EX LIBRIS

NAME

PARTIAL DIFFERENTIAL EQUATIONS AND BOUNDARY-VALUE PROBLEMS WITH APPLICATIONS

International Series in Pure and Applied Mathematics

Ahlfors: *Complex Analysis*
Bender and Orszag: *Advanced Mathematical Methods for Scientists and Engineers*
Boas: *Invitation to Complex Analysis*
Buck: *Advanced Calculus*
Colton: *Partial Differential Equations*
Conte and deBoor: *Elementary Numerical Analysis: An Algorithmic Approach*
Edelstein-Keshet: *Mathematical Models in Biology*
Goldberg: *Matrix Theory with Applications*
Hill: *Experiments in Computational Matrix Algebra*
Lewin and Lewin: *An Introduction to Mathematical Analysis*
Morash: *Bridge to Abstract Mathematics: Mathematical Proof and Structures*
Parzynski and Zipse: *Introduction to Mathematical Analysis*
Pinsky: *Partial Differential Equations and Boundary-Value Problems*
　　　　with Applications
Pinter: *A Book of Abstract Algebra*
Ralston and Rabinowitz: *A First Course in Numerical Analysis*
Ritger and Rose: *Differential Equations with Applications*
Rudin: *Functional Analysis*
Rudin: *Principles of Mathematical Analysis*
Rudin: *Real and Complex Analysis*
Simmons: *Differential Equations with Applications and Historical Notes*
Small and Hosack: *Calculus: An Integrated Approach*
Vanden Eynden: *Elementary Number Theory*
Walker: *Introduction to Abstract Algebra*

Churchill-Brown Series

Complex Variables and Applications
Fourier Series and Boundary Value Problems
Operational Mathematics

PARTIAL DIFFERENTIAL EQUATIONS AND BOUNDARY-VALUE PROBLEMS WITH APPLICATIONS

Second Edition

Mark A. Pinsky

Professor of Mathematics
Northwestern University

With an Appendix by Alfred Gray
on Using Mathematica

McGraw-Hill, Inc.

New York   St. Louis   San Francisco   Auckland   Bogotá   Caracas   Hamburg
Lisbon   London   Madrid   Mexico   Milan   Montreal   New Delhi
Paris   San Juan   São Paulo   Singapore   Sydney   Tokyo   Toronto

This book was set in Times Roman by General Graphic Services, Inc.
The editors were Karen M. Hughes, Richard Wallis, and Jack Maisel;
the production supervisor was Leroy A. Young.
The cover was designed by Rafael Hernandez.
New drawings were done by Fine Line Illustrations, Inc.
R. R. Donnelley & Sons Company was printer and binder.

**PARTIAL DIFFERENTIAL EQUATIONS AND BOUNDARY-VALUE PROBLEMS
WITH APPLICATIONS**

Mathematica is a registered trademark of Wolfram Research, Inc., P.O. Box 6059,
Champaign, IL 61821.

2 3 4 5 6 7 8 9 0 DOC DOC 9 0 9 8 7 6 5 4 3 2 1

ISBN 0-07-050128-9

Library of Congress Cataloging-in-Publication Data
Pinsky, Mark A., (date).
    Partial differential equations and boundary-value problems with
applications / Mark A. Pinsky; with an appendix by Alfred Gray on
''Using Mathematica''. — 2nd ed.
        p.   cm. — (International series in pure and applied
mathematics)
    Rev. ed. of: Introduction to partial differential equations with
applications, c1984.
    Includes bibliographical references and index.
    ISBN 0-07-050128-9
    1. Differential equations, Partial.   2. Boundary-value problems.
I. Pinsky, Mark A., (date).   Introduction to partial differential
equations with applications.   II. Title.   III. Series.
QA374.P55   1991
515'.353—dc20   90-43977

ABOUT THE AUTHOR

Mark A. Pinsky is professor of mathematics at Northwestern University, Evanston, Illinois. He received his Ph.D. degree in mathematics from MIT while holding an NSF Graduate Fellowship. He has held visiting positions at Stanford University, the University of Chicago, and the University of Paris VI. Since 1976 he has been a full professor at Northwestern. He has published more than sixty research articles and four edited volumes. Recently he co-edited "Geometry of Random Motion," volume 73 in the AMS series *Contemporary Mathematics*. From 1977 to 1979 he was Director of the Integrated Science Program at Northwestern, as a result of which he thought about writing for students. He has been the recipient of research grants from NSF and other government agencies. He is a Member and Fellow of the Institute of Mathematical Statistics. He also belongs to the American Mathematical Society, the Mathematical Association of America, and is listed in *Who's Who in America*.

To the memory of Julian H. Blau,
friend and teacher,
1917–1987

CONTENTS

PREFACE

The purpose of this book is to provide an introduction to partial differential equations for students who have finished calculus through ordinary differential equations. The book provides physical motivation, mathematical method, and physical application. While the first and last are the raison d'être for the mathematics, I have chosen to stress the systematic solution algorithms, based on the methods of separation of variables and Fourier series and integrals. My goal has been to achieve a lucid and mathematically correct approach without becoming excessively involved in analysis per se. For example, I have stressed the interpretation of various solutions in terms of asymptotic behavior (for the heat equation) and geometry (for the wave equation).

Chapter 0 contains a brief introduction to the entire subject and some technical material that is used frequently throughout the book. I suggest that Secs. 0.1 through 0.3 could be covered with little difficulty, while Secs. 0.5 and 0.6 are available for later reference. Section 0.4 on Sturm-Liouville theory is heavy going and does not have to be covered in its entirety.

Chapters 1 to 4 contain the basic material on Fourier series and boundary-value problems in rectangular, cylindrical, and spherical coordinates. Bessel and Legendre functions are developed in Chaps. 3 and 4 for those instructors who wish a self-contained development of this material. Instructors who do not wish to use this material on boundary-value problems should cover only Secs. 3.1 and 4.1 in Chaps. 3 and 4. These sections contain several interesting boundary-value problems that can be solved without the use of Bessel or Legendre functions.

Chapter 5 develops Fourier transforms and applies them to solve problems in unbounded regions. This material, which may be treated immediately following Chap. 2 if desired, uses real-variable methods. The student is referred to a subsequent course for complex-variable methods.

The student who has finished all the material through Chap. 5 will have a good working knowledge of the classical methods of solution. To complement

xix

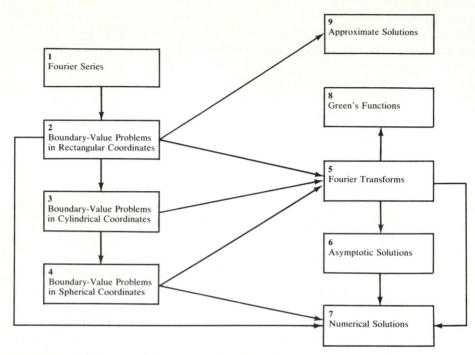

Logical Dependence of Chapters

these basic techniques, I have added chapters on asymptotic solutions (Chap. 6), numerical solutions (Chap. 7), Green's functions (Chap. 8), and approximate solutions (Chap. 9) for instructors who may have additional time or wish to omit some of the earlier material. The flowchart on this page plots various paths through the book.

Chapters 1 and 2 form the heart of the book. These begin with the theory of Fourier series, including a complete discussion of convergence, Parseval's theorem, and the Gibbs phenomenon. We work with the class of piecewise smooth functions, which are infinitely differentiable except at a finite number of points, where all derivatives have left and right limits. Despite the generous dose of theory, it is expected that the student will learn to compute Fourier coefficients and to use Parseval's theorem to estimate the mean square error in approximating a function by the partial sum of its Fourier series.

Chapter 2 takes up the systematic study of the wave equation and the heat equation. It begins with steady-state and time-periodic solutions of the heat equation in Sec. 2.1, including applications to heat transfer and to geophysics, and follows this with the study of initial-value problems in Secs. 2.2 and 2.3, which are treated by a five-stage method. This systematic breakdown allows the student to separate the steady-state solution from the transient solution (found by the separation-of-variables algorithm) and to verify the uniqueness and asymptotic behavior of the solution, as well as to compute the relaxation time. I have found that students can easily appreciate and understand this method, which combines mathematical precision and clear physical interpretation. The five-stage method is used throughout the book, in Secs. 2.5, 3.4, and 4.1. Chapter 2 also includes the wave equation for the vibrating string (Sec. 2.4), solved both by Fourier series and by the d'Alembert formula. Both methods have advantages and disadvantages, which are discussed in detail. My derivations of both the wave equation and the heat equation are from the three-dimensional viewpoint, which I feel is less artificial and more elegant than many treatments that begin with a one-dimensional formulation.

Following Chap. 2, there is a wide choice in the direction of the course. Those instructors who wish to give a complete treatment of boundary-value problems in cylindrical and spherical coordinates, including Bessel and Legendre functions, will want to cover all of Chaps. 3 and 4. Other instructors may ignore this material completely and proceed directly to Chap. 5, Fourier Transforms. An intermediate path might be to cover Secs. 3.1 and/or 4.1, which treat (respectively) Laplace's equation in polar coordinates and spherically symmetric solutions of the heat equation in three dimensions. Neither topic requires any special functions beyond those encountered in trigonometric Fourier series.

Chapter 5 treats Fourier transforms using the complex exponential notation. This is a natural extension of the complex form of the Fourier series, which is covered in Sec. 1.5. Using the Fourier transform, I reduce the heat, Laplace, wave, and telegraph equations to ordinary differential equations with constant coefficients, which can be solved by elementary methods. In many cases these Fourier representations of the solutions can be rewritten as explicit representations (by what is usually known as the Green function method). The method of images for solving problems on a semi-infinite axis is naturally developed here. The Green's functions methods are developed more systematically in Chap. 8, which is new to the second edition. After preparing the one-dimensional case, I give a self-contained treatment of the explicit representation of the solution of Poisson's equation in two and three dimensions.

Throughout the book I emphasize the asymptotic analysis of *series* solutions of boundary-value problems. Chapter 6 gives an elementary account of asymptotic analysis of *integrals,* in particular the Fourier integral representations of the solutions obtained in Chap. 5. The methods include integration by parts, Laplace's method, and the method of stationary phase. These culminate in a detailed asymptotic analysis of the telegraph equation, which illustrates the group velocity of a wave packet.

No introduction to partial differential equations would be complete without some discussion of approximate solutions and numerical methods. Chapter 7 gives the student some working knowledge of the finite difference solution of the heat equation and Laplace's equation in one and two space dimensions. Chapter 9, which is new to the second edition, introduces some perturbational and variational methods which can be used to produce formulas for approximate solutions (which can be implemented numerically). These problems include the vibrating string with nonconstant density, which is treated by "second-order perturbation theory." The material on variational methods first relates differential equations to variational problems and then outlines some direct methods which may be used to arrive at approximate solutions.

Another important addition to the second edition is the Appendix, Using *Mathematica*™, for which we are indebted to Alfred Gray. Computer-assisted methods have become increasingly important in recent years and are now available for the first time in a book at this level.

Each section of the book contains numerous worked examples and a set of exercises. An effort has been made to keep these exercises at a uniform level of difficulty; occasionally more difficult exercises are marked with an asterisk (*). Solutions have been provided for nearly 425 of the 635 exercises in the text.

This book was developed from course notes for Mathematics C91-1 in the Integrated Science Program at Northwestern University. The course has been taught to college juniors since 1977; Chaps. 1 to 5 are covered in a 10-week quarter. I am indebted to my colleagues Leonard Evens, Robert Speed, Paul Auvil, Gene Birchfield, and Mark Ratner for providing valuable suggestions on the mathematics and its applications. The first draft was written in collaboration with Michael Hopkins. The typing was done by Vicki Davis and Julie Mendelson. The solutions were compiled with the assistance of Mark Scherer. Valuable technical advice was further provided by Edward Reiss and Stuart Antman.

In the preparation of the second edition I received valuable comments and suggestions from James W. Brown, Charles Holland, Robert Pego, Mei-Chang Shen, Clark Robinson, Nancy Stanton, Athanassios Tzavaras, David Kapov, and Dennis Kosterman. I also acknowledge the reviewing services of the following individuals: William O. Bray, University of Maine; William E. Fitzgibbon, University of Houston; Peter J. Gingo, University of Akron; Mohammad Kazemi, University of North Carolina-Charlotte; Gilbert N. Lewis, Michigan Technological University; Geoffrey Martin, University of Toledo; Norman Meyers, University of Minnesota–Minneapolis; Allen C. Pipkin, Brown University; R. E. Showalter, University of Texas–Austin; and Grant V. Welland, University of Missouri—St. Louis.

In the preparation of the first edition I was encouraged by John Corrigan of the McGraw-Hill College Division. More recently I have benefited from the editorial services of Karen M. Hughes and Richard Wallis, whose promptness and efficiency have been most appreciated.

Mark A. Pinsky

PARTIAL DIFFERENTIAL EQUATIONS AND
BOUNDARY-VALUE PROBLEMS
WITH APPLICATIONS

CHAPTER
0

PRELIMINARIES

INTRODUCTION

This chapter serves two purposes; it introduces the subject and reviews techniques from previous courses in calculus and ordinary differential equations.

In the first two sections, we introduce some of the classical partial differential equations in two independent variables. In Sec. 0.3 we introduce the notion of orthogonal functions; these arise naturally from separated solutions with boundary conditions and lead, in particular, to the theory of Fourier series of Chap. 1. Section 0.4 contains a detailed treatment of Sturm-Liouville theory. Special cases of this theory arise throughout later chapters and are often handled directly. Hence this material need not be covered in its entirety.

Section 0.5 constitutes a review of some important topics in the theory of infinite series, usually treated in courses of advanced calculus. Section 0.6 gives a corresponding review of the necessary ideas and techniques concerning ordinary differential equations which will be used in later work.

0.1 PARTIAL DIFFERENTIAL EQUATIONS

Sources in Classical Physics

Many laws of physics are expressed mathematically as differential equations. The reader who has studied elementary mechanics is aware of Newton's second law of motion, which expresses the acceleration of a system in terms of the forces on the system. In the case of one or more point particles, this translates to a system of ordinary differential equations when the force law is known.

1

For example, a spring with Hooke's law of elastic restoration gives rise to the linear equation of the harmonic oscillator, which is well studied in elementary courses. A system of particles interacting through several springs gives rise to a system of linear differential equations. Coulomb's law of attraction between gravitational bodies gives rise to a more complicated nonlinear system of ordinary differential equations. Generally speaking, whenever we have a finite number of point particles, the mathematical model is a system of ordinary differential equations, where *time* plays the role of the independent variable and the positions and velocities of the particles function as dependent variables.

In the case of a continuous medium, the positions and velocities of the system are infinite in number and we can no longer use a system of ordinary differential equations to describe the motion. For example, a piano string of length L, which is set into motion by a hammer, has continuously many coordinates corresponding to the points of the interval $0 \leq x \leq L$. The position of the xth coordinate is denoted $y(x; t)$, signifying the net displacement from equilibrium at the time instant t. The velocity of this coordinate is then the first partial derivative $(\partial y/\partial t)(x; t)$, and the acceleration is the second partial derivative $(\partial^2 y/\partial t^2)(x; t)$. Newton's second law of motion can be used to express the forces in terms of the spatial derivatives $\partial y/\partial x$ and $\partial^2 y/\partial x^2$. In Sec. 2.4 we show that if we ignore gravity and assume small displacements, this force is expressed as $T \, \partial^2 y/\partial x^2$, where the positive constant T represents the tension of the string. The intuitive meaning of this is the following: If a given section of the string follows a straight line, then $\partial^2 y/\partial x^2 = 0$ and the force is zero along that section, signifying a state of equilibrium. If another section is concave upward, then $\partial^2 y/\partial x^2 > 0$ along that section and the force is upward, signifying the elastic restoration of the spring; similarly for $\partial^2 y/\partial x^2 < 0$, the force is downward. (The intuitive appeal is even stronger if we replace the string by an equally spaced system of beads, which interact according to elastic restoration.) Newton's second law of motion then translates to the mathematical statement

$$\rho \, \frac{\partial^2 y}{\partial t^2} = T \, \frac{\partial^2 y}{\partial x^2}$$

where ρ is the mass density of the string. This is the *equation of the vibrating string;* it is studied in detail in Sec. 2.4 and throughout later sections. It is normally solved in conjunction with *initial conditions,* representing the position and velocity of the string at the instant $t = 0$, and *boundary conditions,* which require that the ends of the string at $x = 0$, $x = L$ be held fixed during the course of the motion. The resultant mathematical problem is called an *initial-boundary value problem* for the equation of the vibrating string.

Another important physical phenomenon is *heat conduction.* This theory, which was first systematically studied by J. B. Fourier, is derived from direct macroscopic considerations without regard to Newton's laws of mechanics. In the simplest version of this theory, the function $u(x; t)$ represents the temperature of a rod at time t measured at position x. The temporal rate of change of temperature is given by the first partial derivative $\partial u/\partial t$. The *heat equation* is the statement that

$$\frac{\partial u}{\partial t} = K \frac{\partial^2 u}{\partial x^2}$$

This is derived in Sec. 2.1. The positive constant K, called the *diffusivity*, measures the ability of the rod to conduct heat; the larger K is, the faster that heat is conducted throughout the medium. The second derivative $\partial^2 u/\partial x^2$ measures the amount by which the temperature is out of equilibrium; for example, at a local minimum $\partial^2 u/\partial x^2 \geq 0$, and the heat equation states that the temperature will increase. This agrees with intuition, since we expect that heat conduction should be a process of equilibration. Similarly at a local maximum $\partial^2 u/\partial x^2 \leq 0$, and the equation predicts a decrease in temperature. More generally, the second derivative measures the local rate of disequilibrium of temperature—the amount by which $u(x; t)$ differs from the average of its neighbors, to the first order of approximation. Put more simply, the heat equation predicts that hot places will get cooler and cool places will get hotter, in relation to neighboring places.

The heat equation is usually solved in conjunction with initial conditions and boundary conditions. These correspond to the temperature at $t = 0$ and the temperature conditions at the extremities of the rod. These problems are explored in detail throughout this book, beginning in Sec. 2.1.

Equations in Two Variables

From a purely mathematical point of view, one can consider partial differential equations which involve arbitrary combinations of the first and higher partial derivatives of a function of several variables. Such problems have been studied by mathematicians, and some results were obtained. In this book we confine our attention to linear equations of the second order, as defined below. This includes most of the important equations from mathematical physics. In this section we illustrate some of the general principles of operation in the case of two independent variables.

A function of two variables is denoted

$$u = u(x, y)$$

and its partial derivatives by

$$u_x = \frac{\partial u}{\partial x} \qquad u_y = \frac{\partial u}{\partial y}$$

$$u_{xx} = \frac{\partial^2 u}{\partial x^2} \qquad u_{xy} = \frac{\partial^2 u}{\partial x \, \partial y} \qquad u_{yy} = \frac{\partial^2 u}{\partial y^2}$$

For example, if $u(x, y) = x^2 + e^y$, then $u_x = 2x$, $u_y = e^y$, $u_{xx} = 2$, $u_{xy} = 0$, $u_{yy} = e^y$.

A *linear second-order partial differential equation* is an equation of the form

$$\boxed{Au_{xx} + 2Bu_{xy} + Cu_{yy} + Du_x + Eu_y + Fu = G}$$ (0.1.1)

where A, B, C, D, E, F, G are functions of (x, y). The equation is called *homogeneous* if $G = 0$. Some common examples which occur frequently are

1. $u_{xx} + u_{yy} = 0$ (Laplace's equation)
2. $u_{tt} - c^2 u_{xx} = 0$ (the wave equation)
3. $u_t - Ku_{xx} = 0$ (the heat equation)
4. $u_{tt} - c^2 u_{xx} + 2\beta u_t + \alpha u = 0$ (the telegraph equation)
5. $u_{xx} + u_{yy} = G$ (Poisson's equation)

In Chap. 2 we give a derivation of the heat equation in three dimensions, which includes the Laplace and Poisson equations listed here. The wave equation is also derived in Chap. 2.

The first four of these are homogeneous, while the last is nonhomogeneous. Notice that we have used t in place of y in examples 2, 3, and 4 because of the physical interpretation of time.

Superposition Principle and Subtraction Principle

It is not possible to write the general solution of a partial differential equation in a closed form, as we have done for many linear second-order ordinary differential equations. Therefore it is very important to develop methods for combining known solutions, when available. For homogeneous equations we have the following simple rule for combining known solutions:

> **Superposition principle for homogeneous equations.** If $u_1, \ldots, u_n$ are solutions of the same homogeneous linear second-order partial differential equation and $c_1, \ldots, c_n$ are constants, then $c_1 u_1 + \cdots + c_n u_n$ is also a solution of the equation.

Indeed, this results from the fact that the derivatives are *linear* operations.

$$(c_1 u_1 + \cdots + c_n u_n)_x = c_1 (u_1)_x + \cdots + c_n (u_n)_x$$

$$(c_1 u_1 + \cdots + c_n u_n)_{xx} = c_1 (u_1)_{xx} + \cdots + c_n (u_n)_{xx}$$

with similar statements for the other partial derivatives which occur in the equation. When we combine these, we obtain the superposition principle.

For example, it may be checked that $e^{-kx} \cos ky$ is a solution of Laplace's equation for any constant k. Therefore, by the superposition principle, the function

$$u(x, y) = e^{-x} \cos y + 2e^{-\pi x} \cos \pi y - e^{-6x} \cos 6y$$

is also a solution of Laplace's equation.

The superposition principle does not apply to nonhomogeneous equations. For example, if u_1, u_2 are both solutions of Poisson's equation $u_{xx} + u_{yy} = 1$, then $u_1 + u_2$ is a solution of a *different* equation, namely, $u_{xx} + u_{yy} = 2$. Nevertheless there is an important general principle which allows us to relate nonhomogeneous equations to homogeneous equations.

> **Subtraction principle for nonhomogeneous equations.** If u_1, u_2 are solutions of the same nonhomogeneous equation (0.1.1), then $u_1 - u_2$ is a solution of the associated homogeneous equation
>
> $$Au_{xx} + 2Bu_{xy} + Cu_{yy} + Du_x + Eu_y + Fu = 0$$

For example, if u_1, u_2 are both solutions of Poisson's equation, then $u_1 - u_2$ is a solution of Laplace's equation.

The subtraction principle allows us to find the general solution of a nonhomogeneous equation once we know a particular solution of the nonhomogeneous equation and a general solution of the homogeneous equation. Suppose, for example, that we have to solve Poisson's equation $u_{xx} + u_{yy} = 1$ with some definite boundary condition. We observe that the function $v(x, y) = \frac{1}{4}(x^2 + y^2)$ is a solution of Poisson's equation. Therefore, by the subtraction principle, $u - v$ is a solution of Laplace's equation with some new boundary condition. Therefore we have reduced the solution of Poisson's equation to the solution of Laplace's equation.

Classification of Second-Order Equations

It is impossible to formulate a general existence theorem which applies to all linear second-order partial differential equations. Instead of speaking of the general solution, which is customary for ordinary differential equations, it is more natural to specify a solution in terms of certain *boundary conditions*. The exact type of boundary conditions which are natural depends on the equation. For this purpose we classify the equations into three types, depending on the second-order coefficients A, B, C:

> If $AC - B^2 > 0$, the equation is *elliptic*.
> If $AC - B^2 < 0$, the equation is *hyperbolic*.
> If $AC - B^2 = 0$, the equation is *parabolic*.

For example, the equations of Poisson and Laplace are elliptic, while the wave and telegraph equations are hyperbolic. The heat equation is parabolic. The detailed mathematical theory of this classification is beyond the scope of this book. Nevertheless we indicate below types of boundary conditions which are natural for each of the three types.

If the equation is elliptic, we may solve the *Dirichlet problem,* which amounts to specifying the values of $u(x, y)$ on the boundary of a bounded plane region with smooth boundary. Consider the physical problem of determining the electrostatic potential function u in a cylinder where the charge density ρ is specified in the interior and the boundary is required to be an equipotential surface. This leads to the Dirichlet problem

$$u_{xx} + u_{yy} = -\rho \qquad x^2 + y^2 < R^2$$

$$u(x, y) = C \qquad x^2 + y^2 = R^2$$

The constant surface potential C may be taken to be zero without loss of generality.

If the equation is hyperbolic or parabolic, it is natural to solve the *Cauchy problem*. For examples 2 and 4 this amounts to specifying the solution $u(x; t)$ and its time derivative $u_t(x; t)$ on the line $t = 0$. It may also be necessary to specify boundary conditions at the end of a finite interval or, in the case of infinite intervals, to restrict the growth of $u(x; t)$ when x becomes large. As an example of a Cauchy problem, in Chap. 2 we derive the following equations for the vibrating string with fixed ends:

$$u_{tt} - c^2 u_{xx} = 0 \qquad t > 0, 0 < x < L$$

$$u(x, 0) = f_1(x) \qquad 0 < x < L$$

$$u_t(x, 0) = f_2(x) \qquad 0 < x < L$$

$$u(0; t) = 0 \qquad u(L; t) = 0 \qquad t > 0$$

The initial conditions f_1, f_2 correspond, respectively, to the initial position and velocity of the vibrating string; the position is given by $u(x; t)$ at time t at the xth coordinate of the string. The boundary conditions signify that both ends of the string at $x = 0$ and $x = L$ are fixed for all time. In Chap. 2 we give several methods for solving this initial-boundary-value problem.

EXERCISES 0.1

In Exercises 1 to 5, classify each of the second-order equations as elliptic, hyperbolic, or parabolic.

1. $u_{xx} + 6u_{xy} - u_{yy} + 2u_x - u_y = 0$
2. $u_{xx} + xu_{yy} = 0$
3. $u_{xy} = 0$
4. $u_{xx} + 4u_{yy} = 0$
5. $(1 - x^2)u_{xx} + yu_{yy} = 0$

In Exercises 6 to 10, verify that the indicated function satisfies the given partial differential equation.

6. $u(x, y) = \sin kx \sinh ky$ (k constant), $u_{xx} + u_{yy} = 0$
7. $u(x; t) = \sin kx \, e^{-k^2 t}$ (k constant), $u_{xx} - u_t = 0$
8. $u(x, y) = f(x) + g(y)$ (f, g arbitrary), $u_{xy} = 0$
9. $u(x, y) = f(x + y) + g(x - y)$ (f, g arbitrary), $u_{xx} - u_{yy} = 0$
10. $u(x; t) = e^{-x} \cos(t - x)$, $2u_t - u_{xx} = 0$

11. State and prove a suitable form of the superposition principle for linear homogeneous ordinary differential equations of the form $a(t)y'' + b(t)y' + c(t)y = 0$.
12. State and prove a suitable form of the subtraction principle for linear nonhomogeneous ordinary differential equations of the form $a(t)y'' + b(t)y' + c(t)y = d(t)$.

0.2 SEPARATION OF VARIABLES

A fundamental technique for obtaining solutions of partial differential equations is the method of *separation of variables*. This means that we look for particular solutions in the form $u(x, y) = f_1(x)f_2(y)$ and try to obtain ordinary differential equations for $f_1(x)$ and $f_2(y)$. These equations contain a parameter called the *separation constant*. The function $u(x, y)$ is called a *separated solution*.

Example: Separated Solutions of Laplace's Equation

Consider, for example, Laplace's equation $u_{xx} + u_{yy} = 0$. If we let $u(x, y) = f_1(x)f_2(y)$ and substitute, we obtain

$$f_1''(x)f_2(y) + f_1(x)f_2''(y) = 0$$

Dividing by $f_1(x)f_2(y)$ (assumed to be nonzero), we obtain

$$\frac{f_1''(x)}{f_1(x)} + \frac{f_2''(y)}{f_2(y)} = 0$$

The first term depends only on x, while the second term depends only on y. This can happen only if both terms are constant. Thus we must have the ordinary differential equations

$$\frac{f_1''(x)}{f_1(x)} = \lambda \qquad \frac{f_2''(y)}{f_2(y)} = -\lambda$$

where λ is the *separation constant*. These equations may be written in the more standard form

$$f_1''(x) - \lambda f_1(x) = 0$$

$$f_2''(y) + \lambda f_2(y) = 0$$

If $\lambda > 0$, the general solutions of these equations are

$$f_1(x) = A_1 e^{x\sqrt{\lambda}} + A_2 e^{-x\sqrt{\lambda}}$$

$$f_2(y) = A_3 \cos y\sqrt{\lambda} + A_4 \sin y\sqrt{\lambda}$$

If $\lambda = 0$, the general solutions of these equations are

$$f_1(x) = A_1 x + A_2$$

$$f_2(y) = A_3 y + A_4$$

If $\lambda < 0$, the general solutions of these equations are

$$f_1(x) = A_1 \cos x\sqrt{-\lambda} + A_2 \sin x\sqrt{-\lambda}$$

$$f_2(y) = A_3 e^{y\sqrt{-\lambda}} + A_4 e^{-y\sqrt{-\lambda}}$$

To summarize, we have found the following separated solutions of Laplace's equation:

$$u(x, y) = \begin{cases} (A_1 e^{x\sqrt{\lambda}} + A_2 e^{-x\sqrt{\lambda}})(A_3 \cos y\sqrt{\lambda} + A_4 \sin y\sqrt{\lambda}) & \lambda > 0 \\ (A_1 x + A_2)(A_3 y + A_4) & \lambda = 0 \\ (A_1 \cos x\sqrt{-\lambda} + A_2 \sin x\sqrt{-\lambda})(A_3 e^{y\sqrt{-\lambda}} + A_4 e^{-y\sqrt{-\lambda}}) & \lambda < 0 \end{cases}$$

To derive these, we assumed that $u(x, y) \neq 0$. Having now obtained the explicit forms, we can verify independently that in each case $u(x, y)$ satisfies Laplace's equation.

Real and Complex Separated Solutions in General

In looking for separated solutions of a partial differential equation, it is often convenient to allow the functions $f_1(x)$ and $f_2(y)$ to be complex-valued. The following proposition shows that the real and imaginary parts of such solutions will again satisfy the partial differential equation.

Proposition 0.2.1. Let $u(x, y) = v_1(x, y) + iv_2(x, y)$ be a complex-valued solution of the partial differential equation

$$Au_{xx} + 2Bu_{xy} + Cu_{yy} + Du_x + Eu_y + Fu = G$$

where the coefficients A, B, C, D, E, F, G are real-valued functions of (x, y). Then $v_1(x, y) = \text{Re } u(x, y)$ satisfies the partial differential equation, and $v_2(x, y) = \text{Im } u(x, y)$ satisfies the associated homogeneous equation with $G = 0$.

Proof. The operation of partial differentiation is linear; thus

$$u_x = (v_1)_x + i(v_2)_x$$

$$u_{xx} = (v_1)_{xx} + i(v_2)_{xx}$$

with similar expressions for u_y, u_{yy}, and u_{xy}. Substituting these into the partial differential equation and separating the real and imaginary parts yield the result. ∎

We illustrate this technique with the example of Laplace's equation. It is easily verified that the function $u(x, y) = e^{x+iy} = e^x e^{iy}$ satisfies Laplace's equation, since $u_{xx} = u$, $u_{yy} = -u$. To take the real and imaginary parts, we write $u(x, y) = e^x(\cos y + i \sin y) = e^x \cos y + ie^x \sin y$. Since Laplace's equation is homogeneous, we conclude that both $e^x \cos y$ and $e^x \sin y$ are solutions of

Laplace's equation. (This agrees with the previous method, using separation of variables, in the case where $\lambda = 1$.)

Complex-valued separated solutions may always be found if functions A, B, C, D, E, F, which occur in the equation, are independent of (x, y); in this case we speak of a *partial differential equation with constant coefficients*, whose solutions may be written as exponential functions.

Proposition 0.2.2. Consider the homogeneous partial differential equation

$$Au_{xx} + 2Bu_{xy} + Cu_{yy} + Du_x + Eu_y + Fu = 0$$

Suppose that A, B, C, D, E, F are real constants. Then there exist complex separated solutions of the form

$$u(x, y) = e^{\alpha x}e^{\beta y}$$

for appropriate choices of the complex numbers α, β.

Proof. We first note that the ordinary rules for differentiating $e^{\alpha x}$ are valid for complex-valued functions. For example, if $\alpha = a + ib$,

$$\frac{d}{dx}(e^{\alpha x}) = \frac{d}{dx}[e^{ax}(\cos bx + i \sin bx]$$

$$= ae^{ax}\cos bx - be^{ax}\sin bx + i(ae^{ax}\sin bx + be^{ax}\cos bx)$$

$$= e^{ax}(a + ib)(\cos bx + i \sin bx)$$

$$= (a + ib)e^{(a+ib)x}$$

$$= \alpha e^{\alpha x}$$

Similarly, $(d^2/dx^2)(e^{\alpha x}) = \alpha^2 e^{\alpha x}$, with similar expressions for d/dy, d^2/dy^2. Applying this to $u(x, y) = e^{\alpha x}e^{\beta y}$, we have $u_x = \alpha u$, $u_{xx} = \alpha^2 u$, $u_y = \beta u$, $u_{yy} = \beta^2 u$, $u_{xy} = \alpha\beta u$. Substituting these into the partial differential equation, we must have

$$(A\alpha^2 + 2B\alpha\beta + C\beta^2 + D\alpha + E\beta + F)e^{\alpha x}e^{\beta y} = 0$$

But $e^{\alpha x}e^{\beta y} \neq 0$; therefore we obtain a solution if and only if α, β satisfy the quadratic equation

$$\boxed{A\alpha^2 + 2B\alpha\beta + C\beta^2 + D\alpha + E\beta + F = 0}$$

For a given value of β, we may solve this equation for α to obtain in general two roots α_1, α_2. Alternately we may fix α and solve for β to obtain in general two roots β_1, β_2. This proves the proposition. ∎

Examples of Separated Solutions with No Conditions

Example 0.2.1. Find separated solutions of the equation $u_{xx} - u_t = 0$ in the form $u(x, t) = e^{i\mu x}e^{\beta t}$ with μ, β real.

Solution. Substituting $u(x, t) = e^{i\mu x}e^{\beta t}$ in the equation yields the quadratic equation $-\mu^2 - \beta = 0$. Thus $\beta = -\mu^2$, and we have the separated solutions

$$u(x, t) = e^{i\mu x}e^{-\mu^2 t}$$

$$= \cos \mu x\ e^{-\mu^2 t} + i(\sin \mu x\ e^{-\mu^2 t})$$

Therefore we have the real-valued separated solutions $\sin \mu x\ e^{-\mu^2 t}$ and $\cos \mu x\ e^{-\mu^2 t}$. ●

In the above example, the solutions tend to zero when time t tends to infinity. In some problems we may wish to obtain a solution which oscillates in time to represent a periodic disturbance.

Example 0.2.2. Find separated solutions of the equation $u_{xx} - u_t = 0$ in the form $u(x, t) = e^{\alpha x}e^{i\omega t}$, where ω is real and positive.

Solution. Substituting $u(x, t) = e^{\alpha x}e^{i\omega t}$ into equation $u_t - u_{xx} = 0$ yields the quadratic equation $\alpha^2 - i\omega = 0$. This equation has two solutions which may be obtained as follows. Writing the complex number i in the polar form $i = e^{i\pi/2}$, we have the two square roots $i^{1/2} = \pm e^{i\pi/4} = \pm(1 + i)/\sqrt{2}$. Therefore the solutions of the quadratic equation are $\alpha = \pm(1 + i)\sqrt{\omega/2}$. The separated solutions are

$$u(x, t) = \begin{cases} \exp[x(1 + i)\sqrt{\omega/2}] \exp i\omega t = \exp(x\sqrt{\omega/2})] \exp[i(\omega t + x\sqrt{\omega/2})] \\ \\ \exp[-x(1 + i)\sqrt{\omega/2}] \exp i\omega t = \exp[-x\sqrt{\omega/2}) \exp[i(\omega t - x\sqrt{\omega/2})] \end{cases}$$

Taking the real and imaginary parts, we have the real-valued solutions

$$u(x, t) = \begin{cases} e^{x\sqrt{\omega/2}} \cos(\omega t + x\sqrt{\omega/2}) \\ e^{x\sqrt{\omega/2}} \sin(\omega t + x\sqrt{\omega/2}) \\ e^{-x\sqrt{\omega/2}} \cos(\omega t - x\sqrt{\omega/2}) \\ e^{-x\sqrt{\omega/2}} \sin(\omega t - x\sqrt{\omega/2}) \end{cases}$$

These real-valued solutions are no longer in the separated form $f_1(x)f_2(t)$. But because they arise as the real and imaginary parts of complex separated solutions, we refer to them as *quasi-separated solutions*. ●

If some of the coefficients A, B, C, D, E, F are not constant, we will no longer have separated solutions in the form of exponential functions. Even worse, the equation may not admit *any* nonconstant separated solutions, for example, $u_x + (x + y)u_y = 0$. Nevertheless, various classes of equations can still be solved by the separation of variables, for example, any equation of the form

$$A(x)u_{xx} + C(y)u_{yy} + D(x)u_x + E(y)u_y = 0$$

Assuming $u(x, y) = f_1(x)f_2(y)$, we must have

$$A(x)f_1''(x)f_2(y) + C(y)f_1(x)f_2''(y) + D(x)f_1'(x)f_2(y) + E(y)f_1(x)f_2'(y) = 0$$

Dividing by $f_1(x)f_2(y)$, we have

$$\left[A(x)\frac{f_1''(x)}{f_1'(x)} + D(x)\frac{f_1'(x)}{f_1(x)}\right] + \left[C(y)\frac{f_2''(y)}{f_2(y)} + E(y)\frac{f_2'(y)}{f_2(y)}\right] = 0$$

The terms in the first set of brackets depend only on x, while those in the second set depend only on y; therefore both are constant, and we have reduced the problem to ordinary differential equations.

Example 0.2.3. Find separated solutions of the equation $x^2 u_{xx} + 2xu_x + u_{yy} = 0$.

Solution. We let $u(x, y) = f_1(x)f_2(y)$ and obtain

$$\frac{x^2 f_1''(x) + 2xf_1'(x)}{f_1(x)} + \frac{f_2''(y)}{f_2(y)} = 0$$

The separated equations are

$$x^2 f_1''(x) + 2xf_1'(x) - \lambda f_1(x) = 0$$

$$f_2''(y) + \lambda f_2(y) = 0$$

The first equation is a form of Euler's equidimensional equation,* whose solutions are of the form $f_1(x) = x^\gamma$ with $\gamma(\gamma - 1) + 2\gamma - \lambda = 0$. The roots are $\gamma_1 = \frac{1}{2} + \sqrt{\lambda + \frac{1}{4}}$, $\gamma_2 = \frac{1}{2} - \sqrt{\lambda + \frac{1}{4}}$ if $\lambda + \frac{1}{4} \geq 0$. In the case where $\lambda + \frac{1}{4} < 0$, we let $c_\lambda = \sqrt{-\lambda - \frac{1}{4}}$. With this notation the roots are $\gamma_1 = -\frac{1}{2} + ic_\lambda$, $\gamma_2 = -\frac{1}{2} - ic_\lambda$. Therefore

$$f_1(x) = \begin{cases} A_1 x^{\gamma_1} + A_2 x^{\gamma_2} & \text{if } \lambda + \frac{1}{4} > 0 \\ A_1 |x|^{-1/2} + A_2 |x|^{-1/2} \log |x| & \text{if } \lambda + \frac{1}{4} = 0 \\ |x|^{-1/2}[A_1 \cos (c_\lambda \log |x|) + A_2 \sin (c_\lambda \log |x|)] & \text{if } \lambda + \frac{1}{4} < 0 \end{cases}$$

The equation $f_2''(y) + \lambda f_2(y) = 0$ is solved as before in the cases where $\lambda > 0$, $\lambda = 0$, $\lambda < 0$. Therefore we have the separated solutions

$$|x|^{-1/2}[A_1 \cos (c_\lambda \log |x|) + A_2 \sin (c_\lambda \log |x|)]$$
$$\times (A_3 e^{y\sqrt{\lambda}} + A_4 e^{-y\sqrt{-\lambda}}) \qquad \lambda < -\tfrac{1}{4}$$

$$(A_1 |x|^{-1/2} + A_2 |x|^{-1/2} \log |x|)(A_3 e^{y/2} + A_4 e^{-y/2}) \qquad \lambda = -\tfrac{1}{4}$$

$$(A_1 |x|^{\gamma_1} + A_2 |x|^{\gamma_2})(A_3 e^{y\sqrt{-\lambda}} + A_4 e^{-y\sqrt{-\lambda}}) \qquad -\tfrac{1}{4} < \lambda < 0$$

$$(A_1 + A_2 x^{-1})(A_3 + A_4 y) \qquad \lambda = 0$$

$$(A_1 |x|^{\gamma_1} + A_2 |x|^{\gamma_2})(A_3 \cos y\sqrt{\lambda} + A_4 \sin y\sqrt{\lambda}) \qquad \lambda > 0 \qquad ●$$

---

* See Sec. 0.6 for a discussion of Euler's equation.

Examples of Separated Solutions with Boundary Conditions

In many problems we need separated solutions that satisfy certain additional conditions, which are suggested by the physics of the problem. They may be in the form of *boundary conditions* or conditions of *boundedness*.

Example 0.2.4. Find the complex separated solutions of the wave equation $u_{tt} - c^2 u_{xx} = 0$, which are bounded in the form $|u(x; t)| \leq M$ for some constant M and all t, $-\infty < t < \infty$.

Solution. Taking $u(x; t) = e^{ax+bt}$ and substituting in the equation, we have $b^2 - c^2 a^2 = 0$; thus $b = \pm ca$. The separated solutions are of the form $u(x; t) = e^{ax} e^{cat}$, $e^{ax} e^{-cat}$. This solution is bounded for all t if and only if a is pure imaginary, $a = ik$ for k real. Thus the solutions are $u(x; t) = e^{ik(x+ct)}$, $e^{ik(x-ct)}$. The real (quasi-separated) solutions are $\cos k(x + ct)$, $\cos k(x - ct)$, $\sin k(x + ct)$, $\sin k(x - ct)$.   ●

Example 0.2.5. Find the separated solutions of the wave equation $u_{tt} - c^2 u_{xx} = 0$, which satisfy the boundary condition $u(0; t) = 0$, $u(L; t) = 0$.

Solution. Assuming the separated form $u(x; t) = f(x)g(t)$, we have $f(x)g''(t) - c^2 f''(x)g(t) = 0$. Thus $f''(x) + \lambda f(x) = 0$, $g''(t) + \lambda c^2 g(t) = 0$. The boundary conditions require $f(0) = 0$, $f(L) = 0$; thus $f(x) = C \sin(n\pi x/L)$, $g(t) = A \cos(n\pi ct/L) + B \sin(n\pi ct/L)$ for constants A, B, C. The separated solutions are $u(x; t) = [A \cos(n\pi ct/L) + B \sin(n\pi ct/L)] \sin(n\pi x/L)$, $n = 1, 2, \ldots$.   ●

Example 0.2.6. Find the separated solutions of Laplace's equation $u_{xx} + u_{yy} = 0$, which satisfy the boundary condition $u(0, y) = 0$, $u(L, y) = 0$, $u(x, 0) = 0$.

Solution. From the discussion at the beginning of this section, the separated solutions are of three types, depending on the separation constant λ. If $\lambda > 0$, we must have $0 = u(0, y) = A_1 + A_2$, while $0 = u(L, y) = A_1 e^{L\sqrt{\lambda}} + A_2 e^{-L\sqrt{\lambda}}$. This system of simultaneous equations for A_1, A_2 has the unique solution $A_1 = A_2 = 0$, and therefore $\lambda > 0$ does not produce any separated solutions which satisfy the boundary conditions. In the case where $\lambda = 0$, we must have $0 = u(0, y) = A_2$ and $0 = u(L, y) = A_1 L + A_2$; this system of simultaneous equations for A_1, A_2 also has the unique solution $A_1 = A_2 = 0$, and therefore $\lambda = 0$ does not produce any separated solutions which satisfy the boundary conditions. Finally, if $\lambda < 0$, we must have $0 = u(0, y) = A_1$, $0 = u(L, y) = A_1 \cos L\sqrt{-\lambda} + A_2 \sin L\sqrt{-\lambda}$. This system of simultaneous equations for A_1, A_2 has a nonzero solution if and only if $\sin L\sqrt{-\lambda} = 0$, which is satisfied if and only if $L\sqrt{-\lambda} = n\pi$ for some $n = 1, 2, 3, \ldots$. To satisfy the boundary condition $u(x, 0) = 0$, we must have $0 = A_3 + A_4$. Summarizing, we have obtained the following separated solutions of Laplace's equation satisfying the boundary conditions:

$$u(x, y) = A_3 \sin \frac{n\pi x}{L} \left[\exp \frac{n\pi y}{L} - \exp \left(-\frac{n\pi y}{L} \right) \right] \qquad n = 1, 2, \ldots$$

Using the hyperbolic function $\sinh \theta = \frac{1}{2}e^\theta - \frac{1}{2}e^{-\theta}$, we may write these in the more succinct form

$$u(x, y) = 2A_3 \sin \frac{n\pi x}{L} \sinh \frac{n\pi y}{L} \qquad n = 1, 2, \ldots$$

●

EXERCISES 0.2

1. Find the separated equations satisfied by $f_1(x)$, $f_2(y)$ for the following partial differential equations:

(a) $u_{xx} - 2u_{yy} = 0$ (b) $u_{xx} + u_{yy} + 2u_x = 0$

(c) $x^2 u_{xx} - 2yu_y = 0$ (d) $u_{xx} + u_x + u_y - u = 0$

2. Verify that the following are separated solutions of Laplace's equation:

(a) $u(x, y) = e^{3x} \cos 3y$ (b) $u(x, y) = e^{-x} \sin y$

(c) $u(x, y) = e^{2(x+iy)}$ (d) $u(x, y) = (3x + 2)(4y + 3)$

3. Which of the following are solutions of Laplace's equation?

(a) $u(x, y) = e^x \cos 2y$ (b) $u(x, y) = e^x \cos y + e^y \cos x$

(c) $u(x, y) = e^x e^y$ (d) $u(x, y) = (3x + 2)e^y$

In Exercises 4 to 8, find the separated solutions of the indicated equations.

4. $u_{xx} + 2u_x + u_{yy} = 0$

5. $u_{xx} + u_{yy} + 3u = 0$

6. $x^2 u_{xx} + xu_x + u_{yy} = 0$

7. $u_{xx} - u_{yy} + u = 0$

8. $u_{xx} + yu_y + u = 0$

9. Suppose that $u(x, y) = f_1(x)f_2(y)$ is a solution of the equation $u_x + (x + y)u_y = 0$. Show that $f_1(x)$ and $f_2(y)$ are both constant. [*Hint:* Show first that $f_1'/f_1 + (x + y)(f_2'/f_2) = 0$ and deduce that f_1'/f_1 is a linear function of x. From this deduce that yf_2'/f_2 and f_2'/f_2 are both constant and therefore $f_2'/f_2 = 0, f_1'/f_1 = 0$.]

10. Find the separated solutions of Laplace's equation $u_{xx} + u_{yy} = 0$ which satisfy the boundary conditions $u_x(0, y) = 0$, $u_x(L, y) = 0$, $u(x, 0) = 0$.

11. Find the separated solutions of Laplace's equation $u_{xx} + u_{yy} = 0$ which satisfy the boundary conditions $u(0, y) = 0$, $u(L, y) = 0$ and the boundedness condition $|u(x, y)| \le M$ for $y > 0$, where M is a constant independent of (x, y).

12. Find separated solutions of the heat equation $u_t - u_{xx} = 0$ which satisfy the boundary conditions $u(0; t) = 0$, $u(L; t) = 0$.

13. Find separated solutions of the heat equation $u_t - u_{xx} = 0$ which satisfy the boundary conditions $u(0; t) = 0$, $u_x(L; t) = 0$.

14. Find separated solutions of the heat equation $u_t - u_{xx} = 0$ which satisfy the boundary conditions $u_x(0; t) = 0$, $u_x(L; t) = 0$.

0.3 ORTHOGONAL FUNCTIONS

Separated solutions of partial differential equations with boundary conditions often lead to systems of *orthogonal functions*, by way of the *Sturm-Liouville eigenvalue problem*. These two topics are treated in this section and the next. The treatment is entirely self-contained, making no explicit reference to separated solutions of partial differential equations.

Inner Product Space of Functions

The notions of dot product, distance, orthogonality, and projection, which are familiar for vectors in three dimensions, can also be formulated for real-valued functions on an interval $a \leq x \leq b$. The basic notion is the *inner product* of two functions $\varphi(x)$, $\psi(x)$ on the interval $a \leq x \leq b$. This is defined by the integral

$$\langle \varphi, \psi \rangle = \int_a^b \varphi(x)\psi(x) \, dx$$

For example, on the interval $0 \leq x \leq 1$, we have $\langle x, e^{x^2} \rangle = \int_0^1 xe^{x^2} \, dx = \frac{1}{2}(e - 1) = 0.86$, to two decimals.

The analogy between the inner product just defined and the three-dimensional dot product is apparent if we think of the integral as a "continuous sum" of the pointwise products $\varphi(x)\psi(x)$, a generalization of the three-dimensional dot product formula $\mathbf{v} \cdot \mathbf{w} = v_1 w_1 + v_2 w_2 + v_3 w_3$.

The inner product is *linear* and *homogeneous* in both slots. This means that for any functions φ_1, φ_2, ψ_1, ψ_2 and any real number a,

$$\langle \varphi_1, \psi_1 + \psi_2 \rangle = \langle \varphi_1, \psi_1 \rangle + \langle \varphi_1, \psi_2 \rangle$$

$$\langle \varphi_1 + \varphi_2, \psi_1 \rangle = \langle \varphi_1, \psi_1 \rangle + \langle \varphi_2, \psi_1 \rangle$$

$$\langle a\varphi_1, \psi_1 \rangle = a\langle \varphi_1, \psi_1 \rangle$$

$$\langle \varphi_1, a\psi_1 \rangle = a\langle \varphi_1, \psi_1 \rangle$$

Two functions φ, ψ are *orthogonal* by definition if

$$\langle \varphi, \psi \rangle = 0$$

For more than two functions, we say that $\varphi_1, \ldots, \varphi_n$ are orthogonal if $\langle \varphi_i, \varphi_j \rangle = 0$ for $i \neq j$. For example, $\sin x$ and $\cos x$ are orthogonal on the interval $0 \leq x \leq \pi$ since

$$\int_0^\pi \sin x \cos x \, dx = \frac{1}{2} \sin^2 x \Big|_0^\pi = 0$$

In general, the notion of orthogonality just defined is an abstract concept, which formally generalizes the perpendicularity of vectors in three dimensions. There is no implication that the graphs of functions $\varphi(x)$ and $\psi(x)$ intersect at 90°, for example!

The *norm* of a function is the nonnegative number $\|\varphi\|$, defined by

$$\|\varphi\|^2 = \langle \varphi, \varphi \rangle$$

For example, on the interval $0 \leq x \leq \pi$,

$$\|\sin x\|^2 = \int_0^\pi \sin^2 x \, dx = \int_0^\pi \frac{1}{2}(1 - \cos 2x) \, dx = \frac{1}{2}\pi$$

The *distance* between φ and ψ is defined by $d(\varphi, \psi) = \|\varphi - \psi\|$.

For vectors in three-dimensional space, we have the dot product formula $\mathbf{v} \cdot \mathbf{w} = \|\mathbf{v}\| \|\mathbf{w}\| \cos \theta$, where θ is the angle between vectors $\mathbf{v}$ and $\mathbf{w}$ and $\|\mathbf{v}\|$, $\|\mathbf{w}\|$ are the lengths of the respective vectors. We now establish a corresponding fact for nonzero functions $\varphi(x)$, $\psi(x)$, defined on an interval $a \leq x \leq b$. This is known as the *Schwarz inequality*. By the linearity and homogeneity of the inner product, we have, for any real number t,

$$\|\varphi + t\psi\|^2 = \|\varphi\|^2 + 2t\langle \varphi, \psi \rangle + t^2\|\psi\|^2$$

This quadratic function of t is nonnegative and has a minimum at $t = t_0$, where $2\langle \varphi, \psi \rangle + 2t_0\|\psi\|^2 = 0$; at this point the value of the function is nonnegative and given explicitly by

$$\|\varphi\|^2 + 2 \left(-\frac{\langle \varphi, \psi \rangle}{\|\psi\|^2} \right) \langle \varphi, \psi \rangle + \left(-\frac{\langle \varphi, \psi \rangle}{\|\psi\|^2} \right)^2 \|\psi\|^2 = \|\varphi\|^2 - \frac{\langle \varphi, \psi \rangle^2}{\|\psi\|^2}$$

We have proved the *Schwarz inequality*

$$\boxed{\langle \varphi, \psi \rangle^2 \leq \|\varphi\|^2\|\psi\|^2}$$

From this we have, for any two functions φ, ψ,

$$\begin{aligned}
\|\varphi + \psi\|^2 &= \|\varphi\|^2 + 2\langle \varphi, \psi \rangle + \|\psi\|^2 \\
&\leq \|\varphi\|^2 + 2\|\varphi\| \|\psi\| + \|\psi\|^2 \\
&= (\|\varphi\| + \|\psi\|)^2
\end{aligned}$$

We have thus proved the *triangle inequality*

$$\boxed{\|\varphi + \psi\| \leq \|\varphi\| + \|\psi\|}$$

Projection of a Function onto a Finite Orthogonal Set

We now discuss minimizing properties of orthogonal functions. Let $(\varphi_1, \ldots, \varphi_n)$ be a set of orthogonal functions with $\|\varphi_i\| \neq 0$ for $1 \leq i \leq n$. If f is an arbitrary function, we compute the minimum of

$$D(c_1, \ldots, c_n) = \|f - (c_1\varphi_1 + \cdots + c_n\varphi_n)\|^2$$

where $c_1, \ldots, c_n$ range over all real values.

Proposition 0.3.1. The minimum is attained precisely for

$$\bar{c}_i = \frac{\langle f, \varphi_i \rangle}{\|\varphi_i\|^2} \qquad 1 \leq i \leq n$$

Proposition 0.3.2. The coefficients $\bar{c}_1, \ldots, \bar{c}_n$ satisfy *Bessel's inequality*

$$\bar{c}_1^2\|\varphi_1\|^2 + \cdots + \bar{c}_n^2\|\varphi_n\|^2 \leq \|f\|^2$$

Proposition 0.3.3. The minimum distance $d_{\min}$ is given by

$$d_{\min}^2 = D(\bar{c}_1, \ldots, \bar{c}_n) = \|f\|^2 - \bar{c}_1^2\|\varphi_1\|^2 - \cdots - \bar{c}_n^2\|\varphi_n\|^2$$

The function $\bar{c}_1\varphi_1 + \cdots + \bar{c}_n\varphi_n$ is called the *projection* of f on the orthogonal set $(\varphi_1, \ldots, \varphi_n)$; $\bar{c}_i$ is called the *i*th *Fourier coefficient* of f.

The proof of these formulas can be done by calculus. Indeed, we use linearity and homogeneity of the inner product to write

$$D(c_1, \ldots, c_n) = \|f\|^2 - 2 \sum_{i=1}^{n} c_i \langle f, \varphi_i \rangle + \sum_{i=1}^{n} c_i^2 \|\varphi_i\|^2$$

Then we have the partial derivatives

$$\frac{\partial D}{\partial c_1} = 2c_1\|\varphi_1\|^2 - 2\langle f, \varphi_1 \rangle$$

$$\vdots$$

$$\frac{\partial D}{\partial c_n} = 2c_n\|\varphi_n\|^2 - 2\langle f, \varphi_n \rangle$$

These derivatives are zero precisely when

$$c_1\|\varphi_1\|^2 = \langle f, \varphi_1 \rangle, \ldots, c_n\|\varphi_n\|^2 = \langle f, \varphi_n \rangle$$

The matrix of second derivatives is positive; hence we have a minimum, which is the absolute minimum of D. This proves the first statement. To prove the others, we multiply the ith equation above by $2\bar{c}_i$ and sum for $1 \le i \le n$. Thus

$$\sum_{i=1}^{n} 2\bar{c}_i^2\|\varphi_i\|^2 = \sum_{i=1}^{n} 2\bar{c}_i \langle f, \varphi_i \rangle$$

Therefore, $D(\bar{c}_1, \ldots, \bar{c}_n) = \|f\|^2 - \bar{c}_1^2\|\varphi_1\|^2 - \cdots - \bar{c}_n^2\|\varphi_n\|^2$. This is the square of a distance, hence nonnegative, and we have proved Propositions 0.3.2 and 0.3.3.   ■

Example 0.3.1. Find the projection of the function $\cos(\pi x/2)$ on the orthogonal set $(1, x, x^2 - \frac{1}{3})$ on the interval $-1 \le x \le 1$, and compute $d_{\min}$.

Solution. The solution may be written in the form

$$s(x) = \bar{c}_0 + \bar{c}_1 x + \bar{c}_2(x^2 - \tfrac{1}{3})$$

where the Fourier coefficients $\bar{c}_0, \bar{c}_1, \bar{c}_2$ are computed from the equations

$$\bar{c}_0 \int_{-1}^{1} dx = \int_{-1}^{1} \cos \frac{\pi x}{2} \, dx$$

$$\bar{c}_1 \int_{-1}^{1} x^2 \, dx = \int_{-1}^{1} x \cos \frac{\pi x}{2} \, dx$$

$$\bar{c}_2 \int_{-1}^{1} \left(x^2 - \frac{1}{3} \right)^2 dx = \int_{-1}^{1} \left(x^2 - \frac{1}{3} \right) \cos \frac{\pi x}{2} \, dx$$

The first is straightforward since

$$\int_{-1}^{1} \cos \frac{\pi x}{2} \, dx = \frac{2}{\pi} \sin \frac{\pi x}{2} \Big|_{-1}^{1} = \frac{4}{\pi}$$

thus

$$\bar{c}_0 = \frac{2}{\pi}$$

The next is also easy since the function $x \cos (\pi x/2)$ is odd; thus $\bar{c}_1 = 0$. To perform the final integral, we write

$$\int_{-1}^{1} x^2 \cos \frac{\pi x}{2} \, dx = \frac{2}{\pi} \int_{-1}^{1} x^2 \, d \left(\sin \frac{\pi x}{2} \right)$$

$$= \frac{2x^2}{\pi} \sin \frac{\pi x}{2} \Big|_{-1}^{1} - \frac{4}{\pi} \int_{-1}^{1} x \sin \frac{\pi x}{2} \, dx$$

$$= \frac{4}{\pi} + \frac{8}{\pi^2} \int_{-1}^{1} x \, d \left(\cos \frac{\pi x}{2} \right)$$

$$= \frac{4}{\pi} + \frac{8}{\pi^2} \left(x \cos \frac{\pi x}{2} \Big|_{-1}^{1} - \int_{-1}^{1} \cos \frac{\pi x}{2} \, dx \right)$$

$$= \frac{4}{\pi} - \frac{32}{\pi^3}$$

Combining this with the previous integral, we have

$$\int_{-1}^{1} \left(x^2 - \frac{1}{3} \right) \cos \frac{\pi x}{2} \, dx = \frac{4}{\pi} - \frac{32}{\pi^3} - \frac{4}{3\pi}$$

$$= \frac{8\pi^2 - 96}{3\pi^3}$$

But $\int_{-1}^{1} (x^2 - \frac{1}{3})^2 \, dx = \int_{-1}^{1} (x^4 - \frac{2}{3}x^2 + \frac{1}{9}) \, dx = \frac{2}{5} - (\frac{2}{3})^2 + \frac{2}{9} = \frac{8}{45}$. Therefore $\bar{c}_2 = \frac{45}{8}(8\pi^2 - 96)(3\pi^3) = 15(\pi^2 - 12)/\pi^3$. Thus the required orthogonal projection is

$$s(x) = \frac{2}{\pi} + \frac{15(\pi^2 - 12)}{\pi^3} \left(x^2 - \frac{1}{3} \right)$$

For example, to four decimal places of accuracy, we have

$$s(0) = (0.6366) + \tfrac{1}{3}(1.0306) = 0.9801$$

$$s(1) = (0.6366) - \tfrac{2}{3}(1.0306) = -0.0505$$

$$s(\tfrac{1}{2}) = (0.6366) + \tfrac{1}{12}(1.0306) = 0.7225$$

$$s(\tfrac{1}{3}) = (0.6366) + \tfrac{2}{9}(1.0306) = 0.8656$$

$$s(\tfrac{2}{3}) = (0.6366) - \tfrac{1}{9}(1.0306) = 0.5221$$

It is instructive to compare these with the corresponding values of $\cos(\pi x/2)$, which are 1.0000, 0, 0.7071, 0.8667, 0.5000. To compare $d_{\min}$, we have, to four decimals,

$$\bar{c}_0^2 = \left(\frac{2}{\pi}\right)^2 = 0.4053$$

$$\|1\|^2 = \int_{-1}^{1} dx = 2$$

$$\bar{c}_2^2 = \left[\frac{15(\pi^2 - 12)}{\pi^3}\right]^2 = 1.0622$$

$$\|x^2 - \tfrac{1}{3}\|^2 = \int_{-1}^{1}(x^2 - \tfrac{1}{3})^2\, dx = \tfrac{2}{5} - \tfrac{4}{9} + \tfrac{2}{9} = 0.1778$$

$$\left\|\cos\frac{\pi x}{2}\right\|^2 = \int_{-1}^{1}\cos^2\frac{\pi x}{2}\, dx = \frac{1}{2}\int_{-1}^{1}(1 + \cos \pi x)\, dx = 1$$

thus, to four decimals,

$$d_{\min}^2 = 1 - (0.4053)(2) - (1.0622)(0.1778) = 0.0004$$

and, to two decimals, $d_{\min} = 0.02$. ●

Gram-Schmidt Orthogonalization Process

Orthogonal sets of functions are easily manufactured from arbitrary sets of functions by the so-called Gram-Schmidt procedure. Suppose that we are given a set of functions $(\varphi_1, \ldots, \varphi_n)$ not necessarily orthogonal. To avoid redundance, we suppose *linear independence,* meaning that there are no relations of the form $c_1\varphi_1 + \cdots + c_n\varphi_n = 0$ among the $(\varphi_1, \ldots, \varphi_n)$ other than the trivial relation where $c_1 = 0, \ldots, c_n = 0$. In particular, we have $\|\varphi_i\| \neq 0$ for $1 \leq i \leq n$. The orthogonalized functions are constructed by taking $\psi_1 = \varphi_1$ and

$$\psi_i = \varphi_i - \text{proj}(\varphi_i; \psi_1, \ldots, \psi_{i-1}) \qquad 2 \leq i \leq n$$

In detail, we have

$$\psi_1 = \varphi_1$$

$$\psi_2 = \varphi_2 - \frac{\langle\varphi_2, \psi_1\rangle}{\langle\psi_1, \psi_1\rangle}\psi_1$$

$$\psi_3 = \varphi_3 - \frac{\langle\varphi_3, \psi_2\rangle}{\langle\psi_2, \psi_2\rangle}\psi_2 - \frac{\langle\varphi_3, \psi_1\rangle}{\langle\psi_1, \psi_1\rangle}\psi_1$$

$$\vdots$$

$$\psi_n = \varphi_n - \sum_{i=1}^{n-1}\frac{\langle\varphi_n, \psi_i\rangle}{\langle\psi_i, \psi_i\rangle}\psi_i$$

The functions $\psi_1, \ldots, \psi_n$ are orthogonal. Furthermore the sets $(\varphi_1, \ldots, \varphi_n)$ and $(\psi_1, \ldots, \psi_n)$ have the same *linear span; that is,* any function of the form $f = c_1\varphi_1 + \cdots + c_n\varphi_n$ can be written in the form $d_1\psi_1 + \cdots + d_n\psi_n$ for appropriate $d_1, \ldots, d_n$, and conversely.

Example 0.3.2. Let $\varphi_1 = 1$, $\varphi_2 = x$, $\varphi_3 = x^2$ for $0 \le x \le 1$. Apply the Gram-Schmidt procedure to find the orthogonal functions ψ_1, ψ_2, ψ_3.

Solution. We have $\psi_1 = \varphi_1 = 1$, $\langle \varphi_2, \psi_1 \rangle = \int_0^1 x \, dx = \frac{1}{2}$, $\langle \psi_1, \psi_1 \rangle = 1$. Thus $\psi_2 = x - \frac{1}{2}$. The remaining inner products are

$$\langle \varphi_3, \psi_2 \rangle = \int_0^1 x^2(x - \tfrac{1}{2}) \, dx = \tfrac{1}{4} - \tfrac{1}{2}(\tfrac{1}{3}) = \tfrac{1}{12}$$

$$\langle \psi_2, \psi_2 \rangle = \int_0^1 (x - \tfrac{1}{2})^2 \, dx = \tfrac{1}{3} - \tfrac{1}{2} + \tfrac{1}{4} = \tfrac{1}{12}$$

$$\langle \varphi_3, \psi_1 \rangle = \int_0^1 x^2 \, dx = \tfrac{1}{3}$$

Thus $\psi_3 = x^2 - (x - \frac{1}{2}) - \frac{1}{3} = x^2 - x + \frac{1}{6}$. The orthogonal functions are 1, $x - \frac{1}{2}$, $x^2 - x + \frac{1}{6}$, $0 \le x \le 1$. ●

Orthonormal Sets of Functions

The formulas for the Fourier coefficients and the minimum distance become especially simple when the functions $\varphi_1, \ldots, \varphi_n$ are *orthonormal.* This means that $\langle \varphi_i, \varphi_j \rangle = 0$ for $i \ne j$ and $\langle \varphi_i, \varphi_i \rangle = 1$, $1 \le i \le n$. Thus we have for orthonormal functions

$$\bar{c}_i = \langle f, \varphi_i \rangle \qquad 1 \le i \le n$$

$$d^2_{\min} = D(\bar{c}_1, \ldots, \bar{c}_n) = \|f\|^2 - (\bar{c}_1^2 + \cdots + \bar{c}_n^2)$$

If $(\varphi_1, \ldots, \varphi_n)$ is an orthogonal set of functions, we obtain an orthonormal set by replacing φ_i by $\varphi_i/\|\varphi_i\|$, $1 \le i \le n$.

Example 0.3.3. Let $\varphi_1 = 1$, $\varphi_2 = \sin x$, $\varphi_3 = \cos x$ for $-\pi < x < \pi$. Find the corresponding orthonormal set.

Solution. We have

$$\|\varphi_1\|^2 = \int_{-\pi}^{\pi} dx = 2\pi$$

$$\|\varphi_2\|^2 = \int_{-\pi}^{\pi} \sin^2 x \, dx = \pi$$

$$\|\varphi_3\|^2 = \int_{-\pi}^{\pi} \cos^2 x \, dx = \pi$$

Therefore $1/\sqrt{2\pi}$, $(\cos x)/\sqrt{\pi}$ and $(\sin x)/\sqrt{\pi}$ form an orthonormal set. ●

In many problems we are given an *infinite* orthonormal set

$$(\varphi_n)_{n \geq 1} = (\varphi_1, \varphi_2, \ldots)$$

To study these, we apply the above procedure to the finite orthonormal set $(\varphi_1, \ldots, \varphi_n)$. The Fourier coefficients are

$$\overline{c}_i = \langle f, \varphi_i \rangle \qquad 1 \leq i \leq n$$

which do not depend on n. Furthermore, we have Bessel's inequality for each n

$$\sum_{i=1}^{n} \overline{c}_i^2 \leq \|f\|^2 \qquad n = 1, 2, \ldots$$

This is valid for every $n = 1, 2, \ldots$; hence the infinite series $\sum_{i=1}^{\infty} \overline{c}_i^2$ converges, and we have

$$\boxed{\sum_{i=1}^{\infty} \overline{c}_i^2 \leq \|f\|^2}$$

which is the general form of Bessel's inequality, valid for an infinite orthonormal set (φ_n).

Parseval's Theorem and Mean Square Convergence

In many cases of interest Bessel's inequality is an equality, namely,

$$\boxed{\sum_{i=1}^{\infty} \overline{c}_i^2 = \|f\|^2}$$

This statement, known as *Parseval's theorem,* is not true for an arbitrary function. For example, the set of functions $\pi^{-1/2} (\sin nx, \cos nx)_{n \geq 1}$ is an orthonormal set for $-\pi \leq x \leq \pi$. The function $f(x) = 1$ has all Fourier coefficients zero; indeed $\int_{-\pi}^{\pi} \sin nx\, dx = 0 = \int_{-\pi}^{\pi} \cos nx\, dx$, $n \geq 1$. Yet $\|f\|^2 = \int_{-\pi}^{\pi} 1\, dx = 2\pi$. In this case Bessel's inequality is the statement $0 = \sum_{i=1}^{\infty} \overline{c}_i^2 < \|f\|^2 = 2\pi$.

The validity of Parseval's theorem for a particular function f is intimately related to the *mean square convergence* of the Fourier series $\sum_{i=1}^{\infty} \overline{c}_i \varphi_i$, that is,

$$\lim_{n \to \infty} \left\| f - \sum_{i=1}^{n} \overline{c}_i \varphi_i \right\|^2 = 0$$

We state the result as follows.

> **Proposition 0.3.4.** Let $(\varphi_n)_{n \geq 1}$ be an orthonormal set and f an arbitrary function. Parseval's theorem is true if and only if we have mean square convergence of the Fourier series $\sum_{i=1}^{\infty} \overline{c}_i \varphi_i$.

Proof. Let $\bar{c}_i = \langle f, \varphi_i \rangle$ be the ith Fourier coefficient of f. Then by expanding the inner product and using orthonormality on the left side, we have

$$\left\| f - \sum_{i=1}^{n} \bar{c}_i \varphi_i \right\|^2 = \|f\|^2 - 2 \sum_{i=1}^{n} \bar{c}_i \langle f, \varphi_i \rangle + \sum_{i=1}^{n} \bar{c}_i^2$$

$$= \|f\|^2 - \sum_{i=1}^{n} \bar{c}_i^2$$

Letting $n \to \infty$, we see that the right side tends to zero if and only if Parseval's theorem is valid. The left side tends to zero (by definition) if and only if we have mean square convergence. Therefore the proposition is proved. ∎

If Parseval's theorem holds for all functions f in a suitable class of functions, we say that the orthonormal set is *complete* with respect to that class. For example, in Chap. 1 we show that the trigonometric system $\{1/\sqrt{2\pi}, (\sin nx)/\sqrt{\pi}, (\cos nx)/\sqrt{\pi}\}_{n \geq 1}$ is complete with respect to the class of differentiable functions which are 2π-periodic. In the theory of the Lebesgue integral, it is shown that the same trigonometric system is complete with respect to the class of functions f with $\int_{-\pi}^{\pi} |f(x)|^2 \, dx < \infty$.

Complex Inner Product

For complex-valued functions, the inner product and norm are defined by

$$\langle \varphi, \psi \rangle = \int_a^b \varphi(x) \, \overline{\psi(x)} \, dx$$

$$\|\varphi\| = \langle \varphi, \varphi \rangle^{1/2} \geq 0$$

The inner product is linear and homogeneous in the sense that if $\varphi_1, \varphi_2, \psi_1, \psi_2$ are complex-valued functions and a, b, c, d are complex constants, we have

$$\langle a\varphi_1 + b\varphi_2, c\psi_1 + d\psi_2 \rangle = a\bar{c} \langle \varphi_1, \psi_1 \rangle + a\bar{d} \langle \varphi_1, \psi_2 \rangle + b\bar{c} \langle \varphi_2, \psi_1 \rangle + b\bar{d} \langle \varphi_2, \psi_2 \rangle$$

Orthogonality of complex functions is defined by $\langle \varphi, \psi \rangle = 0$.

We record for future use the complex form of the Schwarz inequality.

Proposition 0.3.5. If $\varphi(x), \psi(x), a < x < b$, are complex-valued functions, then

$$\left| \int_a^b \varphi(x) \, \overline{\psi(x)} \, dx \right|^2 \leq \int_a^b |\varphi(x)|^2 \, dx \int_a^b |\psi(x)|^2 \, dx$$

If equality holds and both $\varphi(x)$ and $\psi(x)$ are continuous functions, then $C_1\varphi(x) = C_2\psi(x)$ for some (complex) constants C_1, C_2.

Proof. Let the inner product be written in polar form: $\langle \varphi, \psi \rangle = \int_a^b \varphi(x)\overline{\psi(x)} \, dx = Re^{i\theta}$, where $R = |\int_a^b \varphi(x)\overline{\psi(x)} \, dx|$. Then $\langle \varphi e^{-i\theta}, \psi \rangle = R$ is real. The polynomial $G(t_1, t_2) = \|t_1\varphi e^{-i\theta} - t_2\psi\|^2$ is nonnegative and may be written as $G(t_1, t_2) = t_1^2\|\varphi\|^2 - 2t_1 t_2 R + t_2^2\|\psi\|^2$. Hence the discriminant of the coefficients is nonnegative: $\|\varphi\|^2\|\psi\|^2 -$

$R^2 \geq 0$, which is the required inequality. To deal with the case of equality, we may write the polynomial in the form

$$G(t_1, t_2) = t_1^2\|\varphi\|^2 - 2t_1t_2\|\varphi\|\|\psi\| + t_2^2\|\psi\|^2$$

$$= (t_1\|\varphi\| - t_2\|\psi\|)^2$$

$$= \|t_1\varphi e^{-i\theta} - t_2\psi\|^2$$

If both $\|\varphi\|$ and $\|\psi\|$ are zero, then $\varphi(x) \equiv 0$, $\psi(x) \equiv 0$, and the required equality is true for any values of the complex constants C_1, C_2. Otherwise, one of them is nonzero, say $\|\psi\| \neq 0$. Taking $t_1 = 1$, $t_2 = \|\varphi\|/\|\psi\|$, we have the required equality with $C_1 = 1$, $C_2 = t_2 e^{i\theta}$. If $\|\varphi\| \neq 0$, we obtain the required equality with $C_2 = 1$, $C_1 = \|\psi\|e^{-i\theta}/\|\varphi\|$. ∎

EXERCISES 0.3

1. Let $\varphi_1 = 1$, $\varphi_2 = x$, $\varphi_3 = x^2$ on the interval $0 \leq x \leq 1$. Find the following inner products:
   (a) $\langle \varphi_1, \varphi_2 \rangle$
   (b) $\langle \varphi_1, \varphi_3 \rangle$
   (c) $\|\varphi_1 - \varphi_2\|^2$
   (d) $\|2\varphi_1 + 3\varphi_2\|^2$

2. Which of the following pairs of functions are orthogonal on the interval $0 \leq x \leq 1$?

$$\varphi_1 = \sin 2\pi x \qquad \varphi_2 = x \qquad \varphi_3 = \cos 2\pi x \qquad \varphi_4 = 1$$

3. Apply the Gram-Schmidt procedure to obtain orthogonal functions beginning with the functions $\varphi_1 = 1$, $\varphi_2 = x$, $\varphi_3 = x^2$, for $-1 \leq x \leq 1$.

4. Let $\bar{f} = \bar{c}_1 \varphi_1 + \cdots + \bar{c}_n \varphi_n$ be the projection of f on the orthogonal set $(\varphi_1, \ldots, \varphi_n)$. Show that $f - \bar{f}$ is orthogonal to each of the functions $\varphi_1, \ldots, \varphi_n$.

5. Find the projection of the function $\sin \pi x$ on the orthogonal set $(1, x - \frac{1}{2})$ on the interval $0 \leq x \leq 1$, and compute the minimum distance d_{min}.

6. Find the projection of the function $f(x) = \cos^2 x$ on the orthogonal set $(1, \cos x, \cos 2x)$ on the interval $-\pi \leq x \leq \pi$.

7. Let $\varphi_1(x) = 1$, $\varphi_2(x) = x/|x|$, $\varphi_3(x) = x^2 - \frac{1}{3}$ for $-1 \leq x \leq 1$.
   (a) Show that φ_1, φ_2, and φ_3 form an orthogonal set.
   (b) Find the projection of $f(x) = x$ on this orthogonal set and compute the minimum distance d_{min}.

8. Let $(\varphi_1, \varphi_2, \varphi_3)$ be an orthonormal set of functions on the interval $-1 \leq x \leq 1$, and let f be any function of the form $f(x) = a_1\varphi_1(x) + a_2\varphi_2(x) + a_3\varphi_3(x)$.
   (a) Show that $\|f\|^2 = a_1^2 + a_2^2 + a_3^2$.
   (b) Show that $\langle f, \varphi_1 \rangle = a_1$, $\langle f, \varphi_2 \rangle = a_2$, $\langle f, \varphi_3 \rangle = a_3$.

9. Let $(\varphi_1, \varphi_2, \varphi_3)$ be an orthonormal set of functions on the interval $-1 \leq x \leq 1$, and let $f(x) = a_1\varphi_1(x) + a_2\varphi_2(x) + a_3\varphi_3(x)$, $g(x) = b_1\varphi_1(x) + b_2\varphi_2(x) + b_3\varphi_3(x)$.
   (a) Show that $\langle f, g \rangle = a_1b_1 + a_2b_2 + a_3b_3$.
   (b) Discuss the relation with the three-dimensional dot product formula.

10. Define the angle between two nonzero functions φ, ψ by the formula $\cos \theta = \langle \varphi, \psi \rangle/(\|\varphi\| \|\psi\|)$, $0 \leq \theta \leq \pi$.
    (a) If φ and ψ are orthogonal, show that $\theta = \pi/2$.
    (b) If φ and ψ are proportional, show that $\theta = 0$ or $\theta = \pi$.

(c) If $\theta = 0$ or π, does it follow that φ and ψ are necessarily proportional? (*Hint:* Compute $\|\varphi - c\psi\|^2$ and write it as a perfect square.)

(d) Compute θ if $\varphi(x) = 1$, $\psi(x) = x$ for $0 \le x \le 1$.

11. Find the orthonormal set corresponding to the orthogonal set found in Exercise 3.

0.4 STURM-LIOUVILLE THEORY†

Formulation of the Problem

Orthogonal functions arise naturally from second-order differential equations with boundary conditions. It is our purpose here to collect some useful facts about these equations, whose theory was developed by J. Sturm and J. Liouville in the mid-nineteenth century.

The independent variable is x, with $a \le x \le b$, and the unknown function is $\varphi(x)$, a real-valued function. The general Sturm-Liouville equation is

$$\boxed{(s\varphi')' + (\lambda\rho - q)\varphi = 0} \tag{0.4.1}$$

where $s(x)$, $\rho(x)$, and $q(x)$ are given functions defined for $a \le x \le b$; λ is an unknown constant, called the *eigenvalue parameter;* and ρ is the *weight function.* Equation (0.4.1) will be solved subject to certain *boundary conditions,* which will be explained.

For example, the constant coefficient equation $\varphi'' + \lambda\varphi = 0$ is a Sturm-Liouville equation with $s(x) = 1$, $\rho(x) = 1$, $q(x) = 0$. The (Bessel) equation $(x\varphi')' + \lambda(x\varphi) = 0$ is a Sturm-Liouville equation with $s(x) = x$, $\rho(x) = x$, $q(x) = 0$.

Equation (0.4.1) is in *self-adjoint form.* Many other equations can be written in this form by appropriate choices of functions $s(x)$, $\rho(x)$, $q(x)$. For example, any equation of the form $a(x)\varphi'' + b(x)\varphi' + [\lambda c(x) - d(x)]\varphi = 0$ can be written in the self-adjoint form (0.4.1) by taking

$$s(x) = \exp\left[\int^x \frac{b(u)}{a(u)} \, du\right]$$

$$\rho(x) = s(x)\frac{c(x)}{a(x)} \qquad q(x) = s(x)\frac{d(x)}{a(x)}$$

The functions $s(x)$, $\rho(x)$, $\varphi(x)$ are supposed to be continuous for $a \le x \le b$, with $s(x) > 0$ and $\rho(x) > 0$ for $a < x < b$. If $s(a)\rho(a) = 0$ or $s(b)\rho(b) = 0$, we speak of a *singular Sturm-Liouville problem.* Otherwise, if both $s(a)\rho(a) \ne 0$ and $s(b)\rho(b) \ne 0$, we speak of a *regular Sturm-Liouville problem.* For example, $\varphi'' + \lambda\varphi = 0$ is a regular Sturm-Liouville problem; the Bessel equation $x\varphi' +$

---

† This section, which treats theoretical material, may be omitted by some instructors and developed when necessary for the solution of boundary-value problems in Chaps. 2, 3, and 4. As a bare minimum, one may cover the section on orthogonality and the ensuing examples through Proposition 0.4.3.

$\lambda(x\varphi) = 0$ is a singular Sturm-Liouville problem if $0 < x < 1$ and a regular Sturm-Liouville problem if $1 < x < 2$. If $-1 < x < 1$, the Bessel equation is not of the Sturm-Liouville type since the function $s(x) = 0$ at $x = 0$, which is inside the interval $-1 < x < 1$.

A solution of equation (0.4.1) is defined to be a pair (φ, λ) with φ not identically zero; φ is the *eigenfunction* and λ is the *eigenvalue*. For example, $\varphi(x) = \sin mx$ is an eigenfunction of the Sturm-Liouville equation $\varphi'' + \lambda\varphi = 0$ with eigenvalue $\lambda = m^2$.

Orthogonality of the Eigenfunctions

Sturm-Liouville equations with boundary conditions provide a rich source of orthogonal functions. Each Sturm-Liouville equation has its own orthogonality relation, depending on the *weight function* ρ. For example, if $\varphi'' + \lambda\varphi = 0$, the weight function is $\rho(x) = 1$, whereas for the Bessel equation $(x\varphi')' + \lambda(x\varphi) = 0$ the weight function is $\rho(x) = x$. In general, we say that φ_1, φ_2 are *orthogonal with respect to the weight function* ρ if $\int_a^b \rho(x)\varphi_1(x)\varphi_2(x)\,dx = 0$. This is equivalent to the statement that $\varphi_1\sqrt{\rho}$ and $\varphi_2\sqrt{\rho}$ are orthogonal in the sense of Sec. 0.3.

Proposition 0.4.1. Let (φ_1, λ_1) and (φ_2, λ_2) be solutions of the Sturm-Liouville equation (0.4.1), satisfying the boundary condition

$$s(b)[\varphi_1(b)\varphi_2'(b) - \varphi_1'(b)\varphi_2(b)] = s(a)[\varphi_1(a)\varphi_2'(a) - \varphi_1'(a)\varphi_2(a)] \qquad (0.4.2)$$

Then

$$(\lambda_1 - \lambda_2) \int_a^b \varphi_1(x)\varphi_2(x)\rho(x)\,dx = 0$$

Thus, if $\lambda_1 \neq \lambda_2$, then φ_1 and φ_2 are orthogonal with respect to the weight function ρ.

Proof. We write the Sturm-Liouville equation satisfied by φ_1 as

$$(s\varphi_1')' + (\lambda_1\rho - q)\varphi_1 = 0$$

We multiply this equation by φ_2 and integrate the resulting equation for $a < x < b$. Thus

$$\int_a^b \varphi_2(s\varphi_1')' + \int_a^b \varphi_2(\lambda_1\rho - q)\varphi_1 = 0$$

The first integral can be integrated by parts; thus

$$\varphi_2 s\varphi_1' \Big|_a^b - \int_a^b \varphi_2's\varphi_1' + \int_a^b \varphi_2(\lambda_1\rho - q)\varphi_1 = 0$$

We apply the same method to the Sturm-Liouville equation for φ_2 and get

$$\varphi_1 s\varphi_2' \Big|_a^b - \int_a^b \varphi_1's\varphi_2' + \int_a^b \varphi_1(\lambda_2\rho - q)\varphi_2 = 0$$

When we subtract these and use the boundary conditions (0.4.2), all the terms cancel, except for the final integrals. This yields $(\lambda_1 - \lambda_2) \int_a^b \rho\varphi_1\varphi_2 = 0$, which was to be proved. ∎

We now give some concrete examples of boundary conditions which will occur repeatedly in problems of interest.

Example 0.4.1. *Separable boundary conditions* are of the form

$$\cos \alpha \; \varphi'(a) - \sin \alpha \; \varphi(a) = 0$$

$$\cos \beta \; \varphi'(b) + \sin \beta \; \varphi(b) = 0$$

Here α and β are real constants. If both φ_1 and φ_2 satisfy these boundary conditions, then $\varphi_1\varphi_2' - \varphi_1'\varphi_2 = 0$ at both $x = a$ and $x = b$; therefore (0.4.2) is satisfied.
As examples of separable boundary conditions, we cite the following three examples:

1. $\varphi'(a) = 0$ $\varphi'(b) = 0$ $(\alpha = 0, \beta = 0)$

2. $\varphi(a) = 0$ $\varphi(b) = 0$ $(\alpha = \pi/2, \beta = \pi/2)$

3. $\varphi'(a) = 0$ $\varphi'(b) + \varphi(b) = 0$ $(\alpha = 0, \beta = \pi/4)$

Example 0.4.2. *Periodic boundary conditions* are of the form

$$\varphi(a) = \varphi(b)$$

$$\varphi'(a) = \varphi'(b)$$

where we suppose that $s(a) = s(b)$. Again if φ_1, φ_2 both satisfy these conditions, then $s(\varphi_1\varphi_2' - \varphi_1'\varphi_2)|_a^b = 0$. Hence, orthogonality is ensured by these periodic boundary conditions. We will see that periodic boundary conditions arise from problems in cylindrical and spherical coordinates where the azimuthal angle can be regarded as a periodic variable.

Example 0.4.3. *Singular Sturm-Liouville problems* exist if $s(a) = s(b) = 0$. In this case the orthogonality relation $s(\varphi_1\varphi_2' - \varphi_1'\varphi_2)|_a^b = 0$ is automatically satisfied by any two functions φ_1, φ_2 which have continuous derivatives on the interval $a \le x \le b$. In Chap. 4 we see that this is the case for the Legendre equations, which arise from problems in spherical coordinates.

The computations which were used in the proof of Proposition 0.4.1 can be abstracted in a more systematic fashion by defining the *differential operator*

$$L\varphi = (s\varphi')' - q\varphi$$

This is a linear operator, in the sense that for any two functions φ_1, φ_2 and any two constants c_1, c_2 we have $L(c_1\varphi_1 + c_2\varphi_2) = c_1L\varphi_1 + c_2L\varphi_2$. Furthermore, we have *Lagrange's identity*

$$\varphi_2 L\varphi_1 - \varphi_1 L\varphi_2 = (s(\varphi_1'\varphi_2 - \varphi_1\varphi_2'))'$$

for any two functions φ_1, φ_2 for which the second derivatives exist. This identity can be used to give an alternative proof of the orthogonality result in Proposition 0.4.1 as well as the positivity result (Proposition 0.4.5). The proof of Lagrange's identity is left for the exercises.

The following Sturm-Liouville problem comes up repeatedly in our work:

$$\varphi'' + \lambda\varphi = 0 \qquad a < x < b$$

$$\varphi(a) = 0 \qquad \varphi(b) = 0$$

We now give the complete analysis of the eigenfunctions and eigenvalues of this problem. To do this, we must consider separately the cases $\lambda > 0$, $\lambda = 0$, $\lambda < 0$. The general solution is given by

$$\varphi(x) = A \sin x\sqrt{\lambda} + B \cos x\sqrt{\lambda} \qquad \lambda > 0$$

$$\varphi(x) = Ax + B \qquad \lambda = 0$$

$$\varphi(x) = A \sinh x\sqrt{-\lambda} + B \cosh x\sqrt{-\lambda} \qquad \lambda < 0$$

The hyperbolic functions are defined by $\sinh x = \frac{1}{2}(e^x - e^{-x})$, $\cosh x = \frac{1}{2}(e^x + e^{-x})$. These linear combinations of e^x, e^{-x} are often convenient when boundary conditions are specified at $x = 0$. We note that $\cosh x > 1$ for $x \neq 0$, while $\sinh x = 0$ only when $x = 0$.

Substituting the boundary conditions $\varphi(a) = 0$, $\varphi(b) = 0$, we have

$$0 = A \sin a\sqrt{\lambda} + B \cos a\sqrt{\lambda}$$
$$0 = A \sin b\sqrt{\lambda} + B \cos b\sqrt{\lambda} \qquad \lambda > 0$$

$$0 = Aa + B$$
$$0 = Ab + B \qquad \lambda = 0$$

$$0 = A \sinh a\sqrt{-\lambda} + B \cosh a\sqrt{-\lambda}$$
$$0 = A \sinh b\sqrt{-\lambda} + B \cosh b\sqrt{-\lambda} \qquad \lambda < 0$$

This system of equations has a solution $(A, B) \neq (0, 0)$ if and only if the determinant of the coefficients is nonzero. This translates to three separate equations:

$$\sin a\sqrt{\lambda} \cos b\sqrt{\lambda} - \sin b\sqrt{\lambda} \cos a\sqrt{\lambda} = 0 \qquad \lambda > 0$$

$$a - b = 0 \qquad \lambda = 0$$

$$\sinh a\sqrt{-\lambda} \cosh b\sqrt{-\lambda} - \sinh b\sqrt{-\lambda} \cosh a\sqrt{-\lambda} = 0 \qquad \lambda < 0$$

For $\lambda > 0$, we have $0 = \sin (a - b)\sqrt{\lambda}$ which is satisfied if and only if $(a - b)\sqrt{\lambda} = -n\pi$ for some $n = 1, 2, \ldots$. The eigenfunction is written $\varphi(x) = \cos a\sqrt{\lambda} \sin x\sqrt{\lambda} - \sin a\sqrt{\lambda} \cos x\sqrt{\lambda} = \sin [n\pi(x - a)/(b - a)]$.

For $\lambda = 0$, there are no solutions with $(A, B) \neq (0, 0)$.

For $\lambda < 0$, we have $0 = \sinh (a - b)\sqrt{-\lambda}$, which is satisfied only if $\lambda = 0$, which is impossible.

To summarize, we have the following proposition.

Proposition 0.4.2. The solutions of the Sturm-Liouville equation $\varphi'' + \lambda\varphi = 0$ with the boundary conditions $\varphi(a) = 0$, $\varphi(b) = 0$ are given by $\lambda_n = [n\pi/(b - a)]^2$, $\varphi_n(x) = \sin [n\pi(x - a)/(b - a)]$.

From Proposition 0.4.1 it follows that these functions are orthogonal on the interval $a < x < b$. The projection of a function $f(x)$, $a < x < b$, on this orthogonal set is called a *Fourier sine series* and is studied in detail in Chap. 1. This series has the form $\Sigma B_n \sin [n\pi(x - a)/(b - a)]$, where $B_n = [2/(b - a)] \int_a^b f(x) \sin [n\pi(x - a)/(b - a)] \, dx$.

A closely related problem is the Sturm-Liouville equation $\varphi'' + \lambda\varphi = 0$ with the periodic boundary conditions $\varphi(a) = \varphi(b)$, $\varphi'(a) = \varphi'(b)$. Repeating the above analysis, we consider separately the cases $\lambda > 0$, $\lambda = 0$, $\lambda < 0$. The solutions of the differential equations have the same form in each case. To satisfy the boundary conditions, we must have

$$A \sin a\sqrt{\lambda} + B \cos a\sqrt{\lambda} = A \sin b\sqrt{\lambda} + B \cos b\sqrt{\lambda}$$
$$\sqrt{\lambda}(A \cos a\sqrt{\lambda} - B \sin a\sqrt{\lambda}) = \sqrt{\lambda} (A \cos b\sqrt{\lambda} - B \sin b\sqrt{\lambda}) \qquad \lambda > 0$$

$$Aa + B = Ab + B$$
$$A = A \qquad \lambda = 0$$

$$A \sinh a\sqrt{-\lambda} + B \cosh a\sqrt{-\lambda} = A \sinh b\sqrt{-\lambda} + B \cosh b\sqrt{-\lambda}$$
$$\sqrt{-\lambda}(A \cosh a\sqrt{-\lambda} + B \sinh b\sqrt{-\lambda}) \qquad \lambda < 0$$
$$= \sqrt{-\lambda}(A \cosh b\sqrt{-\lambda} + B \cosh b\sqrt{-\lambda})$$

This system of equations has a solution $(A, B) \neq (0, 0)$ if and only if the determinant of the coefficients is nonzero. This translates to three separate equations:

$$(\sin a\sqrt{\lambda} - \sin b\sqrt{\lambda})^2 + (\cos a\sqrt{\lambda} - \cos b\sqrt{\lambda})^2 = 0 \qquad \lambda > 0$$

$$A = 0 \qquad \lambda = 0$$

$$(\sinh a\sqrt{-\lambda} - \sinh b\sqrt{-\lambda})^2 - (\cosh a\sqrt{-\lambda} - \cosh b\sqrt{-\lambda})^2 = 0 \qquad \lambda < 0$$

For $\lambda > 0$, we have $\sin a\sqrt{\lambda} = \sin b\sqrt{\lambda}$, $\cos a\sqrt{\lambda} = \cos b\sqrt{\lambda}$, hence $0 = \sin a\sqrt{\lambda} \cos b\sqrt{\lambda} - \cos a\sqrt{\lambda} \sin b\sqrt{\lambda} = \sin (a - b)\sqrt{\lambda}$, which leads to $(a - b)\sqrt{\lambda} = m\pi$ for some $m = 1, 2, \ldots$. But $\cos b\sqrt{\lambda} = \cos a\sqrt{\lambda} = \cos (b\sqrt{\lambda} + m\pi) = \cos b\sqrt{\lambda} (-1)^m$, which requires that m be an even integer, $m = 2n$ for some $n = 1, 2, \ldots$. Therefore λ must be of the form $\lambda = [2n\pi/(b - a)]^2$. We note that for this value of λ the entire system of linear equations is identically zero, so that any pair (A, B) is a solution. In this case we have a *doubly degenerate eigenvalue*.

For $\lambda = 0$, the solution is $A = 0$ with B undetermined, hence $\lambda = 0$ is an eigenvalue with $\varphi(x) = 1$.

For $\lambda < 0$, we use the identities for the hyperbolic functions to write the determinant condition in the form $0 = 2[1 - \cosh (a - b)\sqrt{\lambda}]$. This can happen only if $\lambda = 0$, which is impossible. We summarize the results in the following form.

Proposition 0.4.3. The solutions of the Sturm-Liouville equation $\varphi'' + \lambda\varphi = 0$ with the periodic boundary conditions $\varphi(a) = \varphi(b)$, $\varphi'(a) = \varphi'(b)$ are given by $\lambda_n = [2n\pi/(b - a)]^2$, $\varphi_n(x) = A_n \cos [2n\pi x/(b - a)] + B_n \sin [2n\pi x/(b - a)]$, where $n = 0, 1, \ldots$ and constants A_n, B_n are arbitrary.

In most applications the periodic boundary conditions are applied on an interval of the form $-L < x < L$, thus $b - a = 2L$. From Proposition 0.4.1 it follows that the functions $\varphi_n(x) = A_n \cos (n\pi x/L) + B_n \sin (n\pi x/L)$ are orthogonal on $-L < x < L$. The projection of a function $f(x)$, $-L < x < L$, on this orthogonal set is called a *general Fourier series*, which is studied in detail in Chap. 1. The series has the form

$$A_0 + \sum_{n=1}^{\infty} \left(A_n \cos \frac{n\pi x}{L} + B_n \sin \frac{n\pi x}{L} \right)$$

where

$$A_0 = \frac{1}{2L} \int_{-L}^{L} f(x) \, dx$$

$$A_n = \frac{1}{L} \int_{-L}^{L} f(x) \cos \frac{n\pi x}{L} \, dx \qquad n = 1, 2, \ldots$$

$$B_n = \frac{1}{L} \int_{-L}^{L} f(x) \sin \frac{n\pi x}{L} \, dx \qquad n = 1, 2, \ldots$$

Nondegeneracy of the Eigenvalues

In the case of periodic boundary conditions we have, for each eigenvalue $\lambda_n = (n\pi/L)^2$, two linearly independent eigenfunctions $\sin (n\pi x/L)$ and $\cos (n\pi x/L)$. On the other hand, for Sturm-Liouville problems with separable boundary conditions, there is only one linearly independent eigenfunction φ for each eigenvalue λ. We state and prove this result now.

Theorem 0.4.1. Let φ_1, φ_2 be solutions of the Sturm-Liouville problem

$$(s\varphi')' + (\lambda\rho - q)\varphi = 0 \qquad a < x < b$$

$$\cos \alpha \, \varphi'(a) - \sin \alpha \, \varphi(a) = 0$$

$$\cos \beta \, \varphi'(b) + \sin \beta \, \varphi(b) = 0$$

where

$$\varphi_1(x) \neq 0 \qquad \varphi_2(x) \neq 0 \qquad s(a)s(b) \neq 0$$

Then there exists a constant $C \neq 0$ such that

$$\varphi_1(x) = C\varphi_2(x)$$

The following proof may be omitted without loss of continuity.

Proof. By hypothesis, $s(a) \neq 0$. We consider separately two cases.

Case 1: cos $\alpha = 0$. In this case the boundary condition at $x = a$ is $\varphi(a) = 0$. We claim that $\varphi_1'(a) \neq 0$, $\varphi_2'(a) \neq 0$. Indeed, suppose that $\varphi_1'(a) = 0$. The boundary condition requires $\varphi_1(a) = 0$; hence, from the uniqueness theorem for second-order differential equations, $\varphi_1(x) = 0$, a contradiction. Similarly $\varphi_2'(a) \neq 0$.

Now let

$$\varphi(x) = \varphi_2'(a)\varphi_1(x) - \varphi_1'(a)\varphi_2(x)$$

Then φ is a solution of the Sturm-Liouville equation, and

$$\varphi(a) = \varphi_2'(a)\varphi_1(a) - \varphi_1'(a)\varphi_2(a) = 0$$

$$\varphi'(a) = \varphi_2'(a)\varphi_1'(a) - \varphi_1'(a)\varphi_2'(a) = 0$$

Therefore, $\varphi(x) \equiv 0$ by the uniqueness theorem, which proves that $\varphi_2(x) = [\varphi_2'(a)/\varphi_1'(a)]\varphi_1(x)$. This proves the theorem in case 1, with $C = \varphi_2'(a)/\varphi_1'(a) \neq 0$.

Case 2: cos $\alpha \neq 0$. In this case the boundary condition at $x = a$ can be written in the form $\varphi'(a) = \tan \alpha \, \varphi(a)$. We claim that $\varphi_1(a) \neq 0$, $\varphi_2(a) \neq 0$. Indeed, suppose that $\varphi_1(a) = 0$. The boundary condition requires that $\varphi_1'(a) = 0$; hence, from the uniqueness theorem for second-order differential equations, $\varphi_1(x) \equiv 0$, a contradiction. Similarly $\varphi_2(a) \neq 0$.

Now let

$$\varphi(x) = \varphi_2(a)\varphi_1(x) - \varphi_1(a)\varphi_2(x)$$

Then φ is a solution of the Sturm-Liouville equation, and

$$\varphi(a) = \varphi_2(a)\varphi_1(a) - \varphi_1(a)\varphi_2(a) = 0$$

$$\varphi'(a) = \varphi_2(a)\varphi_1'(a) - \varphi_1(a)\varphi_2'(a)$$

$$= \varphi_2(a) \tan \alpha \, \varphi_1(a) - \varphi_1(a) \tan \alpha \, \varphi_2(a)$$

$$= 0$$

Therefore $\varphi(x) = 0$ by the uniqueness theorem, which proves that $\varphi_2(x) = [\varphi_2(a)/\varphi_1(a)]\varphi_1(x)$. This proves the theorem in case 2, with $C = \varphi_2(a)/\varphi_1(a) \neq 0$. ∎

Positivity of the Eigenvalues

It is frequently important to know whether all eigenvalues λ of a Sturm-Liouville problem are positive or zero. This can often be inferred from the boundary conditions, without computing the eigenvalues.

Proposition 0.4.4. Let (φ, λ) be a solution of the Sturm-Liouville equation (0.4.1) satisfying the conditions

$$s\varphi\varphi' \Big|_a^b = s(b)\varphi(b)\varphi'(b) - s(a)\varphi(a)\varphi'(a) \leq 0 \qquad (0.4.3)$$

Then

$$\lambda \int_a^b \rho\varphi^2 \geq \int_a^b q\varphi^2 + \int_a^b s(\varphi')^2$$

If in addition $q(x) \geq 0$, then all the terms on the right are nonnegative and all eigenvalues are positive or zero.

Proof. To prove this result, we multiply the Sturm-Liouville equation by φ and obtain

$$\varphi(s\varphi')' + (\lambda\rho - q)\varphi^2 = 0$$

We integrate this equation for $a < x < b$. The first term can be integrated by parts and produces

$$s\varphi\varphi' \Big|_a^b - \int_a^b s(\varphi')^2 + \int_a^b (\lambda\rho - q)\varphi^2 = 0$$

Condition (0.4.3) states that the first term is less than or equal to zero, and the stated result follows.   ∎

We can use this result in many cases of interest, for example:

Zero boundary conditions: $\varphi(a) = 0$ $\varphi(b) = 0$

Mixed boundary conditions: $\varphi(a) = 0$ $\varphi'(b) = 0$

Singular Sturm-Liouville problem: $s(a) = 0$ $s(b) = 0$

In each case we may verify that the condition $s\varphi\varphi'|_a^b = 0$ is satisfied. Of course, it is also possible to consider various combinations of these three examples.

The conditions (0.4.3) are sufficient, but not necessary, to guarantee positivity. Consider, for example, the equation $\varphi'' + \lambda\varphi = 0$, with the boundary conditions $\varphi(0) = 0$, $\varphi'(L) = \alpha\varphi(L)$. By a direct computation, the eigenvalues are positive or zero if $0 < \alpha < 1/L$. Yet the boundary conditions do not satisfy (0.4.3) for any value of $\alpha > 0$.

Finally, a word on the physical significance of positivity. In Chap. 2 we study the heat equation, which describes temperature changes in space. A solution of the heat equation is often written in the form

$$u(z; t) = U(z) + \sum_{n=1}^{\infty} \varphi_n(z)e^{-\lambda_n Kt}$$

where $U(z)$ is a steady-state solution and (φ_n, λ_n) are solutions of a Sturm-Liouville problem. If all eigenvalues are positive, that is, $\lambda_n > 0$, then we may ignore the infinite series when $t \to \infty$. In other words, *positivity guarantees that we approach a steady-state solution when $t \to \infty$.*

Completeness of the Eigenfunctions

In applications to boundary-value problems, it is important to know that we can expand an arbitrary smooth function into a convergent series of eigen-

functions of a given Sturm-Liouville problem. We now give some general results, the complete proofs of which are beyond the scope of this book. To be
specific, we assume that the boundary conditions are separable, as in Example
0.4.1.

We make the additional assumption that $s(x) > 0$, $\rho(x) > 0$ for $a \leq x \leq$
b. This is a *regular Sturm-Liouville problem*.

Theorem 0.4.2. There exists an infinite sequence of solutions (φ_n, λ_n) of the regular
Sturm-Liouville problem (0.4.1) with the separable boundary conditions
$\cos \alpha \, \varphi'(a) - \sin \alpha \, \varphi(a) = 0$, $\cos \beta \, \varphi'(b) + \sin \beta \, \varphi(b) = 0$. If $f(x)$, $a \leq x \leq b$,
is a smooth function which satisfies these boundary conditions, then the following
series is uniformly convergent for $a \leq x \leq b$:

$$\sum_{n=1}^{\infty} A_n \varphi_n(x) = f(x) \qquad a \leq x \leq b$$

where the Fourier coefficients are given by the formulas

$$A_n \int_a^b \varphi_n(x)^2 \rho(x) \, dx = \int_a^b f(x) \varphi_n(x) \rho(x) \, dx \qquad n = 1, 2, \ldots$$

It is important to notice that the eigenfunctions φ_n are not unique; a
constant multiple of φ_n is again an eigenfunction with the same eigenvalue.
Even if we require the normalization $\int_a^b \varphi_n^2(x) \rho(x) \, dx = 1$, there is still a possible
ambiguity of a minus sign in the definition of φ_n. Therefore in general we do
not insist that the eigenfunctions be normalized.

Let us give some examples of the Sturm-Liouville expansion.

Example 0.4.4

$$\varphi'' + \lambda \varphi = 0 \qquad \varphi(0) = 0 \qquad \varphi(L) = 0$$

In this case we have the solutions $\varphi_n(x) = \sin(n\pi x/L)$ and $\lambda_n = (n\pi/L)^2$. The
Sturm-Liouville expansion is then called a Fourier sine series, which is studied
in detail in Chap. 1.

Example 0.4.5

$$\varphi'' + \lambda \varphi = 0 \qquad \varphi(0) = 0 \qquad \varphi'(L) = 0$$

In this case we have the solutions $\varphi_n(x) = \sin[(n - \frac{1}{2})\pi x/L]$, $\lambda_n = (n - \frac{1}{2})^2 \pi^2/L^2$
for $n = 1, 2, \ldots$. The Sturm-Liouville expansion is equivalent to a Fourier sine
series on the interval $0 \leq x \leq 2L$, for which the even-numbered coefficients are
all zero.

We do not prove Theorem 0.4.2, since it depends on methods from higher
analysis. Nevertheless, we give a sketch of the main arguments used to obtain
the eigenvalues and eigenfunctions. For a complete treatment of the theory,
the reader is referred to an advanced book on ordinary differential equations.†

---

† For example, G. Birkhoff and G. C. Rota, *Ordinary Differential Equations*, 3d ed., John Wiley,
New York, 1978.

To obtain the eigenvalues and eigenfunctions, the first step is to transform the equation to eliminate the first derivative term. This is done by introducing a new function

$$\psi(x) = \varphi(x)\sqrt{s(x)} \tag{0.4.4}$$

Thus $\psi' = \varphi'\sqrt{s} + \varphi(s'/2\sqrt{s})$, $\psi'' = \varphi''\sqrt{s} + \varphi's'/\sqrt{s} + \varphi(\sqrt{s})''$, and we have the equation for ψ:

$$s\psi'' + [\lambda\rho - q - \sqrt{s}(\sqrt{s})'']\psi = 0 \tag{0.4.5}$$

If ρ/s is constant, as is the case in many problems, we have completed the necessary transformations. Otherwise, a further reduction is necessary, where we change both the dependent and independent variables. This is the *Liouville substitution*

$$\psi(x) = \left(\frac{\rho}{s}\right)^{-1/4} v(z)$$

$$\frac{dz}{dx} = \left(\frac{\rho}{s}\right)^{1/2}$$

This defines a new independent variable z and a new function v. It is left as an exercise to show that the Sturm-Liouville equation now takes the form

$$v'' + [\lambda - \bar{q}(z)]v = 0 \qquad \bar{a} < z < \bar{b} \tag{0.4.6}$$

where $$\bar{a} = z(a) \qquad \bar{b} = z(b)$$

and $\bar{q}$ is a new function, which can be determined.

The second step is to take polar coordinates in the (v, v') plane. This transformation, known as the *Prüfer substitution*, is natural, since we expect that the solutions will exhibit oscillatory behavior for large n. Thus we write

$$v(z) = R(z) \sin \theta(z)$$
$$v'(z) = R(z) \cos \theta(z) \tag{0.4.7}$$

The Sturm-Liouville equation for v translates to the following system of two first-order differential equations for (R, θ):

$$\theta' = \cos^2 \theta + [\lambda - \bar{q}(z)] \sin^2 \theta \tag{0.4.8}$$

$$\frac{R'}{R} = \frac{1}{2} [1 + \lambda - \bar{q}(z)] \sin 2\theta \tag{0.4.9}$$

Given a solution (R, θ) of equations (0.4.8) and (0.4.9), we may define $v(z) = R(z) \cos \theta(z)$ and show that v is a solution of the Sturm-Liouville problem (0.4.6). For a given value of λ, we solve (0.4.8) with the boundary condition satisfied at $z = a$, which is possible in theory. When $z = \bar{b}$, we have the value $\theta = \theta(\bar{b}, \lambda)$, which does not satisfy the boundary condition at $z = \bar{b}$, in general. But the dependence on λ is continuous and increasing. Therefore we may apply the intermediate-value property of continuous functions to conclude that the boundary condition at $z = b$ will be satisfied for certain values $\lambda = \lambda_1, \lambda_2, \ldots$.

This concludes the discussion of completeness. The interested reader will find more information in the reference cited.

Asymptotic Behavior of the Eigenvalues and Eigenfunctions

In many problems it is important to know the behavior of the eigenvalues and eigenfunctions (λ_n, φ_n) when $n \to \infty$. It is natural to expect that the behavior will mimic the simplest case $\varphi'' + \lambda\varphi = 0$ where $\lambda_n = (n\pi/L)^2$, $\varphi_n = \sin(n\pi x/L)$. Indeed, the following result confirms our optimism.

Proposition 0.4.5. Let (φ_n, λ_n) be the normalized eigenfunctions and eigenvalues of the regular Sturm-Liouville problem with zero boundary conditions

$$(s\varphi')' + (\lambda\rho - q)\varphi = 0 \qquad a < x < b$$

$$\varphi(a) = 0$$

$$\varphi(b) = 0$$

$$\int_a^b \varphi(x)^2\rho(x) \, dx = 1$$

Then we have the following asymptotic formulas:

$$\sqrt{\lambda_n} = \frac{n\pi}{\bar{b} - \bar{a}} + O\left(\frac{1}{n}\right) \qquad\qquad n \to \infty$$

$$\pm v_n(z) = \sqrt{\frac{2}{\bar{b} - \bar{a}}} \, \sin\frac{n\pi(z - \bar{a})}{\bar{b} - \bar{a}} + O\left(\frac{1}{n}\right) \qquad n \to \infty$$

$$\pm v_n'(z) = n\pi \sqrt{\frac{2}{(\bar{b} - \bar{a})^3}} \, \cos\frac{n\pi(z - \bar{a})}{\bar{b} - \bar{a}} + O(1) \qquad n \to \infty$$

(The $\pm$ sign ambiguity is due to the lack of a natural choice of sign for the eigenfunctions.)

This result will allow us to show that many of the formal series solutions of boundary-value problems in Chaps. 1 to 4 are indeed rigorous solutions. For example, in Sec. 3.4 we encounter the series solution

$$u(x; t) = \sum_{n=1}^{\infty} A_n\varphi_n(x)e^{-\lambda_n K t} \tag{0.4.10}$$

where φ_n is a solution of Bessel's equation with $\varphi_n(R_1) = 0 = \varphi_n(R_2)$. On the interval $R_1 < x < R_2$, this is a regular Sturm-Liouville problem with $s(x) = x = \rho(x)$, $q = 0$. Thus we have the asymptotic behavior $\sqrt{\lambda_n} \sim n\pi/(R_2 - R_1)$, $\sqrt{x}\varphi_n = O(1)$, $(\sqrt{x}\varphi_n)' = O(n)$, $n \to \infty$. This information may be applied to the series for $u(x; t)$ to show that the series for u, u_x, u_{xx}, u_t all converge uniformly when $t > 0$, and thus u is a solution of the heat equation. The completeness of the eigenfunctions shows that the initial condition is satisfied.

EXERCISES 0.4

1. Write each of the following equations in Sturm-Liouville form (0.4.1) and identify the weight function $\rho(x)$.
   (a) $x^2\varphi'' + \lambda\varphi = 0, 1 < x < 2$
   (b) $\sin x \, \varphi'' + \cos x \, \varphi' + \lambda \sin x \, \varphi = 0, 0 < x < \pi$
   (c) $(x\varphi')' + (\lambda - 1/x^2)\varphi = 0, 0 < x < 4$
   (d) $\varphi'' - x\varphi' + \lambda\varphi = 0, -10 < x < 10$
   (e) $\varphi'' - \varphi' + \lambda\varphi = 0, 3 < x < 5$
   (f) $(x\varphi)'' + \lambda x\varphi = 0, 0 < x < 2$

2. Find the solutions (φ_n, λ_n) of the Sturm-Liouville problem

$$\varphi'' + \frac{1}{x}\varphi' + \frac{\lambda}{x^2}\varphi = 0 \quad 1 < x < 2$$

$$\varphi(1) = 0 \qquad \varphi(2) = 0$$

   (*Hint:* The change of variable $x = e^z$ may be used to reduce to an equation with constant coefficients.)

3. Which of the following boundary conditions satisfy the orthogonality conditions (0.4.2)?
   (a) $s(x) = 1, 0 < x < \pi, \varphi(0) = 0, \varphi(\pi) = 0$
   (b) $s(x) = x, 0 < x < \pi, \varphi(0) = 0, \varphi'(\pi) = (0)$
   (c) $s(x) = \sin x, 0 < x < \pi, \varphi(0) = \varphi'(0), \varphi(\pi) = 0$
   (d) $s(x) = e^{-x^2}, -10 < x < 10, \varphi(-10) = \varphi'(10), \varphi'(-10) = \varphi(10)$
   (e) $s(x) = e^{-x}, 0 < x < 10, \varphi(0) = \varphi(10), \varphi'(0) = \varphi'(10)$
   (f) $s(x) = x^2, -5 < x < 5, \varphi(-5) = 0, \varphi'(5) = \varphi(5)$

4. Find the eigenvalues and eigenfunctions of the Sturm-Liouville problem

$$\varphi'' + \lambda\varphi = 0 \qquad 0 < x < L$$

   with the following boundary conditions:
   (a) $\varphi'(0) = 0, \varphi'(L) = 0$
   (b) $\varphi(0) = 0, \varphi'(L) = 0$
   (c) $\varphi(0) = \varphi(L), \varphi'(0) = \varphi'(L)$
   (d) $\varphi(0) = -\varphi(L), \varphi'(0) = -\varphi'(L)$
   (e) $\varphi(0) = 0, \varphi(L) + \varphi'(L) = 0$
   (f) $\varphi(0) - \varphi'(0) = 0, \varphi'(L) = 0$

5. Which of the following boundary conditions satisfy the positivity criterion (0.4.3)?
   (a) $s(x) = 1, 0 < x < \pi, \varphi(0) = 0, \varphi'(\pi) = 0$
   (b) $s(x) = 1 - x^2, -1 < x < 1, \varphi(1) = \varphi'(-1), \varphi'(1) = \varphi(-1)$
   (c) $s(x) = e^{-x^2}, -5 < x < 5, \varphi(5) = -\varphi(-5), \varphi'(5) = -\varphi'(-5)$
   (d) $s(x) = \sin x, 0 < x < \pi, \varphi(0) = \varphi'(0), \varphi(\pi) = -\varphi'(\pi)$
   (e) $s(x) = e^{-x}, 0 < x < 10, \varphi(0) = 0, \varphi'(10) = 0$
   (f) $s(x) = x^2, 0 < x < 6, \varphi(0) = \varphi'(6), \varphi(6) = 0$

6. For which of the following Sturm-Liouville problems can we assert, on the basis of Proposition 0.4.5, that all eigenvalues are positive or zero?

(a) $\varphi'' + \dfrac{1}{x} \varphi' + \left(\lambda - \dfrac{10}{x^2} \right) \varphi = 0,\ 0 < x < 5,\ \varphi(5) = 0$

(b) $x\varphi'' + \varphi' + \left(\lambda + \dfrac{4}{x^2} \right) \varphi = 0,\ 0 < x < 5,\ \varphi(5) = 0$

(c) $\varphi'' - x\varphi' + \lambda\varphi = 0,\ -10 < x < 10,\ \varphi(10) = 0,\ \varphi(-10) = 0$
(d) $\varphi'' - x^2\varphi' + \lambda\varphi = 0,\ -10 < x < 10,\ \varphi(10) = \varphi(-10),\ \varphi'(10) = \varphi'(-10)$
(e) $\varphi'' - x\varphi' + \lambda\varphi = 0,\ -10 < x < 10,\ \varphi(10) = \varphi(-10),\ \varphi'(10) = \varphi'(-10)$
(f) $x^2\varphi'' + x\varphi' + \lambda\varphi = 0,\ 0 < x < \pi,\ \varphi(\pi) = 0$

7. Consider the Sturm-Liouville problem

$$\varphi'' + \lambda\varphi = 0 \qquad 0 < x < 1$$

$$\varphi(0) = 0 \qquad \varphi'(1) = \varphi(1)$$

By examining the graph of the hyperbolic tangent, show that all the eigenvalues are positive. Is the positivity criterion (0.4.3) satisfied?

8. Consider the Bessel equation

$$\varphi'' + \dfrac{d - 1}{x} \varphi' + \left(\lambda - \dfrac{m^2}{x^2} \right) \varphi = 0$$

(a) Show that the weight function is $\rho(x) = x^{d-1}$.
(b) Show that the transformed equation (0.4.5) is

$$\psi'' + \left[\lambda - \dfrac{m^2}{x^2} - \dfrac{(d - 1)(d - 3)}{4x^2} \right] \psi = 0$$

(c) Use this to solve Bessel's equation if $d = 3$, $m = 0$.

9. Let $(R(z),\ \theta(z))$ be solutions of (0.4.8) and (0.4.9), where $\bar{q}(z)$ is a given function. Define $v(z)$ by (0.4.7) and show that v satisfies (0.4.6).

10. Suppose that (φ_n, λ_n) are the solutions of a regular Sturm-Liouville problem. Prove *Parseval's equation*

$$\int_a^b f(x)^2\rho(x)\,dx = \sum_{n=1}^{\infty} \dfrac{[\int_a^b f(x)\varphi_n(x)\rho(x)\,dx]^2}{\int_a^b \varphi_n(x)^2\rho(x)\,dx}$$

valid for any smooth function which satisfies the separated boundary conditions.

11. Consider the Sturm-Liouville problem

$$\varphi'' + \lambda\varphi = 0 \qquad 0 < x < L$$

$$\varphi(0) = 0 \qquad \varphi'(L) = \varphi(L)$$

(a) Show that the eigenvalues and eigenfunctions are determined from the equations

$$\tan L\sqrt{\lambda_n} = \sqrt{\lambda_n} \qquad \varphi_n = \sin x\sqrt{\lambda_n}$$

(b) By examining the graph of the tangent function, prove the asymptotic relation

$$\sqrt{\lambda_n} = \left(n + \dfrac{1}{2} \right) \dfrac{\pi}{L} + O\left(\dfrac{1}{n} \right) \qquad n \to \infty$$

(c) Use part (b) to show that the normalized eigenfunctions satisfy

$$\pm \varphi_n(x) = \sqrt{\frac{2}{L}} \sin \frac{n\pi x}{L} + O\left(\frac{1}{n}\right) \qquad n \to \infty$$

12. Let (φ_n, λ_n) be the normalized eigenfunctions and eigenvalues of the Sturm-Liouville problem

$$\varphi'' + \frac{1}{x}\varphi' + \lambda\varphi = 0 \qquad 0 < R_1 < x < R_2$$

$$\varphi(R_1) = 0 \qquad \varphi(R_2) = 0$$

Use Proposition 0.4.6 to show that

$$\sqrt{\lambda_n} = \frac{n\pi}{R_2 - R_1} + O\left(\frac{1}{n}\right) \qquad n \to \infty$$

$$\pm \varphi_n(x) = \sqrt{\frac{2}{x(R_2 - R_1)}} \sin \frac{n\pi(x - R_1)}{R_2 - R_1} + O\left(\frac{1}{n}\right) \qquad n \to \infty$$

13. Let $u(x; t)$ be defined by (0.4.10). Show that u_x, u_{xx}, u_t exist if $A_n = O(1)$, $n \to \infty$.
14. Consider the Sturm-Liouville problem

$$\varphi'' + \lambda\varphi = 0 \qquad 0 < x < L$$

$$\left. \begin{array}{l} \varphi'(0) = h\varphi(0) \\ \varphi'(L) + h\varphi(L) = 0 \end{array} \right\} \quad h \geq 0$$

(a) Show that the eigenvalues are obtained as solutions of the transcendental equation

$$2h\sqrt{\lambda} \cos L\sqrt{\lambda} + (h^2 - \lambda) \sin L\sqrt{\lambda} = 0$$

(b) If $Lh < \pi/2$, show that $L\sqrt{\lambda_1} < \pi/2$.
(c) If $Lh \geq \pi/2$, show that $\pi/2 \leq L\sqrt{\lambda_1} < \pi$.
(d) Show that $\sqrt{\lambda_{n+1}} \sim n\pi/L$ when $n \to \infty$.
15. Prove the Lagrange identity $(s(\varphi_1'\varphi_2 - \varphi_1\varphi_2'))' = \varphi_2 L\varphi_1 - \varphi_1 L\varphi_2$, where L is the Sturm-Liouville operator $L\varphi = (s\varphi')' - q\varphi$ and φ_1, φ_2 are twice-differentiable functions. (*Hint:* Apply the product rule of differentiation to $s\varphi_1'\varphi_2$, then switch the roles of φ_1, φ_2 and subtract.)
16. Use Lagrange's identity to give an alternate proof of Proposition 0.4.4.
17. Use Lagrange's identity to give an alternate proof of Proposition 0.4.5. (*Hint:* Take $\varphi_2 = 1$.)

0.5   REVIEW OF INFINITE SERIES

Numerical Series

An *infinite series* is an expression of the form

$$\sum_{n=1}^{\infty} a_n$$

where a_n is a sequence of numbers. For example, $\sum_{n=1}^{\infty} 1/n^2$, $\sum_{n=1}^{\infty} (-1)^n/n$, and $\sum_{n=1}^{\infty} 1/3^n$ are familiar examples of infinite series. The summation index n is a dummy variable, so that we make no distinction between $\sum_{n=1}^{\infty} a_n$, $\sum_{m=1}^{\infty} a_m$, and $\sum_{p=1}^{\infty} a_p$.

The *convergence* of the infinite series $\sum_{n=1}^{\infty} a_n$ is formulated in terms of the *partial sums* s_n, defined by $s_n = a_1 + \cdots + a_n$. If $\lim_{n \to \infty} s_n = a$, then we say that the series $\sum_{n=1}^{\infty} a_n$ *converges*, and we write

$$a = \sum_{n=1}^{\infty} a_n$$

If $\lim_{n \to \infty} s_n$ does not exist, then we say that the series *diverges*. If two series $\sum_{n=1}^{\infty} a_n$ and $\sum_{n=1}^{\infty} b_n$ both converge, then the series $\sum_{n=1}^{\infty} (a_n + b_n)$ converges and $\sum_{n=1}^{\infty} (a_n + b_n) = \sum_{n=1}^{\infty} a_n + \sum_{n=1}^{\infty} b_n$. If the series $\sum_{n=1}^{\infty} a_n$ converges and c is any constant, then the series $\sum_{n=1}^{\infty} ca_n$ converges and $\sum_{n=1}^{\infty} ca_n = c \sum_{n=1}^{\infty} a_n$.

Examples of infinite series which converge are

$$\sum_{n=1}^{\infty} \frac{1}{n^2} \qquad \sum_{n=1}^{\infty} \frac{1}{2^n} \qquad \sum_{n=1}^{\infty} e^{-n^2}$$

If the terms a_n decrease too slowly or oscillate too erratically, the infinite series will diverge. Examples are

$$\sum_{n=1}^{\infty} \frac{1}{n} \qquad \sum_{n=1}^{\infty} (-1)^n \qquad \sum_{n=1}^{\infty} \frac{1}{n} \log (1 + n)$$

Tests for convergence are especially simple for series of positive terms $a_n > 0$. In this case the series $\sum_{n=1}^{\infty} a_n$ converges if and only if the partial sums remain bounded, $s_n \leq M$ for some constant M and all $n = 1, 2, \ldots$. A useful criterion for the convergence of series of positive terms is the following *integral test*.

Integral test. Let $a_n = \varphi(n)$, where $\varphi(x)$ is a positive function of $x > 0$, such that $\varphi(x)$ decreases to zero when $x \to \infty$. Then the series $\sum_{n=1}^{\infty} a_n$ converges if and only if the improper integral $\int_1^{\infty} \varphi(x)\, dx$ converges.

Example 0.5.1. Determine for which values of $p > 0$ the series $\sum_{n=1}^{\infty} 1/n^p$ converges.

Solution. We take $\varphi(x) = 1/x^p$, a positive function of $x > 0$, which decreases to zero when $x \to \infty$. The associated improper integral is $\int_1^{\infty} dx/x^p = \lim_{M \to \infty} \int_1^M dx/x^p$. If $p \neq 1$, we have $\int_1^M dx/x^p = (1 - M^{1-p})/(1 - p)$, and thus the limit exists if $p > 1$; the limit does not exist if $p < 1$. If $p = 1$, we have $\int_1^M dx/x = \log M$, and the limit does not exist. Therefore the series $\sum_{n=1}^{\infty} 1/n^p$ converges if and only if $p > 1$. ●

For series whose terms decrease rapidly, one may deduce convergence from the ratio test or the root test, stated as follows.

Ratio test. If there are numbers r, N with $0 \leq r < 1$, $N > 1$ such that $|a_{n+1}/a_n| \leq r$ for all $n > N$, then the series $\sum_{n=1}^{\infty} a_n$ converges.

Root test. If there are numbers r, N with $0 \leq r < 1$, $N > 1$ such that $|a_n|^{1/n} \leq r$ for all $n > N$, then the series $\sum_{n=1}^{\infty} a_n$ converges.

In particular, if $\lim_{n \to \infty} |a_{n+1}/a_n|$ exists and is less than 1, then the ratio test applies. Similarly, if $\lim_{n \to \infty} |a_n|^{1/n}$ exists and is less than 1, the root test applies.

Example 0.5.2. Test the series $\sum_{n=1}^{\infty} x^n/(n!)^2$ and $\sum_{n=1}^{\infty} x^n/n^n$ for convergence.

Solution. In the first case, we have $a_n = x^n/(n!)^2$ with $a_{n+1}/a_n = x/(n + 1)^2$, $\lim_{n \to \infty} (a_{n+1}/a_n) = 0$, and we have convergence by virtue of the ratio test (the root test would be difficult to apply in this case). When $a_n = x^n/n^n$, we have $|a_n|^{1/n} = |x|/n$ with $\lim_{n \to \infty} |a_n|^{1/n} = 0$ and we conclude convergence from the root test. The ratio test could also be used in this case.   ●

Series of Functions: Uniform Convergence

Another source of convergent series is Taylor's theorem with remainder

$$f(x) - f(x_0) = \sum_{n=1}^{N} \frac{(x - x_0)^n}{n!} f^{(n)}(x_0) + \frac{1}{N!} \int_{x_0}^{x} (x - t)^N f^{(N+1)}(t) \, dt$$

We suppose that the function f is *smooth*, that is, that all derivatives $f^{(n)}$ exist and are continuous functions. If these derivatives satisfy the condition $|f^{(n)}(x)| \leq n! c^n$ where $|x - x_0| < 1/c$ for some constant c and $n = 1, 2, \ldots$, then the integral term tends to zero when $N \to \infty$ and we have the *Taylor series*

$$f(x) - f(x_0) = \sum_{n=1}^{\infty} \frac{(x - x_0)^n}{n!} f^{(n)}(x_0) \qquad |x - x_0| < \frac{1}{c}$$

For example,

$$\frac{x}{1 - x} = \sum_{n=1}^{\infty} x^n \qquad\qquad |x| < 1$$

$$\log x = \sum_{n=1}^{\infty} \frac{(-1)^{n-1}(x - 1)^n}{n} \qquad |x - 1| < 1$$

$$e^x - 1 = \sum_{n=1}^{\infty} \frac{x^n}{n!}$$

The last series for e^x converges for all x.

We now consider more general series of functions, of the form

$$\sum_{n=1}^{\infty} u_n(x)$$

The Taylor series just considered are all examples of these. The functions $u_n(x)$ are supposed to be defined on a common interval $a \leq x \leq b$. Let $f_n(x)$ be the partial sum $f_n(x) = u_1(x) + \cdots + u_n(x)$.

Definition. The series $\sum_{n=1}^{\infty} u_n(x)$ *converges pointwise* if for each x, $a \leq x \leq b$, the series $\sum_{n=1}^{\infty} u_n(x)$ converges. We denote the limit by $f(x) = \sum_{n=1}^{\infty} u_n(x)$.

Definition. The series $\sum_{n=1}^{\infty} u_n(x)$ *converges uniformly*, $a \leq x \leq b$, if the series converges pointwise and there exists a sequence of constants ϵ_n such that

$$|f_n(x) - f(x)| \leq \epsilon_n \qquad a \leq x \leq b$$

$$\lim_{n \to \infty} \epsilon_n = 0$$

If the series $\sum_{n=1}^{\infty} u_n(x)$ converges uniformly, then it also converges pointwise, but the converse statement is not true. For example, if $u_n(x) = nxe^{-nx} - (n + 1)xe^{-(n+1)x}$, then the series $\sum_{n=1}^{\infty} u_n(x)$ converges pointwise to $f(x) = xe^{-x}$ for $0 \leq x \leq 1$. But the convergence is not uniform. Indeed we have $f_n(x) = xe^{-x} - nxe^{-nx}$ for all x, $0 \leq x \leq 1$. By taking $x = 1/n$, we have $f_n(1/n) = (1/n)e^{-(1/n)} - e^{-1}$. Thus $|f_n(x) - f(x)|$ has the value $e^{-1} = 0.368 \ldots$ for $x = 1/n$. This contradicts the possibility of a sequence of constants ϵ_n with $\lim_{n \to \infty} \epsilon_n = 0$, so that $|f_n(x) - f(x)| \leq \epsilon_n$ for *all* x, $0 \leq x \leq 1$.

To prove the uniform convergence of a series of functions, we may use the following important criterion.

Weierstrass M test. Let $\sum_{n=1}^{\infty} u_n(x)$ be a series of functions defined for $a \leq x \leq b$. Suppose that there exists a sequence of constants M_n such that

$$|u_n(x)| \leq M_n \qquad a \leq x \leq b$$

and that $\sum_{n=1}^{\infty} M_n$ converges. Then the series $\sum_{n=1}^{\infty} u_n(x)$ converges uniformly for $a \leq x \leq b$.

Example 0.5.3. Show that the Taylor series $\sum_{n=1}^{\infty} x^n/n!$ converges uniformly for $-1 \leq x \leq 1$.

Solution. We have $u_n(x) = x^n/n!$ and $|u_n(x)| \leq 1/n!$ for $-1 \leq x \leq 1$. But the series $\sum 1/n!$ converges. Therefore, the Taylor series $\sum_{n=1}^{\infty} x^n/n!$ is uniformly convergent for $-1 \leq x \leq 1$. ●

The Weierstrass M test includes the *test for absolute convergence* as a special case. We illustrate with an example.

Example 0.5.4. Show that the series $\sum_{n=1}^{\infty} (-1)^n/n^2$ is convergent.

Solution. This series of numbers can be regarded as a special case of a series of functions with $u_n(x) = (-1)^n/n^2$. We have $|u_n(x)| \leq 1/n^2$, which is the general term of a convergent series. ●

More generally, if $\sum_{n=1}^{\infty} a_n$ is a series of constants and $|a_n| \leq M_n$, where $\sum_{n=1}^{\infty} M_n$ is a convergent series of positive terms, then the series $\sum_{n=1}^{\infty} a_n$ is convergent.

The Weierstrass M test gives *sufficient* conditions for the uniform convergence of a series of functions. For example, the series $\sum_{n=1}^{\infty} (-1)^n x^n / n$ does not satisfy these conditions, but it is uniformly convergent for $0 \leq x \leq 1$. (See Exercise 22.)

Uniform convergence can be used to justify many operations with infinite series. We restrict our attention to series of continuous functions $\sum_{n=1}^{\infty} u_n(x)$. We have the following propositions, which give conditions for continuity, integration, and differentiation of a uniformly convergent series of continuous functions. These results are used in subsequent chapters, when we prove that certain series are solutions of differential equations.

Proposition 0.5.1. Suppose that $f(x) = \sum_{n=1}^{\infty} u_n(x)$ is a uniformly convergent series for $a \leq x \leq b$, where $u_n(x)$ is a continuous function for each $n = 1, 2, \ldots$ and $a \leq x \leq b$. Then $f(x)$ is a continuous function, $a \leq x \leq b$.

Proposition 0.5.2. Suppose that $f(x) = \sum_{n=1}^{\infty} u_n(x)$ is a uniformly convergent series for $a \leq x \leq b$, where $u_n(x)$ is a continuous function. Then $\sum_{n=1}^{\infty} \int_a^b u_n(x) \, dx$ is a convergent series of constants, and $\int_a^b f(x) \, dx = \sum_{n=1}^{\infty} \int_a^b u_n(x) \, dx$.

Proposition 0.5.3. Suppose that $f(x) = \sum_{n=1}^{\infty} u_n(x)$ is a uniformly convergent series for $a \leq x \leq b$, where $u_n(x)$ is a continuous function with a continuous derivative $u_n'(x)$ and the series $\sum_{n=1}^{\infty} u_n'(x)$ converges uniformly for $a \leq x \leq b$. Then the function $f(x)$ has a continuous derivative, which is given by $f'(x) = \sum_{n=1}^{\infty} u_n'(x)$.

Example 0.5.5. Show that the function $f(x) = \sum_{n=1}^{\infty} (\cos nx)/2^n$ is continuous for $-\pi \leq x \leq \pi$ and that $f'(x)$ and $f''(x)$ are continuous for $-\pi \leq x \leq \pi$.

Solution. By the Weierstrass M test with $M_n = 1/2^n$, we see that the given series is uniformly convergent, $-\pi \leq x \leq \pi$. Each function $(\cos nx)/2^n$ is continuous; hence $f(x)$ is continuous by Proposition 0.5.1. The differentiated series is $\sum_{n=1}^{\infty} (-n \sin nx)/2^n$, which is also uniformly convergent by the Weierstrass M test with $M_n = n/2^n$, the general term of a convergent series of constants. Therefore by Proposition 0.5.3, $f(x)$ has a continuous derivative $f'(x)$, which is given by $f'(x) = \sum_{n=1}^{\infty} (-n \sin nx)/2^n$. To study $f''(x)$, we note that the differentiated series for f' is $\sum_{n=1}^{\infty} (-n^2 \cos nx)/2^n$, which is uniformly convergent by the Weierstrass M test with $M_n = n^2/2^n$, the general term of a convergent series of constants. Therefore by Proposition 0.5.3, $f'(x)$ has a continuous derivative $f''(x)$, which is given by the differentiated series $f''(x) = \sum_{n=1}^{\infty} (-n^2 \cos nx)/2^n$.   ●

Example 0.5.6. Beginning with the geometric series $x/(1 - x) = \sum_{n=1}^{\infty} x^n$, $-1 < x < 1$, show that

$$\sum_{n=1}^{\infty} nx^n = \frac{x}{(1 - x)^2} \qquad -1 < x < 1$$

$$\sum_{n=1}^{\infty} n^2 x^n = \frac{x(1 + x)}{(1 - x)^3} \qquad -1 < x < 1$$

Solution. Let $u_n(x) = x^n$, and let r be any number with $0 < r < 1$. By the Weierstrass M test, the series $\sum_{n=1}^{\infty} u_n(x)$ is uniformly convergent for $-r \le x \le r$ with $M_n = r^n$, the general term of a convergent series of constants. The differentiated series is $\sum_{n=1}^{\infty} nx^{n-1}$, which is also uniformly convergent for $-r \le x \le r$ by the Weierstrass M test with $M_n = nr^{n-1}$, the general term of a convergent series. On the other hand, from calculus, $(d/dx)[x/(1 - x)] = 1/(1 - x)^2$. Therefore from Proposition 0.5.3 we have $1/(1 - x)^2 = \sum_{n=1}^{\infty} nx^{n-1}$ for $-r \le x \le r$. But r was any number with $0 < r < 1$, and hence this equation is true for all x with $-1 < x < 1$. Multiplying by x gives the required result $x/(1 - x)^2 = \sum_{n=1}^{\infty} nx^n$ for $-1 < x < 1$.

We now apply Proposition 0.5.3 to this series, noting that the differentiated series $\sum_{n=1}^{\infty} n^2 x^{n-1}$ is again uniformly convergent by the Weierstrass M test for $-r \le x \le r$ with $M_n = n^2 r^{n-1}$, the general term of a convergent series of constants. But the derivative can be computed by calculus as $(d/dx)[x/(1 - x)^2] = (1 + x)/(1 - x)^3$. Multiplication by x gives the required result: $\sum_{n=1}^{\infty} n^2 x^n = x(1 + x)/(1 - x)^3$. ●

Often we need to deal with functions that are defined by integrals. The following statement is listed for future reference.

Proposition 0.5.4. Suppose that $f(x) = \int_c^d g(x, y)\, dy$, where the function $g(x, y)$ is continuous for $a \le x \le b$, $c \le y \le d$. Then $f(x)$, $a \le x \le b$, is a continuous function. If, in addition, the partial derivative $(\partial g/\partial x)(x, y)$ exists and is extensible to a continuous function for $a \le x \le b$, $c \le y \le d$, then $f'(x)$ exists for $a \le x \le b$ and is given by the integral formula $f'(x) = \int_c^d (\partial g/\partial x)(x, y)\, dy$.

If a power series $\sum_{n=1}^{\infty} a_n x^n$ converges for $x = x_0$, then it also converges for any x with $|x| < |x_0|$, and by applying the Weierstrass M test, it converges uniformly in any closed subinterval of $-|x_0| < x < |x_0|$ and hence is a continuous function in this region. The following result ensures that the continuity of the function extends to the endpoint of the interval. We normalize to the interval $-1 < x < 1$.

Abel's theorem. Suppose that the power series $\sum_{n=1}^{\infty} a_n x^n$ converges for $x = 1$. Then $\lim_{x \uparrow 1} \sum_{n=1}^{\infty} a_n x^n = \sum_{n=1}^{\infty} a_n$.

Proof. We transform the power series to a related series whose terms have more explicit behavior. Writing $\sum_{n=1}^{\infty} a_n x^n = (1 - x) \sum_{n=1}^{\infty} b_n x^n$, we must have $b_n = a_1 + \cdots + a_n$, and by hypothesis $\lim_{n \to \infty} b_n = B := \sum_{n=1}^{\infty} a_n$. For any N we can write

$$(1 - x) \sum_{n=1}^{\infty} b_n x^n = (1 - x) \sum_{n=1}^{N} b_n x^n + Bx^{N+1} + (1 - x) \sum_{n=N+1}^{\infty} (b_n - B)x^n$$

where we have used the explicit evaluation $\sum_{n=N+1} x^n = x^{N+1}/(1 - x)$ in the second term. For any preassigned ϵ there is an $N = N(\epsilon)$ such that $|b_n - B| < \epsilon$ for $n > N$. If we now let x approach 1, the first term tends to zero, the second term approaches B, while the third term remains less than ϵ. Hence the $\limsup_{x \uparrow 1} |\sum_{n=1}^{\infty} a_n x^n - B| \le \epsilon$ for any $\epsilon > 0$. Since this holds for any $\epsilon > 0$, the indicated limsup must be zero, completing the proof. ■

Double Series

We now consider *double series*. These are of the form

$$\sum_{m,n=1}^{\infty} a_{mn}$$

where the a_{mn} are real numbers defined for $m = 1, 2, \ldots$ and $n = 1, 2, \ldots$. The convergence of a double series is defined in terms of the partial sums

$$s_{mn} = \sum_{i \le m, \, j \le n} a_{ij}$$

If

$$\lim_{\substack{m \to \infty \\ n \to \infty}} s_{mn} = a$$

when the indices $m, n \to \infty$ in any order whatever, then we write

$$a = \sum_{m,n=1}^{\infty} a_{mn}$$

as the sum of the series.

A form of the integral test is applicable to double series of positive terms. Let $a_{mn} = \varphi(m, n)$ where the function $\varphi(x, y)$ is positive and decreasing in variables x and y. Then the convergence or divergence of the double series $\sum_{m,n=1}^{\infty} a_{mn}$ is equivalent to the convergence or divergence of the double integral $\int_1^{\infty} \int_1^{\infty} \varphi(x, y) \, dx \, dy$.

Example 0.5.7. For which values of $p > 0$ is the double series

$$\sum_{m,n=1}^{\infty} \frac{1}{(m^2 + n^2)^{p/2}}$$

convergent?

Solution. We have $\varphi(x, y) = [1/(x^2 + y^2)]^{p/2}$. To study the associated double integral, we take polar coordinates $x = r \cos \theta$, $y = r \sin \theta$, with $dx \, dy = r \, dr \, d\theta$. Thus we examine $\int_0^{\pi/2} \int_1^{\infty} (1/r^p) r \, dr \, d\theta$. This double integral is convergent if and only if $p - 1 > 1$, that is, $p > 2$. We conclude that the double series $\sum_{m,n=1}^{\infty} 1/(m^2 + n^2)^{p/2}$ is convergent if and only if $p > 2$.   ●

The Weierstrass M test and the properties of uniform convergence may also be generalized to double series of functions of one or more variables, for example, the series $f(x, y) = \sum_{m,n=1}^{\infty} u_{m,n}(x, y)$. We may use the properties of uniform convergence to show that $f(x, y)$ has partial derivatives f_x, f_y, which are continuous functions of (x, y) in a rectangular region $a \le x \le b$, $c \le y \le d$.

Big-*O* Notation

Definition. Let $f(t)$, $g(t)$ be two functions defined for $t > 0$. We write

$$f(t) = O(g(t)) \qquad t \to \infty$$

if there exist constants $M > 0$ and $T > 0$ such that $|f(t)| \le Mg(t)$ for $t > T$. Similarly, if $\{a_n\}$ and $\{b_n\}$ are sequences defined for $n = 1, 2, \ldots$, we say that $a_n = O(b_n)$ if there exist constants $M > 0$ and $N > 0$ such that $|a_n| \le Mb_n$ for $n > N$.

For example, we have

$$\frac{2t^2}{1 + t^3} = O\left(\frac{1}{t}\right) \qquad t \to \infty$$

$$\frac{\sin t}{1 + t^2} = O\left(\frac{1}{t^2}\right) \qquad t \to \infty$$

$$e^{-t} = O\left(\frac{1}{t^5}\right) \qquad t \to \infty$$

$$\frac{t^2}{1 + t^2} = O(1) \qquad t \to \infty$$

Each of these can be proved by using facts about the specific functions. In case 1 we have $1 + t^3 \ge t^3$ for $t > 0$, so that $2t^2/(1 + t^3) \le 2t^2/t^3 = 2/t$; hence the definition is satisfied with $M = 2$, $T = 1$. Likewise in case 2, $|\sin t| \le 1$, $(1 + t^2) \ge t^2$, so that the definition is satisfied with $M = 1$, $T = 1$. In case 3 we may use the power series $e^t = 1 + \Sigma_1^\infty t^n/n!$. All the terms are positive when $t > 0$; hence $e^t > t^5/5!$ or $e^{-t} < 5!/t^5$. Therefore we may take $M = 5!$, $T = 1$. Finally we note that $f(t) = O(1)$, as in case 4, is equivalent to the statement that $|f(t)| \le M$ for $t > T$. In this case we say that $f(t)$ remains *bounded* when $t \to \infty$.

If $f(t) = O(g(t))$, $t \to \infty$, it does not follow that $g(t) = O(f(t))$, $t \to \infty$. For example, $e^{-t} = O(1/t^5)$, $t \to \infty$, but $1/t^5 \ne O(e^{-t})$, $t \to \infty$. Similarly we write

$$f(t) = O(g(t)) \qquad t \to t_0$$

if there exist positive constants $M > 0$ and $\delta > 0$ such that $|f(t)| \le Mg(t)$ for $0 < |t - t_0| < \delta$. For example, $\sin t = O(t)$, $t \to 0$, and $\sec t = O(1/|t - \pi/2|)$, $t \to \pi/2$.

Many statements about infinite series can be expressed using these notations. For example, the series $\Sigma_{n=1}^\infty a_n$ converges if $a_n = O(1/n^p)$ for some $p > 1$.

Often we encounter series which depend on a parameter t, for example, the series $\Sigma_{n=1}^\infty e^{-nt}$. This series converges for each $t > 0$ since $\Sigma_{n=1}^\infty e^{-nt} = \Sigma_{n=1}^\infty (e^{-t})^n = e^{-t}/(1 - e^{-t})$. If $t > 1$, $e^{-t} < e^{-1} < 0.38$, and thus $\Sigma_{n=1}^\infty e^{-nt} <$

$[1/(1 - 0.38)]e^{-t} < 1.62e^{-t}$. We have proved that

$$\sum_{n=1}^{\infty} e^{-nt} = O(e^{-t}) \qquad t \to \infty$$

This result can be applied to the series $\sum_{n=1}^{\infty} a_n e^{-nt}$ where $\{a_n\}$ is any sequence of constants with $a_n = O(1)$, $n \to \infty$. Thus we have $\sum_{n=1}^{\infty} a_n e^{-nt} = O(e^{-t})$, $t \to \infty$. This may also be true for certain sequences $\{a_n\}$ with a_n tending to infinity. For example, if $a_n = n^2$, we use the result of Example 0.5.6.

$$\frac{x(1 + x)}{(1 - x)^3} = \sum_{n=1}^{\infty} n^2 x^n \qquad -1 < x < 1$$

Taking $x = e^{-t}$, we have

$$\sum_{n=1}^{\infty} n^2 e^{-nt} = e^{-t} \frac{1 + e^{-t}}{(1 - e^{-t})^3} = O(e^{-t}) \qquad t \to \infty$$

We have proved that

$$\sum_{n=1}^{\infty} n^2 e^{-nt} = O(e^{-t}) \qquad t \to \infty$$

This is used in Chap. 2.

The companion "little-o" notation

$$f(t) = o(g(t)) \qquad t \to \infty$$

means that $f(t)/g(t) \to 0$ when $t \to \infty$. Clearly whenever $f(t) = o(g(t))$, $t \to \infty$, then $f(t) = O(g(t))$, $t \to \infty$, but not conversely. For example, $(1 + t)^5 = O(t^5)$, $t \to \infty$, but $(1 + t)^5 \neq o(t^5)$, $t \to \infty$. The statement that the function f is differentiable at x can be written as the two little-o statements $f(x \pm 1/t) = f(x) \pm (1/t)f'(x) + o(1/t)$, $t \to \infty$. Of course, if f also has a second derivative, then we can write the more accurate big-O statements $f(x + 1/t) = f(x) \pm (1/t)f'(x) + O(1/t^2)$, $t \to \infty$. This illustration reveals why the little-o notation is used less in our applied work: in almost every case of interest we can write a more informative big-O statement using information available.

EXERCISES 0.5

1. Apply the integral test to determine the convergence or divergence of each of the following series of positive terms:

   (a) $\displaystyle\sum_{n=1}^{\infty} \frac{1}{(n + 1) \log (n + 1)}$ (b) $\displaystyle\sum_{n=1}^{\infty} \frac{1}{n(\log 2n)^2}$

   (c) $\displaystyle\sum_{n=1}^{\infty} \frac{\log n}{n}$ (d) $\displaystyle\sum_{n=1}^{\infty} n e^{-n^2}$

2. Find the Taylor series of the following functions:
   (a) $f(x) = \sin x$, $x_0 = 0$
   (b) $f(x) = \log [(1 + x)/(1 - x)]$, $x_0 = 0$

(c) $f(x) = \cosh x = \frac{1}{2}(e^x + e^{-x})$, $x_0 = 0$

(d) $f(x) = xe^{x^2}$, $x_0 = 0$

3. Let $f(x) = 0$ for $x \le 0$ and $f(x) = e^{-1/x}$ for $x > 0$.

   (a) Compute $f'(x)$, $f''(x)$.

   (b) Show by induction that, for $x > 0$, the nth derivative is of the form $f^{(n)}(x) = e^{-1/x}P_{2n}(1/x)$, where P_{2n} is a polynomial of degree $2n$.

   (c) Deduce that $f^{(n)}(0) = 0$ for $n = 1, 2, 3, \ldots$

   (d) Discuss the validity of the Taylor expansion

$$f(x) - f(0) = \sum_{n=1}^{\infty} \frac{x^n f^{(n)}(0)}{n!}$$

4. Apply the Weierstrass M test to verify the uniform convergence of the following series:

   (a) $\displaystyle\sum_{n=1}^{\infty} \frac{x^n}{2^n}$, $-1 \le x \le 1$

   (b) $\displaystyle\sum_{n=1}^{\infty} \frac{1}{n^2} \sin nx$, $-\pi \le x \le \pi$

   (c) $\displaystyle\sum_{n=1}^{\infty} e^{-n^2} \cos nx$, $-\pi \le x \le \pi$

   (d) $\displaystyle\sum_{n=1}^{\infty} \frac{n^4}{2^n} \cos n^2 x$, $0 \le x \le \pi$

5. Which of the series in Exercise 4 can be differentiated term by term according to Proposition 0.5.3?

6. Generalize Example 0.5.6 to find the sum of the series

$$\sum_{n=1}^{\infty} n^3 x^n \qquad \sum_{n=1}^{\infty} n^4 x^n \qquad \text{for } -1 < x < 1$$

7. (a) Show that $1/(1 + x^2) = 1 + \sum_{n=1}^{\infty} (-1)^n x^{2n}$ for $-1 < x < 1$.

   (b) Show that this series is uniformly convergent for $0 \le x \le r$, where $0 < r < 1$.

   (c) Use Proposition 0.5.2 to show that

$$\tan^{-1} x = x + \sum_{n=1}^{\infty} \frac{(-1)^n x^{2n+1}}{2n + 1} \qquad \text{for } 0 < x < 1$$

8. (a) Derive the finite identity

$$\frac{1}{1 + x^2} = 1 - x^2 + x^4 - \cdots + (-1)^n x^{2n} + \frac{(-1)^{n+1} x^{2n+2}}{1 + x^2}$$

   (b) Show that

$$\frac{\pi}{4} = \tan^{-1} 1 = 1 - \frac{1}{3} + \frac{1}{5} - \cdots + \frac{(-1)^n}{2n + 1} + (-1)^{n+1} \int_0^1 \frac{t^{2n+2}}{1 + t^2} dt$$

   (c) Show that $\int_0^1 t^{2n+2}/(1 + t^2) \, dt \le 1/(2n + 3)$.

   (d) Conclude that $\pi/4 = 1 - \frac{1}{3} + \frac{1}{5} - \cdots = \sum_{n=1}^{\infty} (-1)^{n+1}/(2n - 1)$.

9. Which of the following double series are convergent, according to the integral test?

   (a) $\displaystyle\sum_{m,n=1}^{\infty} \frac{1}{m^2 + n^2}$

   (b) $\displaystyle\sum_{m,n=1}^{\infty} e^{-(m^2+n^2)}$

   (c) $\displaystyle\sum_{m,n=1}^{\infty} (m^2 + n^2)e^{-(m+n)}$

   (d) $\displaystyle\sum_{m,n=1}^{\infty} \frac{1}{m^2 n^2}$

10. Find constants $M > 0$, $T > 0$ such that $|f(t)| \leq Mg(t)$ for $t > T$ if
    (a) $f(t) = t^2$, $g(t) = e^t$
    (b) $f(t) = t^{10}$, $g(t) = e^{t/2}$
    (c) $f(t) = (\sin t)(\log t)$, $g(t) = t^{1/2}$
    (d) $g(t) = 100/t$, $f(t) = e^{-t/50}$

11. For each of the following functions, it is true that $f(t) = O(g(t))$, $t \to \infty$. Find suitable constants $M > 0$ and $T > 0$ in each case.
    (a) $f(t) = 3t^3 + 3t^2 + 5$, $g(t) = t^3$
    (b) $f(t) = t^2 + 4t - 2$, $g(t) = t^4$
    (c) $f(t) = (t^3 + 4t)/(t^2 + 2)$, $g(t) = t$
    (d) $f(t) = t^{100}$, $g(t) = e^{t/4}$
    (e) $f(t) = \sin^2 t$, $g(t) = |\sin t|$
    (f) $f(t) = e^{-0.01t}$, $g(t) = t^{-25}$

12. (a) Show that $\lim_{n \to \infty} n^2 x e^{-nx} = 0$ for $0 \leq x \leq 1$.
    (b) Show that $\lim_{n \to \infty} \int_0^1 n^2 x e^{-nx}\, dx = 1$.
    (c) Find a series of functions $u_n(x)$ for which

    $$\int_0^1 \left[\sum_{n=1}^{\infty} u_n(x) \right] dx \neq \sum_{n=1}^{\infty} \left[\int_0^1 u_n(x)\, dx \right]$$

13. If $f_1(t) = O(g(t))$, $f_2(t) = O(g(t))$, $t \to \infty$, show that $f_1(t) + f_2(t) = O(g(t))$, $t \to \infty$.

14. If $f_1(t) = O(g(t))$, $f_2(t) = O(g(t))$, $t \to \infty$, show that $f_1(t)f_2(t) = O(g(t)^2)$, $t \to \infty$.

15. If $f_1(t) = O(g(t))$, $f_2(t) = O(g(t))$, $t \to \infty$, is it true that $f_1(t)/f_2(t) = O(1)$, $t \to \infty$?

16. Show that $\log (1 + t) = \log t + O(1/t)$, $t \to \infty$.

17. Show that $(1 + t)^5 = t^5 + O(t^4)$, $t \to \infty$.

18. Show that $\sum_{n=1}^{\infty} ne^{-nt} = O(e^{-t})$, $t \to \infty$.

19. Use mathematical induction to prove that for any $p = 1, 2, \ldots$, $-1 < x < 1$,

    $$\sum_{n=1}^{\infty} n^p x^n = \frac{x Q_p(x)}{(1 - x)^{p+1}}$$

    where Q_p is a polynomial of degree $p - 1$ ($Q_1 = 1$, $Q_2 = 1 + x$, $Q_3 = x^2 + 4x + 1$).

20. Let $x = e^{-t}$ in Exercise 19 and show that for any $p = 1, 2, \ldots$, $\sum_{n=1}^{\infty} n^p e^{-nt} = O(e^{-t})$, $t \to \infty$.

21. Let $\{a_n\}$ be a sequence with $a_n = O(n^p)$ for some $p = 1, 2, \ldots$. Show that $\sum_{n=1}^{\infty} a_n e^{-nt} = O(e^{-t})$, $t \to \infty$.

22. (a) Derive the identity $1/(1 + t) = 1 - t + t^2 - \cdots + (-t)^{n-1} + (-t)^n/(1 + t)$.
    (b) Integrate this to obtain the identity

    $$\ln (1 + x) = x - \frac{x^2}{2} + \cdots + \frac{(-x)^{n-1}}{n} + (-1)^n \int_0^x \frac{t^n}{1 + t}\, dt$$

    for any $x \geq 0$.
    (c) Show that $\int_0^x [t^n/(1 + t)]\, dt \leq 1/(n + 1)$ for $0 \leq x \leq 1$.
    (d) Conclude that the series $\sum_{n=1}^{\infty} (-x)^n/n$ converges uniformly for $0 \leq x \leq 1$.

0.6 REVIEW OF ORDINARY DIFFERENTIAL EQUATIONS

General Properties of Second-Order Linear Equations

The reader has surely encountered examples of ordinary differential equations in calculus courses. For example, the equation $y' = ky$, which has the general solution $y(t) = Ce^{kt}$, governs exponential growth and decay. The equation $y'' + \omega^2 y = 0$, which has the general solution $y(t) = A \cos \omega t + B \sin \omega t$, governs simple harmonic motion with angular frequency ω. In this section we give a brief account of the necessary facts and techniques of ordinary differential equations, which is used in our subsequent work.

A *second-order linear ordinary differential equation* is of the form

$$a(t)y'' + b(t)y' + c(t)y = f(t) \tag{0.6.1}$$

where $a(t)$, $b(t)$, $c(t)$, $f(t)$ are given functions. If these functions are constant, $a(t) = a$, $b(t) = b$, $c(t) = c$, then we speak of an *equation with constant coefficients*. The equation is said to be *homogeneous* if $f(t) = 0$. The following fact is of great theoretical importance.

> **Existence theorem.** Let $a(t)$, $b(t)$, $c(t)$ be continuous functions on the interval $t_1 \leq t \leq t_2$ with $a(t) \neq 0$. The general solution of the homogeneous equation $a(t)y'' + b(t)y' + c(t)y = 0$ is of the form $y(t) = C_1 y_1(t) + C_2 y_2(t)$, where C_1, C_2 are arbitrary constants and the solutions $y_1(t)$, $y_2(t)$ satisfy the condition $y_1(t)y_2'(t) - y_1'(t)y_2(t) \neq 0$ for $t_1 \leq t \leq t_2$.

The proof of this theorem may be found in textbooks on ordinary differential equations. Our interest lies in finding the solutions when possible and in using the theorem to obtain other important information. For example, we may solve the *initial-value problem* to find a solution of the homogeneous equation $a(t)y'' + b(t)y' + c(t)y = 0$ satisfying $y(t_0) = y_0$, $y'(t_0) = y_1$. Since the general solution is of the form $y(t) = C_1 y_1(t) + C_2 y_2(t)$, the constants C_1, C_2 must satisfy the simultaneous equations $y_0 = C_1 y_1(t_0) + C_2 y_2(t_0)$, $y_1 = C_1 y_1'(t_0) + C_2 y_2'(t_0)$. The determinant of this system is $y_1(t_0)y_2'(t_0) - y_1'(t_0)y_2(t_0)$, which is nonzero from the existence theorem. Therefore we can determine the constants C_1, C_2 and obtain the solution of the initial-value problem.

Another important fact is expressed as follows.

> **Uniqueness theorem.** If $y(t)$, $\bar{y}(t)$ are two solutions of the differential equation $a(t)y'' + b(t)y' + c(t)y = f(t)$ for $t_1 \leq t \leq t_2$, with $y(t_0) = \bar{y}(t_0)$, $y'(t_0) = \bar{y}'(t_0)$ for some t_0, $t_1 \leq t_0 \leq t_2$, then $y(t) = \bar{y}(t)$ for all t, $t_1 \leq t \leq t_2$.

For example, $y(t) = 0$ is the unique solution of the homogeneous equation $a(t)y'' + b(t)y' + c(t)y = 0$ satisfying $y(t_0) = 0$, $y'(t_0) = 0$.

Constant-Coefficient Second-Order Homogeneous Equations

We now outline the procedure for finding the general solution of a second-order homogeneous linear differential equation with *constant* coefficients viz. $ay'' + by' + cy = 0$, where a, b, c are real numbers with $a \neq 0$. We try an exponential solution $y(t) = e^{rt}$ and substitute in the equation with $y'(t) = re^{rt}$, $y''(t) = r^2 e^{rt}$; thus $0 = ar^2 e^{rt} + bre^{rt} + ce^{rt} = (ar^2 + br + c)e^{rt}$. We consider three cases:

Case 1. The characteristic equation $ar^2 + br + c = 0$ has distinct real roots r_1, r_2 with $r_1 \neq r_2$.

Case 2. The characteristic equation $ar^2 + br + c = 0$ has equal real roots $r_1 = r_2$.

Case 3. The characteristic equation $ar^2 + br + c = 0$ has complex roots $r = \alpha \pm i\beta$ with $\beta \neq 0$.

From elementary algebra it is clear that this exhausts all possible cases. Indeed, the quadratic formula yields $r = (-b \pm \sqrt{b^2 - 4ac})/(2a)$ so that case 1 corresponds to $b^2 - 4ac > 0$, case 2 to $b^2 - 4ac = 0$, and case 3 to $b^2 - 4ac < 0$. We now give the general solution of the differential equation in each of the three cases.

Case 1. The general solution is $y(t) = C_1 e^{r_1 t} + C_2 e^{r_2 t}$, where C_1, C_2 are arbitrary constants and r_1, r_2 are the roots of the characteristic equation $ar^2 + br + c = 0$.

Case 2. The general solution is $y(t) = C_1 e^{rt} + C_2 t e^{rt}$, where C_1, C_2 are arbitrary constants and r is the root of the characteristic equation $ar^2 + br + c = 0$.

Case 3. The general solution is $y(t) = e^{\alpha t}(C_1 \cos \beta t + C_2 \sin \beta t)$, where C_1, C_2 are arbitrary constants and $\alpha + i\beta$, $\alpha - i\beta$ are the complex roots of the characteristic equation $ar^2 + br + c = 0$.

Example 0.6.1. Find the general solution of the differential equation $y'' + 3y' + 2y = 0$.

Solution. The characteristic equation is $r^2 + 3r + 2 = 0$, with roots $r_1 = -1$, $r_2 = -2$. Therefore we are in case 1, and the general solution is $y(t) = C_1 e^{-t} + C_2 e^{-2t}$. ●

Often, when working with case 1, we may find it convenient to express the solution in terms of hyperbolic functions. Recall the defining equations $\sinh x = \frac{1}{2}(e^x - e^{-x})$, $\cosh x = \frac{1}{2}(e^x + e^{-x})$. These have the properties $\sinh(-x) = -\sinh x$, $\cosh(-x) = \cosh x$, $\sinh 0 = 0$, $\cosh 0 = 1$ and $e^x = \cosh x + \sinh x$, $e^{-x} = \cosh x - \sinh x$. Thus any linear combination of exponential functions may be written in terms of hyperbolic functions. The solution of case 1 is written as $y(t) = C_1(\cosh r_1t + \sinh r_1t) + C_2(\cosh r_2t - \sinh r_2t)$. To verify the solution, we may use the differentiation formulas $(d/dx)(\sinh x) = \cosh x$, $(d/dx)(\cosh x) = \sinh x$.

Example 0.6.2. Solve the initial-value problem $y'' - 4y = 0$, $y(0) = 0$, $y'(0) = 6$.

Solution. The characteristic equation is $r^2 - 4 = 0$, with roots $r_1 = 2$, $r_2 = -2$. The general solution is $y(t) = C_1e^{2t} + C_2e^{-2t} = (C_1 + C_2) \cosh 2t + (C_1 - C_2) \sinh 2t$. Setting $t = 0$ and using the initial conditions, we must have $0 = (C_1 + C_2)$, $6 = 2(C_1 - C_2)$. The required solution is $y(t) = 3 \sinh 2t$. ●

We now illustrate a typical example from case 3.

Example 0.6.3. Solve the initial-value problem $y'' + 2y' + 2y = 0$, $y(0) = 3$, $y'(0) = -1$.

Solution. The characteristic equation is $r^2 + 2r + 2 = 0$, which has the complex roots $r = -1 \pm i$. Therefore we are in case 3, and the general solution is $y(t) = e^{-t}(C_1 \cos t + C_2 \sin t)$. The initial conditions require $3 = C_1$, $-1 = C_2 - C_1$. The solution is $y(t) = e^{-t}(3 \cos t + 2 \sin t)$. ●

Euler's Equation

Once the constant coefficient equation has been found, we can solve other second-order equations by a transformation of the independent variable. This is defined by an increasing function $u = \varphi(t)$ which is twice-differentiable with a twice-differentiable inverse function. A new function is defined by writing $Y(u) = y(t)$, and the derivatives are transformed by using the chain rule in the form $dy/dt = (dY/du)(du/dt)$, $d^2y/dt^2 = (dY/du)(d^2u/dt^2) + (d^2Y/du^2)(du/dt)^2$. When these are expressed in terms of u and substituted in the second-order equation, we obtain a new equation for the transformed function $Y(u)$.

The Euler (equidimensional) equation arises in the important case of the transformation $u = e^t$, with $du/dt = e^t = u$, $d^2u/dt^2 = e^t = u$. The derivatives are transformed as $y'(t) = Y'(u) du/dt = uY'(u)$, and $y''(t) = Y'(u) (d^2u/dt^2) + Y''(u) (du/dt)^2 = uY'(u) + u^2Y''(u)$; and the constant-coefficient equation $ay'' + by' + cy = 0$ is transformed to the equation $a[u^2Y''(u) + uY'(u)] + buY'(u) + cY(u) = 0$. The general form of Euler's equation is

$$\alpha u^2 Y'' + \beta u Y' + \gamma Y = 0 \qquad (0.6.2)$$

The independent variable u is supposed to be positive: $u > 0$.

To solve this, we note that the exponential solution of the constant-coefficient equation is transformed as $y(t) = e^{rt} = (e^t)^r = u^r = Y(u)$. Substitution of the trial solution $Y(u) = u^r$ into the eulerian equation produces the quadratic equation $\alpha r(r - 1) + \beta r + \gamma = 0$, corresponding to the characteristic equation of the constant-coefficient second-order equation. We are led to the following threefold classification:

Proposition. For the Euler equation (0.6.2), one of three situations arises:

1. The indicial equation $\alpha r(r - 1) + \beta r + \gamma = 0$ has two distinct real roots $r_1 \neq r_2$. In this case two linearly independent solutions are given by $Y_1(u) = u^{r_1}$, $Y_2(u) = u^{r_2}$.
2. The indicial equation $\alpha r(r - 1) + \beta r + \gamma = 0$ has a repeated real root r. In this case two linearly independent solutions are given by $Y_1(u) = u^r$, $Y_2(u) = u^r \ln u$.
3. The indicial equation $\alpha r(r - 1) + \beta r + \gamma = 0$ has complex roots $r_1 = a + ib$, $r_2 = a - ib$ with $b > 0$. Then two linearly independent solutions are given by $Y_1(u) = u^a \cos (b \ln u)$, $Y_2(u) = u^a \sin (b \ln u)$.

Example 0.6.4. Find the general solution of the eulerian equation $u^2 Y'' - 3u Y' + 4Y = 0$.

Solution. The indicial equation is $r(r - 1) - 3r + 4 = 0$, which simplifies to $0 = r^2 - 4r + 4 = (r - 2)^2$, which has the repeated root $r = 2$. The general solution is of the form $Y(u) = c_1 u^2 + c_2 u^2 \ln u$, for arbitrary constants c_1, c_2. ●

First-Order Linear Equations

We now briefly discuss the *first-order linear equation,* in general form,

$$b(t)y' + c(t)y = f(t)$$

Equations of this type may be solved by the method of *integrating factors.* This consists of finding a function $\mu(t)$ such that $\mu(t)[y'(t) + b(t)y(t)/c(t)] = [\mu(t)y(t)]'$ for every possible function $y(t)$. This requires that $\mu(t)$ satisfy the auxiliary equation $\mu'(t)b(t) = \mu(t)c(t)$. This may be solved by writing $[\log \mu(t)]' = \mu'(t)/\mu(t) = c(t)/b(t)$; thus $\mu(t) = \exp \int_{t_0}^{t} c(t')/b(t') \, dt'$. Having obtained $\mu(t)$, we can get the solution $y(t)$ by a single integration through $[\mu(t)y(t)]' = \mu(t) f(t)/b(t)$.

Example 0.6.5. Find the general solution of the equation $y' + cy = d$, where c, d are constants with $c \neq 0$.

Solution. An integrating factor is obtained by solving $\mu'(t) = c\mu(t)$; thus $\mu(t) = e^{ct}$. Multiplying the equation by e^{ct}, we have $(e^{ct}y)' = de^{ct}$, $e^{ct}y(t) = (d/c)e^{ct} + C_1$, where C_1 is an arbitrary constant. The general solution is $y(t) = d/c + C_1 e^{-ct}$. ●

Method of Particular Solutions

Returning to second-order equations, we now discuss the nonhomogeneous equation $a(t)y'' + b(t)y' + c(t)y = f(t)$. The general solution is obtained by the *method of particular solutions* as follows: Suppose that $y_0(t)$ is a known solution of the nonhomogeneous equation $a(t)y'' + b(t)y' + c(t)y = f(t)$. Then the general solution is obtained in the form $y(t) = y_0(t) + C_1 y_1(t) + C_2 y_2(t)$, where C_1, C_2 are arbitrary constants and $y_1(t)$, $y_2(t)$ are the solutions of the homogeneous equations discussed above.

We illustrate the method of particular solutions in the case of an equation with constant coefficients.

Example 0.6.6. Find the general solution of the equation $ay'' + by' + cy = f$, where a, b, c, f are constants with $a \neq 0$, $c \neq 0$.

Solution. A particular solution is $y_0(t) = d/c$, since $y_0'(t) = 0$, $y_0''(t) = 0$ for this choice of $y_0(t)$. Thus the general solution of the equation is $y(t) = d/c + C_1 y_1(t) + C_2 y_2(t)$, where C_1, C_2 are arbitrary constants. It will depend on the exact values of a, b, c whether case 1, 2, or 3 applies. ●

Example 0.6.7. Find the general solution of the equation $ay'' + by' = f$, where a, b, f are constants with $a \neq 0$, $b \neq 0$.

Solution. A particular solution is $y_0(t) = ft/b$, since $y_0'(t) = f/b$, $y_0''(t) = 0$ for this choice of $y_0(t)$. To obtain the general solution, we have the characteristic equation $ar^2 + br = 0$, with roots $r = 0$, $-b/a$. The general solution of the equation is $y(t) = ft/b + C_1 + C_2 e^{-bt/a}$. ●

An important method for finding particular solutions is the method of the *variation of parameters*, which we illustrate here for the constant-coefficient second-order equation. Suppose that $v(t)$ is the solution of the homogeneous equation $av'' + bv' + cv = 0$ with the initial conditions $v(0) = 0$, $v'(0) = 1$ (a, b, c constants, $a \neq 0$). Then the formula

$$y(t) = \frac{1}{a} \int_0^t v(t - s) f(s) \, ds$$

provides a particular solution of the equation $ay'' + by' + cy = f(t)$ satisfying the initial conditions $y(0) = 0$, $y'(0) = 1$.

Example 0.6.8. Find the particular solution of the equation $y'' + 4y = e^{t^2}$.

Solution. The method of variation of parameters provides the solution in the form $y(t) = \frac{1}{2} \int_0^t \sin 2(t - s) e^{s^2} \, ds$. ●

Steady-State and Relaxation Time

Many differential equations which occur in applications have solutions $y(t)$ which tend to a limit y_∞ when the time t tends to infinity. This represents an equilibrium state, or *steady state*, of the system. When this happens, it is useful to have a quantitative measure of the time necessary for the system to come within a fixed fraction of the steady state. The *relaxation time* τ is defined by the following limit.

$$\frac{1}{\tau} = -\lim_{t \to \infty} \frac{1}{t} \ln |y(t) - y_\infty|$$

Example 0.6.9. For the differential equation $y' + 3y = 6$ with the initial condition $y(0) = 1$, find the steady state and the relaxation time.

Solution. The solution of the differential equation can be obtained with the integrating factor e^{3t}. Thus $(ye^{3t})' = 6e^{3t}$, $ye^{3t} = 1 + 2(e^{3t} - 1) = 2e^{3t} - 1$. Thus $y(t) = 2 - e^{-3t}$, and the steady state is $y_\infty = 2 = \lim_{t \to \infty} y(t)$. To find the relaxation time, we write $|y(t) - 2| = e^{-3t}$, $\ln |y(t) - 2| = -3t$, $(-1/t) \ln |y(t) - 2| = 3$. Therefore $\tau = \frac{1}{3}$, and we have found the relaxation time. ●

The relaxation time is closely related to the *half-life* T, defined through $|y(T) - y_\infty|/|y(0) - y_\infty| = \frac{1}{2}$. In the previous example $|y(t) - y_\infty|/|y(0) - y_\infty| = e^{-3t}$; therefore the half-life is obtained by solving $e^{-3T} = \frac{1}{2}$ or $T = \frac{1}{3} \ln 2 = 0.23$, to two decimals. This is typical for first-order equations, where the relaxation time and half-life differ by the factor $\ln 2$. We choose to work with the relaxation time because of this absence of the factor $\ln 2$ in the formulas. In general, relaxation time gives a rough measure of the time necessary for the system to come within $1/e \cong 0.37$ of the steady state.

Note that the relaxation time depends on the solution as well as the equation. Consider, for example, the differential equation $y'' - 5y' + 6y = 0$ and the two solutions $y_1(t) = e^{-3t}$ and $y_2(t) = e^{-2t}$. Both solutions have the same steady state $y_\infty = 0$. For $y_1(t)$ we have the relaxation time $\tau_1 = \frac{1}{3}$, whereas for the solution $y_2(t)$ we have the relaxation time $\tau_2 = \frac{1}{2}$.

To obtain a practical estimate of the time necessary to reach the steady state, we may solve the equation $[y(t) - y_\infty]/[y(0) - y_\infty] = q$, a fixed fraction. For $q = 0.01$, we get $e^{-t/\tau} = 0.01$, $t/\tau = \ln 100 = 4.6$. Thus *after 5 units of relaxation time the system comes to within 1 percent of the steady state.*

We now return to the discussion of second-order linear equations.

Power Series Solutions

If the homogeneous equation $a(t)y'' + b(t)y' + c(t)y = 0$ does not have constant coefficients, the solutions $y_1(t)$, $y_2(t)$ cannot be written as elementary functions, in general. The *method of power series* provides a useful algorithm for obtaining solutions for a wide class of second-order equations with nonconstant coefficients. We suppose that the functions $a(t)$, $b(t)$, $c(t)$ have convergent power series expansions about $t = t_0$:

$$a(t) = a_0 + \sum_{n=1}^{\infty} a_n(t - t_0)^n$$

$$b(t) = b_0 + \sum_{n=1}^{\infty} b_n(t - t_0)^n$$

$$c(t) = c_0 + \sum_{n=1}^{\infty} c_n(t - t_0)^n$$

The point $t = t_0$ is called an *ordinary point* if $a_0 = a(t_0) \neq 0$. Under these hypotheses, we can find power series solutions $y(t)$ of the equation $a(t)y'' + b(t)y' + c(t)y = 0$, convergent for $|t - t_0| < \delta$, an interval on which $a(t) \neq 0$. To find these solutions, we assume a solution of the form

$$y(t) = y_0 + \sum_{n=1}^{\infty} y_n(t - t_0)^n$$

Differentiating formally, we have

$$y'(t) = \sum_{n=1}^{\infty} ny_n(t - t_0)^{n-1}$$

$$y''(t) = \sum_{n=1}^{\infty} n(n - 1)y_n(t - t_0)^{n-2}$$

We form the power series for $a(t)y''$, $b(t)y'$, and $c(t)y$. Summing these and equating the coefficients of $1, t - t_0, \ldots$ to zero, we obtain linear equations for the coefficients $y_2, y_3, \ldots$. The coefficients y_0, y_1 are determined by the initial conditions. In this way we obtain the power series solution of the equation.

Example 0.6.10. Find the power series solution of the equation $y'' + ty = 0$ with the initial conditions $y(0) = 1$, $y'(0) = 0$.

Solution. We assume a power series of the form

$$y(t) = 1 + \sum_{n=2}^{\infty} y_n t^n$$

Differentiating, we have

$$y''(t) = \sum_{n=2}^{\infty} n(n - 1)y_n t^{n-2}$$

Thus

$$y'' + ty = 2y_2 + t(1 + 6y_3) + t^2(12y_4) + t^3(y_2 + 20y_5) + \cdots$$
$$+ t^n[y_{n-1} + (n + 2)(n + 1)y_{n+2}] + \cdots$$

Equating each coefficient to zero, we have $y_2 = 0$, $y_3 = -\frac{1}{6}$, $y_4 = 0$, $y_5 = 0$, $y_6 = \frac{1}{180}$, and, in general,

$$y_{3k} = \frac{(-1)^k}{3k(3k - 1)(3k - 3) \cdots 6 \cdot 5 \cdot 3 \cdot 2}$$

and

$$y_{3k+1} = 0 = y_{3k+2}$$

The solution is

$$y(t) = 1 + \sum_{k=1}^{\infty} \frac{(-1)^k t^{3k}}{3k(3k-1)(3k-3)\cdots 6 \cdot 5 \cdot 3 \cdot 2} \qquad \bullet$$

The method of power series can be used to produce the two solutions $y_1(t)$, $y_2(t)$ with $y_1(t)y_2'(t) - y_1'(t)y_2(t) \neq 0$. In fact, we can determine these uniquely by the initial conditions $y_1(t_0) = 1$, $y_1'(t_0) = 0$, $y_2(t_0) = 0$, $y_2'(t_0) = 1$.

We now discuss power series solutions of the equation $a(t)y'' + b(t)y' + c(t)y = 0$ with $a(t_0) = 0$. We say that $t = t_0$ is a *regular singular point* if the functions $(t - t_0)b(t)/a(t)$ and $(t - t_0)^2 c(t)/a(t)$ both have convergent power series expansions about $t = t_0$.

$$\frac{(t-t_0)b(t)}{a(t)} = \beta(t) = \beta_0 + \sum_{n=1}^{\infty} \beta_n(t-t_0)^n$$

$$\frac{(t-t_0)^2 c(t)}{a(t)} = \gamma(t) = \gamma_0 + \sum_{n=1}^{\infty} \gamma_n(t-t_0)^n$$

Thus the equation is written in the form

$$(t-t_0)^2 y'' + (t-t_0)\beta(t)y' + \gamma(t)y = 0$$

When $\beta(t)$ and $\gamma(t)$ are constants, this is the Euler equation (0.6.2), whose typical solutions are of the form $(t - t_0)^r$ for a suitable value of r. In the general case, we look for a power series solution in the form

$$y = (t-t_0)^r \left[1 + \sum_{n=1}^{\infty} y_n(t-t_0)^n \right]$$

$$= (t-t_0)^r + \sum_{n=1}^{\infty} y_n(t-t_0)^{n+r}$$

The exponent r and the coefficients y_1, y_2, . . . are found by substituting in the differential equation and equating the coefficients of $(t - t_0)^r$, $(t - t_0)^{r+1}$, ... Thus we have the formal series

$$y' = r(t-t_0)^{r-1} + \sum_{n=1}^{\infty} (n+r)y_n(t-t_0)^{n+r-1}$$

$$y'' = r(r-1)(t-t_0)^{r-2} + \sum_{n=1}^{\infty} (n+r)(n+r-1)y_n(t-t_0)^{n+r-2}$$

Equating the coefficient of $(t - t_0)^r$ to zero gives the *indicial equation*

$$r(r-1) + \beta_0 r + \gamma_0 = 0$$

This equation has two roots r_1, r_2. If they are real, we take the larger root $r = \frac{1}{2}[1 - \beta_0 + \sqrt{(1 - \beta_0)^2 - 4\gamma_0}]$. With this value of r we equate the coefficients of $(t - t_0)^{r+1}$, $(t - t_0)^{r+2}$, ... to zero to obtain the coefficients $y_1, y_2, \ldots$. The resultant power series is called the *Frobenius solution* of the equation.

Example 0.6.11. Find the Frobenius solution of $t^2y'' + ty = 0$.

Solution. We have $t_0 = 0$, $\beta(t) = 0$, $\gamma(t) = t$. The indicial equation is $r(r - 1) = 0$, with roots $r = 0, 1$. We look for the power series solution in the form $y = t + \sum_{n=1}^{\infty} y_n t^{n+1}$. Thus $y' = 1 + \sum_{n=1}^{\infty} (n + 1)y_n t^n$, $y'' = \sum_{n=1}^{\infty} (n + 1)ny_n t^{n-1}$, $t^2y'' + ty = t^2(2y_1 + 1) + t^3(6y_2 + y_1) + \cdots + t^n[n(n - 1)y_{n-1} + y_{n-2}] + \cdots$. Matching each coefficient to zero, we have $y_1 = -\frac{1}{2}$, $y_2 = 1/(3 \cdot 2 \cdot 2)$, $y_3 = -1/(4 \cdot 3 \cdot 3 \cdot 2 \cdot 2)$, $\ldots$, $y_n = (-1)^n/[(n + 1)(n!)^2]$.
The solution is

$$y = t - \frac{t^2}{2} + \frac{t^3}{3 \cdot 2 \cdot 2} - \frac{t^4}{4 \cdot 3 \cdot 3 \cdot 2 \cdot 2} \cdots$$

$$= \sum_{n=1}^{\infty} (-1)^{n+1} \frac{nt^n}{(n!)^2} \qquad \bullet$$

EXERCISES 0.6

In Exercises 1 to 5, find the general solution of each of the indicated equations.
1. $y'' + 4y = 0$
2. $y'' + 4y' + 4y = 0$
3. $y'' + 2y' - 15y = 0$
4. $y'' + 3y' = 0$
5. $3y'' - 5y' - 2y = 0$

In Exercises 6 to 10, solve the indicated initial-value problems.
6. $y'' + 4y = 0;\ y(0) = 0,\ y'(0) = 2$
7. $y'' + 4y' + 4y = 0;\ y(0) = 1,\ y'(0) = 0$
8. $y'' + 2y' - 15y = 0;\ y(0) = 2,\ y'(0) = 3$
9. $y'' + 3y' = 0;\ y(1) = 0,\ y'(1) = 4$
10. $3y'' - 5y' - 2y = 0;\ y(0) = 1,\ y'(0) = 1$

In Exercises 11 to 15, find the general solution of each of the indicated first-order equations.
11. $y' + 2ty = e^{-t^2}$
12. $ty' + y = 1$
13. $y' + 3y = e^{2t}$
14. $ty' + 4y = t^2$
15. $(\sin t)y' + (\cos t)y = \cos t$

In Exercises 16 to 20, find the general solution of each of the indicated equations, and solve the indicated initial-value problem.
16. $y'' + 4y' + 6y = 2;\ y(0) = 0,\ y'(0) = 2$
17. $y'' - 4y = 2;\ y(0) = 0,\ y'(0) = 0$
18. $y'' + 4y' = 2;\ y(0) = 0,\ y'(0) = 4$

19. $y'' = 0$; $y(0) = 3$, $y'(0) = 4$

20. $ty'' + y' = -1$; $y(1) = 0$, $y'(1) = 0$

In Exercises 21 to 25, find the steady-state solution and the relaxation time for each of the following solutions of differential equations.

21. $y(t) = 3 + 4e^{-2t}$ (solution of $y' + 2y = 6$)

22. $y(t) = 5 + 3e^{-t} + e^{-3t}$ (solution of $y'' + 4y' + 3y = 15$)

23. $y(t) = e^{-t} \cosh t$ (solution of $y'' + 4y' + 4y = 2$)

24. $y(t) = 1 + e^{-2t} \sinh t$ (solution of $y'' + 4y' + 3y = 3$)

25. $y(t) = 4 + e^{-t} + 3te^{-t}$ (solution of $y'' + 2y' + y = 4$)

The following problems deal with power series solutions of second-order equations.

26. Find the power series solution of $y'' + 4ty = 0$ with the initial conditions $y(0) = 0$, $y'(0) = 1$.

27. Find the power series solution of $y'' + 4ty = 0$ with the initial condition $y(0) = 1$, $y'(0) = 0$.

28. Find the power series solution of $y'' + t^2y = 0$ with the initial condition $y(0) = 0$, $y'(0) = 1$.

29. Find the power series solution of $y'' + t^2y = 0$ with the initial condition $y(0) = 1$, $y'(0) = 0$.

30. Find the power series solution of $y'' + (1 + t)y = 0$ with the initial condition $y(0) = 0$, $y'(0) = 1$.

31. Which of the following second-order differential equations has a regular singular point at $t = 0$?

(a) $t^2y'' + 3ty + y = 0$ (b) $t^2y'' + y' + 3ty = 0$

(c) $t^2y'' + t^2y' + y = 0$ (d) $ty'' + y' = 0$

(e) $ty'' + y = 0$ (f) $y'' + ty = 0$

In Exercises 32 to 35, find the indicial equation and the Frobenius solution for the indicated equation, which has a regular singular point at $t = 0$.

32. $t^2y'' + y = 0$

33. $t^2y'' + ty' - y = 0$

34. $t^2y'' + ty' + 3t^2y = 0$

35. $t^2y'' + (1 + 3t^2)y = 0$

FOURIER SERIES

INTRODUCTION

Many of the classical partial differential equations with boundary conditions have separated solutions which involve trigonometric functions and superpositions of them. This leads to the theory of Fourier series, which is developed here in its own right. In this chapter we show how to represent an "arbitrary function" by an infinite sum of trigonometric functions. The theory is developed in a self-contained manner, independent of Sturm-Liouville theory. In particular, we give proofs of all statements regarding orthogonality, pointwise convergence, uniform convergence, and integration or differentiation of Fourier series. Applications to partial differential equations are developed in Chap. 2.

1.1 DEFINITIONS AND EXAMPLES

A *trigonometric series* on $(-L, L)$ is a function of the form

$$A_0 + \sum_{n=1}^{\infty} \left(A_n \cos \frac{n\pi x}{L} + B_n \sin \frac{n\pi x}{L} \right) \tag{1.1.1}$$

where $A_0, A_1, B_1, \ldots$ are constants. This is a series of sines and cosines, whose frequencies are multiples of a basic angular frequency π/L and whose amplitudes are arbitrary. We recognize the functions $\cos(n\pi x/L)$, $\sin(n\pi x/L)$ as the eigenfunctions of the Sturm-Liouville problem $f'' + \lambda f = 0$ with the periodic boundary conditions $f(-L) = f(L)$, $f'(-L) = f'(L)$. We let m, n take the values $0, 1, 2, \ldots$, noting that $\cos 0 = 1$, $\sin 0 = 0$.

Orthogonality Relations

Proposition 1.1.1. We have the orthogonality relations

$$\int_{-L}^{L} \cos \frac{n\pi x}{L} \cos \frac{m\pi x}{L}\, dx = \begin{cases} 0 & n \neq m \\ L & n = m \neq 0 \\ 2L & n = m = 0 \end{cases} \tag{1.1.2}$$

$$\int_{-L}^{L} \sin \frac{n\pi x}{L} \sin \frac{m\pi x}{L}\, dx = \begin{cases} 0 & n \neq m \\ L & n = m \neq 0 \\ 0 & n = m = 0 \end{cases} \tag{1.1.3}$$

$$\int_{-L}^{L} \sin \frac{n\pi x}{L} \cos \frac{m\pi x}{L}\, dx = 0 \qquad \text{all } m, n \tag{1.1.4}$$

To prove these, we use the trigonometric identities

$$\cos \alpha \cos \beta = \tfrac{1}{2}[\cos (\alpha - \beta) + \cos (\alpha + \beta)] \tag{1.1.2\textprime}$$

$$\sin \alpha \sin \beta = \tfrac{1}{2}[\cos (\alpha - \beta) - \cos (\alpha + \beta)] \tag{1.1.3\textprime}$$

$$\sin \alpha \cos \beta = \tfrac{1}{2}[\sin (\alpha - \beta) + \sin (\alpha + \beta)] \tag{1.1.4\textprime}$$

Thus to prove (1.1.2), we have, for $n \neq m$,

$$\int_{-L}^{L} \cos \frac{n\pi x}{L} \cos \frac{m\pi x}{L}\, dx = \frac{1}{2} \int_{-L}^{L} \left[\cos \frac{(n - m)\pi x}{L} + \cos \frac{(n + m)\pi x}{L} \right] dx$$

$$= \frac{L}{2\pi} \left\{ \frac{\sin [(n - m)\pi x/L]}{n - m} \Big|_{-L}^{L} + \frac{\sin [(n + m)\pi x/L]}{n + m} \Big|_{-L}^{L} \right\}$$

$$= 0$$

If $n = m \neq 0$, we have

$$\int_{-L}^{L} \cos^2 \frac{n\pi x}{L}\, dx = \frac{1}{2} \int_{-L}^{L} \left(1 + \cos \frac{2n\pi x}{L} \right) dx$$

$$= \frac{1}{2} \left(2L + \frac{L}{2n\pi} \sin \frac{2n\pi x}{L} \Big|_{-L}^{L} \right)$$

$$= L$$

Finally, if $n = m = 0$, the integral is $2L$. This completes the proof of (1.1.2). The proofs of (1.1.3) and (1.1.4) are left as exercises. ∎

Having established the orthogonality and performed the computation of these integrals, we now define the Fourier series of a function $f(x)$, $-L < x < L$.

Definition of Fourier Coefficients

Definition. Let $f(x)$, $-L < x < L$, be a real-valued function. The *Fourier series* of f is the trigonometric series (1.1.1) where (A_n, B_n) are defined by

$$A_0 = \frac{1}{2L} \int_{-L}^{L} f(x)\, dx \tag{1.1.5}$$

$$A_n = \frac{1}{L} \int_{-L}^{L} f(x) \cos \frac{n\pi x}{L}\, dx \qquad n = 1, 2, \ldots \tag{1.1.6}$$

$$B_n = \frac{1}{L} \int_{-L}^{L} f(x) \sin \frac{n\pi x}{L}\, dx \qquad n = 1, 2, \ldots \tag{1.1.7}$$

These definitions were suggested in Sec. 0.3, where we showed that the minimum of $\|f - \sum_{n=1}^{N} a_n \varphi_n\|^2$ is achieved by choosing $a_1, \ldots, a_n$ as the Fourier coefficients $\langle f, \varphi_n \rangle / \langle \varphi_n, \varphi_n \rangle$.

Even Functions and Odd Functions

To simplify the computation of Fourier series of many functions encountered in practice, we may exploit symmetry arguments. A function $f(x)$, $-L < x < L$, is *even* if $f(-x) = f(x)$, $-L < x < L$. A function $f(x)$, $-L < x < L$, is *odd* if $f(-x) = -f(x)$, $-L < x < L$. For example, $f(x) = x$, $f(x) = x^3$, and $f(x) = \sin x$ are odd functions, whereas $f(x) = x^2$, $f(x) = x^4$, and $f(x) = \cos x$ are even functions. Of course, many functions are neither even nor odd, for example, $f(x) = x + x^2$. The product of two even functions is an even function, the product of an odd function and an even function is an odd function, and the product of two odd functions is an even function. These properties result from the multiplication facts $(+1)(+1) = +1$, $(-1)(+1) = -1$, $(-1)(-1) = +1$. If $f(x)$, $-L < x < L$, is an odd function, the integral $\int_{-L}^{L} f(x)\, dx = 0$. This may be seen in detail by writing

$$\int_{-L}^{0} f(x)\, dx = -\int_{L}^{0} f(-t)\, dt \qquad x = -t,\ dx = -dt$$

$$= \int_{0}^{L} f(-t)\, dt \qquad \int_{0}^{L} = -\int_{L}^{0}$$

$$= -\int_{0}^{L} f(t)\, dt \qquad \text{oddness}$$

But t is a dummy variable of integration; thus

$$\int_{-L}^{L} f(x)\, dx = \int_{-L}^{0} f(x)\, dx + \int_{0}^{L} f(x)\, dx$$

$$= -\int_{0}^{L} f(x)\, dx + \int_{0}^{L} f(x)\, dx$$

$$= 0$$

Similarly, if $f(x)$, $-L < x < L$, is an even function, then $\int_{-L}^{L} f(x)\, dx = 2\int_{0}^{L} f(x)\, dx$.

Proposition 1.1.2. If $f(x)$, $-L < x < L$, is an even function, then $B_n = 0$, $n = 1$, $2, \ldots$. If $f(x)$, $-L < x < L$, is an odd function, then $A_n = 0$, $n = 0, 1, 2, \ldots$.

Proof. To prove these facts, we note that $\sin(n\pi x/L)$ is an odd function and $\cos(\pi x/L)$ is an even function since $\sin(-\theta) = -\sin\theta$, $\cos(-\theta) = \cos\theta$. Now if $f(x)$, $-L < x < L$, is an even function, the product $f(x)\sin(n\pi x/L)$ is an odd function and we have $B_n = 0$. If $f(x)$, $-L < x < L$, is an odd function, the product $f(x)\cos(n\pi x/L)$ is an odd function and we have $A_n = 0$. ∎

Examples

Example 1.1.1. Compute the Fourier series of $f(x) = x$, $-L < x < L$.†

Solution. The function $f(x)$, $-L < x < L$, is an odd function; therefore $A_n = 0$, $n = 0, 1, 2, \ldots$. To compute B_n, we note that $f(x)\sin(n\pi x/L)$ is an even function; thus

$$B_n = \frac{1}{L}\int_{-L}^{L} x \sin\frac{n\pi x}{L}\, dx$$

$$= \frac{2}{L}\int_{0}^{L} x \sin\frac{n\pi x}{L}\, dx$$

We integrate by parts with $u = x$, $dv = \sin(n\pi x/L)\, dx$. Thus

$$B_n = \frac{2}{L}\left(-x\frac{L}{n\pi}\cos\frac{n\pi x}{L}\Big|_{0}^{L} + \frac{L}{n\pi}\int_{0}^{L}\cos\frac{n\pi x}{L}\, dx\right)$$

The last integral is zero, and we have $B_n = -[2L/(n\pi)]\cos n\pi = [2L/(n\pi)](-1)^{n+1}$. Therefore the Fourier series of $f(x) = x$, $-L < x < L$, is

$$\frac{2L}{\pi}\sum_{n=1}^{\infty}\frac{(-1)^{n+1}}{n}\sin\frac{n\pi x}{L} \qquad \bullet$$

Example 1.1.2. Compute the Fourier series of $f(x) = |x|$, $-L < x < L$.

---

† This example also appears in the Appendix on "Using Mathematica."

Solution. The function $f(x)$, $-L < x < L$, is an even function; therefore $B_n = 0$, $n = 1, 2, \ldots$. To compute A_n, we note that the product $f(x) \cos (n\pi x/L)$ is an even function; thus, for $n \neq 0$,

$$A_n = \frac{1}{L} \int_{-L}^{L} |x| \cos \frac{n\pi x}{L} \, dx$$

$$= \frac{2}{L} \int_{0}^{L} x \cos \frac{n\pi x}{L} \, dx$$

We integrate by parts with $u = x$, $dv = \cos (n\pi x/L) \, dx$. Thus

$$A_n = \frac{2}{L} \left(x \frac{L}{n\pi} \sin \frac{n\pi x}{L} \Big|_0^L - \frac{L}{n\pi} \int_0^L \sin \frac{n\pi x}{L} \, dx \right)$$

The first term is zero at both endpoints $x = 0$, $x = L$, while the integral can be evaluated as $\int_0^L \sin (n\pi x/L) \, dx = [L/(n\pi)][1 - (-1)^n]$. Thus we have $A_n = -[2L/(n^2\pi^2)][1 - (-1)^n]$ for $n \neq 0$. For $n = 0$, we have $A_0 = \frac{1}{2} \int_0^L x \, dx = L/2$. Therefore the Fourier series of $f(x) = |x|$, $-L < x < L$, is

$$\frac{L}{2} - \frac{2L}{\pi^2} \sum_{n=1}^{\infty} \frac{1 - (-1)^n}{n^2} \cos \frac{n\pi x}{L}$$

This can also be written as $L/2 - (4L/\pi^2) \sum_{m=1}^{\infty} \cos [(2m - 1)\pi x/L]/(2m - 1)^2$ by writing $n = 2m - 1$ and noting that $1 - (-1)^n = 0$ if n is even and $1 - (-1)^n = 2$ if n is odd. ●

We show in Sec. 1.2 that these Fourier series are convergent and that the equation

$$f(x) = A_0 + \sum_{n=1}^{\infty} \left(A_n \cos \frac{n\pi x}{L} + B_n \sin \frac{n\pi x}{L} \right)$$

is valid for $-L < x < L$. We illustrate this graphically for the above two examples. To do this, we define the *partial sum of order N* of a trigonometric series as the function

$$f_N(x) = A_0 + \sum_{n=1}^{N} \left(A_n \cos \frac{n\pi x}{L} + B_n \sin \frac{n\pi x}{L} \right)$$

In Figs. 1.1.1 and 1.1.2, we give the partial sums for the Fourier series of the two previous examples.

The method of these two examples may be extended to compute the Fourier series of any polynomial $f(x) = c_0 + c_1 x + \cdots + c_k x^k$. To do this, it is sufficient to handle each term separately and to integrate by parts. Thus we have the reduction formulas

$$\int_{-L}^{L} x^k \sin \frac{n\pi x}{L} \, dx = -\frac{Lx^k}{n\pi} \cos \frac{n\pi x}{L} \Big|_{-L}^{L} + \frac{Lk}{n\pi} \int_{-L}^{L} x^{k-1} \cos \frac{n\pi x}{L} \, dx$$

$$\int_{-L}^{L} x^k \cos \frac{n\pi x}{L} \, dx = \frac{Lx^k}{n\pi} \sin \frac{n\pi x}{L} \Big|_{-L}^{L} - \frac{Lk}{n\pi} \int_{-L}^{L} x^{k-1} \sin \frac{n\pi x}{L} \, dx$$

Proceeding inductively, we can compute the necessary integrals.

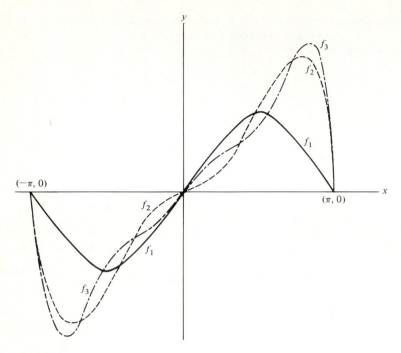

FIGURE 1.1.1
Graphs of the partial sums $f_N(x)$ for $N = 1, 2, 3$ of the Fourier series of $f(x) = x$, $-\pi < x < \pi$.

If a function $f(x)$, $-L < x < L$, can be written as a finite trigonometric sum, then its Fourier series is that trigonometric sum. For example, the Fourier series of $f(x) = \sin^2 x$, $-\pi < x < \pi$, can be obtained from the observation that $\sin^2 x = \frac{1}{2}(1 - \cos 2x)$; thus $B_n = 0$ for all n, while $A_0 = \frac{1}{2}$, $A_2 = -\frac{1}{2}$, and $A_n = 0$ for $n = 1, 3, 4, 5, \ldots$. It is not necessary to perform any integrations to find the Fourier series in this case.

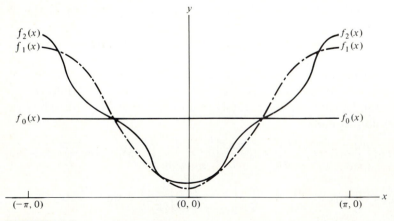

FIGURE 1.1.2
Graphs of the partial sums $f_N(x)$ for $N = 0, 1, 2$ of the Fourier series of $f(x) = |x|$, $-\pi < x < \pi$.

Periodic Functions

Definition. A function $f(x)$, $-\infty < x < \infty$, is *2L-periodic* if

$$f(x + 2L) = f(x) \qquad -\infty < x < \infty$$

For example, $\sin (n\pi x/L)$ and $\cos (n\pi x/L)$ are 2L-periodic for $n = 1, 2, \ldots,$ since

$$\sin \frac{n\pi}{L} (x + 2L) = \sin \left(\frac{n\pi x}{L} + 2n\pi \right) = \sin \frac{n\pi x}{L}$$

$$\cos \frac{n\pi}{L} (x + 2L) = \cos \left(\frac{n\pi x}{L} + 2n\pi \right) = \cos \frac{n\pi x}{L}$$

The sum, difference, and product of any two 2L-periodic functions are again 2L-periodic. Therefore any convergent trigonometric series defines a 2L-periodic function $f(x)$, $-\infty < x < \infty$. Conversely, we can speak of the Fourier series of a 2L-periodic function $f(x)$, $-\infty < x < \infty$, by restricting x to $-L < x < L$ and computing the Fourier series as we have just done.

Example 1.1.3. Compute the Fourier series of the 2L-periodic function $f(x) = -1$ if $(2n - 1)L < x < 2nL$, $f(x) = 1$ if $2nL < x < (2n + 1)L$, $n = 0, \pm 1, \pm 2, \ldots$.

Solution. Here f is an odd function; thus $A_n = 0$, $B_n = (2/L) \int_0^L \sin (n\pi x/L) \, dx = (2/L)[L/(n\pi)][1 - (-1)^n]$. Therefore $(2/\pi) \sum_{n=1}^{\infty} [1 - (-1)^n] \times \sin [(n\pi x/L)/n]$ is the Fourier series.   ●

Fourier Sine and Cosine Series

We are given a function f defined on $(0, L)$, and we desire a Fourier series representation. To get this, we extend f to the interval $(-L, L)$ and then compute the Fourier coefficients. It turns out that there are two natural ways of doing this. These two methods give rise to the Fourier sine and cosine series.

One way of extending f is to define a new function f_O by

$$f_O(x) = \begin{cases} f(x) & 0 < x < L \\ -f(-x) & -L < x < 0 \\ 0 & x = 0 \end{cases}$$

f_O is called the *odd extension* of f to $(-L, L)$. It is an odd function, and therefore its Fourier coefficients are given as follows:

$$A_n = 0 \qquad n = 0, 1, \ldots,$$

$$B_n = \frac{1}{L} \int_{-L}^{L} f_O(x) \sin \frac{n\pi x}{L} \, dx = \frac{2}{L} \int_0^L f(x) \sin \frac{n\pi x}{L} \, dx$$

Therefore we have the *Fourier sine series*

$$\sum_{n=1}^{\infty} B_n \sin \frac{n\pi x}{L} \qquad 0 \le x \le L$$

where

$$B_n = \frac{2}{L} \int_0^L f(x) \sin \frac{n\pi x}{L} \, dx$$

Another way of extending f to the interval $(-L, L)$ is to define

$$f_E(x) = \begin{cases} f(x) & 0 < x < L \\ f(-x) & -L < x < 0 \\ 0 & x = 0 \end{cases}$$

Here f_E is called the *even extension* of f to $(-L, L)$. It is an even function defined on the interval $(-L, L)$. [Of course, we could define $f_E(0) = \lim_{x \to 0} f(x)$, if this limit exists. The definition $f_E(0) = 0$ is completely arbitrary.] The Fourier coefficients of f_E are as follows:

$$B_n = 0 \qquad n = 1, 2, \ldots$$

$$A_0 = \frac{1}{2L} \int_{-L}^{L} f_E(x) \, dx = \frac{1}{L} \int_0^L f(x) \, dx$$

$$A_n = \frac{1}{L} \int_{-L}^{L} f_E(x) \cos \frac{n\pi x}{L} \, dx = \frac{2}{L} \int_0^L f(x) \cos \frac{n\pi x}{L} \, dx$$

Therefore we have the *Fourier cosine series*

$$A_0 + \sum_{n=1}^{\infty} A_n \cos \frac{n\pi x}{L} \qquad 0 \le x \le L$$

where

$$A_0 = \frac{1}{L} \int_0^L f(x) \, dx$$

and

$$A_n = \frac{2}{L} \int_0^L f(x) \cos \frac{n\pi x}{L} \, dx$$

The Fourier sine series may also be described as the orthogonal series associated with the Sturm-Liouville problem $f'' + \lambda f = 0$ with boundary conditions $f(0) = 0$, $f(L) = 0$. Likewise, the Fourier cosine series may be described as the orthogonal series associated with the boundary conditions $f'(0) = 0$, $f'(L) = 0$.

Example 1.1.4. Compute the Fourier sine series of $f(x) = 1, 0 < x < L$.

Solution. We have

$$B_n = \frac{2}{L} \int_0^L \sin \frac{n\pi x}{L} \, dx = -\frac{2L}{Ln\pi} \cos \frac{n\pi x}{L} \Big|_0^L = \frac{2}{n\pi} [1 - (-1)^n]$$

The Fourier sine series is

$$\frac{2}{\pi} \sum_{n=1}^{\infty} \frac{1 - (-1)^n}{n} \sin \frac{n\pi x}{L}$$ ●

We now give an alternate method for computing the Fourier sine series of certain polynomials. Let $f(x)$, $0 \le x \le L$, be a function with $f(0) = 0$, $f(L) = 0$, and $f''(x)$ continuous for $0 \le x \le L$. Then

$$B_n = \frac{2}{L} \int_0^L f(x) \sin \frac{n\pi x}{L}\, dx$$

$$= \frac{2}{n\pi} f(x) \cos \frac{n\pi x}{L} \Big|_L^0 + \frac{2}{n\pi} \int_0^L f'(x) \cos \frac{n\pi x}{L}\, dx$$

The first term is zero, and the second term can be integrated again by parts with the result

$$B_n = -\left(\frac{L}{n\pi}\right)^2 \frac{2}{L} \int_0^L f''(x) \sin \frac{n\pi x}{L}\, dx$$

Therefore the Fourier sine series of $f(x)$, $0 < x < L$, is obtained from the Fourier sine series of $f''(x)$, $0 < x < L$, by multiplication of the nth term of the series by $-[L/(n\pi)]^2$.

Example 1.1.5. Find the Fourier sine series of $f(x) = x^3 - L^2 x$, $0 < x < L$.

Solution. The function satisfies $f(0) = 0$, $f(L) = 0$ with $f''(x) = 6x$. The Fourier sine series of $6x$ is $(12L/\pi) \sum_1^\infty (-1)^{n+1} \sin [(n\pi x/L)/n]$. Therefore, the Fourier sine series of $f(x)$ is $(12L^3/\pi^3) \sum_1^\infty (-1)^n \sin [(n\pi x/L)/n^3]$.   ●

EXERCISES 1.1

In Exercises 1 to 10, compute the Fourier series of the indicated functions.
1. $f(x) = x^2$, $-L < x < L$
2. $f(x) = x^3$, $-L < x < L$
3. $f(x) = |x|^3$, $-L < x < L$
4. $f(x) = e^x$, $-L < x < L$ (See also the Appendix on ''Using Mathematica.'')
5. $f(x) = \sin^2 2x$, $-\pi < x < \pi$
6. $f(x) = \cos^3 x$, $-\pi < x < \pi$
7. $f(x) = \begin{cases} 0 & -L < x < 0 \\ 1 & 0 \le x < L \end{cases}$
8. $f(x) = \begin{cases} 0 & -L < x < 0 \\ x & 0 \le x < L \end{cases}$
9. $f(x) = \begin{cases} 0 & -\pi < x < 0 \\ \sin x & 0 \le x < \pi \end{cases}$
10. $f(x) = \sinh x = \frac{1}{2}(e^x - e^{-x})$, $-L < x < L$
11. Prove the orthogonality relations (1.1.3). [*Hint:* Use the trigonometric identities (1.1.3)'.]

12. Prove the orthogonality relations (1.1.4). [*Hint:* Use the trigonometric identities (1.1.4)'.] Can these be deduced from Sturm-Liouville theory if $m = n$?

13. Prove the following facts about even and odd functions:
    (*a*) The product of two even functions is even.
    (*b*) The product of two odd functions is even.
    (*c*) The product of an even function and an odd function is odd.
    (*d*) Which of statements (*a*), (*b*), (*c*) remains true if the word "product" is replaced by "sum"?

14. Let f be an arbitrary function. Show that there exist an odd function f_1 and an even function f_2 such that $f = f_1 + f_2$.

15. Which of the following functions are even, odd, or neither?
    (*a*) $f(x) = x^3 - 3x$ (*b*) $f(x) = x^2 + 4$
    (*c*) $f(x) = \cos 3x$ (*d*) $f(x) = x^3 - 3x^2$
    (*e*) $f(x) = \sin x - 3x^5$ (*f*) $f(x) = |x| \sin x$
    (*g*) $f(x) = x^2 - \cos x$ (*h*) $f(x) = \cos^3 x$

16. Find the Fourier sine series for the following functions:
    (*a*) $f(x) = x, 0 \le x \le L$ (*b*) $f(x) = x^2, 0 \le x \le L$
    (*c*) $f(x) = e^x, 0 \le x \le L$ (*d*) $f(x) = x^3, 0 \le x \le L$
    (*e*) $f(x) = \sin x, 0 \le x \le L$ (*f*) $f(x) = \cos x, 0 \le x \le L$

17. Find the Fourier cosine series for the functions in Exercise 16.

18. Let $f(x), -L < x < L$, be an odd function which satisfies the symmetry condition

$$f(L - x) = f(x)$$

    Show that $A_n = 0$ for all n

                  $B_n = 0$ for all even n

19. Let $f(x), -L < x < L$, be an odd function which satisfies the symmetry condition

$$f(L - x) = -f(x)$$

    Show that $A_n = 0$ for all n

                  $B_n = 0$ for all odd n

20. A function f on $(0, \pi/2)$ is to be expanded into a Fourier series

$$f(x) = A_0 + \sum_{n=1}^{\infty} (A_n \cos nx + B_n \sin nx)$$

    By extending f to $(-\pi, \pi)$ in four different ways, give four different prescriptions for finding the Fourier coefficients $\{A_n\}_{n=0}^{\infty}$, $\{B_n\}_{n=1}^{\infty}$. [*Hint:* There are two choices for extending f to $(0, \pi)$ and two more choices for further extending f to $(-\pi, \pi)$.]

21. Illustrate the expansions of Exercise 20 with $f(x) = 1, 0 < x < \pi/2$. Find the four different Fourier series.

For each of the functions in Exercises 22 to 29, state whether it is periodic and find the smallest period.

22. $f(x) = \sin \pi x$
23. $f(x) = \sin 2x + \sin 3x$
24. $f(x) = \sin 4x + \cos 6x$

25. $f(x) = \sin x + \sin \pi x$

26. $f(x) = x - [x]$ ($[x] =$ integer part of x)

27. $f(x) = \tan x$

28. $f(x) = \sum_{n=0}^{\infty} (-1)^n x^{2n}/(2n)!$

29. $f(x) = \sin x^2$

30. Compute the Fourier sine series of $f(x) = x^2 - Lx$, $0 < x < L$.

31. Compute the Fourier sine series of $f(x) = x^4 - 2Lx^3 + L^3x$, $0 < x < L$.

32. Let $f(x)$, $-L < x < L$, be an even function. Show that

$$\int_{-L}^{L} f(x) \, dx = 2 \int_{0}^{L} f(x) \, dx$$

33. Show that the derivative of an even function is an odd function.

34. Show that the derivative of an odd function is an even function.

1.2 CONVERGENCE OF FOURIER SERIES

In this section we discuss the validity of the equation

$$f(x) = A_0 + \sum_{n=1}^{\infty} \left(A_n \cos \frac{n\pi x}{L} + B_n \sin \frac{n\pi x}{L} \right)$$

where A_n, B_n are the Fourier coefficients of the function $f(x)$, $-L < x < L$. For simplicity in writing, we take $L = \pi$ in the exposition; all results obtained can be transformed to the interval $(-L, L)$ by the change of variable $x' = \pi x/L$.

Piecewise Continuous Functions

Recall that a function f is *continuous* at x if $\lim_{y \to x} f(y) = f(x)$. Not all Fourier series converge, even if we impose the restriction that their functions be continuous. In fact, there exist continuous functions on $[-\pi, \pi]$ whose Fourier series diverge at an infinite number of points! We therefore need to focus our attention on another class of functions, the so-called piecewise smooth functions. We first define the concept of piecewise continuous functions.

Definition. A function $f(x)$, $a < x < b$, is *piecewise continuous* if there is a finite set of points $a = x_0 < x_1 < \cdots < x_p < x_{p+1} = b$ such that

$$f \text{ is continuous at } x \qquad x \neq x_i, i = 1, \ldots, p \qquad (1.2.1)$$

$$\lim_{\substack{\epsilon \to 0 \\ \epsilon > 0}} f(x_i + \epsilon) \text{ exists} \qquad i = 0, \ldots, p \qquad (1.2.2)$$

$$\lim_{\substack{\epsilon \to 0 \\ \epsilon > 0}} f(x_i - \epsilon) \text{ exists} \qquad i = 1, \ldots, p + 1 \qquad (1.2.3)$$

The limit (1.2.2) is denoted $f(x_i + 0)$ and is called the *right limit*. Likewise, the limit (1.2.3) is denoted $f(x_i - 0)$ and is called the *left limit*. These are supposed to be finite.

Piecewise Smooth Functions

Definition. A function $f(x)$, $a < x < b$, is said to be *piecewise smooth* if f and all its derivatives are piecewise continuous.

Of course, we assume that the subdivision points $x_0 < x_1 < \cdots < x_{p+1}$ are the same for f and all its derivatives. With this definition, the derivative of a piecewise smooth function is again piecewise smooth.

If $f(x)$, $a < x < b$, is piecewise smooth, then $f'(x)$ exists except for $x = x_1, \ldots, x_p$. This is the *piecewise derivative* of f. Many of the usual operations with ordinary derivatives are valid for piecewise derivatives; the sum, difference, and product rules are valid except at the finite collection of points $(x_1, \ldots, x_p)$. The quotient rule is also valid unless the denominator is zero. The fundamental theorem of calculus must be modified for piecewise smooth functions to the form

$$f(b - 0) - f(a + 0) = \int_a^b f'(x)\, dx + \sum_{i=1}^{p} [f(x_i + 0) - f(x_i - 0)]$$

Indeed, on each interval (x_i, x_{i+1}) we may apply the ordinary fundamental theorem of calculus in the form

$$f(x_{i+1} - 0) - f(x_i + 0) = \int_{x_i}^{x_{i+1}} f'(x)\, dx$$

Adding these equations for $i = 0, 1, \ldots, p$ gives the result.

If the piecewise smooth function $f(x)$, $a < x < b$, is also *continuous*, then the fundamental theorem of calculus may be applied in its usual form

$$f(b - 0) - f(a + 0) = \int_a^b f'(x)\, dx$$

With these rules in mind, we may operate freely with piecewise smooth functions.

Example 1.2.1

$$f(x) = |x| \qquad -\pi < x < \pi$$

We take $x_0 = -\pi$, $x_1 = 0$, $x_2 = \pi$. Here f is continuous on the entire interval; f' is piecewise continuous, with $f'(0 + 0) = 1$, $f'(0 - 0) = -1$. All higher derivatives are zero; hence f is piecewise smooth on $(-\pi, \pi)$. ●

Example 1.2.2

$$f(x) = \begin{cases} x^2 & -\pi < x < 0 \\ x^2 + 1 & 0 \le x < \pi \end{cases}$$

In this example f is continuous, with the exception of the point $x = 0$, where we have $f(0 + 0) = 1$ and $f(0 - 0) = 0$. All higher derivatives are piecewise continuous on $(-\pi, \pi)$, so f is piecewise smooth on $(-\pi, \pi)$. ●

Example 1.2.3

$$f(x) = x|x| \qquad -\pi < x < \pi$$

In this case f and f' are continuous. And f'' is continuous everywhere except at $x = 0$, where we have $f''(0 + 0) = 2$ and $f''(0 - 0) = -2$. All higher derivatives are zero; thus f is piecewise smooth on $(-\pi, \pi)$.   ●

Example 1.2.4

$$f(x) = x^2 \sin\frac{1}{x} \qquad -\pi < x < \pi \qquad f(0) = 0$$

Function f is continuous on $(-\pi, \pi)$, and f' is continuous on $(-\pi, \pi)$ with the exception of the point $x = 0$. However, $f'(0 + 0)$ and $f'(0 - 0)$ do not exist (see Exercise 2), so f is piecewise continuous but *it is not piecewise smooth*.   ●

Example 1.2.5

$$f(x) = \frac{1}{x^2 - \pi^2} \qquad -\pi < x < \pi$$

Although in this case f is continuous on $(-\pi, \pi)$, it is not piecewise continuous on $(-\pi, \pi)$ since $f(-\pi + 0)$ and $f(\pi - 0)$ are not finite. In particular, f is not piecewise smooth.   ●

When working with piecewise smooth functions, we can omit the definition of $f(x)$ at the subdivision points $x_0, x_1, \ldots, x_{p+1}$. This causes no difficulty in the discussion of Fourier series, since the Fourier coefficients A_n, B_n are defined as integrals, which are insensitive to the value of $f(x)$ at a finite number of points. More precisely, if $f_1(x) = f_2(x)$, except for $x = x_0, x_1, \ldots, x_{p+1}$, then $\int_a^b f_1(x)\, dx = \int_a^b f_2(x)\, dx$. Therefore we see that the *Fourier coefficients do not depend on any of the numbers $f(x_0), \ldots, f(x_{p+1})$.*

Suppose f is piecewise smooth on (a, b). We define the *periodic extension* of f by setting

$$f(x + n(b - a)) = f(x) \qquad \text{where } x \in (a, b)$$

and n is an integer (positive or negative).

It is left as an exercise to show that the periodic extension of f is piecewise smooth on any open interval and that it is periodic with period $b - a$. It is also left as an exercise to show that

$$\int_c^d f(x)\, dx = \int_a^b f(x)\, dx \qquad \text{if } d - c = b - a$$

where f is *any* periodic function of period $b - a$.

Finally, notice that we now have the convention

$$f(a - 0) \equiv f(b - 0)$$

and

$$f(b + 0) \equiv f(a + 0)$$

Let $f(x)$, $-L < x < L$, be a piecewise smooth function, and let $f(x)$, $-\infty < x < \infty$, be the periodic extension of f; $\bar{f}$ is a $2L$-periodic function with $\bar{f}(x) = f(x)$ for $-L < x < L$. For any $\epsilon > 0$, we have by periodicity $\bar{f}(L + \epsilon) = \bar{f}(-L + \epsilon) = f(-L + \epsilon)$. Letting ϵ approach zero, we see that $\bar{f}(L + 0) = \bar{f}(-L + 0) = f(-L + 0)$. Similarly, $\bar{f}(-L - 0) = f(L - 0)$. The average value of $\bar{f}$ at $x = L$, which is defined as $\frac{1}{2}[\bar{f}(L + 0) + \bar{f}(L - 0)]$, can therefore be written as $\frac{1}{2}[f(L - 0) + f(-L + 0)]$. Similarly, $\frac{1}{2}[\bar{f}(-L + 0) + \bar{f}(-L - 0)] = \frac{1}{2}[f(-L + 0) + f(L - 0)]$. We now state the main result of this section.

Convergence Theorem

Theorem 1.2.1. Let f be piecewise smooth on $(-\pi, \pi)$. Then the Fourier series of f converges for all x to the value $\frac{1}{2}[\bar{f}(x + 0) + \bar{f}(x - 0)]$, where $\bar{f}$ is the periodic extension of f.

The restriction to the interval $(-\pi, \pi)$ is of no significance. It has been made here so that instead of writing $\cos(m\pi x/L)$ and $\sin(m\pi x/L)$, we may write $\cos mx$ and $\sin mx$.

Before proceeding with the proof, we need two lemmas.

Two Useful Lemmas

Lemma 1 (Riemann).   If f and f' are piecewise continuous on (a, b), then

1. $\lim\limits_{\lambda \to \infty} \int_a^b f(x) \sin \lambda x \, dx = 0$

2. $\lim\limits_{\lambda \to \infty} \int_a^b f(x) \cos \lambda x \, dx = 0$

Proof. Only 1 will be proved; 2 is established in an almost identical manner. First we write

$$\int_a^b f(x) \sin \lambda x \, dx = \sum_{i=0}^{p} \int_{x_i}^{x_{i+1}} f(x) \sin \lambda x \, dx$$

It remains to show that

$$\lim_{\lambda \to \infty} \int_{x_i}^{x_{i+1}} f(x) \sin \lambda x \, dx = 0$$

For this we integrate by parts, with $u = f(x)$, $dv = \sin \lambda x \, dx$. Thus

$$\int_{x_i}^{x_{i+1}} f(x) \sin \lambda x \, dx = \frac{-f(x) \cos \lambda x}{\lambda}\bigg|_{x_i}^{x_{i+1}} + \frac{1}{\lambda} \int_{x_i}^{x_{i+1}} f'(x) \cos \lambda x \, dx$$

Each of these tends to zero when $\lambda \to \infty$. This completes the proof of Lemma 1.

∎

We wish to examine the limit as $N \to \infty$ of

$$f_N(x) = A_0 + \sum_{m=1}^{N} (A_m \cos mx + B_m \sin mx)$$

Using the definitions of A_0, A_m, B_m given in Sec. 1.1, formulas (1.1.5) to (1.1.7), we have

$$f_N(x) = \frac{1}{2\pi} \int_{-\pi}^{\pi} f(t) \, dt + \sum_{m=1}^{N} \frac{1}{\pi} \int_{-\pi}^{\pi} f(t)(\cos mt \cos mx + \sin mt \sin mx) \, dt$$

$$= \frac{1}{2\pi} \int_{-\pi}^{\pi} f(t) \, dt + \sum_{m=1}^{N} \frac{1}{\pi} \int_{-\pi}^{\pi} f(t) \cos m(t - x) \, dt$$

$$= \frac{1}{\pi} \int_{-\pi}^{\pi} f(t) \left[\frac{1}{2} + \sum_{m=1}^{N} \cos m(t - x) \right] dt$$

Clearly it would be useful to be able to write

$$\frac{1}{2} + \sum_{m=1}^{N} \cos m(t - x)$$

in a more compact form. Therefore we use a second lemma.

Lemma 2

$$\frac{1}{2} + \cos \alpha + \cdots + \cos N\alpha = \frac{\sin (N + \frac{1}{2})\alpha}{2 \sin \frac{1}{2}\alpha} \qquad \alpha \text{ real}, \ \alpha \neq 0, \ \pm 2\pi, \ \ldots$$

Proof. Setting $S = \frac{1}{2} + \cos \alpha + \cdots + \cos N\alpha$, we have

$$S \sin \alpha = \frac{1}{2} \sin \alpha + \sin \alpha \cos \alpha + \cdots + \sin \alpha \cos N\alpha$$

From the addition formulas

$$\sin (a + b) = \sin a \cos b + \cos a \sin b$$

$$\sin (a - b) = \sin a \cos b - \cos a \sin b$$

we have

$$\sin a \cos b = \frac{1}{2}[\sin (a + b) + \sin (a - b)]$$

$$= \frac{1}{2}[\sin (a + b) - \sin (b - a)]$$

so that

$$S \sin \alpha = \frac{1}{2}[\sin \alpha + \sin 2\alpha - 0 + \sin 3\alpha - \sin \alpha$$

$$+ \cdots + \sin (N + 1)\alpha - \sin (N - 1)\alpha]$$

$$= \frac{1}{2}[\sin N\alpha + \sin (N + 1)\alpha]$$

To complete the proof, we average the addition formulas as follows:

$$\frac{1}{2}[\sin (a + b) + \sin (a - b)] = \sin a \cos b$$

Setting $a + b = (N + 1)\alpha$, $a - b = N\alpha$, we have $a = (N + \frac{1}{2})\alpha$, $b = \frac{1}{2}\alpha$, so that

$$\tfrac{1}{2}[\sin N\alpha + \sin (N + 1)\alpha] = \sin (N + \tfrac{1}{2})\alpha \cos \tfrac{1}{2}\alpha$$

and

$$S = \frac{\sin (N + \tfrac{1}{2})\alpha \cos \tfrac{1}{2}\alpha}{\sin \alpha}$$

Substituting the identity $\sin \alpha = 2 \sin \frac{1}{2}\alpha \cos \frac{1}{2}\alpha$ completes the proof of Lemma 2. (For a shorter proof of Lemma 2, using complex numbers, see Exercise 13 at the end of this section.) ∎

In view of Lemma 2, we can write

$$f_N(x) = \frac{1}{\pi} \int_{-\pi}^{\pi} f(t) \frac{\sin (N + \tfrac{1}{2})(t - x)}{2 \sin \tfrac{1}{2}(t - x)} \, dt$$

This form is preferable because it makes no mention of the Fourier coefficients A_n, B_n.

Dirichlet Kernel

To proceed further, we make the definition

$$D_N(u) = \begin{cases} \dfrac{\sin (N + \tfrac{1}{2})u}{2\pi \sin (u/2)} & u \neq 0, \pm 2\pi, \ldots \\[2ex] \dfrac{2N + 1}{2\pi} & u = 0, \pm 2\pi, \ldots \end{cases}$$

D_N is the *Dirichlet kernel,* an even, 2π-periodic function. From Lemma 2 we see that

$$\int_0^{\pi} D_N(u) \, du = \tfrac{1}{2} = \int_{-\pi}^{0} D_N(u) \, du$$

From Fig. 1.2.1 we see that $D_N(u)$ behaves roughly as a periodic function with period $2\pi/N$, except in the neighborhood of $u = 0$, $\pm 2\pi$, where it is peaked.

Proof of Convergence

To complete the proof of Theorem 1.2.1, we extend f to $\bar{f}$, a 2π-periodic function. Therefore the product $D_N(t - x)\bar{f}(t)$ is also a 2π-periodic function of t for each x. We now write

$$f_N(x) = \int_{-\pi}^{\pi} \bar{f}(t) D_N(t - x) \, dt$$

$$= \int_{-\pi-x}^{\pi-x} \bar{f}(x + u) D_N(u) \, du \qquad t - x = u$$

$$= \int_{-\pi}^{\pi} \bar{f}(x + u)D_N(u)\, du \qquad \text{periodicity}$$

$$= \left\{ \int_{-\pi}^{0} + \int_{0}^{\pi} \right\} \bar{f}(x + u)D_N(u)\, du$$

We analyze the two integrals separately and show that

$$\lim_{N \to \infty} \int_{0}^{\pi} \bar{f}(x + u)D_N(u)\, du = \tfrac{1}{2}\bar{f}(x + 0)$$

$$\lim_{N \to \infty} \int_{-\pi}^{0} \bar{f}(x + u)D_N(u)\, du = \tfrac{1}{2}\bar{f}(x - 0)$$

from which the result will follow. We carry out the analysis of only the first integral in detail, for the second is identical in every respect. Let $g(u) :=$ $[\bar{f}(x + u) - \bar{f}(x + 0)]/u$. Then

$$\int_{0}^{\pi} [\bar{f}(x + u) - \bar{f}(x + 0)]D_N(u)\, du = \frac{1}{\pi} \int_{0}^{\pi} g(u)U(u) \sin (N + \tfrac{1}{2})u\, du$$

where
$$U(u) = \frac{u}{2 \sin (u/2)} \qquad u \neq 0$$

$$U(0) = 1$$

Using L'Hospital's rule, we see that the function $U(u)$ is continuous and has a continuous derivative, $-\pi \leq u \leq \pi$. Similarly, we can use L'Hospital's rule

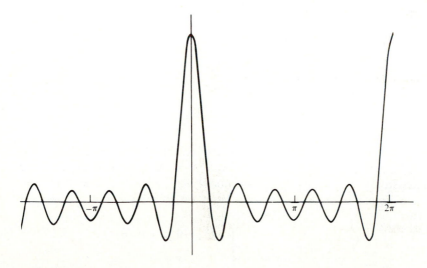

FIGURE 1.2.1
The Dirichlet kernel $D_N(u)$ for $N = 5$.

to compute the limits

$$\lim_{u \downarrow 0} g(u) = \bar{f}'(x + 0)$$

$$\lim_{u \downarrow 0} g'(u) = \tfrac{1}{2}\bar{f}''(x + 0)$$

Therefore $g(u)$ is piecewise continuous with a piecewise continuous derivative. But $U(u)$ has a continuous derivative, and therefore the product $g(u)U(u)$ also has a piecewise continuous derivative. Applying Lemma 1, we have proved that

$$\lim_{N \to \infty} \int_0^\pi g(u)U(u) \sin (N + \tfrac{1}{2})u \, du = 0$$

Writing this in terms of $\bar{f}$, we have

$$\lim_{N \to \infty} \int_0^\pi \bar{f}(x + u)D_N(u) \, du = \bar{f}(x + 0) \int_0^\pi D_N(u) \, du = \tfrac{1}{2}\bar{f}(x + 0)$$

which was to be proved.

An examination of the graph of $D_N(u)$ (Fig. 1.2.1) helps one to see an intuitive motivation of the proof. Since $\int_{-\pi}^\pi D_N(u) \, du = 1$, it follows from the graph that as N gets large, the area tends to concentrate around $u = 0$, so that $\int_{-\pi}^\pi f(u)D_N(u) \, du$ tends to pick off the values of $f(u)$ near $u = 0$. Thus,

$$\lim_{N \to \infty} \int_{-\pi}^\pi f(u)D_N(u) \, du = \tfrac{1}{2}[f(0 + 0) + f(0 - 0)]$$

for functions $f(x)$ which are piecewise smooth.

Having proved the convergence of the Fourier series, we can now obtain many useful conclusions. Referring to the first two examples in Sec. 1.1, we have the convergent Fourier series

$$2 \sum_{n=1}^\infty \frac{(-1)^{n+1}}{n} \sin nx = x \qquad -\pi < x < \pi$$

$$\frac{\pi}{2} - \frac{2}{\pi} \sum_{n=1}^\infty \frac{1 - (-1)^n}{n^2} \cos nx = |x| \qquad -\pi < x < \pi$$

Both of these examples are continuous functions, from which it follows that $f(x + 0) = f(x - 0) = f(x)$ for all x, $-\pi < x < \pi$. Note, however, that the periodic extension is not necessarily continuous e.g., $f(x) = x$, $-\pi < x < \pi$.

As an example of a discontinuous function, we have the convergent Fourier series

$$\frac{2}{\pi} \sum_{n=1}^\infty \frac{1 - (-1)^n}{n} \sin nx = \begin{cases} 1 & 0 < x < \pi \\ 0 & x = 0 \\ -1 & -\pi < x < 0 \end{cases}$$

These can also be used to obtain various numerical series. Taking $x = 0$

in the Fourier series for $|x|$, we have $0 = \pi/2 - (2/\pi)(2 + \frac{2}{9} + \frac{2}{25} + \cdots)$, $\pi^2/8 = 1 + \frac{1}{9} + \frac{1}{25} + \cdots$. Similarly, taking $x = \pi/2$ in the third example, we obtain $\pi/4 = 1 - \frac{1}{3} + \frac{1}{5} - \cdots$.

EXERCISES 1.2

1. Which of the following functions are piecewise smooth?
   (a) $f(x) = |x|^{3/2}$, $-2 < x < 2$
   (b) $f(x) = [x] - x$, $0 < x < 3$ and $[x] =$ integer part of x
   (c) $f(x) = x^4 \sin(1/x)$, $-1 < x < 1$
   (d) $f(x) = e^{-(1/x^2)}$, $-1 < x < 1$
2. Let $f(x) = x^2 \sin(1/x)$.
   (a) Show that $\lim_{x\to 0} f(x) = 0$.
   (b) Graph $f(x)$.
   (c) Show that $f'(0 + 0)$ does not exist by considering $f'(h)$ as $h \to 0$ through the values $1/(2n\pi)$ and $1/[(2n + 1)\pi]$, $n = 1, 2, \ldots$.
3. Let f and g be piecewise smooth on (a, b).
   (a) Show that $f + g$ is piecewise smooth on (a, b).
   (b) Show that $f \cdot g$ is piecewise smooth on (a, b).
   (c) What restrictions must be made on g in order that f/g be piecewise smooth on (a, b)?
4. Let f be the periodic extension of a function that is piecewise smooth on (a, b).
   (a) Show that f is piecewise smooth on (c, d), $c < d$.
   (b) Show that f is periodic with period $b - a$.
   (c) Show that
   $$\int_c^d f(x)\, dx = \int_a^b f(x)\, dx \qquad \text{if } d - c = b - a$$
5. Show that
   $$U(u) = \begin{cases} \dfrac{u}{2 \sin \frac{1}{2}u} & u \neq 0,\ -\pi \leq u \leq \pi \\[2mm] 1 & u = 0 \end{cases}$$
   is continuous and has a continuous derivative for $-\pi \leq u \leq \pi$. (*Hint:* Use L'Hospital's rule.)
6. Let $f(x)$, $a < x < b$, be a piecewise smooth function. Let $g(u) = [f(x + u) - f(x + 0)]/u$ for $u \neq 0$. Show that $g(0 + 0) = f'(x + 0)$, $g(0 - 0) = f'(x - 0)$. (*Hint:* Use L'Hospital's rule.)
7. Let $g(u)$ be defined as in Exercise 6. Show that $g'(0 + 0) = \frac{1}{2}f''(x + 0)$, $g'(0 - 0) = \frac{1}{2}f''(x - 0)$.
8. Prove that $D_N(u)$ is even and periodic with period 2π.
9. Use Lemma 1 and the properties of the Dirichlet kernel to compute the following limits:
   (a) $\displaystyle\lim_{N\to\infty} \int_{-\pi/2}^{\pi/2} D_N(u)\, du$
   (b) $\displaystyle\lim_{N\to\infty} \int_{0}^{\pi/2} D_N(u)\, du$
   (c) $\displaystyle\lim_{N\to\infty} \int_{-\pi/6}^{\pi/6} D_N(u)\, du$
   (d) $\displaystyle\lim_{N\to\infty} \int_{\pi/2}^{\pi} D_N(u)\, du$

10. What is the maximum value of $D_N(u)$, $-\pi \le u \le \pi$?

11. Find all solutions of $D_N(u) = 0$.

12. Find all solutions of $D'_N(u) = 0$.

13. There is another way of establishing Lemma 2. Recall that $e^{ix} = \cos x + i \sin x$.

(a) Show that

$$\cos x = \frac{e^{ix} + e^{-ix}}{2} \qquad \sin x = \frac{e^{ix} - e^{-ix}}{2i}$$

(b) Prove Lemma 2, using part (a) and the fact that

$$1 + r + \cdots + r^n = \frac{r^{n+1} - 1}{r - 1} \qquad r \ne 1$$

*14. This exercise establishes the formula

$$\int_0^\infty \frac{\sin x}{x} \, dx = \frac{\pi}{2}$$

(a) Let

$$f(u) = \frac{1}{2 \sin (u/2)} - \frac{1}{u} \qquad u \ne 0$$

$$f(0) = 0$$

Show that f, f' are continuous on $(0, \pi)$. (The only trouble occurs at $u = 0$. Use L'Hospital's rule to show that the appropriate limits are finite.)

(b) Conclude that

$$\lim_{N \to \infty} \int_0^\pi \left[\frac{1}{2 \sin (u/2)} - \frac{1}{u} \right] \sin (N + \tfrac{1}{2})u \, du = 0$$

(c) Hence show that

$$\lim_{N \to \infty} \int_0^\pi D_N(u) \, du = \lim_{N \to \infty} \int_0^\pi \frac{\sin (N + \tfrac{1}{2})u}{u} \, du$$

(d) Make the appropriate substitution in the second definite integral and recall the appropriate facts about $D_N(u)$ to conclude that

$$\lim_{N \to \infty} \int_0^{(N+1/2)\pi} \frac{\sin x}{x} \, dx = \frac{\pi}{2}$$

(e) If $(N - \tfrac{1}{2})\pi \le X \le (N + \tfrac{1}{2})\pi$, show that

$$\int_0^X \frac{\sin x}{x} \, dx = \int_0^{(N+1/2)\pi} \frac{\sin x}{x} \, dx + \epsilon_X$$

where $|\epsilon_X| \le 1/(N - \tfrac{1}{2})$. Conclude that the improper integral converges to $\pi/2$.

15. (a) Set $x = \pi/2$ in the Fourier series for $f(x) = x$, $-\pi < x < \pi$, to obtain the formula

$$\frac{\pi}{4} = 1 - \frac{1}{3} + \frac{1}{5} - \frac{1}{7} \cdots$$

(b) Set $x = \pi/4$ in the series in part (a) to obtain

$$\frac{\pi}{4} = \sqrt{2} \left(1 + \frac{1}{3} - \frac{1}{5} - \frac{1}{7} \cdots \right) - \left(1 - \frac{1}{3} + \frac{1}{5} - \frac{1}{7} \cdots \right)$$

(c) Conclude from part (b) that

$$\frac{\pi}{2\sqrt{2}} = 1 + \frac{1}{3} - \frac{1}{5} - \frac{1}{7} + \frac{1}{9} + \frac{1}{11} - \frac{1}{13} - \frac{1}{15} \cdots .$$

(d) If we set $x = \pi$ in the series in (a), we find that the series sums to zero. Why does this not contradict $f(x) = x$?

16. (a) Show that

$$x^2 = \frac{\pi^2}{3} - 4 \cos x + \cos 2x - \frac{4}{9} \cos 3x + \cdots + (-1)^m \frac{4}{m^2} \cos mx + \cdots$$

for $-\pi \le x \le \pi$.

(b) Setting $x = 0$ in (a), find the sum

$$1 - \frac{1}{4} + \frac{1}{9} - \frac{1}{16} \cdots = \sum_{m=1}^{\infty} (-1)^{m+1} \frac{1}{m^2}$$

(c) What is

$$\sum_{m=1}^{\infty} \frac{1}{m^2}$$

[*Hint:* Set $x = \pi$ in part (a).]

(d) What is

$$\sum_{m \text{ odd}} \frac{1}{m^2}$$

[*Hint:* Add (b) and (c).]

17. Let $f(x) = x$, $-\pi < x < \pi$. What is the sum of the Fourier series for $x = -\pi$, $x = \pi$?

18. Let $f(x) = e^x$, $-\pi < x < \pi$. What is the sum of the Fourier series for $x = -\pi$, $x = \pi$?

*19. Let $f(x)$, $g(x)$ be piecewise smooth functions for $a < x < b$. Show that

$$f(b - 0)g(b - 0) - f(a + 0)g(a + 0) = \int_a^b f(x)g'(x)\, dx + \int_a^b f'(x)g(x)\, dx$$

$$- \sum_{i=1}^p [f(x_i + 0)g(x_i + 0) - f(x_i - 0)g(x_i - 0)]$$

*20. Use Exercise 19 to prove the following integration-by-parts formula for piecewise smooth functions:

$$\int_a^b f(x)g'(x)\, dx = fg \Big|_{a+0}^{b-0} - \int_a^b f'(x)g(x)\, dx + \sum_{i=1}^p g(x_i - 0)[f(x_i + 0) - f(x_i - 0)]$$

$$+ \sum_{i=1}^p f(x_i - 0)[g(x_i + 0) - g(x_i - 0)]$$

$$+ \sum_{i=1}^p [f(x_i + 0) - f(x_i - 0)][g(x_i + 0) - g(x_i - 0)]$$

*21. By examining the proof of the main theorem, show that the conclusion is valid if f, f', f'' are piecewise continuous.

22. On the basis of Exercise 21, for which $n \geq 1$, can we assert that the Fourier series of $x^n \sin(1/x)$ is convergent for all x, $-\pi < x < \pi$?

*23. Let $f(x)$, $-\pi < x < \pi$, be a piecewise smooth function with Fourier coefficients A_n, B_n. Apply Exercise 20 with $a = -\pi$, $b = \pi$, $g'(x) = \cos nx$ to find an asymptotic formula for A_n, B_n, $n \to \infty$.

1.3   UNIFORM CONVERGENCE AND THE GIBBS PHENOMENON

We have seen that the Fourier series of a piecewise smooth function converges to the function except at points of discontinuity, where it converges to the average of the function's left and right limits. Since we are interested in approximating functions by partial sums of their Fourier series, it is of interest how the Fourier series converge near a discontinuity, that is, how the partial sums of Fourier series behave near discontinuities of their functions. We turn first to an example.

Example of Gibbs Overshoot†

Consider the function

$$f(x) = \begin{cases} -1 & -\pi \leq x < 0 \\ 1 & 0 \leq x \leq \pi \end{cases}$$

The cosine coefficients are all zero (f is odd), and the sine coefficients are given by

$$B_n = \frac{2}{\pi} \int_0^\pi \sin nx \, dx = \frac{2}{n\pi} [1 - (-1)^n]$$

The partial sum of the Fourier series is therefore

$$f_{2n-1}(x) = \frac{4}{\pi} \left[\sin x + \frac{\sin 3x}{3} + \cdots + \frac{\sin(2n-1)x}{2n-1} \right]$$

From the graph of Fig. 1.3.1 we see that just before the discontinuity the partial sums overshoot the right and left limits and then slope rapidly toward their mean. On the interval $-\pi \leq x \leq \pi$, f_1 has one maximum and one minimum, f_3 has three maxima and three minima, f_5 has five maxima and five minima, etc. We can actually calculate the overshoot by computing the derivative

$$f'_{2n-1}(x) = \frac{4}{\pi} [\cos x + \cos 3x + \cdots + \cos(2n-1)x] \tag{1.3.1}$$

and solving the equation $f'(x) = 0$.

To solve this equation, we multiply (1.3.1) by $\sin x$ and use the identity

$$\sin x \cos kx = \tfrac{1}{2}[\sin(k+1)x - \sin(k-1)x]$$

---

† This example also appears in the Appendix on "Using Mathematica."

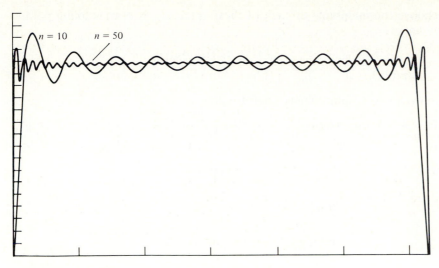

FIGURE 1.3.1
The Gibbs phenomenon for $n = 10$ and $n = 50$.

and get

$$\pi \sin x \, f'_{2n-1}(x) = 2 \left\{ \sin 2x + \sum_{k=1}^{n-1} [\sin 2(k+1)x - \sin 2kx] \right\}$$

$$= 2 \sin 2nx$$

Therefore, the extrema occur at the points

$$2nx = \pm \pi, \ \pm 2\pi, \ \ldots, \ \pm(2n-1)\pi$$

These points are equally spaced in $[-\pi, \pi]$. It is the maximum closest to the discontinuity [that is, when $x = \pi/(2n)$] that is of interest, so we wish to compute

$$f_{2n-1}\left(\frac{\pi}{2n}\right) = \frac{4}{\pi} \left[\sin \frac{\pi}{2n} + \frac{1}{3} \sin \frac{3\pi}{2n} + \cdots + \frac{1}{2n-1} \sin \frac{(2n-1)\pi}{2n} \right]$$

for large n. The technique we use for evaluating this sum consists of rewriting the sum so that it looks like the approximating sum of a riemannian integral and then evaluating the integral. Our answer is exact when $n \to \infty$ and so should give a good approximation for large n.

The function whose integral we will approximate is $g(x) = (\sin x)/x$. Consider the partition of $[0, \pi]$, given by the points $\{x_k\}$, where

$$x_k = \frac{\pi k}{n} \qquad k = 1, \ldots, n$$

$$\Delta x_k = \frac{\pi}{n}$$

If we choose the midpoints of each of these intervals as our sampling points, then we have

$$\frac{\sin [\pi/(2n)]}{\pi/(2n)} \frac{\pi}{n} + \cdots + \frac{\sin [(2n - 1)\pi/(2n)]}{(2n - 1)\pi/(2n)} \frac{\pi}{n} \to \int_0^\pi \frac{\sin x}{x} dx$$

If we rearrange our sum, we see that it equals

$$\frac{2n}{\pi} \frac{\pi}{n} \left\{ \sin \frac{\pi}{2n} + \frac{\sin [3\pi/(2n)]}{3} + \cdots + \frac{\sin [(2n - 1)\pi/(2n)]}{2n - 1} \right\} = \frac{\pi}{2} f_{2n-1} \left(\frac{\pi}{2n} \right)$$

Therefore the *limit of the overshoot* is given by

$$\lim_{n \to \infty} f_{2n-1} \left(\frac{\pi}{2n} \right) = \frac{2}{\pi} \int_0^\pi \frac{\sin x}{x} dx$$

We can approximate the integral numerically as follows:

$$\sin x = x - \frac{x^3}{3!} + \frac{x^5}{5!} - \frac{x^7}{7!} \cdots$$

so

$$\frac{\sin x}{x} = 1 - \frac{x^2}{3!} + \frac{x^4}{5!} - \frac{x^6}{7!} + \cdots$$

and

$$\frac{2}{\pi} \int_0^\pi \frac{\sin x}{x} dx = \frac{2}{\pi} \int_0^\pi \left(1 - \frac{x^2}{3!} + \frac{x^4}{5!} - \frac{x^6}{7!} \right) dx + \cdots$$

$$= \frac{2}{\pi} \left(\pi - \frac{\pi^3}{18} + \frac{\pi^5}{600} - \frac{\pi^7}{35,280} \right) + \cdots$$

$$= 2 - \frac{\pi^2}{9} + \frac{\pi^4}{300} - \frac{\pi^6}{17,640} + \cdots$$

$$= 2 - 1.11 + 0.33 - 0.04 + \cdots$$

$$= 1.18 \quad \text{to two decimal places}$$

   This means that if we stand at any one point we will land on the graph of $f(x)$ in the limit $n \to \infty$. However, if we ride the crest of the worst point possible for each n, then we will never reach the graph of $f(x)$. When $n \to \infty$, we will be left dangling approximately 1.18 units above the origin. This behavior can be described by saying that the partial sums *do not* converge *uniformly* to $f(x)$ (that is, the *entire* curve is not arbitrarily close to the graph of f for sufficiently large n). Rather they converge to the graph indicated in Fig. 1.3.2. This is known as the *Gibbs phenomenon*. Notice that the overshoot of 1.18 is 9 percent of the jump made at the discontinuity. This is characteristic of the overshoot due to any discontinuity in any piecewise smooth function f. In fact, we have the following general fact, whose proof is omitted.

   Let f be piecewise smooth on $(-\pi, \pi)$. Then the amount of overshoot near a

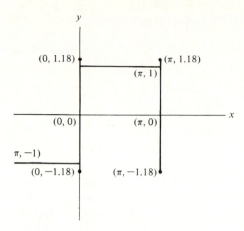

FIGURE 1.3.2

discontinuity, due to the Gibbs phenomenon, is approximately equal to

$$0.09 \, |f(x_0 + 0) - f(x_0 - 0)|$$

for large n.

Uniform and Nonuniform Convergence

In many problems it is important to avoid the Gibbs phenomenon, in other words, to be sure that the function $f(x)$ is well approximated by the partial sum $f_n(x)$ at all points of the interval $-L \le x \le L$. Recall that a sequence of functions $f_n(x)$, $a \le x \le b$, converges *uniformly* to a limit function $f(x)$, $a \le x \le b$, if

$$|f_n(x) - f(x)| \le \epsilon_n \qquad a \le x \le b, \quad n = 1, 2, \ldots$$

where

$$\lim_{n \to \infty} \epsilon_n = 0$$

This is clearly violated in the Gibbs phenomenon, for in the previous example $\lim_{n \to \infty} [f_{2n-1}(\pi/(2n)) - f(\pi/(2n))] = 0.18. \ldots$

Two Theorems on Uniform Convergence

We give two general criteria for uniform convergence. The first can be tested on the series, while the second can be tested on the function.

First criterion for uniform convergence. Let $f(x)$, $-L < x < L$, be a piecewise smooth function. Suppose that the Fourier coefficients A_n, B_n satisfy

$$\sum_{n=1}^{\infty} (|A_n| + |B_n|) < \infty$$

Then the Fourier series converges uniformly.

For example, $\sum_{n=1}^{\infty} (\sin nx)/n^2$ is a uniformly convergent Fourier series.

Second criterion for uniform convergence. Let $f(x)$, $-L < x < L$, be a piecewise smooth function. Suppose in addition that

$$f \text{ is continuous} \qquad -L < x < L$$

$$f(-L + 0) = f(L - 0)$$

Then the Fourier series converges uniformly.

For example, $f(x) = |x|$ has a uniformly convergent Fourier series.

Within the class of piecewise smooth functions, these criteria are necessary and sufficient: If the Fourier series of a piecewise smooth function converges uniformly, then f is continuous, $f(-L + 0) = f(L - 0)$, and $\sum_{n=1}^{\infty} (|A_n| + |B_n|) < \infty$. Once we leave the domain of piecewise smooth functions, life becomes much more complicated; for example, the Fourier series $\sum_{n=2}^{\infty} (\sin nx)/(n \log n)$ is known to be uniformly convergent,† but it does not satisfy the first criterion. Of course, the sum of this series must be a *continuous* function by the general properties of uniform convergence.

Differentiation of Fourier Series

We now give a general criterion for differentiating a Fourier series.

Proposition 1.3.1. Let $f(x)$, $-L < x < L$, be a continuous piecewise smooth function with $f(L - 0) = f(-L + 0)$. Then

$$\frac{1}{2}[f'(x + 0) + f'(x - 0)] = \sum_{n=1}^{\infty} \frac{n\pi}{L}\left(B_n \cos \frac{n\pi x}{L} - A_n \sin \frac{n\pi x}{L}\right)$$

For example, suppose that we want to compute the Fourier series of $f(x) = x^2$, $-\pi < x < \pi$. The Fourier series of this even function is of the form $A_0 + \sum_{n=1}^{\infty} A_n \cos nx$, where the A_n are to be determined. From the proposition we may write

$$2x = -\sum_{n=1}^{\infty} nA_n \sin nx$$

But from Example 1.1.1 we know that

$$2x = 4 \sum_{n=1}^{\infty} (-1)^{n+1} \frac{\sin nx}{n}$$

Therefore $A_n = 4(-1)^n/n^2$ for $n = 1, 2, \ldots$. To compute A_0, we must return to the definition $A_0 = [1/(2\pi)] \int_{-\pi}^{\pi} x^2 \, dx = \pi^2/3$. Therefore we have the Fourier series $x^2 = \pi^2/3 + 4 \sum_{n=1}^{\infty} [(-1)^n/n^2] \cos nx$, $-\pi < x < \pi$.

---

† A. Zygmund, *Trigonometrical Series*, Cambridge University Press, 1959, vol. 1, p. 182.

Integration of Fourier Series

The following proposition shows that a Fourier series may be integrated term by term under very general conditions. Since the proof is instructive, we present the details.

Proposition 1.3.2. Let $f(x)$, $-\pi < x < \pi$, be a piecewise smooth function with Fourier series

$$A_0 + \sum_{n=1}^{\infty} (A_n \cos nx + B_n \sin nx)$$

Then

$$\int_{x_0}^{x} f(u)\, du = A_0(x - x_0) + \sum_{n=1}^{\infty} \left[\frac{A_n}{n} (\sin nx - \sin nx_0) \right.$$

$$\left. + \frac{B_n}{n} (\cos nx_0 - \cos nx) \right] \qquad -\pi \leq x_0 < x \leq \pi$$

Proof. Let $F(x) = \int_{-\pi}^{x} [f(u) - A_0]\, du$. The function F is continuous and piecewise smooth with $F(-\pi) = F(\pi)$. Therefore by the basic convergence theorem we have

$$F(x) = \bar{A}_0 + \sum_{n=1}^{\infty} (\bar{A}_n \cos nx + \bar{B}_n \sin nx) \qquad -\pi \leq x \leq \pi$$

where $\bar{A}_n$, $\bar{B}_n$ are the Fourier coefficients of F. To compute these, we have, for $n \neq 0$,

$$\bar{A}_n = \frac{1}{\pi} \int_{-\pi}^{\pi} F(x) \cos nx\, dx$$

$$= \frac{1}{\pi} \int_{-\pi}^{\pi} \cos nx \left\{ \int_{-\pi}^{x} [f(u) - A_0]\, du \right\} dx$$

$$= \frac{1}{\pi} \int_{-\pi}^{\pi} [f(u) - A_0] \left(\int_{u}^{\pi} \cos nx\, dx \right) du$$

$$= -\frac{1}{\pi} \int_{-\pi}^{\pi} [f(u) - A_0] \frac{\sin nu}{n}\, du$$

$$= -\frac{B_n}{n}$$

In the same fashion, we have

$$\bar{B}_n = \frac{1}{\pi} \int_{-\pi}^{\pi} F(x) \sin nx\, dx$$

$$= \frac{1}{n\pi} \int_{-\pi}^{\pi} [f(u) - A_0](\cos nu - \cos n\pi)\, du$$

$$= \frac{A_n}{n}$$

Recalling the definition of $F(x)$, we have proved that

$$\int_{-\pi}^{x} f(u)\, du = A_0(x + \pi) + \bar{A}_0$$

$$+ \sum_{n=1}^{\infty} \frac{1}{n} (A_n \sin nx - B_n \cos nx) \qquad -\pi \le x \le \pi$$

If we replace x by x_0 and subtract the result, then $\bar{A}_0$ cancels, and we have proved the stated result. ∎

EXERCISES 1.3

1. Let

$$f_{2n-1}(x) = \frac{4}{\pi} \left[\sin x + \frac{1}{3} \sin 3x + \cdots + \frac{1}{2n-1} \sin (2n-1)x \right]$$

Show that

$$f_{2n-1}\left(\frac{k\pi}{2n}\right) \to \frac{2}{\pi} \int_{0}^{k\pi} \frac{\sin x}{x}\, dx \qquad k = 1, 2, \ldots$$

[*Hint:* Write the sum for $f_{2n-1}(k\pi/(2n))$ as the approximating sum for an appropriate riemannian integral.]

2. Estimate the integral $\int_{0}^{k\pi} (\sin x)/x \, dx$ for $k = 2, 3, 4$.

3. Let $f(x)$, $-L < x < L$, be a piecewise smooth function. Show that the first criterion for uniform convergence follows from the Weierstrass M test (Sec. 0.5).

4. Let $f(x)$, $-L < x < L$, be a piecewise smooth function. Show that $A_n = O(1/n)$, $B_n = O(1/n)$ when $n \to \infty$.

5. Let $f(x)$, $-L < x < L$, be a piecewise smooth function. Let A_n', B_n' be the Fourier coefficients of f'.

$$A_n' = \frac{1}{L} \int_{-L}^{L} f'(x) \cos \frac{n\pi x}{L}\, dx$$

$$B_n' = \frac{1}{L} \int_{-L}^{L} f'(x) \sin \frac{n\pi x}{L}\, dx$$

If f is continuous and $f(-L + 0) = f(L - 0)$, show that

$$A_n' = \frac{n\pi}{L} B_n \qquad B_n' = -\frac{n\pi}{L} A_n$$

6. Let $f(x)$, $-L < x < L$, be a continuous piecewise smooth function with $f(-L + 0) = f(L - 0)$. Use Exercises 4 and 5 to show that $A_n = O(1/n^2)$, $B_n = O(1/n^2)$ when $n \to \infty$.

7. Let $f(x)$, $-L < x < L$, be a continuous piecewise smooth function with $f(-L + 0) = f(L - 0)$. Use Exercise 6 to prove the second criterion for uniform convergence.

8. Use Exercise 5 and the main convergence theorem of Sec. 1.2 to prove the proposition on differentiating a Fourier series.

9. Let $f(x) = \sum_{n=1}^{\infty} e^{-n^2\pi/L^2} \sin(n\pi x/L)$ be the Fourier series of a piecewise smooth function. Show that

$$f'(x) = \sum_{n=1}^{\infty} \frac{n\pi}{L} e^{-n^2\pi/L^2} \cos\frac{n\pi x}{L}$$

$$f''(x) = -\sum_{n=1}^{\infty} \left(\frac{n\pi}{L}\right)^2 e^{-n^2\pi/L^2} \sin\frac{n\pi x}{L}$$

10. Consider the Fourier series of $f(x) = x$ found in Example 1.1.1. By formally differentiating the series at $x = 0$, show that it is not valid to differentiate a Fourier series term by term, even if the function is differentiable.

11. Consider the Fourier series of $f(x) = x$ found in Example 1.1.1. By integrating this series, find a series for x^2.

12. Integrate the series of Exercise 11 and compare the result with Example 1.1.5.

13. Among the series for x, x^2, $x^3 - L^2x$ found in exercises 10 to 12, which are uniformly convergent?

*14. Let $f(x) = x$, $-\pi < x < \pi$. Find the maximum of the partial sum $f_N(x)$, and verify the presence of Gibb's phenomenon.

1.4   PARSEVAL'S THEOREM AND MEAN SQUARE ERROR

Having developed the convergence properties of Fourier series, we now turn to some concrete computations which show how Fourier series may be used in various problems.

Statement and Proof of Parseval's Theorem

The key to these applications is *Parseval's theorem,* a form of the pythagorean theorem, which is valid in the setting of Fourier series. Thus, let $f(x)$, $-L < x < L$, be a piecewise smooth function with Fourier series

$$A_0 + \sum_{n=1}^{\infty} \left(A_n \cos\frac{n\pi x}{L} + B_n \sin\frac{n\pi x}{L}\right)$$

Parseval's theorem states that

$$\boxed{\frac{1}{2L} \int_{-L}^{L} f(x)^2 \, dx = A_0^2 + \tfrac{1}{2} \sum_{n=1}^{\infty} (A_n^2 + B_n^2)} \qquad (1.4.1)$$

The left side represents the mean square of the function $f(x)$, $-L < x < L$. The right side represents the sum of the squares of the Fourier components in the various coordinate directions $\cos(n\pi x/L)$, $\sin(n\pi x/L)$.

The proof of Parseval's theorem is especially simple if the piecewise smooth function is also continuous with $f(-L + 0) = f(L - 0)$. In that case we multiply the uniformly convergent Fourier series by $f(x)$ to obtain

$$f(x)^2 = A_0 f(x) + \sum_{n=1}^{\infty} \left[A_n f(x) \cos \frac{n\pi x}{L} + B_n f(x) \sin \frac{n\pi x}{L} \right]$$

This series is also uniformly convergent, and we may integrate term by term for $-L < x < L$, with the result

$$\int_{-L}^{L} f(x)^2 \, dx = A_0 \int_{-L}^{L} f(x) \, dx + \sum_{n=1}^{\infty} \left[A_n \int_{-L}^{L} f(x) \cos \frac{n\pi x}{L} \right.$$
$$\left. + B_n \int_{-L}^{L} f(x) \sin \frac{n\pi x}{L} \right]$$

On the right we recognize the integrals which define the Fourier coefficients A_0, A_n, B_n. Dividing both sides by $2L$, we obtain equation (1.4.1), the desired form of Parseval's theorem in this case. The proof in the general case may be found in Exercises 21 to 29 at the end of this section. ∎

Application to Mean Square Error

Our first application of Parseval's theorem is to the *mean square error* σ_N^2, defined by

$$\sigma_N^2 = \frac{1}{2L} \int_{-L}^{L} [f(x) - f_N(x)]^2 \, dx$$

This number measures the average amount by which the partial sum $f_N(x)$ differs from $f(x)$. The Fourier series of $f(x) - f_N(x)$ is

$$\sum_{N+1}^{\infty} \left(A_n \cos \frac{n\pi x}{L} + B_n \sin \frac{n\pi x}{L} \right)$$

and therefore, by Parseval's theorem, we have

$$\frac{1}{2L} \int_{-L}^{L} [f(x) - f_N(x)]^2 \, dx = \frac{1}{2} \sum_{N+1}^{\infty} (A_n^2 + B_n^2)$$

and the formula

$$\sigma_N^2 = \frac{1}{2} \sum_{N+1}^{\infty} (A_n^2 + B_n^2) \tag{1.4.2}$$

The mean square error is half the sum of the squares of the remaining Fourier coefficients. This formula shows, in particular, that the mean square error tends to zero when N tends to infinity.

Example 1.4.1. Let $f(x) = |x|$, $-\pi < x \, \pi$. Find the mean square error, and give an asymptotic estimate when $N \to \infty$.

Solution. We have $B_n = 0$, $A_{2m} = 0$, $A_{2m-1} = -4/[\pi(2m - 1)^2]$, so that

$$\sigma_{2N-1}^2 = \sigma_{2N}^2 = \frac{1}{2} \sum_{n=2N+1}^{\infty} A_n^2$$

$$= \frac{1}{2} \sum_{m=N+1}^{\infty} \left[\frac{4}{\pi(2m - 1)^2} \right]^2 \qquad n = 2m - 1$$

$$= \frac{8}{\pi^2} \sum_{m=N+1}^{\infty} \frac{1}{(2m - 1)^4}$$

Although we cannot make a closed-form evaluation of this series, we can still make a useful *asymptotic* estimate. To do this, we compare the sum with the integral

$$\frac{8}{\pi^2} \int_N^{\infty} \frac{1}{(2x - 1)^4} \, dx = \frac{4}{3\pi^2} \frac{1}{(2N - 1)^3}$$

Figure 1.4.1 shows the comparison of a sum with an integral. This gives us the useful asymptotic statement

$$\sigma_N^2 = O(N^{-3}) \qquad N \to \infty \qquad \bullet$$

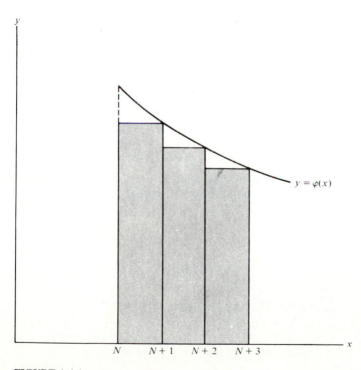

FIGURE 1.4.1

Illustrating the relation $\sum_{m=N+1}^{\infty} \varphi(m) \le \int_N^{\infty} \varphi(x) \, dx$.

Example 1.4.2. Let $f(x) = x$, $-\pi < x < \pi$. Find the mean square error, and give an asymptotic estimate when $N \to \infty$.

Solution. We have $A_m = 0$, $B_m = (-1)^{m-1}(2/m)$, and therefore

$$\sigma_N^2 = \frac{1}{2} \sum_{m=N+1}^{\infty} \frac{4}{m^2} = 2 \sum_{m=N+1}^{\infty} \frac{1}{m^2}$$

To obtain a useful asymptotic estimate of this sum, we compare with the integral

$$2 \int_N^\infty \frac{dx}{x^2} = \frac{2}{N}$$

so that

$$\sigma_N^2 = O(N^{-1}) \qquad N \to \infty \qquad \bullet$$

Application to the Isoperimetric Theorem

We now give an application of Fourier series to geometry, the *isoperimetric theorem*. Suppose that we have a smooth closed curve in the xy plane which encloses an area A and has perimeter P. We will prove that

$$P^2 \geq 4\pi A$$

with equality if and only if the curve is a circle.

To do this, suppose that the curve is described by parametric equations $x = x(t)$, $y = y(t)$ where $-\pi \leq t \leq \pi$. The functions $x(t)$, $y(t)$ are supposed smooth and satisfy the normalization $x(-\pi) = x(\pi)$, $y(-\pi) = y(\pi)$ because the curve is closed. From calculus, the perimeter and area are given by the formulas

$$P = \int_{-\pi}^{\pi} \sqrt{(x')^2 + (y')^2}\, dt \qquad A = \int_{-\pi}^{\pi} xy'\, dt$$

where $x' = dx/dt$, $y' = dy/dt$. By changing the functions $x(t)$, $y(t)$, we may suppose that $(x')^2 + (y')^2$ is constant (see Exercise 20); in fact,

$$(x')^2 + (y')^2 = \frac{P^2}{4\pi^2}$$

Now we introduce the convergent Fourier series

$$x(t) = a_0 + \sum_{n=1}^{\infty} (a_n \cos nt + b_n \sin nt) \qquad -\pi \leq t \leq \pi$$

$$y(t) = c_0 + \sum_{n=1}^{\infty} (c_n \cos nt + d_n \sin nt) \qquad -\pi \leq t \leq \pi$$

Since the functions $x(t)$, $y(t)$ are supposed smooth, we also have the convergent Fourier series

$$x'(t) = \sum_{n=1}^{\infty} n(-a_n \sin nt + b_n \cos nt) \qquad -\pi \le t \le \pi$$

$$y'(t) = \sum_{n=1}^{\infty} n(-c_n \sin nt + d_n \cos nt) \qquad -\pi \le t \le \pi$$

Applying Parseval's theorem, we have

$$\frac{P^2}{2\pi} = \int_{-\pi}^{\pi} [x'(t)^2 + y'(t)^2] \, dt = \pi \sum_{n=1}^{\infty} n^2(a_n^2 + b_n^2 + c_n^2 + d_n^2)$$

$$A = \int_{-\pi}^{\pi} x(t)y'(t) \, dt$$

$$= \frac{1}{4} \int_{-\pi}^{\pi} \{[x(t) + y'(t)]^2 - [x(t) - y'(t)]^2\} \, dt$$

$$= \pi \sum_{n=1}^{\infty} n(a_n d_n - b_n c_n)$$

Performing the necessary algebraic steps, we have

$$\frac{P^2}{2\pi} - 2A = \pi \sum_{n=1}^{\infty} [n(a_n - d_n)^2 + n(b_n + c_n)^2 + n(n - 1)(a_n^2 + b_n^2 + c_n^2 + d_n^2)]$$

The right side is a sum of squares with positive coefficients; thus $P^2/(2\pi) - 2A \ge 0$. If the sum is zero, then all the terms are zero, in particular, $a_n^2 + b_n^2 + c_n^2 + d_n^2 = 0$ for $n > 1$ and $a_1 - d_1 = 0$, $b_1 + c_1 = 0$. This means that

$$x(t) = a_0 + a_1 \cos t - c_1 \sin t \qquad -\pi \le t \le \pi$$

$$y(t) = c_0 + c_1 \cos t + a_1 \sin t \qquad -\pi \le t \le \pi$$

which is the equation of a circle of radius $\sqrt{a_1^2 + c_1^2}$ with center (a_0, c_0). The proof is complete. ∎

EXERCISES 1.4

Find the mean square errors for the Fourier series of the functions in Exercises 1 to 3.

1. $f(x) = \begin{cases} 1 & 0 < x < \pi \\ 0 & x = 0 \\ -1 & -\pi < x < 0 \end{cases}$

2. $f(x) = x^2, \ -\pi \le x \le \pi$

3. $f(x) = \sin 10x, \ -\pi < x < \pi$

4. Write out Parseval's theorem for the Fourier series of Exercise 1.

5. Write out Parseval's theorem for the Fourier series of Exercise 2.

6. Show that, in Exercise 1, $\sigma_N^2 = O(N^{-1})$, $N \to \infty$.

7. Show that, in Exercise 2, $\sigma_N^2 = O(N^{-3})$, $N \to \infty$.

8. Let $f(x) = x(\pi - x)$, $0 \le x \le \pi$.
   (a) Compute the Fourier sine series of f.
   (b) Compute the Fourier cosine series of f.
   (c) Find the mean square error incurred by using N terms of each series, and find asymptotic estimates when $N \to \infty$.
   (d) Which series gives a better mean square approximation of f?

9. Let $f(x)$, $g(x)$, $-L \le x \le L$, be piecewise smooth functions with Fourier series

$$f(x) = A_0 + \sum_{n=1}^{\infty} \left(A_n \cos \frac{n\pi x}{L} + B_n \sin \frac{n\pi x}{L} \right)$$

$$g(x) = C_0 + \sum_{n=1}^{\infty} \left(C_n \cos \frac{n\pi x}{L} + D_n \sin \frac{n\pi x}{L} \right)$$

Show that

$$\frac{1}{2L} \int_{-L}^{L} f(x)g(x)\, dx = A_0 C_0 + \frac{1}{2} \sum_{n=1}^{\infty} (A_n C_n + B_n D_n)$$

Note that this formula corresponds to the dot product formula

$$(a_1\mathbf{i} + b_1\mathbf{j} + c_1\mathbf{k}) \cdot (a_2\mathbf{i} + b_2\mathbf{j} + c_2\mathbf{k}) = a_1 a_2 + b_1 b_2 + c_1 c_2$$

for vectors in the three-dimensional vector space $\mathbf{R}^3$.

10. Let $f(x) = \cos ax/\sin a\pi$, $-\pi \le x \le \pi$, where $0 < a < \frac{1}{2}$.
   (a) Find the Fourier series of f.
   (b) Give an asymptotic estimate for the mean square error incurred in approximating f by the first N terms of the Fourier series.
   *(c) Apply Parseval's theorem to obtain the following integral formula:

$$\sum_{n=-\infty}^{\infty} (a^2 - n^2)^{-2} = \frac{\pi}{2} (a \sin a\pi)^{-2} \int_{-\pi}^{\pi} \cos^2 ax\, dx$$

   *(d) Prove that $\sum_{n=1}^{\infty} n^{-4} = \pi^4/90$. (*Hint:* Make a three-term Taylor expansion of Exercise 10c in powers of a, and identify the coefficients.)

11. Let $\varphi(x)$ be defined for $x > 0$ with $\varphi(x) > 0$, $\varphi(x) < 0$, and let the integral $\int_1^{\infty} \varphi(x)\, dx$ be convergent.
   (a) Show that

$$\int_{N+1}^{\infty} \varphi(x)\, dx \le \sum_{N+1}^{\infty} \varphi(n) \le \int_{N}^{\infty} \varphi(x)\, dx$$

   (b) Deduce from this that

$$-\varphi(N + 1) \le \sum_{N+1}^{\infty} \varphi(n) - \int_{N}^{\infty} \varphi(x)\, dx \le 0$$

12. Let $\varphi(x) = 1/x^s$ where $s > 1$.
   (a) Use Exercise 11 to show that

$$\frac{-1}{(N + 1)^s} \le \sum_{N+1}^{\infty} \frac{1}{n^s} - \frac{1}{s - 1} \frac{1}{N^{s-1}} \le 0$$

(b) Show that this may be written in the form

$$\sum_{N+1}^{\infty} \frac{1}{n^s} = \frac{1}{(s-1)N^{s-1}} \left[1 + O\left(\frac{1}{N}\right)\right] \qquad N \to \infty$$

13. Let σ_N^2 be the mean square error in the Fourier series of $f(x) = x$, $-\pi < x < \pi$. Use Exercise 12 to show that $\sigma_N^2 = (1/N)[1 + O(1/N)]$, $N \to \infty$.

14. Let $\varphi(x) = 1/P(x)$ where $P(x)$ is a polynomial of degree s, $s > 1$. Modify Exercise 11b to show that

$$\sum_{N+1}^{\infty} \varphi(n) = \int_N^{\infty} \varphi(x)\, dx \left[1 + O\left(\frac{1}{N}\right)\right] \qquad N \to \infty$$

15. Let σ_N^2 be the mean square error in the Fourier series of $f(x) = |x|$, $-\pi < x < \pi$. Use Exercise 14 to find an asymptotic estimate of the form $\sigma_N^2 = (C/N^s)[1 + O(1/N)]$, $N \to \infty$ for appropriate constants C, s.

16. Let $\varphi(x) = e^{-x}$, $x > 0$. Discuss the validity of the asymptotic estimate

$$\sum_{N+1}^{\infty} \varphi(n) = \int_N^{\infty} \varphi(x)\, dx \, [1 + O(1/N)] \qquad N \to \infty$$

17. Compute the ratio P^2/A for an equilateral triangle.

18. Compute the ratio P^2/A for a square.

19. Compute the ratio P^2/A for a regular polygon of n sides, and compare with the isoperimetric theorem.

*20. Let $x(t)$, $y(t)$ be smooth functions, $-\pi \le t \le \pi$, with $(x')^2 + (y')^2 \ne 0$. Let $s(t) = \int_{-\pi}^{t} \sqrt{(x')^2 + (y')^2}$, $P = s(\pi)$, $\bar{t} = -\pi + 2\pi s/P$, $\bar{x}(\bar{t}) = x(t)$, $\bar{y}(\bar{t}) = y(t)$. Show that $-\pi \le \bar{t} \le \pi$ and $(d\bar{x}/d\bar{t})^2 + (d\bar{y}/d\bar{t})^2 = P^2/(4\pi^2)$.

The following exercises are designed to lead to the proof of Parseval's theorem for an arbitrary piecewise smooth function $f(x)$, $-L < x < L$ (suggested by Professor N. Stanton).

21. Let $f(x)$, $-L < x < L$, be a piecewise smooth function. Show that for each $\epsilon > 0$, there is a continuous piecewise smooth function $f^*(x)$, $L < x < L$, with $f^*(-L + 0) = f^*(L - 0)$ such that $[1/(2L)] \int_{-L}^{L} [f(x) - f^*(x)]^2\, dx < \epsilon^2$. [*Hint:* Across each subdivision point x_i $(1 \le i \le p)$ replace $f(x)$ by a linear function on the interval $x_i - h < x < x_i + h$, where h is chosen in terms of ϵ, p and the maximum of $|f(x)|$, $-L < x < L$.

22. Show that the Fourier coefficients of f and f^* satisfy

$$(A_0 - A_0^*)^2 + \tfrac{1}{2} \sum_{n \ge 1} [(A_n - A_n^*)^2 + (B_n - B_n^*)^2] < \epsilon^2$$

(*Hint:* Use Bessel's inequality.)

23. Develop a suitable triangle inequality for the norm $\sqrt{A_0^2 + \tfrac{1}{2}\Sigma(A_n^2 + B_n^2)}$ to prove that, for the Fourier coefficients introduced above, we have

$$\sqrt{A_0^2 + \tfrac{1}{2}\Sigma_{n\ge1}(A_n^2 + B_n^2)} \ge \sqrt{A_0^{*2} + \tfrac{1}{2}\Sigma_{n\ge1}(A_n^{*2} + B_n^{*2})} - \epsilon$$

24. Use Parseval's theorem for the function $f^*(x)$, $-L < x < L$, to prove that

$$\sqrt{A_0^2 + \tfrac{1}{2}\Sigma_{n\ge1}(A_n^2 + B_n^2)} \ge \sqrt{[1/(2L)]\int_{-L}^{L} f(x)^2\, dx} - 2\epsilon$$

25. Combine Exercise 24 with Bessel's inequality to conclude Parseval's theorem for the piecewise smooth function $f(x)$, $-L < x < L$.

The following exercises give an alternate treatment of the results obtained in Exercises 22 to 25.

26. Let $f(x)$, $-L < x < L$, be a piecewise smooth function, and let $f^*(x)$, $-L < x < L$, be the continuous function constructed in Exercise 21. Use Proposition 0.3.1 to show that $\|f - f_N\| \le \|f - f_N^*\|$, where f_N is the Nth partial sum of the Fourier series of the function $f(x)$, $-L < x < L$, and f_N^* is the corresponding Nth partial sum for the function f^*.

27. Use the triangle inequality from Sec. 0.3 to show that $\|f - f_N^*\| \le \|f - f^*\| + \|f^* - f_N^*\|$.

28. Show that there is an integer N_0 such that for $N \ge N_0$ we have $\|f - f_N^*\| < 2\epsilon$. (*Hint:* Use the proposition relating mean square convergence with Parseval's theorem from Sec. 0.3, coupled with the Parseval theorem already proved for the function $f^*(x)$, $-L < x < L$.)

29. Conclude Parseval's theorem for the piecewise smooth function $f(x)$, $-L < x < L$.

1.5   COMPLEX FORM OF FOURIER SERIES

Fourier Series and Fourier Coefficients

It is often useful to rewrite the formulas of Fourier series using complex numbers. To do this, we begin with De Moivre's formula

$$e^{i\theta} = \cos \theta + i \sin \theta \tag{1.5.1}$$

and the immediate consequences $\cos \theta = \frac{1}{2}(e^{i\theta} + e^{-i\theta})$, $\sin \theta = (e^{i\theta} - e^{-i\theta})/2i$. We apply these to a Fourier series

$$f(x) = A_0 + \sum_{n=1}^{\infty} \left(A_n \cos \frac{n\pi x}{L} + B_n \sin \frac{n\pi x}{L} \right)$$

$$= A_0 + \frac{1}{2} \sum_{n=1}^{\infty} [(A_n - iB_n)e^{in\pi x/L} + (A_n + iB_n)e^{-in\pi x/L}]$$

Therefore we let $\alpha_n = \frac{1}{2}(A_n - iB_n)$, $n = 1, 2, \ldots$; $\alpha_n = \frac{1}{2}(A_{-n} + iB_{-n})$, $n = -1, -2, \ldots$; and $\alpha_0 = A_0$. With this convention the Fourier series assumes the form

$$f(x) = \sum_{-\infty}^{\infty} \alpha_n e^{in\pi x/L} \tag{1.5.2}$$

To obtain integral formulas for the coefficients α_n, we use (1.1.6) and (1.1.7):

$$2\alpha_n = A_n - iB_n$$

$$= \frac{1}{L} \int_{-L}^{L} f(x) \times \left(\cos \frac{n\pi x}{L} - i \sin \frac{n\pi x}{L} \right) dx \qquad n = 1, 2, \ldots$$

$$= \frac{1}{L} \int_{-L}^{L} f(x) e^{-in\pi x/L} \, dx$$

with a corresponding formula for the plus sign. When $n = 0$, equation (1.1.5) shows that α_0 is given appropriately. Thus we have

$$\boxed{\alpha_n = \frac{1}{2L} \int_{-L}^{L} f(x) e^{-in\pi x/L} \qquad n = 0, \pm 1, \pm 2, \ldots} \qquad (1.5.3)$$

Parseval's Theorem in Complex Form

Finally we retrieve the appropriate form of Parseval's theorem. To do this, multiply (1.5.2) by $f(x)$ and integrate on $(-L, L)$. The result is

$$\boxed{\frac{1}{2L} \int_{-L}^{L} f(x)^2 \, dx = \sum_{-\infty}^{\infty} |\alpha_n|^2} \qquad (1.5.4)$$

Applications of Complex Notation

The functions $e^{in\pi x/L}$ satisfy an orthogonality relation, which may be written in the form

$$\int_{-L}^{L} e^{in\pi x/L} e^{-im\pi x/L} \, dx = \begin{cases} 0 & n \neq m \\ 2L & n = m \end{cases}$$

These may be proved by using De Moivre's formula and the orthogonality of the trigonometric functions $\cos (n\pi x/L)$, $\sin (n\pi x/L)$. Knowing these orthogonality relations, we can develop the complex form of Fourier series in its own right, without reference to the original formulas of Sec. 1.1.

The theory of Fourier series may also be extended to *complex-valued functions* $f(x)$, $-L < x < L$. These are of the form $f(x) = f_1(x) + if_2(x)$, where f_1, f_2 are real-valued functions. The Fourier coefficients are defined by the same formulas $\alpha_n = [1/(2L)] \int_{-L}^{L} f(x) e^{in\pi x/L} \, dx$. If both f_1, f_2 are piecewise smooth functions, then the complex Fourier series converges for all x to $\frac{1}{2}[\bar{f}(x + 0) + \bar{f}(x - 0)]$, where $\bar{f}$ is the periodic extension of the piecewise smooth function $f(x)$, $-L < x < L$. This convergence is understood as the limit of the sum $\sum_{-N}^{N}$ when N tends to infinity.

The Fourier coefficients of a real-valued function are characterized by the relation

$$\alpha_{-n} = \bar{\alpha}_n$$

where the bar indicates the *complex conjugate* of a complex number: if $c = a + ib$, then $\bar{c} = a - ib$.

To simplify the computation of complex Fourier series, we indicate some formulas which are often used. If $c = a + ib$ is a complex number, the exponential function $e^{cx} = e^{ax}e^{ibx} = e^{ax}(\cos bx + i \sin bx)$. From this we have $(d/dx)(e^{cx}) = ae^{ax} \cos bx - be^{ax} \sin bx + i(ae^{ax} \sin bx + be^{ax} \cos bx) = (a + ib)e^{ax}(\cos bx + i \sin bx) = ce^{cx}$. Hence the differentiation formula

$$\frac{d}{dx} e^{cx} = ce^{cx}$$

is valid for any complex number c.

Example 1.5.1. Compute the complex Fourier series of $f(x) = e^{ax}$, $-\pi < x < \pi$, where a is a real number.

Solution. The Fourier coefficients are given by the formula

$$\alpha_n = \frac{1}{2\pi} \int_{-\pi}^{\pi} e^{ax}e^{-inx} \, dx = \frac{1}{2\pi} \int_{-\pi}^{\pi} e^{(a-in)x} \, dx$$

Noting that $(d/dx)(e^{(a-in)x}) = (a - in)e^{(a-in)x}$, we have

$$\alpha_n = \frac{1}{2\pi} \frac{1}{a - in} (e^{(a-in)\pi} - e^{(a-in)(-\pi)})$$

$$= \frac{1}{2\pi} \frac{1}{a - in} (-1)^n (e^{a\pi} - e^{-a\pi})$$

$$= \frac{1}{\pi} \sinh a\pi \frac{(-1)^n (a + in)}{a^2 + n^2}$$

The complex Fourier series of $f(x) = e^{ax}$, $-\pi < x < \pi$, is

$$\frac{1}{\pi} \sinh a\pi \sum_{-\infty}^{\infty} \frac{(-1)^n (a + in)}{a^2 + n^2} e^{inx} \qquad \bullet$$

As our next application of complex Fourier series, we compute the Fourier series of

$$f(x) = \cos^m x \qquad -\pi < x < \pi$$

If we were to use the real form of Fourier series, we would encounter many cumbersome trigonometric identities. With the complex approach, we avoid these. We begin with the identity

$$\cos x = \tfrac{1}{2}(e^{ix} + e^{-ix})$$

We expand the mth power, using the binomial theorem

$$(e^{ix} + e^{ix})^m = \sum_{j=0}^{m} \binom{m}{j} e^{ijx} e^{-i(m-j)x}$$

Therefore,

$$\cos^m \dot{x} = \left(\frac{1}{2}\right)^m \sum_{j=0}^{m} \binom{m}{j} e^{i(2j-m)x}$$

This is the complex form of the Fourier series for $\cos^m x$. As a by-product, we can obtain some useful integrals. To do this, we multiply the previous equation by e^{-inx} and integrate for $-\pi < x < \pi$. By orthogonality all the integrals are zero except when $2j - m - n = 0$, in which case the integral is 2π. In particular, $m + n$ must be even. Therefore we have

$$\frac{1}{2\pi} \int_{-\pi}^{\pi} (\cos x)^m e^{-inx}\, dx = \begin{cases} 0 & m + n \text{ odd} \\ \left(\frac{1}{2}\right)^m \binom{m}{j} & 0 \le m + n = 2j \le 2m \end{cases}$$

The Fourier series for $\cos^m x$ can also be written in a real form, to obtain familiar trigonometric identities. It is simpler to consider separately the cases of m even and m odd. Thus, if $m = 2k + 1$, we can group the terms of the Fourier series in pairs: $j = 0$ with $j = m$, $j = 1$ with $j = m - 1$, etc. To each pair, we apply De Moivre's theorem with the result

$$\cos^{2k+1} x = \left(\frac{1}{2}\right)^{2k+1} \left[2 \cos (2k + 1)x + \cdots + 2\binom{2k + 1}{k} \cos x \right]$$

In particular, this gives the identities

$$\cos^3 x = \tfrac{1}{4}(\cos 3x + 3 \cos x)$$

$$\cos^5 x = \tfrac{1}{16}(\cos 5x + 5 \cos 3x + 10 \cos x)$$

If m is even, we group the terms $j = 0$ with $j = m$, etc., as before and finish with one ungrouped term in the middle. Applying De Moivre's theorem again, we have, with $m = 2k$,

$$\cos^{2k} x = \left(\frac{1}{2}\right)^{2k} \left[2 \cos 2kx + \cdots + 2\binom{2k}{k - 1} \cos 2x + \binom{2k}{k} \right]$$

In particular, we retrieve the identities

$$\cos^2 x = \tfrac{1}{2}(\cos 2x + 1)$$

$$\cos^4 x = \tfrac{1}{8}(\cos 4x + 4 \cos 2x + 3)$$

EXERCISES 1.5

1. Verify that the orthogonality relations hold, in the form

$$\int_{-L}^{L} e^{in\pi x/L} e^{-im\pi x/L} \, dx = \begin{cases} 0 & n \neq m \\ 2L & n = m \end{cases}$$

2. Use the formulas in Exercise 1 to prove equation (1.5.3) from (1.5.2). You may assume that the series (1.5.2) converges uniformly for $-L < x < L$.

3. Use the complex form to find the Fourier series of $f(x) = e^x$, $-L < x < L$.

4. Let $0 < r < 1$, $f(x) = 1/(1 - re^{ix})$, $-\pi < x < \pi$. Find the Fourier series of f. (*Hint:* First expand f as a power series in r.)

5. Use Exercise 4 to derive the real formulas

$$\frac{1 - r \cos x}{1 + r^2 - 2r \cos x} = 1 + \sum_{n=1}^{\infty} r^n \cos nx \qquad 0 \le r < 1$$

$$\frac{r \sin x}{1 + r^2 - 2r \cos x} = \sum_{n=1}^{\infty} r^n \sin nx \qquad 0 \le r < 1$$

6. Use Exercise 5 to derive the formula

$$1 + 2 \sum_{n=1}^{\infty} r^n \cos nx = \frac{1 - r^2}{1 + r^2 - 2r \cos x} \qquad 0 < r < 1$$

7. Show that the convergence theorem, Theorem 1.2.1, can be written in complex form as

$$\tfrac{1}{2}[\overline{f}(x + 0) + \overline{f}(x - 0)] = \lim_{N \to \infty} \sum_{-N}^{N} \alpha_n e^{in\pi x/L}$$

8. Show that the unrestricted double limit

$$\lim_{M,N \to \infty} \sum_{-M}^{N} \alpha_n e^{in\pi x/L}$$

does not exist in general.
(*Hint:* Try Example 1.1.3 at $x = 0$.)

BOUNDARY-VALUE PROBLEMS IN RECTANGULAR COORDINATES

INTRODUCTION

In this chapter we apply Fourier series to some typical problems of classical physics. These include the heat equation, the wave equation, and Laplace's equation in rectangular coordinates. The methods of separation of variables and the superposition principle enable us to solve boundary-value problems and initial-value problems in the form of infinite series of separated solutions. The rigorous validity and asymptotic behavior are studied in detail.

2.1 THE HEAT EQUATION

We consider a three-dimensional body in which heat can flow freely. By $u(x, y, z; t)$ we denote the temperature measured at the point (x, y, z) at instant t. We suppose that $u(x, y, z; t)$ is a smooth function of $(x, y, z; t)$, and we proceed to determine a partial differential equation for u.

Fourier's Law of Heat Conduction

To derive the heat equation, we consider a solid material which occupies a portion of three-dimensional space. A basic quantity of importance is the *heat current density* $\mathbf{q}(\mathbf{x}; t)$. This vector quantity represents the rate of heat flow per unit time at the point $\mathbf{x} = (x, y, z)$. *Fourier's law* states that

$$\boxed{\mathbf{q} = -k \operatorname{grad} u}$$

where k is the thermal conductivity of the material. From calculus we know that grad u points in the direction of the maximum increase of u. Since heat is expected to flow from warmer to cooler regions, we insert the minus sign in Fourier's law. Thus $\mathbf{q}$ points in the direction of maximum *decrease* of u, and $|\mathbf{q}|$ is the rate of heat flow in that direction. If $\mathbf{n}$ is any unit vector, the scalar quantity $\mathbf{q} \cdot \mathbf{n}$ is called the *heat flux* in direction $\mathbf{n}$. It measures the rate of heat flow per unit time per unit area across a plane with normal vector $\mathbf{n}$.

Derivation of the Heat Equation

During a small time interval $(t, t + \Delta t)$, heat flows through the material and may also be generated by internal sources, at a rate $s(\mathbf{x}; t)$. Therefore the amount of heat which enters any region R of the material in time interval $(t, t + \Delta t)$ is, to first order in Δt, given by

$$Q = \left(-\iint_{\partial R} \mathbf{q} \cdot \mathbf{n} \, dS + \iiint_R s \, dV \right) \Delta t + O(|\Delta t|^2)$$

where $\mathbf{n}$ is the outward-pointing normal vector and the minus sign is in front of the surface integral because $\mathbf{q} \cdot \mathbf{n} \, dS$ is the density of heat flowing *out* of the surface element dS per unit time.

However, this heat Q has the effect of raising the temperature by the amount $u_t \, \Delta t$, to first order in Δt. Therefore we can write

$$Q = \left(\iiint_R c\rho u_t \, dV \right) \Delta t + O(|\Delta t)|^2)$$

where c is the heat capacity per unit mass and ρ is the mass density of the material. Equating these, dividing by Δt, and letting $\Delta t \to 0$, we have the *continuity equation*

$$\boxed{\iiint_R c\rho u_t \, dV = -\iint_{\partial R} \mathbf{q} \cdot \mathbf{n} \, dS + \iiint_R s \, dV}$$

This equation is valid for any region, no matter how large or small. In particular, we take a region R about the point (x, y, z), divide by the volume, and take the limit when the diameter of the region tends to zero. The surface integral can be handled by the divergence theorem

$$\iint_{\partial R} \mathbf{q} \cdot \mathbf{n} \, dS = \iiint_R (\operatorname{div} \mathbf{q}) \, dV$$

and we obtain the differential form of the continuity equation:

$$c\rho u_t = -\text{div } \mathbf{q} + s$$

Combining this with Fourier's law, we have

$$\boxed{c\rho u_t = \text{div } (k \text{ grad } u) + s}$$

This is the general form of the heat equation.

In most problems k is independent of $\mathbf{x}$, and we can bring it outside and thus obtain the heat equation in the form

$$u_t = K \text{ div } (\text{grad } u) + r = K\nabla^2 u + r$$

where $K = k/(c\rho)$ and $r = s/(c\rho)$ are the renormalized conductivity and source terms, respectively. Here K is called the *thermal diffusivity* of the material. The *laplacian* of a function u is defined by

$$\nabla^2 u = \text{div } (\text{grad } u) = u_{xx} + u_{yy} + u_{zz}$$

Remark. We can derive the heat equation without using the divergence theorem, by the following direct argument. Let R be the rectangular box defined by the inequalities $x_1 \le x \le x_2$, $y_1 \le y \le y_2$, $z_1 \le z \le z_2$, and let q^x, q^y, q^z be the components of the heat current density vector. Then

$$\iint_{\partial R} \mathbf{q} \cdot \mathbf{n} \, dS = \int_{z_1}^{z_2} \int_{y_1}^{y_2} [q^x(x_2, y, z) - q^x(x_1, y, z)] \, dy \, dz$$

$$+ \int_{z_1}^{z_2} \int_{x_1}^{x_2} [q^y(x, y_2, z) - q^y(x, y_1, z)] \, dx \, dz$$

$$+ \int_{y_1}^{y_2} \int_{x_1}^{x_2} [q^z(x, y, z_2) - q^z(x, y, z_1)] \, dx \, dy$$

We must show that

$$\frac{1}{(x_2 - x_1)(y_2 - y_1)(z_2 - z_1)} \iint_{\partial R} \mathbf{q} \cdot \mathbf{n} \, dS$$

tends to $(q_x^x + q_y^y + q_z^z)(x_1, y_1, z_1)$ when $x_2 \to x_1$, $y_2 \to y_1$, $z_2 \to z_1$. To do this, we consider each of the three integrals separately. For the first integral we have to examine

$$\frac{1}{(y_2 - y_1)(z_2 - z_1)} \int_{z_1}^{z_2} \int_{y_1}^{y_2} \frac{q^x(x_2, y, z) - q^x(x_1, y, z)}{x_2 - x_1} \, dy \, dz$$

When $x_2 \to x_1$, the integrand tends to $q_x^x(x_1, y, z)$, a continuous function. When $y_2 \to y_1$, $z_2 \to z_1$, the resulting integral tends to $q_x^x(x_1, y_1, z_1)$. The same result is obtained if we first let $y_2 \to y_1$, $z_2 \to z_1$. The second integral, where q^x is replaced by q^y, tends to $q_y^y(x_1, y_1, z_1)$ when $x_2 \to x_1$, $y_2 \to y_1$, $z_2 \to z_1$ in any order. Similarly for the third integral. This proves that $\iint_R \mathbf{q} \cdot \mathbf{n} \, dS$, divided by the volume of the box R, tends to div $\mathbf{q}$ when the sides tend to zero, in any order. Referring to the continuity equation and letting $x_2 \to x_1$, $y_2 \to y_1$, $z_2 \to z_1$, we have proved that $c\rho u_t(x_1, y_1, z_1) = -\text{div } \mathbf{q}(x_1, y_1, z_1) + s(x_1, y_1, z_1)$, which was to be shown.

Boundary Conditions

The heat equation describes the flow of heat within the solid material. To completely determine the time evolution of temperature, we must also consider boundary conditions of various forms. For example, if the material is in contact with an ice-water bath, it is natural to suppose that $u = 32°F$ on the boundary. Alternately, we can imagine that the heat flux across the boundary is given; therefore by Fourier's law the appropriate boundary condition is of the type $\nabla u \cdot \mathbf{n} = a$, a given function on the boundary. For example, an insulated surface would necessitate $\nabla u \cdot \mathbf{n} = 0$ on the boundary. A third type of boundary condition results from Newton's law of cooling, written in the form

$$\mathbf{q} \cdot \mathbf{n} = h(u - T)$$

The heat flux across the boundary is proportional to the difference between the temperature u of the body and the temperature T of the surrounding medium.

Steady-State Solutions

An important class of solutions of the heat equation is the *steady-state solutions*. This means that $\partial u/\partial t = 0$ or that u is a function of (x, y, z), independent of t. Thus we must have $K\nabla^2 u + r = 0$, a form of Poisson's equation. If in addition there are no internal sources of heat, then $r = 0$ and u satisfies Laplace's equation $\nabla^2 u = 0$. We restate this as follows.

> **Proposition 2.1.1.** Steady-state solutions of the heat equation, with no internal heat sources, are solutions of Laplace's equation.

Thus, Laplace's equation is a special case of the heat equation.

> **Example 2.1.1.** Find the steady-state solution of the heat equation $u_t = K\nabla^2 u$ in the slab $0 < z < L$ and satisfying the boundary conditions $u(x, y, 0) = T_1$, $(\partial u/\partial z + hu)(x, y, L) = 0$, where T_1, h are positive constants.
>
> *Solution.* Laplace's equation is $u_{xx} + u_{yy} + u_{zz} = 0$. Since the boundary conditions are independent of (x, y), we look for the solution in the form $u(x, y, z) = U(z)$, independent of (x, y). Thus U must satisfy $U''(z) = 0$, whose general solution is $U(z) = A + Bz$. The boundary condition at $z = 0$ requires $T_1 = A$, while the boundary condition at $z = L$ requires $B + h(A + BL) = 0$. Thus $B(1 + hL) = -hA = -hT_1$, and the solution is $U(z) = T_1 - hT_1z/(1 + hL)$.   ●

In many problems it is important to compute the flux through the faces of the slab. From our earlier discussion, the flux is given by $-k\nabla u \cdot \mathbf{n}$, where $\mathbf{n} = (0, 0, 1)$ for the upper face and $\mathbf{n} = (0, 0, -1)$ for the lower face. Thus in

Example 2.1.1, the flux from the upper face is $-k\,\partial U/\partial z = khT_1/(1 + hL)$, while the flux from the lower face is $k\,\partial U/\partial z = -khT_1/(1 + hL)$.

We now consider an example with internal heat sources.

Example 2.1.2. Find the steady-state solution of the heat equation $u_t = K\nabla^2 u + r$ in the slab $0 < z < L$ and satisfying the boundary conditions $u(x, y, 0) = T_1$, $(\partial u/\partial z + hu)(x, y, L) = 0$, where r, K, h, T_1 are positive constants. Find the flux through the upper and lower faces.

Solution. The boundary conditions are independent of (x, y); hence we look for the solution in the form $u(x, y, z) = U(z)$, independent of (x, y). Thus U must satisfy $KU''(z) + r = 0$, whose general solution is $U(z) = -rz^2/(2K) + A + Bz$. The boundary condition at $z = 0$ requires $T_1 = A$, while the boundary condition at $z = L$ requires $-rL/K + B + h[-rL^2/(2K) + A + BL] = 0$. Thus $B(hL + 1) = rL/K + hrL^2/(2K) - hT_1$. The solution is $U(z) = -rz^2/(2K) + T_1 + Bz$, where $B(1 + b) = (rL/K)(1 + \frac{1}{2}b) - hT_1$, and the *Biot modulus* b is defined as $b = hL$. The flux through the upper face is $-kU'(L) = krL/K - kB$. The flux through the lower face is $kU'(0) = kB$. ●

In some cases the steady-state solution is not uniquely determined by the boundary conditions. For example, the heat equation $u_t = K\nabla^2 u$ with the boundary conditions $u_z(x, y, 0) = 0$, $u_z(x, y, L) = 0$ has the solution $U(z) = A$ for *any* constant A. This phenomenon of nonuniqueness is equivalent to the statement that $\lambda = 0$ is an eigenvalue of the Sturm-Liouville problem with the associated homogeneous boundary conditions. Indeed, if we have two different steady-state solutions $U_1(z)$, $U_2(z)$ with the same nonhomogeneous boundary conditions, then the difference $U(z) = U_1(z) - U_2(z)$ is a nonzero solution of the homogeneous equation $U''(z) = 0$, satisfying the homogeneous boundary conditions. This is exactly the statement that $\lambda = 0$ is an eigenvalue of the Sturm-Liouville problem with these homogeneous boundary conditions. We come back to this point in Sec. 2.3.

Time-Periodic Solutions†

Another important class of solutions of the heat equation is the *periodic solutions*. These correspond to a stationary regime, where the solution exists for all time, $-\infty < t < \infty$. Typically the solution is specified by a boundary condition and a condition of boundedness. We illustrate with the following problem from geophysics.

The temperature at the surface of the earth is a given periodic function of time, and we seek the temperature z units below the surface. We assume that there are no internal heat sources and that the thermal diffusivity is a constant throughout the earth.

---

† This class of solutions is also treated in the Appendix on "Using Mathematica."

To formulate this problem, we suppose that the earth is flat and that the surface is given by the equation $z = 0$. (In Chap. 4 we show that the flat earth is a valid approximation for shallow depths.) The temperature on the surface is independent of location and depends only on time. Therefore we must solve the problem

$$u_t = Ku_{zz} \qquad z > 0, \ -\infty < t < \infty$$

$$u(0; t) = \varphi(t) \qquad -\infty < t < \infty$$

where φ is periodic with period T. In addition, we require that the temperature be bounded

$$|u(z; t)| \leq M$$

since we do not expect the temperature variations within the earth to exceed the variations on the surface.

To solve this problem, first we look for separated solutions of the form

$$u(z; t) = f(z)g(t)$$

Since the heat equation has real coefficients, the real and imaginary parts of a complex-valued solution are again solutions. Thus we may allow $f(z)$, $g(t)$ to be complex-valued. Substituting into the heat equation, we have

$$\frac{Kf''(z)}{f(z)} = \frac{g'(t)}{g(t)}$$

Both sides must be a constant, which we call $-\lambda$. Thus we have the ordinary differential equations

$$g'(t) + \lambda g(t) = 0$$

$$f''(z) + \frac{\lambda}{K} f(z) = 0$$

The first equation has the solution $g(t) = e^{-\lambda t}$. Since we require bounded solutions for $-\infty < t < \infty$, λ must be pure imaginary, $\lambda = i\beta$ with β real. To solve the second equation, we try $f(z) = e^{\gamma z}$. Thus we must have $\gamma^2 e^{\gamma z} + (\lambda/K)e^{\gamma z} = 0$, yielding the quadratic equation

$$\gamma^2 + \frac{i\beta}{K} = 0$$

In the case where $\beta > 0$, this has two solutions

$$\gamma = \pm(-1 + i) \sqrt{\frac{\beta}{2K}}$$

Since we require bounded solutions for $z > 0$, we must take the solution with Re $\gamma < 0$, that is, the plus sign. Therefore we have the *separated solutions, in complex form,*

$$e^{-i\beta t}e^{(-1+i)z\sqrt{\beta/(2K)}}$$

Taking the real and imaginary parts, we have the real solutions

$$e^{-cz}\cos(\beta t - cz)$$
$$c = \sqrt{\frac{\beta}{2K}}$$
$$e^{-cz}\sin(\beta t - cz)$$

(If $\beta < 0$, it can be shown that no new solutions are obtained.) We refer to these as the *quasi-separated solutions*.

To solve the original problem, we suppose that the boundary temperature has been expanded as a Fourier series

$$\varphi(t) = A_0 + \sum_{n=1}^{\infty}\left(A_n\cos\frac{2n\pi t}{T} + B_n\sin\frac{2n\pi t}{T}\right)$$

We take $\beta_n = 2n\pi/T$, $c_n = \sqrt{n\pi/(KT)}$ in the quasi-separated solutions just developed to obtain the solution in the form

$$u(z; t) = A_0 + \sum_{n=1}^{\infty}e^{-c_n z}[A_n\cos(\beta_n t - c_n z) + B_n\sin(\beta_n t - c_n z)]$$

To verify that this is indeed a rigorous solution to the original problem, we may suppose that A_n, B_n are bounded by some constant. Then it may be shown that the formal series for u_z, u_{zz}, u_t converge uniformly, and hence u indeed satisfies the heat equation.

Example 2.1.3. Solve the heat equation $u_t = Ku_{zz}$ for $z > 0$, $-\infty < t < \infty$, with the boundary condition

$$u(0; t) = A_0 + A_1\cos\frac{2\pi t}{T}$$

where A_0, A_1, T are positive constants. Graph the solution as a function of t for $z\sqrt{\pi/(KT)} = 0$, $\pi/2$, π, $3\pi/2$, 2π and $0 \le t \le T$.

Solution. Referring to the general solution just obtained, we let $B_n = 0$ for $n \ge 1$ and $A_n = 0$ for $n \ge 2$. The solution is

$$u(z; t) = A_0 + A_1 e^{-c_1 z}\cos\left(\frac{2\pi t}{T} - z\sqrt{\frac{\pi}{KT}}\right)$$

In Fig. 2.1.1 we plot the temperature as a function of time for the depths indicated.
●

Applications to Geophysics

This theory can be used to study the seasonal variations of temperature within the earth. For $z = 0$, the maximum of $u(z; t)$ is assumed at $t = 0$, $\pm T$, $\pm 2T$, ... For $z = \sqrt{\pi KT}$, $u(z; t)$ assumes its *minimum* value for the same times, $t = 0$,

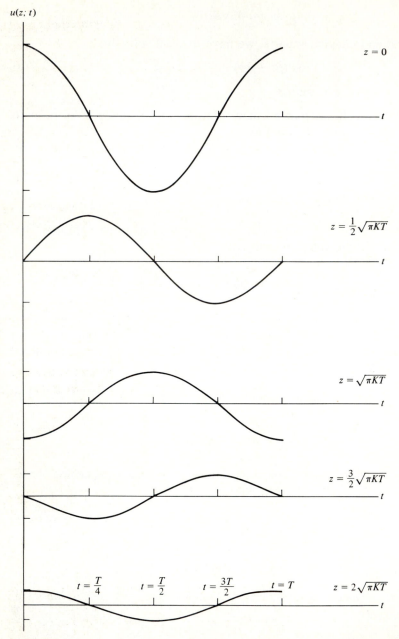

FIGURE 2.1.1

104

$\pm T, \pm 2T, \ldots$. Stated differently, when it is summer on the earth's surface, it is winter at a depth of $z = \sqrt{\pi KT}$.

Example 2.1.4. Suppose that $K = 2 \times 10^{-3}$ cm²/s, $T = 3.15 \times 10^7$ s. Find the depth necessary for a change from summer to winter.

Solution. We have $\sqrt{\pi KT} = 4.45 \times 10^2$ cm. Therefore, when it is summer on the earth's surface, it is winter at a depth of 4.4 m. ●

This theory can also be used to estimate the thermal diffusivity of the earth. To do this, we define the *amplitude variation* of the solution $u(z; t)$ as

$$A(z) = \max_{-\infty < t < \infty} u(z; t) - \min_{-\infty < t < \infty} u(z; t)$$

By measuring $A(z)$ at different depths, we may determine the diffusivity K. Indeed, using the solution obtained in Example 2.1.3, we have $\max u(z; t) = A_0 + A_1 e^{-c_1 z}$, $\min u(z; t) = A_0 - A_1 e^{-c_1 z}$, and thus $A(z) = 2A_1 e^{-c_1 z}$ and $A(z)/A(0) = e^{-c_1 z}$. Let z_1 be the depth for which $e^{-c_1 z} = \frac{1}{2}$. Since $c_1 = \sqrt{\pi/(KT)}$, we have $\sqrt{\pi/(KT)}z_1 = \ln 2$, $K = \pi z_1^2/[T(\ln 2)^2]$.

Example 2.1.5. Estimate the thermal diffusivity of the earth if the summer-winter amplitude variation decreases by a factor of 2 at a depth of 1.3 m.

Solution. We take $T = (365)(24)(3600) = 3.15 \times 10^7$ s, $z_1 = 1.3$ m. Thus

$$K = \frac{\pi(1.3)^2}{(3.15 \times 10^7)(0.69)^2} = 3.5 \times 10^{-7} \text{ m}^2/\text{s}$$ ●

EXERCISES 2.1

1. Find the steady-state solution of the heat equation $u_t = K\nabla^2 u$ in the slab $0 < z < L$ satisfying the boundary conditions $u(x, y, 0) = T_1$, $u(x, y, L) = T_2$ where T_1, T_2 are positive constants.
2. For the solution found in Exercise 1, find the flux through the upper face $z = L$.
3. Find the steady-state solution of the heat equation $u_t = K\nabla^2 u$ in the slab $0 < z < L$ satisfying the boundary conditions $(\partial u/\partial z)(x, y, 0) = \Phi$, $u(x, y, L) = T_0$ where Φ, T_0 are positive constants.
4. Find the steady-state solution of the heat equation $u_t = K\nabla^2 u$ in the slab $0 < z < L$ satisfying the boundary conditions $[k(\partial u/\partial z) - h(u - T_0)](x, y, 0) = 0$, $[k(\partial u/\partial z) + h(u - T_0)](x, y, 0) = 0$.
5. Find the steady-state solution of the heat equation $u_t = K\nabla^2 u - \beta(u - T_3)$ in the slab $0 < z < L$ satisfying the boundary conditions $u(x, y, 0) = T_1$, $u(x, y, L) = T_2$ where T_1, T_2, T_3, β are positive constants.
6. Find the steady-state solution of the heat equation $u_t = K\nabla^2 u + r$ in the slab $0 < z < L$ satisfying the boundary conditions $(\partial u/\partial z)(x, y, 0) = 0$, $u(x, y, L) = T_1$ where K, r, T_1 are positive constants. Find the flux through the face $z = L$.

7. Find the steady-state solution of the heat equation $u_t = K\nabla^2 u + r$ in the slab $0 < z < L$ satisfying the boundary conditions $u(x, y, 0) = T_1$, $u(x, y, L) = T_2$, where K, r, T_1, T_2 are positive constants. If $T_1 = T_2$, show that the flux across the plane $z = \frac{1}{2}L$ is zero.

*8. Find the steady-state solution of the heat equation $u_t = K\nabla^2 u + r(z)$ in the slab $0 < z < L$ satisfying the boundary condition $u(x, y, 0) = 0$, $u(x, y, L) = 0$ where $r(z) = r_0$ for $L/3 < z < 2L/3$, $r(z) = 0$ for $0 < z < L/3$ and $2L/3 < z < L$, and r_0 and K are positive constants. (*Hint:* Although u is not smooth, it may be supposed that u and u_z are both continuous.)

9. A wall of thickness 25 cm has outside temperature $-10°C$ and inside temperature $18°C$. The conductivity is $k = 0.0016$ cal/(s·cm·°C), and there are no internal heat sources. Find the steady-state heat flux through the outer wall per unit area.

10. Find the solution of the heat equation $u_t = K\nabla^2 u$ in the half-space $z > 0$ for $-\infty < t < \infty$ and satisfying the conditions $|u(z; t)| \leq M$, $u(0; t) = A_0 + A_1 \cos 2\pi t/T_1 + A_2 \cos 2\pi t/T_2$, where A_0, A_1, A_2, T_1, T_2 are positive constants.

11. Let $u(z; t) = e^{-cz} \cos (\beta t - cz)$, where β, c are constants. Show that u satisfies the heat equation $u_t = Ku_{zz}$ if and only if $c^2 = \beta/(2K)$.

In Exercises 12 to 14 it is required to find the solution of the heat equation in the slab $0 < z < L$, where one face is maintained at temperature zero. Thus we have the boundary-value problem

$$u_t = Ku_{zz} \qquad\qquad 0 < z < L, \ -\infty < t < \infty$$

$$u(0; t) = A_0 + A_1 \cos \frac{2\pi t}{T} \qquad -\infty < t < \infty$$

$$u(L; t) = 0 \qquad\qquad -\infty < t < \infty$$

12. Find all complex separated solutions satisfying the heat equation and of the form $u(z; t) = e^{\gamma z}e^{i\beta t}$, where β is positive.

13. By taking the real and imaginary parts of the complex-valued solutions found in Exercise 12, show that we have the quasi-separated solutions

$$u(z; t) = e^{cz} \cos (\beta t + cz) \qquad u(z; t) = e^{-cz} \cos (\beta t - cz)$$

$$u(z; t) = e^{cz} \sin (\beta t + cz) \qquad u(z; t) = e^{-cz} \sin (\beta t - cz)$$

where $c = \sqrt{\beta/(2K)}$.

*14. By taking suitable linear combinations of the quasi-separated solutions found in Exercise 13 and a steady-state solution, solve the boundary-value problem in the slab $0 < z < L$.

15. Suppose that the daily temperature variation at the earth's surface is a periodic function $\varphi(t) = A_0 + A_1 \cos (2\pi t/T)$. Find the depth necessary for a change from maximum to minimum daily temperature if $K = 2 \times 10^{-3}$ cm²/s and $T = 24 \times 3600$ s.

16. Find the bounded solution of the heat equation $u_t = Ku_{zz}$ for $z > 0$, $-\infty < t < \infty$, satisfying the boundary condition $u(0; t) = 1$ for $0 < t < \frac{1}{2}T$, $u(0; t) = -1$ for $\frac{1}{2}T < t < T$ and $u(0; t)$ is periodic with period T.

17. Find the bounded solution of the heat equation $u_t = Ku_{zz}$ for $z > 0$, $-\infty < t < \infty$, satisfying the boundary condition $u_z(0; t) = A_1 \cos \beta t$, where β and A_1 are positive constants.

18. Find the bounded solution of the heat equation $u_t = Ku_{zz}$ for $z > 0$, $-\infty < t < \infty$, satisfying the boundary condition $u_z(0; t) - hu(0; t) = A_1 \cos \beta t$, where h, β, A_1 are positive constants.

19. For the solution found in Exercise 14, find the limit of $u(z; t)$ when $L \to \infty$ and compare with Example 2.1.3.

20. Find the steady-state solution of the heat equation $u_t = K\nabla^2 u + r$ in the slab $0 < z < L$ and satisfying the boundary condition $u_z(0; t) = h[u(0; t) - T_1]$, $u_z(L; t) = -h[u(L; t) - T_2]$, where r, h, T_1, T_2 are positive constants.

21. For what values of the constants K, r, Φ_1, Φ_2 does there exist a steady-state solution of the equation $u_t = K\nabla^2 u + r$ satisfying the boundary conditions $u_z(x, y, 0; t) = \Phi_1$, $u_z(x, y, L; t) = \Phi_2$?

2.2 HOMOGENEOUS BOUNDARY CONDITIONS ON A SLAB

Many problems in mathematical physics and engineering involve a partial differential equation with initial conditions and boundary conditions. In this section we consider the case of homogeneous boundary conditions for the heat equation in the slab $0 < z < L$. In the next section we consider the general nonhomogeneous boundary condition.

A homogeneous boundary condition at $z = 0$ has one of the following forms:

$$u(0; t) = 0 \quad \text{or} \quad u_z(0; t) = 0 \quad \text{or} \quad u_z(0; t) = hu(0; t)$$

where h is a nonzero constant. All three of these may be included in the following succinct form:

$$\cos \alpha \, u_z(0; t) - \sin \alpha \, u(0; t) = 0 \qquad (2.2.1)$$

where $0 \le \alpha \le \pi$. When $\alpha = \pi/2$, we have the first boundary condition $u(0; t) = 0$; when $\alpha = 0$, we have the second boundary condition $u_z(0; t) = 0$; and when $\tan \alpha = h$, we have the third boundary condition $u_z(0; t) = hu(0; t)$. Similarly, the general homogeneous boundary condition at $z = L$ can be written in the form

$$\cos \beta \, u_z(L; t) + \sin \beta \, u(L; t) = 0 \qquad (2.2.2)$$

The constant β is not related to α, in general.

Separated Solutions with Boundary Conditions

We now discuss separated solutions of the heat equation $u_t = Ku_{zz}$ with homogeneous boundary conditions (2.2.1) and (2.2.2). A separated solution of the heat equation is of the form

$$u(z; t) = f(z)g(t)$$

Substituting in the heat equation $u_t = Ku_{zz}$, we obtain

$$f(z)g'(t) = Kf''(z)g(t)$$

Dividing by $Kf(z)g(t)$, we obtain $g'(t)/Kg(t) = f''(z)/f(z)$. The left side depends on t alone, and the right side depends on z alone; therefore each is a constant, which we call $-\lambda$. Thus we have the ordinary differential equations

$$g'(t) + \lambda Kg(t) = 0$$

$$f''(z) + \lambda f(z) = 0$$

The first equation has the solution $g(t) = e^{-\lambda K t}$, which is never zero. To the second equation we must add boundary conditions (2.2.1) and (2.2.2). The product $u(z; t) = f(z)g(t)$ satisfies (2.2.1) if and only if $f(z)$ satisfies the boundary condition $\cos \alpha \, f'(0) - \sin \alpha \, f(0) = 0$. Similarly $u(z; t)$ satisfies (2.2.2) if and only if $f(z)$ satisfies the boundary condition $\cos \beta \, f'(L) + \sin \beta \, f(L) = 0$. This leads us to the following proposition.

> **Proposition 2.2.1.** The separated solutions of the heat equation $u_t = Ku_{zz}$ with boundary conditions (2.2.1) and (2.2.2) are of the form $u_n(z; t) = e^{-\lambda_n K t} f_n(z)$, where $f_n(z)$ is an eigenfunction of the Sturm-Liouville equation $f''(z) + \lambda f(z) = 0$ with the boundary conditions $\cos \alpha \, f'(0) - \sin \alpha \, f(0) = 0$, $\cos \beta \, f'(L) + \sin \beta \, f(L) = 0$. These eigenfunctions satisfy the orthogonality relation $\int_0^L f_n(z)f_m(z) \, dz = 0$ for $m \neq n$.

Our first example corresponds to a slab with both faces maintained at temperature zero.

> **Example 2.2.1.** Find all the separated solutions of the heat equation $u_t = Ku_{zz}$ for $0 < z < L$ and satisfying the boundary conditions $u(0; t) = 0$, $u(L; t) = 0$.

> **Solution.** The associated Sturm-Liouville problem is $f''(z) + \lambda f(z) = 0$ with boundary conditions $f(0) = 0$, $f(L) = 0$. In Sec. 0.4 we found that the solutions of this are $f_n(z) = \sin(n\pi z/L)$, $\lambda_n = (n\pi/L)^2$. Thus we have the separated solutions $u_n(z; t) = \sin(n\pi z/L) \, e^{-(n\pi/L)^2 K t}$, $n = 1, 2, \dots$. ●

Our next example corresponds to a slab with one face insulated and the other face maintained at temperature zero.

> **Example 2.2.2.** Find all the separated solutions of the heat equation $u_t = Ku_{zz}$ for $0 < z < L$ and satisfying the boundary conditions $u(0; t) = 0$, $u_z(L; t) = 0$.

> **Solution.** The associated Sturm-Liouville problem is $f''(z) + \lambda f(z) = 0$ with the boundary conditions $f(0) = 0$, $f'(L) = 0$. Since these boundary conditions satisfy the positivity condition of Proposition 0.4.5, we may restrict attention to $\lambda \geq 0$. For $\lambda = 0$, the general solution of the differential equation is $f(z) = Az + B$. The first boundary condition requires $B = 0$, while the second boundary condition requires $A = 0$. Hence $\lambda = 0$ is not an eigenvalue. For $\lambda > 0$, the general solution of the differential equation is $f(z) = A \sin z\sqrt{\lambda} + B \cos z\sqrt{\lambda}$. The first boundary condition requires that $B = 0$, while the second boundary condition requires that $A \cos L\sqrt{\lambda} = 0$. For a nonzero solution we must take $L\sqrt{\lambda} = (n - \frac{1}{2})\pi$, $n = 1$, $2, \dots$. Therefore the solutions are $f_n(z) = \sin(n - \frac{1}{2})\pi z/L$, $\lambda_n = (n - \frac{1}{2})^2\pi^2/L^2$.

The separated solutions of the heat equation are $u_n(z; t) = \sin\left[(n - \frac{1}{2})\pi z/L\right]$ $\exp\left[-(n - \frac{1}{2})^2\pi^2 Kt/L^2\right]$, $n = 1, 2, \ldots$.　　　　　●

Solution of the Initial-Value Problem in a Slab

Having obtained the separated solutions of the heat equation with homogeneous boundary conditions, we can solve the following *initial-value problem:*

$$u_t = Ku_{zz} \qquad t > 0, 0 < z < L$$

$$\cos\alpha \, u_z(0; t) - \sin\alpha \, u(0; t) = 0 \qquad t > 0$$

$$\cos\beta \, u_z(0; t) + \sin\beta \, u(0; t) = 0 \qquad t > 0$$

$$u(z; 0) = v(z) \qquad 0 < z < L$$

where $v(z)$ is a piecewise smooth function.

To solve this initial-value problem, first we expand $v(z)$ in a series of eigenfunctions of the Sturm-Liouville problem in the form

$$v(z) = \sum_{n=1}^{\infty} A_n f_n(z) \qquad 0 < z < L$$

[If v is discontinuous at z, the series converges to $\frac{1}{2}v(z + 0) + \frac{1}{2}v(z - 0)$.] The formal solution of the initial-value problem is given by the series

$$u(z; t) = \sum_{n=1}^{\infty} A_n f_n(z)e^{-\lambda_n Kt} \qquad (2.2.3)$$

The solution has been written as a superposition of separated solutions of the heat equation satisfying the indicated homogeneous boundary conditions. The Fourier coefficients are obtained from the orthogonality relations by the formulas

$$\int_0^L v(z)f_n(z) \, dz = A_n \int_0^L f_n(z)^2 \, dz \qquad n = 1, 2, \ldots$$

To prove that the formal solution (2.2.3) is a rigorous solution of the heat equation, we must check that, for each $t > 0$, the series for u, u_z, u_{zz}, and u_t is uniformly convergent for $0 \le z \le L$. This can be shown for each type of boundary condition which we consider.

Example 2.2.3. Solve the initial-value problem $u_t = Ku_{zz}$ for $t > 0$, $0 < z < L$, with the boundary conditions $u(0; t) = 0$, $u(L; t) = 0$ and the initial condition $u(z; 0) = 1$.

Solution. The separated solutions of the heat equation satisfying the boundary conditions are $\sin(n\pi z/L) \, e^{-(n\pi/L)^2 Kt}$, $n = 1, 2, \ldots$. To satisfy the initial condition, we must expand the function $v(z) = 1$ in a Fourier sine series. The Fourier coefficients are given by

$$A_n \int_0^L \sin^2 \frac{n\pi z}{L} \, dz = \int_0^L \sin \frac{n\pi z}{L} \, dz = \frac{L}{n\pi} [1 - (-1)^n]$$

Thus $A_n = [2/(n\pi)][1 - (-1)^n]$, and the solution is

$$u(z; t) = \frac{2}{\pi} \sum_1^\infty \frac{1 - (-1)^n}{n} \sin \frac{n\pi z}{L} e^{-(n\pi/L)^2 Kt}$$

For $t > 0$ and $0 \le z \le L$, this series converges uniformly, owing to the exponential factor. Likewise the series for u_z, u_{zz}, and u_t converge uniformly for $0 \le z \le L$ and each $t > 0$. Thus u is a rigorous solution of the heat equation. ⬤

Asymptotic Behavior and Relaxation Time

In Example 2.2.3 we have obtained a *transient solution* of the heat equation, meaning that $u(z; t)$ tends to zero when t tends to infinity. To analyze this more generally, we assume that the boundary conditions are $u(0; t) = 0$, $u(L; t) = 0$ and that the initial condition is $u(z; 0) = v(z)$, a piecewise smooth function. The solution is

$$u(z; t) = \sum_{n=1}^\infty A_n \sin \frac{n\pi z}{L} e^{-(n\pi/L)^2 Kt}$$

where the A_n are the Fourier sine coefficients of the piecewise smooth function $v(z)$, $0 < z < L$. Thus

$$A_n = \frac{2}{L} \int_0^L v(z) \sin \frac{n\pi z}{L} \, dz \qquad \text{and} \qquad |A_n| \le 2M$$

where M is the maximum of $|v(z)|$, $0 < z < L$. Writing $a = \pi^2 K/L^2$ and noting that $|\sin (n\pi z/L)| \le 1$, we have

$$|u(z; t)| \le 2M \sum_{n=1}^\infty e^{-n^2 at}$$

But $n^2 \ge n$ for $n \ge 1$, and thus $e^{-n^2 at} \le e^{-nat} = (e^{-at})^n$. Thus

$$|u(z; t)| \le 2M \sum_{n=1}^\infty (e^{-at})^n$$

$$= 2M \frac{e^{-at}}{1 - e^{-at}}$$

where we have used the formula for the sum of a geometric series $\sum_{n=1}^\infty \gamma^n = \gamma/(1 - \gamma)$, $0 \le \gamma < 1$. When $t \to \infty$, $e^{-at} \to 0$, and we have shown that

$$u(z; t) = O(e^{-at}) \qquad t \to \infty$$

In particular, $u(z; t) \to 0$ when $t \to \infty$; this means that $u(z; t)$ is a transient solution.

We define the *relaxation time* τ by the formula

$$\frac{1}{\tau} = -\lim_{t \to \infty} \frac{1}{t} \ln |u(z; t)|$$

provided that the limit exists and is independent of z, $0 < z < L$. For transient solutions of the heat equation, the relaxation time can be computed explicitly from the first nonzero term of the series solution.

Proposition 2.2.2. For the Sturm-Liouville equation $f''(z) + \lambda f(z) = 0$ with boundary conditions (2.2.1) and (2.2.2), suppose that all eigenvalues λ_n are positive. Then $\sum_{n=1}^{\infty} A_n f_n(z) e^{-\lambda_n K t}$ is a transient solution of the heat equation, and the relaxation time is given by $\tau = 1/(\lambda_1 K)$ if $A_1 \neq 0$.

Example 2.2.4. Compute the relaxation time for the solution $u(z; t) = \sum_1^{\infty} A_n \sin(n\pi z/L) e^{-(n\pi/L)^2 K t}$.

Solution. We write

$$u(z; t) = A_1 \sin \frac{\pi z}{L} e^{-(\pi/L)^2 K t} + \sum_2^{\infty} A_n \sin \frac{n\pi z}{L} e^{-(n\pi/L)^2 K t}$$

From the preceding analysis the last series is $O(e^{-4\pi^2 K t/L^2})$ when $t \to \infty$. If $A_1 \neq 0$, we may write

$$u(z; t) = A_1 \sin \frac{\pi z}{L} e^{-(\pi/L)^2 K t}[1 + O(e^{-3\pi^2 K t/L^2})]$$

By taking natural logarithms, $\ln |u(z; t)| = \ln |A_1| + \ln \sin(\pi z/L) - (\pi/L)^2 K t + O(e^{-3\pi^2 K t/L^2})$. Thus $\lim_{t \to \infty} (1/t) \ln |u(z; t)| = -\pi^2 K/L^2$. We have proved that $\tau = L^2/(\pi^2 K)$ provided that $A_1 \neq 0$. ●

This analysis of relaxation time shows that, for large t, the solution $u(z; t)$ is well approximated by the first term of the series solution. This also can be seen graphically, by plotting the function $z \to u(z; t)$ for various values of t. When t is small, the solution is close to the initial function $v(z)$. As t increases, the solution tends to zero and assumes the shape of a sine curve, corresponding to the first term of the series solution. The graphs in Fig. 2.2.1 plot the solution of the initial-value problem $u_t = 2u_{zz}$ for $0 < z < \pi$, with boundary conditions $u(z; 0) = 0$, $u(\pi; 0) = 0$ and initial conditions $u(z; 0) = 2z$ for $t = 0, 0.005, 0.01, 0.05, 0.1, 0.2, 0.3, 0.5, 0.7$, and 0.8.

Uniqueness of Solutions

We now discuss the *uniqueness* of the solution of the initial-value problem. We have found a solution as a series of separated solutions, but it is conceivable that by another method we might produce a distinct solution of the heat equation with the same initial conditions and boundary conditions. We will prove that this is impossible. To be specific, we take the zero boundary conditions $u(0; t) = 0$, $u(L; t) = 0$.

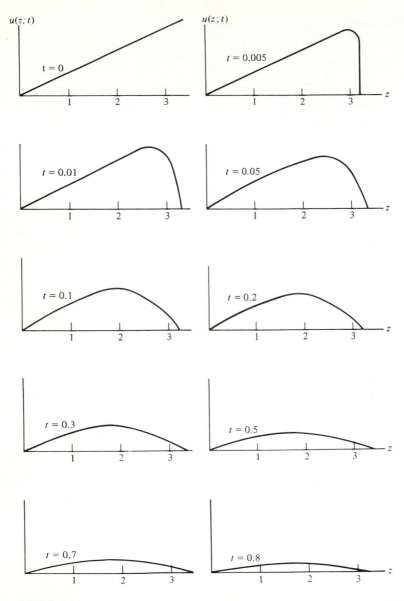

FIGURE 2.2.1

For this purpose, suppose that u_1, u_2 are two solutions with the same initial and boundary conditions, and set $u = u_1 - u_2$. Then u satisfies the heat equation with zero boundary conditions and zero initial conditions. Let

$$w(t) = \tfrac{1}{2} \int_0^L u(z; t)^2 \, dz$$

Then
$$w'(t) = \int_0^L u(z; t)u_t(z; t)\, dz$$

$$= K \int_0^L u(z; t)u_{zz}(z; t)\, dz$$

$$= Ku(z; t)u_z(z; t)\Big|_0^L - K \int_0^L u_z(z; t)^2\, dz$$

where we have used the heat equation to obtain the second equation and integration by parts to obtain the third equation. Using the boundary conditions, we see that the first term in the final equation is zero. Therefore

$$w'(t) = -K \int_0^L u_z(z; t)^2\, dz$$

But K is a positive constant, and $u_z(z; t)^2 \geq 0$, since squares of real numbers are greater than or equal to zero. Thus we have both

$$w'(t) \leq 0 \qquad \text{and} \qquad w(t) \geq 0$$

But $u(z; 0) = 0$, which means that

$$w(0) = 0$$

To complete the proof, we use the fundamental theorem of calculus.

$$w(t) = w(0) + \int_0^t w'(s)\, ds \leq 0$$

Since $w(t) \geq 0$, we are forced to conclude that $w(t) \equiv 0$, which means that $u(z; t) = 0$ for each t, that is, $u_1(z; t) = u_2(z; t)$. Hence we have proved uniqueness of the solution.

The careful reader will notice that we have used the boundary conditions only to show that $uu_z\,|_0^L = 0$. Hence our proof applies also to other boundary conditions, for example, $u_z(0) = 0$, $u_z(L) = 0$.

Examples of Transcendental Eigenvalues

In certain problems we must solve the heat equation $u_t = Ku_{zz}$ with the boundary conditions

$$u(0; t) = 0 \qquad u_z(L; t) + hu(L; t) = 0$$

where h is a *positive* constant. We will see that the eigenvalues are determined from a transcendental equation.

The separated solutions of the problem are of the form $u(z; t) = f(z)g(t)$, where $g'(t) + K\lambda g(t) = 0$, and f is the solution of the Sturm-Liouville problem $f''(z) + \lambda f(z) = 0$ with the boundary conditions $f(0) = 0$, $f'(L) + hf(L) = 0$. The positivity criterion (Proposition 0.4.4) implies that any eigenvalue λ satisfies $\lambda > 0$. Therefore the eigenfunctions which satisfy the boundary conditions are

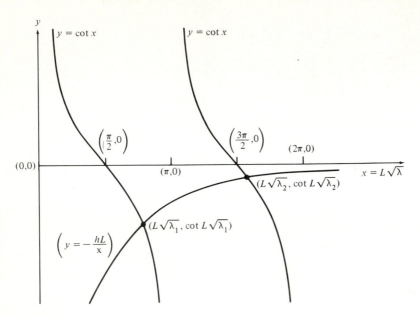

FIGURE 2.2.2
Graphical solution of the transcendental equation $\cot(L\sqrt{\lambda}) = -h/\sqrt{\lambda}$.

of the form $f(z) = B \sin z\sqrt{\lambda}$ with $B \neq 0$. The second boundary condition gives the transcendental equation $\sqrt{\lambda} \cos L\sqrt{\lambda} + h \sin L\sqrt{\lambda} = 0$. Although this cannot be solved in closed form explicitly, a graphical solution can be determined by graphing the cotangent function versus the reciprocal function. Indeed, we must have $\sin L\sqrt{\lambda} \neq 0$ for any solution (otherwise, $\cos L\sqrt{\lambda} = \pm 1$, which does not satisfy the equation), and the transcendental equation may be rewritten in the form $\cot L\sqrt{\lambda} = -hL/(L\sqrt{\lambda})$. From Fig. 2.2.2 it is clear that the smallest eigenvalue λ_1 satisfies $\pi/2 < L\sqrt{\lambda_1} < \pi$ and that the higher eigenvalues satisfy $(n - \frac{1}{2})\pi < L\sqrt{\lambda_n} < n\pi L\sqrt{\lambda_n} \to (n - \frac{1}{2})\pi$ when $n \to \infty$.

To complete the mathematical analysis, we will examine the case when the constant of proportionality is negative, the physical meaning of which is not immediately obvious.

We conclude this section by discussing the heat equation with the homogeneous boundary conditions

$$u(0; t) = 0 \qquad u_z(L, t) = hu(L; t) \tag{2.2.4}$$

where h is a positive constant. Following the previous discussion, we look for separated solutions in the form

$$u(z; t) = f(z)g(t)$$

Substituting in the heat equation, we obtain the ordinary differential equations

$$g'(t) + \lambda K g(t) = 0$$

$$f''(z) + \lambda f(z) = 0$$

with the boundary conditions

$$f(0) = 0 \qquad f'(L) = hf(L)$$

The general solution satisfying the first boundary condition is of the form

$$f(z) = \begin{cases} B \sinh z\sqrt{-\lambda} & \lambda < 0 \\ Bz & \lambda = 0 \\ B \sin z\sqrt{\lambda} & \lambda > 0 \end{cases}$$

For the second boundary condition we consider three cases.

Case 1. $\lambda < 0$, $B\sqrt{-\lambda} \cosh L\sqrt{-\lambda} = hB \sinh L\sqrt{-\lambda}$

Case 2. $\lambda = 0$, $B = hBL$

Case 3. $\lambda > 0$, $B\sqrt{\lambda} \cos L\sqrt{\lambda} = hB \sin L\sqrt{\lambda}$

In each case we desire a nontrivial solution; hence $B \neq 0$. Dividing by B, we can rewrite these three equations.

Case 1. $\lambda < 0$, $\tanh L\sqrt{-\lambda} = \dfrac{L\sqrt{-\lambda}}{Lh}$

Case 2. $\lambda = 0$, $Lh = 1$

Case 3. $\lambda > 0$, $\tan L\sqrt{\lambda} = \dfrac{L\sqrt{\lambda}}{Lh}$

For $\lambda < 0$, we examine the graph of $y = \tanh x$ (Fig. 2.2.3).

If $Lh \leq 1$, we see from the graph that the line $y = x/(Lh)$ does not intersect the curve $y = \tanh x$ for $x > 0$; hence there are no solutions for $\lambda < 0$, $Lh \leq 1$. If $Lh > 1$, the line $y = x/(Lh)$ intersects the curve $y = \tanh x$ in exactly one place $x > 0$; hence there is one solution $\lambda < 0$ if $Lh > 1$.

For $\lambda = 0$, we have a solution if and only if $Lh = 1$.

For $\lambda > 0$, we examine the graph of $y = \tan x$, shown in Fig. 2.2.4. From the graph we see that the line $y = x/(Lh)$ intersects the curve $y = \tan x$ infinitely

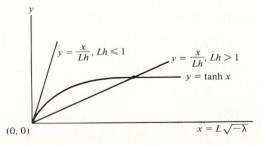

$y = \dfrac{x}{Lh}, Lh \leq 1$

$y = \dfrac{x}{Lh}, Lh > 1$

$y = \tanh x$

$(0, 0)$

$x = L\sqrt{-\lambda}$

FIGURE 2.2.3
Graphical solution of the transcendental equation $\tanh L\sqrt{-\lambda} = \sqrt{-\lambda}/h$.

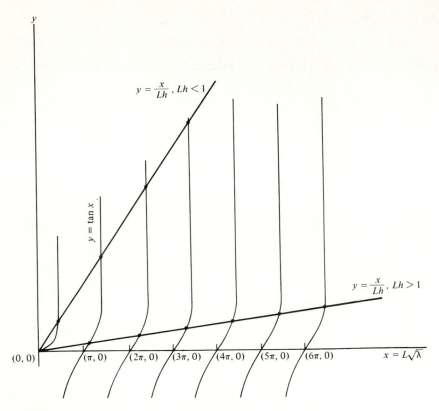

FIGURE 2.2.4
Graphical solution of the transcendental equation $\tan L\sqrt{\lambda} = \sqrt{\lambda}/h$.

many times. If $Lh < 1$, the first such intersection occurs for $0 < L\sqrt{\lambda} < \pi/2$; otherwise, the first intersection occurs for $\pi < L\sqrt{\lambda} < 3\pi/2$.

Summarizing, we have the following:

$Lh < 1$ All separated solutions are of the form

$$u_n(z; t) = B_n \sin z\sqrt{\lambda_n}\, e^{-\lambda_n Kt} \qquad n = 1, 2, \ldots$$

$$\frac{(n-1)^2\pi^2}{L^2} < \lambda_n < \frac{(n-\frac{1}{2})^2\pi^2}{L^2} \qquad \tan L\sqrt{\lambda_n} = \frac{\sqrt{\lambda_n}}{h}$$

$Lh = 1$ There is one separated solution of the form

$$u_1(z; t) = B_1 z \qquad \lambda_1 = 0$$

All other separated solutions are of the form

$$u_n(z; t) = B_n \sin z\sqrt{\lambda_n}\, e^{-\lambda_n Kt} \qquad n = 2, 3, \ldots$$

$$\frac{(n-1)^2\pi^2}{L^2} < \lambda_n < \frac{(n-\frac{1}{2})^2\pi^2}{L^2} \qquad \tan L\sqrt{\lambda_n} = \frac{\sqrt{\lambda_n}}{h}$$

$Lh > 1$ There is one separated solution of the form

$$u_1(z; t) = B_1 \sinh z\sqrt{-\lambda_1}\, e^{-\lambda_1 Kt} \qquad \lambda_1 < 0,\ \tanh L\sqrt{-\lambda_1} = \frac{\sqrt{-\lambda_1}}{h}$$

All other separated solutions are of the form

$$u_n(z; t) = B_n \sin z\sqrt{\lambda_n}\, e^{-\lambda_n Kt} \qquad n = 2, 3, \ldots$$

$$\frac{(n - 1)^2\pi^2}{L^2} < \lambda_n < \frac{(n - \tfrac{1}{2})^2\pi^2}{L^2} \qquad \tan L\sqrt{\lambda_n} = \frac{\sqrt{\lambda_n}}{h}$$

Therefore we have found all the separated solutions of the heat equation with the boundary conditions (2.2.4). We emphasize that the eigenvalues λ_n must be determined graphically or by a numerical method. There is no elementary formula for the solution of the transcendental equation $\tan L\sqrt{\lambda_n} = \sqrt{\lambda_n}/h$.

EXERCISES 2.2

1. Solve the initial-value problem $u_t = Ku_{zz}$ for $t > 0$, $0 < z < L$, with the boundary conditions $u(0; t) = 0$, $u(L; t) = 0$ and the initial condition $u(z; 0) = z$, $0 < z < L$.
2. Solve the initial-value problem $u_t = Ku_{zz}$ for $t > 0$, $0 < z < L$, with the boundary conditions $u(0; t) = 0$, $u(L; t) = 0$ and the initial conditions $u(z; 0) = T$ for $0 < z < \tfrac{1}{2}L$, $u(z; 0) = 0$ for $\tfrac{1}{2}L < z < L$ where T is a positive constant.
3. Solve the initial-value problem $u_t = Ku_{zz}$ for $t > 0$, $0 < z < L$, with the boundary conditions $u(z; 0) = 0$, $u_z(L; t) = 0$ and the initial condition $u(z; 0) = 3 \sin [\pi z/(2L)] + 5 \sin [3\pi z/(2L)]$.
4. Find all the separated solutions of the heat equation $u_t = Ku_{zz}$ satisfying the boundary conditions $u_z(0; t) = 0$, $u_z(L; t) = 0$.
5. Solve the initial-value problem $u_t = Ku_{zz}$ for $t > 0$, $0 < z < L$, with the boundary conditions $u_z(0; t) = 0$, $u_z(L; t) = 0$ and the initial condition $u(z, 0) = z$, $0 < z < L$.
6. Solve the initial-value problem $u_t = Ku_{zz}$ for $t > 0$, $0 < z < L$, with the boundary conditions $u_z(0; t) = 0$, $u_z(L; t) = 0$ and the initial condition $u(z; 0) = 3 + 4\cos (\pi z/L) + 7 \cos (3\pi z/L)$, $0 < z < L$.
7. Consider the heat equation

$$u_t = Ku_{zz} \qquad t > 0, 0 < z < L$$

$$u(0; t) = 0 \qquad u_z(L; t) = -hu(L, t)$$

where $h > 0$. Show that *all* separated solutions are of the form

$$u_n(z; t) = B_n \sin z\sqrt{\lambda_n}\, e^{-\lambda_n Kt}$$

where $\lambda_n > 0$ are the solutions of

$$\tan L\sqrt{\lambda} = -\frac{\sqrt{\lambda}}{h}$$

8. By direct computation, show that

$$\int_0^L \sin z\sqrt{\lambda_n} \, \sin z\sqrt{\lambda_m} \, dz = 0 \qquad n \neq m$$

where $\{\lambda_n\}$ are the solutions obtained in Exercise 7.

9. Solve the initial-value problem $u_t = Ku_{zz}$ for $t > 0$, $0 < z < L$, with the boundary conditions $u(0; t) = 0$, $u_z(L; t) = -hu(L; t)$ and the initial condition $u(z; 0) = 1$, where h is a positive constant.

10. Find all the separated solutions of the heat equation $u_t = Ku_{zz}$ satisfying the boundary conditions $u_z(0; t) = hu(0; t)$, $u_z(L; t) + hu(L; t) = 0$, where h is a positive constant.

11. Solve the heat equation $u_t = Ku_{zz}$ for $t > 0$, $0 < z < L$, with the boundary conditions $u_z(0; t) = hu(0; t)$, $u_z(L; t) + hu(L; t) = 0$ and the initial condition $u(z; 0) = 1$, where h is a positive constant.

12. Find the relaxation time for the solution found in Exercise 1.

13. Find the relaxation time for the solution found in Exercise 2.

14. Find the relaxation time for the solution found in Exercise 3.

15. Find the relaxation time for the solution found in Exercise 9. (It may be expressed in terms of the solution of a certain transcendental equation.)

16. In Exercise 2, suppose that $K = 0.15$ cm²/s, $L = 40$ cm, $T = 100°C$. Compute the relaxation time and $u(20; t)$ for $t = 0.1, 1.0, 10.0, 100$ min.

17. In Exercise 3 suppose that $K = 0.15$ cm²/s, $L = 40$ cm, $T = 100°C$. Compute the relaxation time and $u(20; t)$ for $t = 0.1, 1.0, 10.0, 100$ min.

Exercises 18 to 20 treat heat flow in a circular ring of circumference L.

18. Find all the separated solutions of the heat equation $u_t = Ku_{zz}$ satisfying the periodic boundary conditions $u(0; t) = u(L; t)$, $u_z(0; t) = u_z(L; t)$.

19. Solve the heat equation $u_t = Ku_{zz}$ with the periodic boundary conditions $u(0; t) = u(L; t)$, $u_z(0; t) = u_z(L; t)$ and the initial conditions $u(z; 0) = 100$ if $0 < z < \frac{1}{2}L$ and $u(z; 0) = 0$ if $\frac{1}{2}L < z < L$.

20. For the solution of Exercise 19, find the relaxation time. Compare your result with the relaxation time for heat flow in a slab of width L, with zero boundary conditions, found in Exercise 13.

2.3   INHOMOGENEOUS BOUNDARY CONDITIONS ON A SLAB

In this section we give the complete treatment of initial-value problems for the heat equation in the slab $0 < z < L$, with general boundary conditions. Our analysis is based on a *five-stage method*, which also applies to initial-value problems for the heat equation in other coordinate systems, studied in Chaps. 3 and 4.

Statement of Problem

An initial-value problem for the heat equation in the slab $0 < z < L$ is the following set of four conditions:

$$u_t = Ku_{zz} + r \tag{2.3.1}$$

$$\cos \alpha \, u_z(0; t) - \sin \alpha \, u(0; t) = T_1 \tag{2.3.2}$$

$$\cos \beta \, u_z(L; t) + \sin \beta \, u(L; t) = T_2 \tag{2.3.3}$$

$$u(z; 0) = v(z) \tag{2.3.4}$$

where $v(z)$ is a piecewise smooth function and $\alpha, \beta, r, T_1, T_2, K$ are constants. We seek the solution $u(z; t)$ for all $t > 0$, $0 < z < L$.

The heat equation (2.3.1) is inhomogeneous, owing to the internal source term r. The boundary conditions (2.3.2) and (2.3.3) are inhomogeneous, owing to the constants T_1, T_2.

Five-Stage Method of Solution

STAGE 1: Steady-state solution. We ignore the initial conditions and look for a function $U(z)$ which satisfies the heat equation (2.3.1) and the boundary conditions (2.3.2) and (2.3.3). Thus we must have $KU''(z) + r = 0$, whose general solution is $U(z) = -rz^2/(2K) + A + Bz$. The boundary conditions determine the constants A and B, as we have shown by examples in Sec. 2.1, except in the case where $\lambda = 0$ is an eigenvalue of the associated homogeneous boundary conditions. In this case we take the most general steady-state solution; the ambiguity will be resolved in stage 3.

STAGE 2: Transformation of the problem. Having obtained the steady-state solution, we use the subtraction principle to transform to a heat equation with no internal sources and homogeneous boundary conditions. To do this, we define a new unknown function

$$v(z; t) = u(z; t) - U(z)$$

We have $v_t(z; t) = u_t(z; t)$, $v_z(z; t) = u_z(z; t) - U'(z)$, $v_{zz}(z; t) = u_{zz}(z; t) - U''(z)$. Thus $v_t - Kv_{zz} = u_t - K[u_{zz} - U''(z)] = Ku_{zz} + r - Ku_{zz} - r = 0$. Likewise, v satisfies boundary conditions (2.3.2) and (2.3.3) with $T_1 = 0$, $T_2 = 0$. Thus we have

$$v_t = Kv_{zz} \tag{2.3.1a}$$

$$\cos \alpha \, v_z(0; t) - \sin \alpha \, v(0; t) = 0 \tag{2.3.2b}$$

$$\cos \beta \, v_z(L; t) + \sin \beta \, v(L; t) = 0 \tag{2.3.3c}$$

$$v(z; 0) = v(z) - U(z) \tag{2.3.4d}$$

Thus $v(z; t)$ satisfies a homogeneous equation with homogeneous boundary conditions and a new initial condition. This problem was treated in Sec. 2.2.

STAGE 3: Separation of variables. To determine the new unknown function $v(z; t)$, we use a superposition of separated solutions of the heat equation with homogeneous boundary conditions (2.3.2b) and (2.3.3c)

$$v(z; t) = \sum_{n=1}^{\infty} A_n f_n(z) e^{-\lambda_n Kt}$$

Coefficients A_n are determined by expanding the initial condition $v(z) - U(z)$ in a series of eigenfunctions $\sum_{n=1}^{\infty} A_n f_n(z)$. Equivalently, they may be computed from the integrals

$$\int_0^L [v(z) - U(z)] f_n(z) \, dz = A_n \int_0^L f_n(z)^2 \, dz \qquad n = 1, 2, \ldots$$

The formal solution of the initial-value problem is

$$\boxed{u(z; t) = U(z) + \sum_{n=1}^{\infty} A_n f_n(z) e^{-\lambda_n K t}} \tag{2.3.5}$$

If $\lambda = 0$ is an eigenvalue of the Sturm-Liouville problem for $f''(z) + \lambda f(z) = 0$ with the associated homogeneous boundary conditions, the steady-state solution $U(z)$ is determined uniquely by requiring that $U(z) - v(z)$ be orthogonal to the eigenfunction $f_1(z)$, for which $\lambda_1 = 0$. We take $A_1 = 0$ in this case.

STAGE 4: Verification of the solution. At this point it is appropriate to verify that the formal solution (2.3.5) indeed satisfies the initial-value problem. To illustrate the proof, we assume that $\alpha = \pi/2$, $\beta = \pi/2$ where the eigenfunctions are $f_n(z) = \sin(n\pi z/L)$. Then the series for u, u_z, u_{zz}, u_t is

$$u: \qquad U(z) + \sum_{n=1}^{\infty} A_n \sin \frac{n\pi z}{L} e^{-\lambda_n K t}$$

$$u_z: \qquad U'(z) + \sum_{n=1}^{\infty} A_n \frac{n\pi}{L} \cos \frac{n\pi z}{L} e^{-\lambda_n K t}$$

$$u_{zz}: \qquad U''(z) - \sum_{n=1}^{\infty} A_n \left(\frac{n\pi}{L}\right)^2 \sin \frac{n\pi z}{L} e^{-\lambda_n K t}$$

$$u_t: \qquad - \sum_{n=1}^{\infty} A_n K \left(\frac{n\pi}{L}\right)^2 \sin \frac{n\pi z}{L} e^{-\lambda_n K t}$$

We have $\lambda_n = (n\pi/L)^2$, $|A_n| \leq 2M$, where M is the maximum of $|v(z) - U(z)|$. Therefore, for each $t > 0$, each of these series is uniformly convergent for $0 \leq z \leq L$. Hence the colon signs become equality, and u satisfies the heat equation $u_t = Ku_{zz} + r$, together with the boundary conditions.

The initial conditions are satisfied in the form

$$\lim_{t \downarrow 0} u(z; t) = \tfrac{1}{2} f(z + 0) + \tfrac{1}{2} f(z - 0)$$

To see this, we apply Abel's theorem from Sec. 0.5 to the series (2.3.5) with $x = e^{-\pi^2 K t/L^2}$ leading us to conclude that $\lim_{t \downarrow 0} \sum_{n=1}^{\infty} A_n \sin(n\pi z/L) e^{-\lambda_n K t} = \sum_{n=1}^{\infty} A_n \sin(n\pi z/L)$. This convergent Fourier series has the sum $\tfrac{1}{2} v(z + 0) + \tfrac{1}{2} v(z - 0)$. But $v(z) = f(z) - U(z)$, and the polynomial $U(z)$ is continuous. Thus $U(z) + \sum_{n=1}^{\infty} A_n \sin(n\pi z/L) e^{-\lambda_n K t}$ tends to $\tfrac{1}{2} f(z + 0) + \tfrac{1}{2} f(z - 0)$ when $t \downarrow 0$.

If u_1, u_2 are two solutions of the problem, their difference $u = u_1 - u_2$

must satisfy the heat equation with zero initial and boundary conditions. But we have already shown in Sec. 2.2 that such a function is zero. Hence we have proved the uniqueness of our solution.

STAGE 5: Asymptotic behavior. We now discuss the convergence of $u(z; t)$ to the steady-state solution $U(z)$ when $t \to \infty$. Although this may be obvious on physical grounds, the mathematical analysis has not been given.

To do this, we assume that all eigenvalues are positive, $\lambda_n > 0$ for $n = 1, 2, \ldots$. The basic fact about convergence is

$$\boxed{u(z; t) - U(z) = O(e^{-\lambda_1 K t}) \qquad t \to \infty}$$

We illustrate the proof in the case of zero boundary conditions $u(0; t) = 0$, $u(L; t) = 0$. In this case $\lambda_n = (n\pi/L)^2$, $|A_n| \leq 2M$, and we have

$$|u(z; t) - U(z)| = \left| \sum_{n=1}^{\infty} A_n \sin \frac{n\pi z}{L} e^{-(n\pi/L)^2 K t} \right|$$

$$\leq \sum_{n=1}^{\infty} 2M (e^{-at})^{n^2} \qquad a = \frac{\pi^2 K}{L^2}$$

$$\leq 2M \sum_{n=1}^{\infty} (e^{-at})^n$$

$$= \frac{2M e^{-at}}{1 - e^{-at}}$$

The denominator tends to 1 when $t \to \infty$, and we have proved that $u(z; t) - U(z) = O(e^{-at})$, $t \to \infty$, which was to be shown.

Finally we discuss the *relaxation time*. This is the number τ defined by the equation

$$\frac{1}{\tau} = -\lim_{t \to \infty} \frac{1}{t} \ln |u(z; t) - U(z)|$$

provided that the limit exists and is independent of z, $0 < z < L$.

The basic fact about relaxation time is that

$$\boxed{\tau = \frac{1}{\lambda_1 K}}$$

provided that $A_1 \neq 0$.† We illustrate the computation in the case of zero bound-

---

† If $A_1 = 0$ and $A_2 \neq 0$, then $\tau = 1/(\lambda_2 K)$.

ary conditions $u(0; t) = 0$, $u(L; t) = 0$. In this case

$$u(z; t) - U(z) = A_1 \sin \frac{\pi z}{L} e^{-\pi^2 K t/L^2} + \sum_{n=2}^{\infty} A_n \sin \frac{n \pi z}{L} e^{-(n\pi/L)^2 K t}$$

$$= A_1 \sin \frac{\pi z}{L} e^{-\pi^2 K t/L^2} + O(e^{-4\pi^2 K t/L^2})$$

$$= A_1 \sin \frac{\pi z}{L} e^{-\pi^2 K t/L^2}[1 + O(e^{-3\pi^2 K t/L^2})]$$

$$\ln |u(z; t) - U(z)| = \ln \left| A_1 \sin \frac{\pi z}{L} \right| - \frac{\pi^2 K t}{L^2} + \ln [1 + O(e^{-3\pi^2 K t/L^2})]$$

The final term tends to zero when $t \to \infty$, and the first term is independent of t. Dividing by t, we see that

$$\lim_{t \to \infty} \frac{1}{t} \ln |u(z; t) - U(z)| = - \frac{\pi^2 K}{L^2}$$

Therefore the relaxation time is given by $\tau = L^2/(\pi^2 K)$, provided $A_1 \neq 0$. This makes good physical sense because the larger the conductivity K, the smaller the relaxation time. Likewise, the larger the width of the slab L, the larger the relaxation time. [If $A_1 = 0$ and $A_2 \neq 0$, then $\tau = L^2/(4\pi^2 K)$.]

This concludes the five stages of our analysis of the initial-value problem (2.3.1) to (2.3.4). We use this method repeatedly for solving problems with inhomogeneous boundary conditions.

Examples

Example 2.3.1. Solve the initial-value problem for the heat equation $u_t = K u_{zz}$ with the boundary conditions $u(0; t) = T_1$, $u_z(L; t) = 0$ and the initial condition $u(z; 0) = T_3$ where T_1, T_3 are positive constants.

Solution. We use the five-stage method outlined above.

　　　　Stage 1: Steady-state solution. In this case the steady-state solution satisfies $U_{zz} = 0$, $U(0) = T_1$, $U_z(L) = 0$. The general solution is $U(z) = Az + B$, and the boundary conditions give $B = T_1$, $A = 0$, so that the steady-state solution is

$$U(z) = T_1$$

　　　　Stage 2: Transformation of the problem. Again we set $v(z; t) = u(z; t) - U(z)$. Then $v(z; t)$ satisfies

1'. $v_t = K v_{zz}$
2'. $v(0; t) = 0$, $t > 0$
3'. $v_z(L; t) = 0$, $t > 0$
4'. $v(z; 0) = T_3 - T_1$

The boundary conditions 2' and 3' are *homogeneous;* this means that the super-position principle can be applied.

    Stage 3: Separation of variables. We look for separated solutions

$$v(z; t) = f(z)g(t)$$

that satisfy the heat equation and the homogeneous boundary conditions $v(0; t) = 0$; $v_z(L; t) = 0$. The heat equation requires that

$$\frac{f''(z)}{f(z)} = \frac{g'(t)}{Kg(t)}$$

and hence both sides equal the constant $-\lambda$. Thus

$$g'(t) + K\lambda g(t) = 0$$

$$f''(z) + \lambda f(z) = 0$$

    The first equation requires that

$$g(t) = Ce^{-\lambda Kt}$$

for some constant C. The second equation must be solved by taking into account the boundary conditions. Separating the three cases $\lambda > 0$, $\lambda = 0$, and $\lambda < 0$, we have

1. $f(z) = A \cos z\sqrt{\lambda} + B \sin z\sqrt{\lambda}$ $\lambda > 0$
2. $f(z) = A + Bz$ $\lambda = 0$
3. $f(z) = A \cosh z\sqrt{-\lambda} + B \sinh z\sqrt{-\lambda}$ $\lambda < 0$

    We now apply the boundary conditions. In the case where $\lambda > 0$, this means that $A = 0$, $B\sqrt{\lambda} \cos L\sqrt{\lambda} = 0$, so that for a nonzero solution we must have

$$L\sqrt{\lambda} = (n - \tfrac{1}{2})\pi \qquad n = 1, 2, \ldots$$

In the case where $\lambda = 0$, we must have $A = 0$, $B = 0$, which is a zero solution. Finally, in the case where $\lambda < 0$, the boundary conditions require $A = 0$, $B\sqrt{-\lambda} \cosh \sqrt{-\lambda} = 0$, again a zero solution.

    The separated solutions of the heat equation which satisfy the homogeneous boundary conditions are therefore of the form

$$\sin\left[\left(n - \frac{1}{2}\right)\frac{\pi z}{L}\right] \exp\left\{-\left[\left(n - \frac{1}{2}\right)\frac{\pi}{L}\right]^2 Kt\right\} \qquad n = 1, 2, \ldots$$

    As before, the superposition principle applies and makes any function of the form

$$v(z; t) = \sum_{n=1}^{\infty} A_n \sin\left[\left(n - \frac{1}{2}\right)\frac{\pi z}{L}\right] \exp\left\{-\left[\left(n - \frac{1}{2}\right)\frac{\pi}{L}\right]^2 Kt\right\}$$

a solution of the heat equation with homogeneous boundary conditions. To satisfy the new initial conditions, we set $t = 0$ and obtain

$$T_3 - T_1 = \sum_{n=1}^{\infty} A_n \sin\left[\left(n - \frac{1}{2}\right)\frac{\pi z}{L}\right]$$

This is not a Fourier sine series on $(0, L)$. It can be thought of as a Fourier sine series on $(0, 2L)$ for which the even-numbered sine terms are absent. To determine the coefficients A_n, we use the orthogonality relations, with the result

$$A_n = \frac{2}{L} \int_0^L (T_3 - T_1) \sin\left[\left(n - \frac{1}{2}\right) \frac{\pi z}{L}\right] dz$$

$$= \frac{2}{\pi} \frac{T_3 - T_1}{n - \frac{1}{2}} \qquad n = 1, 2, \ldots$$

Therefore the solution to the original problem is

$$u(z; t) = T_1 + \frac{2}{\pi} (T_3 - T_1) \sum_{n=1}^{\infty} \frac{\sin\left[(n - \frac{1}{2})(\pi z/L)\right]}{n - \frac{1}{2}} \exp\left\{-\left[\left(n - \frac{1}{2}\right) \frac{\pi}{L}\right]^2 Kt\right\}$$

Stage 4: Verification of the solution. As before, it can be verified that $u(z; t)$ satisfies the heat equation, initial conditions, and boundary conditions. We leave the verification as an exercise.

Stage 5: Asymptotic behavior. To determine the asymptotic behavior as $t \to \infty$, we note that $|A_n| \le 4/[\pi(T_3 - T_1)]$, $(n - \frac{1}{2})^2 \ge \frac{1}{2}(n - \frac{1}{2})$, and therefore

$$|u(z; t) - U(z)| \le \frac{4}{\pi} (T_3 - T_1) \sum_{n=1}^{\infty} \exp\left[-\frac{1}{2}\left(n - \frac{1}{2}\right) \frac{\pi^2}{L^2} Kt\right]$$

$$= \frac{(4/\pi)(T_3 - T_1) \exp\left[-\pi^2 Kt/(4L^2)\right]}{1 - \exp\left[-\pi^2 Kt/(2L^2)\right]}$$

Hence we see, as before, that $u(z; t) \to U(z)$ when $t \to \infty$. The relaxation time is given by $\tau = 4L^2/(\pi^2 K)$, provided that $T_1 \ne T_3$. ●

Our next example corresponds to a slab where one face exchanges heat by convection and the other face is maintained at a fixed temperature.

Example 2.3.2. Solve the initial-value problem for the heat equation $u_t = Ku_{zz}$ with the boundary conditions $u(0; t) = T_1$, $u_z(L; t) = -h[u(L; t) - T_2]$ and the initial condition $u(z; 0) = T_3$ where h, T_1, T_2, T_3 are positive constants.

Solution. We use the five-stage method.

Stage 1. We look for the steady-state solution $U(z)$ which satisfies $KU'' = 0$ and the boundary conditions $U(0) = T_1$, $U'(L) = -h[U(L) - T_2]$. The general solution of the equation is $U(z) = A + Bz$. The first boundary condition requires $A = T_1$, while the second requires that $B = -h(A + BL - T_2)$. Solving these, we have the steady-state solution $U(z) = T_1 + hz(T_2 - T_1)/(1 + hL)$.

Stage 2. We use the steady-state solution to transform to a homogeneous equation with homogeneous boundary conditions. Thus setting $v(z; t) = u(z; t) - U(z)$, we have $v_t = Kv_{zz}$ with the boundary conditions $v(0; t) = 0$, $v_z(L; t) + hv(L; t) = 0$ and the initial condition $v(z; 0) = T_3 - U(z)$.

Stage 3. The separated solutions of the heat equation $v_t = Kv_{zz}$ with the homogeneous boundary conditions are of the form $v(z; t) = e^{-\lambda Kt} f(z)$, where λ

is an eigenvalue and $f(z)$ is an eigenfunction of the Sturm-Liouville problem $f''(z) + \lambda f(z) = 0$ with the boundary conditions $f(0) = 0$, $f'(L) + hf(L) = 0$. These boundary conditions satisfy the positivity condition of Sec. 0.4 for Sturm-Liouville systems, and thus we know that $\lambda > 0$. The general solution of the differential equation is $f(z) = A \sin z\sqrt{\lambda} + B \cos z\sqrt{\lambda}$. The boundary condition at $z = 0$ requires $B = 0$, while the boundary condition at $z = L$ requires $A\sqrt{\lambda} \cos L\sqrt{\lambda} + Ah \sin L\sqrt{\lambda} = 0$. We obtain a nonzero solution by taking $A \neq 0$; thus λ must be a solution of the transcendental equation $\sqrt{\lambda} \cos L\sqrt{\lambda} + h \sin L\sqrt{\lambda} = 0$. These eigenvalues may be obtained graphically by examining the graph of the cotangent function (Fig. 2.3.1). Therefore the separated solutions of the heat equation with homogeneous boundary conditions are $\sin z\sqrt{\lambda_n}\, e^{-\lambda_n Kt}$, where the λ_n are determined as before.

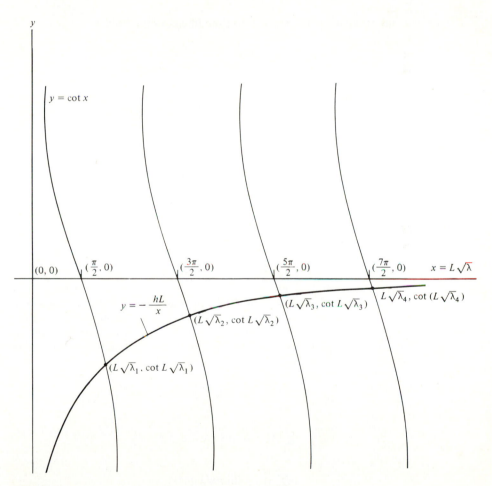

FIGURE 2.3.1
Graphical solution of the transcendental equation $\cot L\sqrt{\lambda} = -h/\sqrt{\lambda}$.

The initial-value problem for $v(z; t)$ is solved by a superposition of separated solutions

$$v(z; t) = \sum_{n=1}^{\infty} A_n \sin z\sqrt{\lambda_n}\, e^{-\lambda_n K t}$$

The Fourier coefficients are obtained from the expansion of

$$T_3 - U(z) = \sum_{n=1}^{\infty} A_n \sin z\sqrt{\lambda_n} \qquad 0 < z < L$$

Using orthogonality, we have the integral formulas

$$\int_0^L [T_3 - U(z)] \sin z\sqrt{\lambda_n}\, dz = A_n \int_0^L \sin (z\sqrt{\lambda_n})^2\, dz \qquad n = 1, 2, \ldots$$

To compute these integrals, we may use the integration formulas

$$\int_0^L \sin z\sqrt{\lambda}\, dz = \frac{1 - \cos L\sqrt{\lambda}}{\sqrt{\lambda}}$$

$$\int_0^L z \sin z\sqrt{\lambda}\, dz = \frac{-L \cos L\sqrt{\lambda}}{\sqrt{\lambda}} + \frac{\sin L\sqrt{\lambda}}{\lambda}$$

$$\int_0^L \sin^2 z\sqrt{\lambda}\, dz = \frac{1}{2}\left(L - \frac{\sin 2L\sqrt{\lambda}}{2\sqrt{\lambda}} \right)$$

By making the necessary substitutions and using the transcendental equation for λ_n, we obtain the Fourier coefficients in the form

$$\tfrac{1}{2}A_n(\lambda_n L + h \sin^2 L\sqrt{\lambda_n}) = (T_3 - T_1)(\sqrt{\lambda_n} + h \sin L\sqrt{\lambda_n})$$
$$+ h(T_1 - T_2) \sin L\sqrt{\lambda_n}$$

The formal solution of the initial-value problem is $u(z; t) = U(z) + \sum_{n=1}^{\infty} A_n \sin z\sqrt{\lambda_n}e^{-\lambda_n K t}$, where the A_n have just been obtained and the λ_n are determined from the transcendental equation $\sqrt{\lambda_n} \cos L\sqrt{\lambda_n} + h \sin L\sqrt{\lambda_n} = 0$.

Stage 4: Verification of the solution. From the graph of the cotangent function, we see that $L\sqrt{\lambda_n} - (n - \tfrac{1}{2})\pi \to 0$ when $n \to \infty$. From the formula for A_n, we see that A_n tends to zero when $n \to \infty$; in particular, $|A_n| \le M$ for some constant M. These estimates permit us to conclude that, for each $t > 0$, the following series are uniformly convergent for $0 \le z \le L$:

$$\sum_{n=1}^{\infty} A_n \sin z\sqrt{\lambda_n}e^{-\lambda_n K t}$$

$$\sum_{n=1}^{\infty} A_n\sqrt{\lambda_n} \cos z\sqrt{\lambda_n}e^{-\lambda_n K t}$$

$$\sum_{n=1}^{\infty} A_n\lambda_n \sin z\sqrt{\lambda_n}e^{-\lambda_n K t}$$

Therefore $u(z; t)$ is a differentiable function, and $u(z; t)$ satisfies the heat equation $u_t = Ku_{zz}$ for $t > 0$, $0 < z < L$.

Stage 5: Asymptotic behavior. When $t \to \infty$, $u(z; t)$ tends to the steady-state solution $U(z)$. The relaxation time is obtained from the first term of the series. Thus, if $T_3 - T_1$ and $T_1 - T_2$ are both positive, we have $\tau = 1/(\lambda_1 K)$ where λ_1 is obtained from the graph of the cotangent function in Fig. 2.3.1.   ●

Temporally Inhomogeneous Problems

The methods used above to study the general boundary-value problem for the one-dimensional heat equation in the temporally homogeneous case also can be used to study problems with explicit time dependence. The most general problem of this type is of the form

$$u_t - Ku_{zz} = r(z; t) \qquad 0 < z < L, t > 0$$

$$\cos \alpha \, u_z(0; t) - \sin \alpha \, u(0; t) = T_1(t) \qquad t > 0$$

$$\cos \beta \, u_z(L; t) + \sin \beta \, u(L; t) = T_2(t) \qquad t > 0$$

$$u(z; 0) = v(z) \qquad 0 < z < L$$

Here $r(z; t)$, $v(z)$, $T_1(t)$, $T_2(t)$ are given functions, assumed to be piecewise smooth in each variable.

To solve a problem of this type, first we consider the case of homogeneous boundary conditions, that is, $T_1(t) \equiv 0$, $T_2(t) \equiv 0$. The solution is sought in the form of a series of eigenfunctions of the homogeneous problem. Thus

$$r(z; t) = \sum_{n=1}^{\infty} r_n(t)\varphi_n(z) \qquad v(z) = \sum_{n=1}^{\infty} v_n\varphi_n(z) \qquad u(z; t) = \sum_{n=1}^{\infty} u_n(t)\varphi_n(z)$$

where $\varphi_n(z)$ are normalized eigenfunctions of the Sturm-Liouville problem for $\varphi'' + \lambda\varphi = 0$ with the associated homogeneous boundary conditions. The expansion coefficients $r_n(t)$, v_n, and $u_n(t)$ are obtained from the orthogonality of the eigenfunctions as the generalized Fourier coefficients: $r_n(t) = \int_0^L r(z; t)\varphi_n(z) \, dz$, $v_n = \int_0^L v(z)\varphi_n(z) \, dz$, $u_n(t) = \int_0^L u(z; t)\varphi_n(z) \, dz$. Substituting the series for $u(z; t)$ in the inhomogeneous heat equation, we have

$$u_t - Ku_{zz} = \sum_{n=1}^{\infty} [u_n'(t) + K\lambda_n u_n(t)]\varphi_n(z) = \sum_{n=1}^{\infty} r_n(t)\varphi_n(z)$$

Therefore we choose $u_n(t)$ as the solution of the ordinary differential equation $u_n'(t) + K\lambda_n u_n(t) = r_n(t)$ with the initial conditions $u_n(0) = v_n$. This ordinary differential equation is easily solved by the method of variation of parameters as

$$u_n(t) = v_n e^{-\lambda_n Kt} + \int_0^t r_n(s)e^{-\lambda_n K(t-s)} \, ds$$

It is not difficult to show that the series obtained converges uniformly for $0 < z < L$ together with the differentiated series and that the function obtained satisfies the equation $u_t - Ku_{zz} = r(z; t)$.

To solve the original problem with $T_1(t)$ and $T_2(t)$ nonzero, we reduce to homogeneous boundary conditions by defining a new function

$$v(z; t) = u(z; t) - [A(t) + zB(t)]$$

In order for $u(z; t)$ to solve the stated equation, we must have

$$v_t - Kv_{zz} = r(z; t) - [A'(t) + zB'(t)]$$

$$v(z; 0) = v(z) - [A(0) + zB(0)]$$

Here the functions $A(t)$, $B(t)$ are chosen so that the linear function $A(t) + zB(t)$ satisfies the inhomogeneous boundary conditions of the given problem. So we must have

$$\cos \alpha \, B(t) - \sin \alpha \, A(t) = T_1(t)$$

$$\cos \beta \, B(t) + (\sin \beta) \, [A(t) + LB(t)] = T_2(t)$$

This system of linear equations has a unique solution if and only if the determinant is nonzero: $\cos \alpha \sin \beta + \sin \alpha \cos \beta + L \sin \alpha \sin \beta \neq 0$. This happens precisely when $\lambda = 0$ is *not* an eigenvalue of the Sturm-Liouville problem for $\varphi'' + \lambda \varphi = 0$ with the associated homogeneous boundary conditions. Assuming this, we can determine $U(z; t)$ and infer that the function $v(z; t)$ so defined satisfies the homogeneous boundary conditions. Therefore we seek to determine $v(z; t)$ as a time-dependent series of eigenfunctions:

$$v(z; t) = \sum_{n=1}^{\infty} v_n(t)\varphi_n(z)$$

This can be done as above, once we determine the right side. We must have

$$v_t - Kv_{zz} = r(z; t) - [A'(t) + zB'(t)]$$

$$v(z; 0) = v(z) - [A(0) + zB(0)]$$

To solve the problem, first we must find the generalized Fourier series of the linear function $a + bz$:

$$a + bz = \sum_{n=1}^{\infty} (a\langle 1, \varphi_n \rangle + b\langle z, \varphi_n \rangle)\varphi_n(z)$$

Replacing $r_n(t)$ and v_n suitably, we are led to the solution

$$v_n(t) = e^{-\lambda_n K t}[v_n - A(0)\langle 1, \varphi_n \rangle - B(0)\langle z, \varphi_n \rangle]$$
$$+ \int_0^t e^{-\lambda_n K(t-s)}[h_n(s) - A'(s)\langle 1, \varphi_n \rangle - B'(s)\langle z, \varphi_n \rangle] \, ds$$

Example 2.3.3. Find the solution of the heat equation $u_t - Ku_{zz} = 0$ with the boundary conditions $u(0; t) = a_0 + b_0 t$, $u(L; t) = a_1 + b_1 t$ and the initial condition $u(z; 0) = 0$.

Solution. In this case the associated homogeneous boundary conditions are $u(0; t) = 0$, $u(L; t) = 0$, for which $\lambda = 0$ is not an eigenvalue. Therefore we can determine the functions $A(t)$ and $B(t)$ uniquely to yield $U(z; t) = a_0 + b_0 t + (z/L)[a_1 - a_0 + (b_1 - b_0)t]$ so that the new function $v(z; t)$ satisfies the inhomogeneous heat equation $v_t - Kv_{zz} = -[b_0 + (z/L)(b_1 - b_0)]$ with $v(z; 0) = -a_0 - (z/L)(a_1 - a_0)$. The appropriate Fourier series is

$$b_0 + (b_1 - b_0) \left(\frac{z}{L}\right) = \frac{2}{\pi} \sum_{n=1}^{\infty} \left\{ \frac{b_0}{n} [1 - (-1)^n] + \frac{b_1 - b_0}{n} (-1)^{n+1} \right\} \sin \frac{n\pi z}{L}$$

The solution coefficients are obtained as

$$v_n(t) = e^{-\lambda_n K t} \frac{2}{\pi} \left\{ -\frac{a_0}{n} [1 - (-1)^n] - (a_1 - a_0)(-1)^{n+1} \right\}$$

$$- \frac{2}{\pi} \frac{1 - e^{-\lambda_n K t}}{\lambda_n K} \left\{ \frac{b_0}{n} [1 - (-1)^n] + \frac{b_1 - b_0}{n} (-1)^{n+1} \right\}$$

The solution of the given problem is $u(z; t) = U(z; t) + \sum_{n=1}^{\infty} v_n(t)\varphi_n(z)$ when we make the appropriate insertions for $U(z; t)$ and $v_n(t)$ from above. ●

EXERCISES 2.3

1. Solve the initial-value problem for the heat equation $u_t = Ku_{zz}$ with the boundary conditions $u(0; t) = T_1$, $u_z(L; t) = \Phi_2$ and the initial condition $u(z; 0) = T_3$, where T_1, Φ_2, T_3 are positive constants. Find the relaxation time.
2. Solve the initial-value problem for the heat equation $u_t = Ku_{zz}$ with the boundary conditions $u_z(0; t) = 0$, $u_z(L; t) + h[u(L; t) - T_2] = 0$ and the initial condition $u(z; 0) = T_3$, where h, T_2, T_3 are positive constants.
3. Solve the initial-value problem for the heat equation $u_t = Ku_{zz} + r$ with the boundary conditions $u(0; t) = T_1$, $u(L; t) = T_2$ and the initial condition $u(z; 0) = T_3$, where r, T_1, T_2, T_3 are positive constants. Find the relaxation time.
4. Consider heat flow in an infinite slab $0 \le z \le L$ with the no-flux boundary conditions $u_z = 0$ at $z = 0$, $z = L$ and the initial condition $u = 273 + 96(2L - 4z)$. Find the solution of this initial-value problem, and determine the relaxation time. (*Warning:* The steady-state solution cannot be determined from the boundary conditions alone.)
5. Solve the initial-value problem for the heat equation $u_t = Ku_{zz}$ with the boundary conditions $u_z(0; t) = \Phi$, $u_z(L; t) = \Phi$ and the initial condition $u(z; 0) = T_3$, where Φ, T_3 are positive constants.
6. Consider heat flow in an infinite slab $0 \le z \le L$ with a source that generates heat at a rate per unit volume that is directly proportional to the distance from the face $z = 0$. Initially the temperature is zero throughout the slab; both faces are maintained at that temperature forever.
   (a) Show that this leads to the equation $u_t = Ku_{zz} + az$ with $u(0; t) = 0$, $u(L; t) = 0$, $u(z; 0) = 0$.
   (b) Find the steady-state solution U, the full solution u, and the relaxation time τ.
7. Consider heat flow in an infinite slab $0 \le z \le L$ which is initially at temperature zero. The face $z = 0$ is maintained at zero temperature, and the temperature at the face $z = L$ increases linearly with time.

(*a*) Show that this leads to the heat equation $u_t = Ku_{zz}$ with $u(0; t) = 0$, $u(L; t) = At$, $u(z; 0) = 0$.

(*b*) Show that the solution may be obtained in the form $u(z; t) = Azt/L + v(z; t)$, where $v(z; t)$ is obtained as a suitable Fourier sine series.

8. If a slender wire is placed in a medium which exchanges heat with the surroundings, the "linear law of surface heat transfer" dictates the equation $u_t = Ku_{zz} - bu$, where b is a positive constant. Assume that the ends of the wire are insulated and that initially the temperature $u = T_0$.

(*a*) Find the steady-state solution of the problem.

(*b*) Find the solution of the initial-boundary-value problem.

[*Hint:* The function $w(z; t) := e^{-bt}u(z; t)$ solves a known problem.]

9. With reference to the treatment of temporally inhomogeneous problems, suppose that the functions $r(z; t)$, $v(z)$, $T_1(t)$, $T_2(t)$ are uniformly bounded for $t > 0$, $0 < z < L$. Prove that the series defining the functions $v(z; t)$, $v_z(z; t)$, $v_{zz}(z; t)$, $v_t(z; t)$ all converge uniformly for $0 < z < L$ when $t > 0$ and that $v(z; t)$ satisfies the appropriate heat equation.

10. Find the solution of the inhomogeneous heat equation

$$u_t - Ku_{xx} = 1 - e^{-t}$$

with the boundary conditions

$$u(0; t) = 0, \ u_x(L; t) = 0$$

and the initial condition

$$u(x; 0) = 100.$$

2.4 THE VIBRATING STRING

In this section we derive and solve the equation of the vibrating string. This equation is the one-dimensional form of the wave equation, which occurs throughout many branches of mathematical physics.

Derivation of the Equation

Imagine a perfectly flexible elastic string, which at rest is stretched between two fixed pegs. For convenience we take a system of rectangular coordinates, so that the pegs are at the points $(0, 0, 0)$ and $(L, 0, 0)$. The points of the string are labeled by a parameter s, $0 \le s \le L$ (see Fig. 2.4.1). The motion of the string is described by a vector function $\mathbf{r}(s; t) = (X(s; t), Y(s; t), Z(s; t))$, which gives the rectangular coordinates of the string point s at time t. Thus the vector $\partial\mathbf{r}/\partial s$ is tangent to the string at point s (Fig. 2.4.2). The vector $\partial\mathbf{r}/\partial t$ is the instantaneous velocity of the string at the point s, while the vector $\partial^2\mathbf{r}/\partial t^2$ is the instantaneous acceleration of the string point s.

We now determine a system of partial differential equations satisfied by the functions $X(s; t)$, $Y(s; t)$, $Z(s; t)$. To do this, we apply Newton's second law of motion, stating that the force on any segment of the string is the time

$(0, 0, 0)$ ⟶ s ⟶ $(L, 0, 0)$

FIGURE 2.4.1
Vibrating string in equilibrium position.

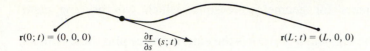

$\mathbf{r}(0; t) = (0, 0, 0)$ $\dfrac{\partial \mathbf{r}}{\partial s}(s; t)$ $\mathbf{r}(L; t) = (L, 0, 0)$

FIGURE 2.4.2
Vibrating string in motion—tangent vector.

derivative of the momentum of that segment. The mass of the string is given by a *mass density function* $\rho(s)$, $0 \le s \le L$. This means that

$$\int_a^b \rho(s)\, ds = \begin{array}{l} \text{mass of segment of} \\ \text{string for which } a \le s \le b \end{array}$$

Likewise, $\displaystyle\int_a^b \rho(s)\, \frac{\partial \mathbf{r}}{\partial t}(s; t)\, ds = \begin{array}{l} \text{momentum of segment of} \\ \text{string for which } a \le s \le b \end{array}$

A typical segment of the string for which $a \le s \le b$ is subject to contact forces exerted at a and b by the rest of the string and to external forces (such as gravity) from the environment. These external forces may be written as a vector function $\mathbf{F}(s; t) = (F_1(s; t), F_2(s; t), F_3(s; t))$, representing the force per unit mass acting on point s of the string. To describe the contact forces, we introduce the *tension* $T(s; t)$. The segment of the string with $b \le s \le L$ exerts a contact force at b, on the segment $a \le s \le b$ of the string, of the form

$$T^+(b; t)\, \frac{(\partial \mathbf{r}/\partial s)(b; t)}{|(\partial \mathbf{r}/\partial s)(b; t)|} = \begin{array}{l} \text{force on segment } a \le s \le b \\ \text{due to segment } b \le s \le L \end{array}$$

This means that the force is directed along the tangent to the string, as illustrated in Fig. 2.4.3. This property is a mathematical statement of the assumption that the string is perfectly flexible.

Similarly, we set

$$-T^-(a; t)\, \frac{(\partial \mathbf{r}/\partial s)(a; t)}{|(\partial \mathbf{r}/\partial s)(a; t)|} = \begin{array}{l} \text{force on segment } a \le s \le b \\ \text{due to segment } 0 \le s \le a \end{array}$$

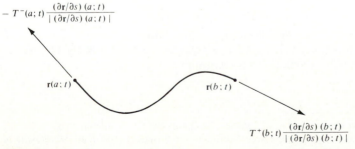

$-T^-(a; t)\dfrac{(\partial \mathbf{r}/\partial s)\,(a; t)}{|\,(\partial \mathbf{r}/\partial s)\,(a; t)\,|}$

$\mathbf{r}(a; t)$

$\mathbf{r}(b; t)$

$T^+(b; t)\dfrac{(\partial \mathbf{r}/\partial s)\,(b; t)}{|\,(\partial \mathbf{r}/\partial s)\,(b; t)\,|}$

FIGURE 2.4.3
Contact forces on vibrating string.

The minus sign enters for geometric reasons, which will become clear momentarily.

The force on the segment $a \le s \le b$ is the external force plus these contact forces. Applying Newton's second law, we have

$$\int_a^b \rho(s) \frac{\partial^2 \mathbf{r}}{\partial t^2} (s; t) \, ds = \int_a^b \rho(s) \mathbf{F}(s; t) \, ds + T^+(b; t) \frac{(\partial \mathbf{r}/\partial s)(b; t)}{|(\partial \mathbf{r}/\partial s)(b; t)|}$$

$$- T^-(a; t) \frac{(\partial \mathbf{r}/\partial s)(a; t)}{|(\partial \mathbf{r}/\partial s)(a; t)|}$$

This is a vector integral equation, equivalent to three scalar integral equations. Now we let $b \to a$. In the limit the integrals vanish, and we conclude that $T^+(a; t) = T^-(a; t)$. We accordingly drop the plus and minus signs and write $T(s; t)$ for the common value, called the *tension* at $(s; t)$.

To write these integral equations as differential equations, we differentiate with respect to b, set $b = s$, and obtain the following differential form of the equation of the vibrating string:

$$\rho(s) \frac{\partial^2 \mathbf{r}}{\partial t^2} (s; t) = \rho(s) \mathbf{F}(s; t) + \frac{\partial}{\partial s} \left[T(s; t) \frac{(\partial \mathbf{r}/\partial s)(s; t)}{|(\partial \mathbf{r}/\partial s)(s; t)|} \right] \qquad (2.4.1)$$

This equation is equivalent to three scalar equations for the four variables X, Y, Z, T. We obtain a determinate system by saying how the tension $T(s; t)$ is influenced by the stretch factor $|(\partial \mathbf{r}/\partial s)(s; t)|$. For each elastic material, there is a well-defined function N expressing this dependence by

$$T(s; t) = N \left(\left| \frac{\partial \mathbf{r}}{\partial s} (s; t) \right|, s \right) \qquad (2.4.2)$$

The equation obtained by substituting (2.4.2) into (2.4.1) is a rigorous consequence of Newton's second law of motion. However, this equation is extremely difficult to solve. Therefore we will make assumptions to obtain simplified equations which may be solved. The alert reader will note that some of the steps we take may be difficult to justify within our treatment. The essential point is that *the solutions of the simplified equations can be shown to be close, in an appropriate sense, to the solutions of the exact equations (2.4.1) and (2.4.2)*. For more information the reader is referred to the book by H. Weinberger† or to the excellent article by S. Antman.‡

---

† H. Weinberger, *A First Course in Partial Differential Equations*, Ginn, Boston, 1965.
‡ S. Antman, "The Equations for the Large Vibrations of Strings," *American Mathematical Monthly*, vol. 87, pp. 359–370, 1980.

Linearized Model

To obtain the simplified equations, we look for solutions which describe *small vibrations*. This means that

$$X(s; t) = s + \epsilon x(s; t)$$

$$Y(s; t) = \epsilon y(s; t)$$

$$Z(s; t) = \epsilon z(s; t) \tag{2.4.3}$$

$$T(s; t) = T_0 + \epsilon T_1(s; t)$$

$$F(s; t) = \epsilon f(s; t)$$

The parameter ϵ may be thought of as a rough measure of the maximum displacement of the string from its neutral position $X = s$, $Y = 0$, $Z = 0$. The expressions T_0 and T_1 may be found by substituting the first three equations (2.4.3) into (2.4.1) and (2.4.2) and discarding higher powers of ϵ. In particular, T_0 is the tension of the string in its neutral position $X = s$, $Y = 0$, $Z = 0$.

With the assumptions (2.4.3), we have $\partial X/\partial s = 1 + \epsilon\, \partial x/\partial s$, $\partial Y/\partial s = \epsilon\, \partial y/\partial s$, $\partial Z/\partial s = \epsilon\, \partial z/\partial s$; thus

$$\left|\frac{\partial \mathbf{r}}{\partial s}\right|^2 = \left(1 + \epsilon \frac{\partial x}{\partial s}\right)^2 + \left(\epsilon \frac{\partial y}{\partial s}\right)^2 + \left(\epsilon \frac{\partial z}{\partial s}\right)^2$$

$$= 1 + 2\epsilon \left(\frac{\partial x}{\partial s}\right) + \epsilon^2 \left[\left(\frac{\partial x}{\partial s}\right)^2 + \left(\frac{\partial y}{\partial s}\right)^2 + \left(\frac{\partial z}{\partial s}\right)^2\right]$$

Taking the square root and using the Taylor expansion $(1 + a)^{1/2} = 1 + \frac{1}{2}a + O(a^2)$, we have

$$\left|\frac{\partial \mathbf{r}}{\partial s}\right| = 1 + \epsilon \frac{\partial x}{\partial s} + O(\epsilon^2)$$

Thus
$$\frac{\partial \mathbf{r}/\partial s}{|\partial \mathbf{r}/\partial s|} = \left(1 + O(\epsilon^2), \epsilon \frac{\partial y}{\partial s} + O(\epsilon^2), \epsilon \frac{\partial z}{\partial s} + O(\epsilon^2)\right)$$

If we insert these into (2.4.1) and use (2.4.3), we obtain the following equations for the x, y, and z components:

$$\epsilon \rho(s) \frac{\partial^2 x}{\partial t^2} = \epsilon \rho(s) f_1(s; t) + \frac{\partial}{\partial s} \{[T_0 + \epsilon T_1(s; t)][1 + O(\epsilon^2)]\}$$

$$\epsilon \rho(s) \frac{\partial^2 y}{\partial t^2} = \epsilon \rho(s) f_2(s; t) + \frac{\partial}{\partial s} \left\{[T_0 + \epsilon T_1(s; t)]\left[\epsilon \frac{\partial y}{\partial s} + O(\epsilon^2)\right]\right\}$$

$$\epsilon \rho(s) \frac{\partial^2 z}{\partial t^2} = \epsilon \rho(s) f_3(s; t) + \frac{\partial}{\partial s} \left\{[T_0 + \epsilon T_1(s; t)]\left[\epsilon \frac{\partial z}{\partial s} + O(\epsilon^2)\right]\right\}$$

Equating the coefficients of ϵ in each of these three equations, we obtain the following simplified equations for the longitudinal and transverse vibrations:

$$\rho(s)\,\frac{\partial^2 x}{\partial t^2} = \rho(s)f_1(s;\,t) + \frac{\partial}{\partial s}\,T_1(s;\,t)$$

$$\rho(s)\,\frac{\partial^2 y}{\partial t^2} = \rho(s)f_2(s;\,t) + T_0\,\frac{\partial^2 y}{\partial s^2}$$

$$\rho(s)\,\frac{\partial^2 z}{\partial t^2} = \rho(s)f_3(s;\,t) + T_0\,\frac{\partial^2 z}{\partial s^2}$$

We are interested in only the transverse vibrations obtained by solving the second and third equations. These are of the same form, the so-called *one-dimensional wave equation*

$$\boxed{\frac{\partial^2 y}{\partial t^2} = f(s;\,t) + \frac{T_0}{\rho(s)}\,\frac{\partial^2 y}{\partial s^2}} \qquad (2.4.4)$$

Example 2.4.1. Suppose that $\rho(s) = \rho$, $f(s;\,t) = g$, independent of $(s;\,t)$. Find the steady-state solution of the wave equation (2.4.4) satisfying the boundary conditions $y(0) = 0$, $y(L) = 0$.

Solution. The function $y(s)$ must satisfy the ordinary differential equation

$$0 = g + \frac{T_0}{\rho}\,y''(s)$$

The general solution of this equation is $y(s) = -[\rho g/(2T_0)]s^2 + As + B$, where A, B are arbitrary constants. Applying the boundary conditions, we have $0 = B$, $0 = -[\rho g/(2T_0)]L^2 + AL + B$. The solution is $y(s) = [\rho g/(2T_0)]\,(Ls - s^2)$.   ●

The Plucked String

Now we turn to time-dependent solutions of the wave equation, specifically the problem of the *plucked string*. We suppose that the string is uniform $[\rho(s) = \rho]$ and that no outside forces are present $[f(s;\,t) = 0]$. We let

$$c^2 = \frac{T_0}{\rho}$$

which has the dimensions of (velocity)^2. The wave equation is now written as $y_{tt} = c^2 y_{ss}$. The initial position of the string is supposed to be given by a function $f_1(s)$, $0 \le s \le L$, while the initial velocity is zero. Thus we have the problem

$$y_{tt} = c^2 y_{ss} \qquad\qquad t > 0,\, 0 < s < L$$

$$y(0;\,t) = 0 = y(L;\,t) \qquad t > 0$$

$$y(s;\,0) = f_1(s) \qquad\qquad 0 < s < L$$

$$y_t(s; 0) = 0 \qquad\qquad 0 < s < L$$

In Example 0.2.5 we showed that the separated solutions which satisfy the boundary conditions and the second initial condition are $\sin(n\pi s/L)\cos(n\pi ct/L)$, $n = 1, 2, \ldots$. Therefore we may obtain a formal solution by the superposition principle as

$$y(s; t) = \sum_{n=1}^{\infty} B_n \cos \frac{n\pi ct}{L} \sin \frac{n\pi s}{L} \qquad (2.4.5)$$

To find the coefficients B_n, we set $t = 0$. This requires that

$$f_1(s) = \sum_{n=1}^{\infty} B_n \sin \frac{n\pi s}{L} \qquad (2.4.6)$$

Therefore B_n is the nth Fourier sine coefficient of f_1:

$$B_n = \frac{2}{L} \int_0^L f_1(s) \sin \frac{n\pi s}{L} \, ds \qquad (2.4.7)$$

If $f_1(s)$, $0 < s < L$, is continuous and piecewise smooth, the Fourier series (2.4.6) converges for all s to the odd periodic extension of f_1, denoted $\bar{f}_1$. Therefore, to solve the problem, we have a simple rule: Given f_1, compute B_n from (2.4.7) and substitute into (2.4.5) to solve the problem. This is called the *Fourier representation* of the solution. We illustrate with the problem of the symmetric plucked string.

Example 2.4.2. Solve the vibrating string problem in the case where

$$f_1(s) = \begin{cases} s & 0 < s < L/2 \\ L - s & L/2 < s < L \end{cases}$$

Solution. To compute the Fourier coefficients, we notice that f_1 is even about $s = L/2$, whereas $\sin(n\pi s/L)$ is even (respectively, odd) about $s = L/2$ if n is odd (respectively, even). Therefore $B_n = 0$ if n is even. If n is odd, we have

$$B_n = \frac{2}{L} \int_0^L f_1(s) \sin \frac{n\pi s}{L} \, ds$$

$$= \frac{4}{L} \int_0^{L/2} f_1(s) \sin \frac{n\pi s}{L} \, ds$$

$$= \frac{4}{L} \int_0^{L/2} s \sin \frac{n\pi s}{L} \, ds$$

$$= \frac{4}{L} \frac{Ls}{n\pi} \cos \frac{n\pi s}{L} \Big|_{L/2}^{0} + \frac{4}{n\pi} \int_0^{L/2} \cos \frac{n\pi s}{L} \, ds$$

$$= \frac{4L}{n^2\pi^2} \sin \frac{n\pi}{2}$$

If n is even, $B_n = 0$, while if n is odd, we write $n = 2m - 1$, $\sin(n\pi/2) = (-1)^{m+1}$. Therefore we have solved the problem:

$$y(s; t) = \frac{4L}{\pi^2} \sum_{m=1}^{\infty} \frac{(-1)^{m+1}}{(2m-1)^2} \cos\left[(2m-1)\frac{\pi ct}{L}\right] \sin\left[(2m-1)\frac{\pi s}{L}\right] \quad \bullet$$

The Fourier representation (2.4.5) displays the solution as a Fourier sine series for each time t. The nth term of this series is a purely harmonic vibration of frequency $n\pi c/L$ and amplitude B_n. Hence this form of the solution affords a natural analogy between the vibrating string and an infinite system of harmonic oscillators.

Acoustic Interpretations

In the theory of acoustics, the numbers $\omega_n = n\pi c/L$ are interpreted as the *frequencies* of purely harmonic vibrations. It is a characteristic feature of the vibrating string that these numbers are multiples of a common frequency $\omega_1 = \pi c/L$, which is called the *fundamental frequency*. The higher frequencies are called *overtones,* and they strengthen the quality of the sound.

If the initial position of the string is given by $y(s; 0) = C \sin(n\pi s/L)$, then the string will vibrate at the corresponding frequency $\omega_n = n\pi c/L$ when released from rest. In practice, this initial condition is rarely met; instead we often have initial conditions which result from striking or bowing the string. In these cases both the fundamental frequency and many overtones will be present in the resultant vibration, which is written as a superposition of several purely harmonic vibrations. The resulting impulses which are transmitted are characteristic of the particular stringed instrument. We list below typical solutions for both a piano string and a violin string, corresponding to the fundamental frequency of 440 cycles/s (concert A).

$$y(s; t) = \sum_{n=1}^{8} A_n \sin\frac{n\pi s}{L} \cos\frac{n\pi ct}{L}$$

	Piano	Violin
A_1	1	1
A_2	0.20	1
A_3	0.25	0.45
A_4	0.10	0.50
A_5	0.08	1.00
A_6	0.00	0.03
A_7	0.00	0.03
A_8	0.00	0.03

According to these data, the piano vibration is much closer to a purely harmonic vibration than the violin vibration. To see this numerically, we use the formula for the mean square error, developed in Sec. 1.4; thus

$$\frac{1}{L} \int_0^L \left[y(s; t) - A_1 \sin \frac{\pi s}{L} \cos \frac{\pi c t}{L} \right]^2 ds = \frac{1}{2} \sum_{n=2}^{8} A_n^2 \cos^2 \frac{n \pi c t}{L}$$

In the case of the piano vibration,

$$\sum_{n=2}^{8} A_n^2 = (0.20)^2 + (0.25)^2 + (0.10)^2 + (0.08)^2 = 0.1189$$

whereas in the case of the violin vibration

$$\sum_{n=2}^{8} A_n^2 = 1 + (0.45)^2 + (0.50)^2 + (1.00)^2 + (0.03)^2 + (0.03)^2 + (0.03)^2$$

$$= 2.4552$$

To obtain a meaningful comparison, we define the *fractional mean square error* by

$$\bar{\sigma}^2 = \frac{\sum_{n=2}^{8} A_n^2}{\sum_{n=1}^{8} A_n^2}$$

For the piano vibration we have $\bar{\sigma}^2 = 0.1189/1.1189 = 0.1063$, whereas for the violin vibration we have $\bar{\sigma}^2 = 2.4552/3.4552 = 0.7106$, nearly 7 times as large.†

Explicit (d'Alembert) Representation

We now return to the mathematical discussion.

The Fourier representation (2.4.5) also has some disadvantages, which we now discuss. On the one hand, it is difficult to verify that $y(s; t)$ actually is a solution of the wave equation. Consider, for example, the computation of the second time derivative y_{tt} of the solution obtained in Example 2.4.2. Proceeding formally, we encounter the series

$$\frac{4c^2}{L} \sum_{m=1}^{\infty} (-1)^{m+1} \cos \left[(2m - 1) \frac{\pi c t}{L} \right] \sin \left[(2m - 1) \frac{\pi s}{L} \right]$$

We have lost the factor $1/(2m - 1)^2$, which ensured convergence of the series for $y(s; t)$. The new series converges for no value of t.

A second disadvantage of the Fourier representation is that it provides little geometric insight into the motion of the vibrating string. We expect that

---

† Data obtained from E. Donnell Blackham, "The Physics of the Piano," *Scientific American*, December 1965.

an initial disturbance will be propagated as some sort of wave motion, but the Fourier representation does not show this directly.

To overcome these difficulties, we rewrite the solution (2.4.5) in a form which avoids the Fourier coefficients B_n. To do this, we use the trigonometric identity $\sin \alpha \cos \beta = \frac{1}{2}[\sin (\alpha + \beta) + \sin (\alpha - \beta)]$ and proceed formally:

$$y(s; t) = \sum_{n=1}^{\infty} B_n \sin \frac{n\pi s}{L} \cos \frac{n\pi c t}{L}$$

$$= \frac{1}{2} \sum_{n=1}^{\infty} B_n \left\{ \sin \left[\frac{n\pi}{L} (s + ct) \right] + \sin \left[\frac{n\pi}{L} (s - ct) \right] \right\}$$

We recognize the Fourier sine series for $\bar{f}_1(s + ct)$, $\bar{f}_1(s - ct)$, where $\bar{f}_1$ is the odd $2L$-periodic extension of $f_1(s)$, $0 < s < L$. Therefore we have the *explicit representation*

$$\boxed{y(s; t) = \frac{1}{2}[\bar{f}_1(s + ct) + \bar{f}_1(s - ct)]} \tag{2.4.8}$$

In physical terms, we have written $y(s; t)$ as a sum of two traveling waves, one moving to the right with speed c and the other moving to the left with speed c. This will enable us to obtain graphical representations of the solution.*

Using the representation (2.4.8), we now verify that $y(s; t)$ satisfies the wave equation. Indeed, whenever $\bar{f}'_1, \bar{f}''_1$ exist, we have

$$y_t = \frac{1}{2}c\bar{f}'_1(s + ct) - \frac{1}{2}c\bar{f}'_1(s - ct)$$

$$y_{tt} = \frac{1}{2}c^2\bar{f}''_1(s + ct) + \frac{1}{2}c^2\bar{f}''_1(s - ct)$$

$$y_s = \frac{1}{2}\bar{f}'_1(s + ct) + \frac{1}{2}\bar{f}'_1(s - ct)$$

$$y_{ss} = \frac{1}{2}\bar{f}''_1(s + ct) + \frac{1}{2}\bar{f}''_1(s - ct)$$

Clearly $y_{tt} = c^2 y_{ss}$. To check the boundary conditions, we have

$$y(0; t) = \frac{1}{2}[\bar{f}_1(ct) + \bar{f}_1(-ct)] = 0$$

since $\bar{f}_1$ is odd, and

$$y(L; t) = \frac{1}{2}[\bar{f}_1(L + ct) + \bar{f}_1(L - ct)]$$

$$= \frac{1}{2}[\bar{f}_1(L + ct) - \bar{f}_1(-L + ct)]$$

$$= \frac{1}{2}[\bar{f}_1(L + ct) - \bar{f}_1(L + ct)]$$

$$= 0$$

where we have used the oddness and $2L$ periodicity of $\bar{f}_1$. The initial condition $y(s; 0) = \bar{f}_1(s)$ is satisfied everywhere, while $y_t(s; 0) = 0$ wherever $\bar{f}'_1$ is defined. This completes the validation of the solution.

---

* See also the Appendix on "Using Mathematica."

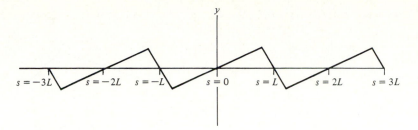

FIGURE 2.4.4
The odd, $2L$-periodic extension of $f_1(s)$, $0 < s < L$.

The second application of (2.4.8) is to obtain a picture of the motion of the vibrating string. We now illustrate this in the case of the asymmetric plucked string, where

$$f_1(s) = \begin{cases} s(L - s_0) & 0 < s < s_0 \\ s_0(L - s) & s_0 < s < L \end{cases}$$

When we extend f_1 as an odd periodic function, we obtain a graph, depicted in Fig. 2.4.4. The extended function satisfies $\bar{f}_1(-s) = -\bar{f}_1(s)$, $\bar{f}_1(s + 2L) = \bar{f}_1(s)$ for any s. We substitute this into (2.4.8) and get

$$y(s; t) = \tfrac{1}{2}[\bar{f}_1(s + ct) + \bar{f}_1(s - ct)]$$

To obtain a picture of the motion of the plucked string, we must average the left and right translates of $\bar{f}_1$. Since the average of two linear functions is again linear, it suffices to plot five points in the interval $0 \le s \le L$: the two endpoints where the string is fixed, the interior point where $\bar{f}_1(s + ct) = \bar{f}_1(s - ct)$, and the two interior points where $\bar{f}'_1(s + ct)$ changes sign.

The diagrams in Fig. 2.4.5 give a motion picture of the plucked string during the half-period $0 \le t \le L/c$. For convenience we take $s_0 = 3L/4$. At $t = 0$, we have the odd periodic function $\bar{f}_1(s)$ with vertices at $A(s = 3L/4)$, $B(s = 5L/4)$, and $C(s = -3L/4)$. These points, which determine the discontinuities of $\bar{f}'_1$, move along the axis and are labeled $A_\pm$, $B_\pm$, $C_\pm$. For $t = L/(8c)$, we have vertices A_+, A_- as the only vertices with $0 \le s \le L$. When $t = L/(4c)$, A_- has arrived at the middle of the interval, and the plucked string is symmetric about $s = L/2$. The vertex A_+ is replaced by B_-. When $L/(4c) \le t \le 3L/(4c)$, part of the string goes below the axis, until it reaches the symmetric configuration shown at $t = 3L/(4c)$. When $t = 3L/(4c)$, the A_- vertex disappears from the interval and is replaced by C_+; this results in the symmetric configuration completely below the axis. Finally, when $t = L/c$, the string has completed one-half of its period and is congruent to its initial position with vertex at $s = L/4$.

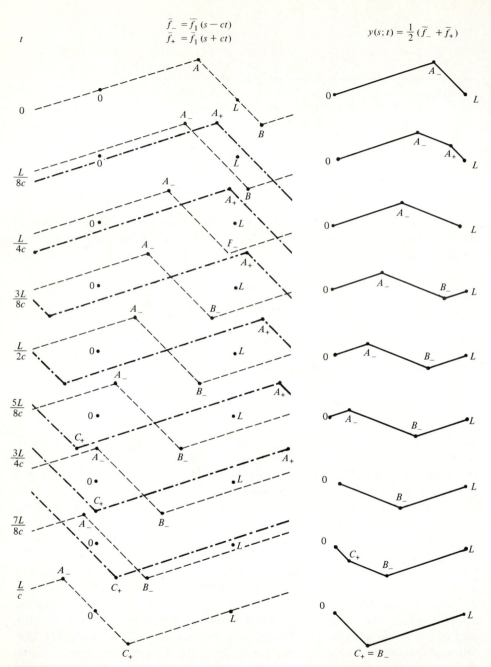

FIGURE 2.4.5
Successive positions of the asymmetric plucked string.

140

The Struck String

We now solve the problem of the vibrating string, starting from equilibrium, with nonzero initial velocity. Thus we must solve

$$y_{tt} = c^2 y_{ss} \qquad t > 0, 0 < s < L$$

$$y(0; t) = 0 = y(L; t) \qquad t > 0$$

$$y(s; 0) = 0 \qquad 0 < s < L$$

$$y_t(s; 0) = f_2(s) \qquad 0 < s < L$$

The initial velocity profile $f_2(s)$ is unspecified for the moment.

To solve this problem, we begin with the separated solutions which satisfy the wave equation, the boundary conditions, and the first initial condition. These are of the form

$$\sin \frac{n\pi s}{L} \sin \frac{n\pi ct}{L}$$

To satisfy the second initial condition, we try a superposition of these:

$$y(s; t) = \sum_{n=1}^{\infty} B_n \sin \frac{n\pi s}{L} \sin \frac{n\pi ct}{L} \qquad (2.4.9)$$

This is the Fourier representation of the solution. The coefficients B_n are determined by differentiating this series with respect to t and setting $t = 0$; thus

$$y_t(s; t) = \sum_{n=1}^{\infty} B_n \sin \left(\frac{n\pi s}{L} \frac{n\pi c}{L} \right) \cos \frac{n\pi ct}{L}$$

Setting $t = 0$, we must have

$$y_t(s; 0) = f_2(s) = \sum_{n=1}^{\infty} \frac{n\pi c}{L} B_n \sin \frac{n\pi s}{L} \qquad (2.4.10)$$

In other words, $(n\pi c/L)B_n$ is the nth Fourier sine coefficient of $f_2(s)$, $0 < s < L$:

$$\frac{n\pi c}{L} B_n = \frac{2}{L} \int_0^L f_2(s) \sin \frac{n\pi s}{L} \, ds$$

If $f_2(s)$, $0 < s < L$, is continuous and piecewise smooth, the Fourier series (2.4.10) converges for all s to the odd periodic extension of f_2, denoted $\bar{f}_2$. This completes the Fourier representation of the solution.

To obtain an explicit representation, we apply the trigonometric identity $\sin \alpha \sin \beta = \frac{1}{2}[\cos (\alpha - \beta) - \cos (\alpha + \beta)]$ to the Fourier representation. Thus

$$y(s; t) = \frac{1}{2} \sum_{n=1}^{\infty} B_n \left\{ \cos \left[\frac{n\pi}{L} (s - ct) \right] - \cos \left[\frac{n\pi}{L} (s + ct) \right] \right\}$$

We use calculus to rewrite this as

$$y(s; t) = \frac{1}{2} \sum_{n=1}^{\infty} B_n \frac{n\pi}{L} \int_{s-ct}^{s+ct} \sin \frac{n\pi z}{L} \, dz$$

$$= \frac{1}{2} \int_{s-ct}^{s+ct} \left(\sum_{n=1}^{\infty} B_n \frac{n\pi}{L} \sin \frac{n\pi z}{L} \right) dz$$

where we have formally interchanged the summation and integration. Aside from the factor c, we recognize the formula in parentheses as the Fourier sine series for $\bar{f}_2$, the odd $2L$-periodic extension of $f_2(s)$, $0 < s < L$. Thus we have the *explicit representation*

$$y(s; t) = \frac{1}{2c} \int_{s-ct}^{s+ct} \bar{f}_2(z) \, dz \qquad (2.4.11)$$

This formula defines a solution of the wave equation and satisfies the boundary conditions and both initial conditions.

In particular cases, this formula can be used to graph the solution of the wave equation.

Example 2.4.3. Graph the solution of the vibrating string if

$$f_2(s) = \begin{cases} 0 & 0 < s < L/4 \\ 1 & L/4 < s < 3L/4 \\ 0 & 3L/4 < s < L \end{cases}$$

Solution. To graph the solution, first we extend f_2 as an odd $2L$-periodic function. Next, from formula (2.4.11), we have $y_s(s; t) = [1/(2c)][\bar{f}_2(s + ct) - \bar{f}_2(s - ct)]$. This quantity is 0, $\pm 1/(2c)$, $\pm 1/c$. For each t, $y(s; t)$ is linear on each segment on which y_s is constant. Finally, we have the set of graphs shown in Fig. 2.4.6 of the solution for $t = 0$, $L/(8c)$, $L/(4c)$, $3L/(8c)$, $L/(2c)$. ●

The explicit solutions just obtained can be combined to obtain a solution of the wave equation for the general initial conditions $y(s; 0) = f_1(s)$, $y_t(s; 0) = f_2(s)$. For this purpose, consider the function

$$y(s; t) = \frac{1}{2} [f_1(s + ct) + f_1(s - ct)] + \frac{1}{2c} \int_{s-ct}^{s+ct} f_2(z) \, dz \qquad (2.4.12)$$

Suppose that f_1, f_1', f_1'', f_2, f_2' are continuous functions. Then

$$y_t = \frac{c}{2} [f_1'(s + ct) - f_1'(s - ct)] + \frac{1}{2} [f_2(s + ct) + f_2(s - ct)]$$

$$y_{tt} = \frac{c^2}{2} [f_1''(s + ct) + f_1''(s - ct)] + \frac{c}{2} [f_2'(s + ct) - f_2'(s - ct)]$$

$$y_s = \frac{1}{2}[f_1'(s + ct) + f_1'(s - ct)] + \frac{1}{2c}[f_2(s + ct) - f_2(s - ct)]$$

$$y_{ss} = \frac{1}{2}[f_1''(s + ct) + f_1''(s - ct)] + \frac{1}{2c}[f_2'(s + ct) - f_2'(s - ct)]$$

We observe that $y(s; 0) = f_1(s)$, $y_t(s; 0) = f_2(s)$, and $y(s; t)$ satisfies the wave equation $y_{tt} = c^2 y_{ss}$ for all (s, t). We have made no use of boundary conditions or periodicity considerations. This general solution is called *d'Alembert's solution* of the wave equation. We summarize the result as a proposition.

Proposition. Let f_1, f_2 be continuous functions with continuous derivatives f_1', f_1'', f_2'. The d'Alembert formula (2.4.12) gives a solution of the wave equation $y_{tt} = c^2 y_{ss}$ valid for all $t > 0$, $-\infty < s < \infty$, and satisfies the initial conditions $y(s; 0) = f_1(s)$, $y_t(s; 0) = f_2(s)$.

The careful reader will note that Examples 2.4.2 and 2.4.3 do not satisfy the hypotheses of the proposition. Therefore this proposition, although mathematically rigorous, excludes examples of physical interest. To improve this

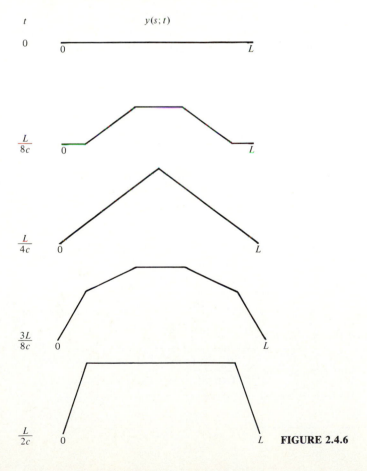

FIGURE 2.4.6

point of the theory, mathematicians have extended the concept of *solution* to include functions $y(s; t)$ for which some of the indicated partial derivatives may fail to exist. This concept of *weak solution* is explained by Weinberger.†

Inhomogeneous Equation

In the remainder of this section, we consider a vibrating string with a time-dependent external force, using the method of Fourier series. As our first model we consider the problem

$$y_{tt} - c^2 y_{ss} = g(s) \cos \omega t \qquad 0 < s < L, t > 0$$

$$y(0; t) = 0 = y(L; t) \qquad t > 0$$

where $g(s)$, $0 < s < L$, is a piecewise smooth function and ω is a positive constant.

We look for a particular solution in the form

$$y(s; t) = \sum_{n=1}^{\infty} A_n(t) \sin \frac{n\pi s}{L}$$

(The general solution can be found by adding a solution of the homogeneous equation, which has already been discussed.) To find the coefficient functions $A_n(t)$, we substitute in the differential equation and use the Fourier sine expansion of $g(s)$:

$$g(s) = \sum_{n=1}^{\infty} B_n \sin \frac{n\pi s}{L}$$

$$\sum_{n=1}^{\infty} \left[A_n''(t) + \left(\frac{n\pi c}{L} \right)^2 A_n(t) \right] \sin \frac{n\pi s}{L} = \cos \omega t \sum_{n=1}^{\infty} B_n \sin \frac{n\pi s}{L}$$

We choose $A_n(t)$ to be solutions of the ordinary differential equations

$$A_n''(t) + \left(\frac{n\pi c}{L} \right)^2 A_n(t) = B_n \cos \omega t$$

If $\omega \neq n\pi c/L$ for any n, a particular solution is

$$A_n(t) = A_n \cos \omega t$$

$$A_n \left[-\omega^2 + \left(\frac{n\pi c}{L} \right)^2 \right] = B_n$$

The formal solution of the problem is

$$y(s; t) = \cos \omega t \sum_{n=1}^{\infty} \frac{B_n \sin (n\pi s/L)}{(n\pi c/L)^2 - \omega^2}$$

---

† Weinberger, op. cit., p. 15.

The solution is a periodic function of time with the same period. The series for $y(s; t)$ converges uniformly for $0 \leq s \leq L$. If $g(s)$ is continuous and satisfies the boundary conditions, the differentiated series for y_s, y_{ss}, y_t, y_{tt} also converge uniformly for $0 \leq s \leq L$ and $y(s; t)$ is a solution of the problem.

Example 2.4.4. Find a particular solution of the problem $y_{tt} - c^2 y_{ss} = A \cos \omega t$, satisfying the boundary conditions $y(0; t) = 0 = (L; t)$, where A, ω are positive constants with $\omega \neq n \pi c/L$.

Solution. In this case we use the Fourier sine series expansion of the constant function

$$1 = \frac{2}{\pi} \sum_{n=1}^{\infty} \frac{1 - (-1)^n}{n} \sin \frac{n \pi s}{L} \qquad 0 < s < L$$

Thus $B_n = [2/(n\pi)][1 - (-1)^n]$, and the solution is

$$y(s; t) = \frac{2A}{\pi} \cos \omega t \sum_{n=1}^{\infty} \frac{1 - (-1)^n}{n[(n\pi c/L)^2 - \omega^2]} \sin \frac{n \pi s}{L}$$

The series for y_s, y_{ss}, y_t, y_{tt} converges uniformly for $\delta \leq s \leq L - \delta$ for any $\delta > 0$, and this implies that $y(s; t)$ is a solution of the equation.   ●

Example 2.4.5. Find a particular solution of the problem $y_{tt} - c^2 y_{ss} = A \cos \omega t$ satisfying the boundary conditions $y(0; t) = 0 = y(L; t)$, where A, ω are positive constants with $\omega = N \pi c/L$ for some integer N.

Solution. We look for a particular solution in the form

$$y(s; t) = \sum_{n=1}^{\infty} A_n(t) \sin \frac{n \pi s}{L}$$

Repeating the analysis of the previous cases, we choose $A_n(t)$ as a solution of the ordinary differential equation

$$A_n''(t) + \left(\frac{n \pi c}{L}\right)^2 A_n(t) = \frac{2}{\pi} A \frac{1 - (-1)^n}{n} \cos \omega t$$

If $n \neq N$, this is solved as before.

    To solve the equation for $n = N$, we use the following observation. For any $\omega' \neq \omega$,

$$\left(\frac{d^2}{dt^2} + \omega^2\right) \cos \omega' t = [\omega^2 - (\omega')^2] \cos \omega' t$$

while

$$\left(\frac{d^2}{dt^2} + \omega^2\right) \cos \omega t = 0$$

Therefore

$$\left(\frac{d^2}{dt^2} + \omega^2\right) \frac{\cos \omega' t - \cos \omega t}{\omega^2 - (\omega')^2} = \cos \omega' t$$

Taking the limit $\omega' \to \omega$, we have

$$\lim_{\omega' \to \omega} \frac{\cos \omega' t - \cos \omega t}{\omega^2 - (\omega')^2} = \frac{t}{2\omega} \sin \omega t$$

Thus we have the solution

$$
y(s; t) = A \frac{1 - (-1)^N}{N} \frac{t}{2\omega} \sin \omega t \sin \frac{N\pi s}{L}
$$

$$
+ \frac{2}{\pi} A \cos \omega t \sum_{n \neq N} \frac{\sin (n\pi s/L)}{(n\pi c/L)^2 - \omega^2} \frac{1 - (-1)^n}{n}
$$

This solution is not a periodic function of time, but oscillates with increasing amplitude as time progresses. This is the phenomenon of *resonance*. ●

EXERCISES 2.4

1. Consider the initial-value problem for the symmetric plucked string

$$
y_{tt} = c^2 y_{ss} \qquad\qquad t > 0, 0 < s < L
$$

$$
y(0; t) = 0 = y(L; t) \qquad t > 0
$$

$$
y(s; 0) = \begin{cases} s & 0 < s \leq L/2 \\ L - s & L/2 < s < L \end{cases}
$$

$$
y_t(s; 0) = 0 \qquad\qquad 0 < s < L
$$

Make a graphical representation of the solution for $ct = L/4, L/2, 3L/4, L$. At what time is $y(s; t) = 0$ for all $0 < s < L$?

2. Let $y(s; t) = \sum_{n=1}^{\infty} B_n \cos (n\pi ct/L) \sin (n\pi s/L)$ be a solution of the vibrating string problem. Suppose that the string is further constrained at its midpoint, so that $y(L/2; t) = 0$ for all t. What condition does this impose on the coefficients B_n?

3. Let $y(s; t) = \sum_{n=1}^{\infty} B_n \cos (n\pi ct/L) \sin (n\pi s/L)$ be a solution of the vibrating string problem. Suppose that the string is constrained so that $y(L/3; t) = 0$ for all t. What condition does this impose on the coefficients B_n?

4. The *energy* of a vibrating string of tension T_0 and density $\rho = m/L$ is defined by

$$
E = \frac{1}{2} \int_0^L (\rho y_t^2 + T_0 y_s^2) \, ds
$$

Let

$$
y(s; t) = \sum_{n=1}^{\infty} (\tilde{A}_n \cos \omega_n t + \tilde{B}_n \sin \omega_n t) \sin \frac{n\pi s}{L}
$$

be a solution of the wave equation $\omega_n = n\pi c/L$. Use Parseval's theorem to write E as an infinite series involving $\tilde{A}_n$, $\tilde{B}_n$, and verify the law of conservation of energy.

5. Let $y(s; t)$ be a solution of the wave equation $y_{tt} = c^2 y_{ss}$ satisfying the boundary conditions $y(0; t) = 0 = y(L; t)$. By differentiating under the integral sign, show directly that $dE/dt = 0$, where E is the energy defined in Exercise 4.

6. Consider the initial-value problem for the plucked string of Exercise 2.4.2. Compute the total energy corresponding to normal modes $n \neq 1$, and show that this is less than one-half of the total energy.

7. Suppose $f_2(s)$ is an odd 2L-periodic function. Let $y(s; t) = [1/(2c)] \int_{s-ct}^{s+ct} f_2(z) \, dz$. Show that $y(0; t) = 0$, $y(L; t) = 0$ for all t.

8. Let $f(s)$, $-L < s < L$, be a piecewise smooth function. Extend f to a 2L-periodic

function defined for $-\infty < s < \infty$. Show that the resulting function is piecewise smooth on every interval $a < s < b$.

9. Let $f(s)$, $-L \leq s \leq L$, be an odd function. Extend f to a $2L$-periodic function defined for $-\infty < s < \infty$. Show that the resulting function is again odd.

10. Let $f(s)$, $-\infty < s < \infty$, be an odd function with the property that $f(L - s) = f(s)$. Show that $f(s + L/2) + f(s - L/2) = 0$ for all s.

11. Let $f(s)$, $0 \leq s < L$, satisfy $f(s) = f(L - s)$. Let y be the solution of the wave equation $y(s; 0) = f(s)$, $y_t(s; 0) = 0$. Show that $y(s, L/2c) = 0$ for $0 < s < L$.

12. Show that the Fourier series solution obtained in Example 2.4.2 converges uniformly for $0 \leq s \leq L$.

13. Consider the following initial-value problem for the wave equation $y_{tt} = c^2 y_{ss}$ for $t > 0$, $0 < s < L$; $y(0; t) = y(L; t) = 0$ for $t > 0$; $y(s; 0) = 0$ and $y_t(s; 0) = 1$ for $0 < s < L$.
    (a) Find the Fourier representation of the solution.
    (b) Find the explicit representation of the solution, and graph the solution for $t = 0$, $L/(4c)$, $L/(2c)$, $3L/(4c)$, L/c.

Exercises 14 to 16 are designed to review the techniques from calculus which are used in establishing d'Alembert's formula. Recall that the fundamental theorem of calculus states that $(d/dx) \int_0^x f(z)\, dz = f(x)$ for any continuous function f. The chain rule for differentiating composite functions states that $(d/dx)F(G(x)) = F'(G(x))G'(x)$.

14. Let f be a continuous function and set $F(x) = \int_0^x f(z)\, dz$. Show that $\int_{x-ct}^{x+ct} f(z)\, dz = F(x + ct) - F(x - ct)$.

15. Use the chain rule and the fundamental theorem of calculus to show that $(d/dx) \cdot \int_{x-ct}^{x+ct} f(z)\, dz = f(x + ct) - f(x - ct)$.

16. Use the chain rule and the fundamental theorem of calculus to show that $(d/dt) \cdot \int_{x-ct}^{x+ct} f(z)\, dz = cf(x + ct) + cf(x - ct)$.

17. The voltage $u(x; t)$ in a transmission cable is known to satisfy the partial differential equation $u_{tt} + 2au_t + a^2u = c^2 u_{xx}$, where a and c are positive constants. Let $y(x; t) = e^{at}u(x; t)$, and show that y satisfies the wave equation $y_{tt} = c^2 y_{xx}$.

18. Use Exercise 17 and d'Alembert's formula to solve the initial-value problem $u_{tt} + 2au_t + a^2u = c^2 u_{xx}$ for $t > 0$, $-\infty < x < \infty$, with initial conditions $u(x; 0) = g_1(x)$, $u_t(x; 0) = 0$.

19. Use Exercise 17 and d'Alembert's formula to solve the initial-value problem $u_{tt} + 2au_t + a^2u = c^2 u_{xx}$ for $t > 0$, $-\infty < x < \infty$, with initial conditions $u(x; 0) = 0$, $u_t(x; 0) = g_2(x)$.

20. A vibrating string with friction in a periodic force field is described by the equation $y_{tt} + 2ay_t - c^2 y_{ss} = A \cos \omega t$ and the boundary conditions $y(0; t) = 0 = y(L; t)$, where A, a, ω are positive constants. Find a particular solution which is also periodic in time.

21. The longitudinal displacements of a cylindrical bar of natural length L are described by the equation $y_{tt} = a^2 y_{ss}$. At time $t = 0$ both ends are released and left free, leading to the boundary conditions $y_s(0; t) = 0$, $y_s(L; t) = 0$ and the initial conditions $y(s; 0) = f(s)$, $y_t(s; 0) = 0$, where $f(s)$ is the initial longitudinal displacement.
    (a) Find a Fourier representation of the solution of the initial-value problem. (*Hint:* Use a cosine series.)
    (b) Determine the coefficients in the series if $f(s) = cs$.
    (c) Find an explicit (d'Alembert) representation of the solution in the general case.

22. (Refer to Exercise 21.) At time $t = 0$ the end of the bar at $s = 0$ is left free while a constant longitudinal force F_0 is applied at the end $s = L$. This leads to the boundary conditions $y_s(0; t) = 0$, $y_s(L; t) = F_0/E$, where E is *Young's modulus* and the initial conditions are $y(s; 0) = 0$, $y_t(s; 0) = 0$.

   (a) Find a steady-state solution of the equation and the boundary conditions.

   (b) Show that the solution can be written in the form $y(s; t) = Y(s; t) + U(s) + V(t)$, where $U(s)$ is the steady-state solution, V satisfies $V(0) = 0$, $V'(0) = 0$, $V''(t) = $ constant, and Y is a solution of Exercise 21a for the choice $f(s) = -U(s)$.

   (c) Find an explicit (d'Alembert) representation of the solution.

2.5   SOLUTIONS IN A RECTANGLE

In this section we consider boundary-value problems in rectangular coordinates (x, y, z) where more than one of these variables appears in the solution. This is in contrast to the previous sections where the solution depended on z alone. We will solve initial-value problems for the heat equation, boundary-value problems for Laplace's equation, and the wave equation for a vibrating membrane.

Double Fourier Series

The key idea in our work is a *double Fourier series*. To illustrate this, we display a double Fourier sine series

$$\sum_{m,n=1}^{\infty} A_{mn} \sin \frac{m\pi x}{L_1} \sin \frac{n\pi y}{L_2}$$

which may be used in problems involving a rectangle or column, $0 < x < L_1$, $0 < y < L_2$. Similarly, a double Fourier cosine series is of the form $\sum A_{mn} \cdot \cos(m\pi x/L_1) \cos(n\pi y/L_2)$. Clearly we could consider other combinations of these, where we mix sines and cosines for example. All double series of this type are of the form $\sum_{m,n} \varphi_m(x)\psi_n(y)$, where φ_m, ψ_n are the eigenfunctions of a Sturm-Liouville problem. Accordingly the corresponding functions of two variables obey suitable orthogonality relations. For example, in the case of double Fourier sine series, we have

$$\int_0^{L_2} \int_0^{L_1} \sin \frac{m\pi x}{L_1} \sin \frac{n\pi y}{L_2} \sin \frac{m'\pi x}{L_1} \sin \frac{n'\pi y}{L_2} \, dx \, dy$$

$$= \begin{cases} \frac{1}{4}L_1 L_2 & \text{if both } m = m' \text{ and } n = n' \\ 0 & \text{otherwise} \end{cases} \tag{2.5.1}$$

Many functions can be written as sums of multiple Fourier series. In the case of double Fourier sine series in the rectangle $0 < x < L_1$, $0 < y < L_2$, we have the expansion formulas

$$f(x, y) = \sum_{m,n=1}^{\infty} A_{mn} \sin \frac{m\pi x}{L_1} \sin \frac{n\pi y}{L_2} \tag{2.5.2}$$

$$A_{mn} = \frac{4}{L_1 L_2} \int_0^{L_2} \int_0^{L_1} f(x, y) \sin \frac{m\pi x}{L_1} \sin \frac{n\pi y}{L_2} \, dx \, dy$$

If $f(x, y)$ is a smooth function in the rectangle, the series converges to $f(x, y)$ for $0 < x < L_1$, $0 < y < L_2$.

Application to the Heat Equation (Homogeneous Boundary Conditions)

As our first application of double Fourier series, we consider the heat equation in a rectangular column $0 < x < L_1$, $0 < y < L_2$, with the homogeneous boundary conditions that $u = 0$ on all four sides of the column, $x = 0$, $x = L_1$, $y = 0$, $y = L_2$. The separated solutions, which depend on (x, y, t), are of the form

$$u(x, y; t) = f_1(x)f_2(y)g(t)$$

Substituting in the heat equation and dividing by Ku, we have

$$\frac{g'(t)}{Kg(t)} = \frac{f_1''(x)}{f_1(x)} + \frac{f_2''(y)}{f_2(y)}$$

The left side depends on t, and the right side depends on (x, y). Therefore each is a constant, which we call $-\lambda$. Applying the same argument to the right side, we see that both f_1''/f_1 and f_2''/f_2 are constants, to be called $-\mu_1$ and $-\mu_2$, respectively. Therefore we have the ordinary differential equations

$$g'(t) + \lambda Kg(t) = 0$$

$$f_1''(x) + \mu_1 f_1(x) = 0 \tag{2.5.3}$$

$$f_2''(y) + \mu_2 f_2(y) = 0$$

where $\lambda = \mu_1 + \mu_2$. From the boundary conditions we must have $f_1(0) = 0$, $f_1(L_1) = 0$, $f_2(0) = 0$, $f_2(L_2) = 0$. The solutions of these Sturm-Liouville problems are $f_1(x) = \sin(m\pi x/L_1)$, $\mu_1 = (m\pi/L_1)^2$, $f_2(y) = \sin(n\pi y/L_2)$, $\mu_2 = (n\pi/L_2)^2$; we have $g(t) = e^{-\lambda Kt}$, where $\lambda = (m\pi/L_1)^2 + (n\pi/L_2)^2$. Thus we have the separated solutions of the heat equation in the column, with zero boundary conditions:

$$\sin \frac{m\pi x}{L_1} \sin \frac{n\pi y}{L_2} e^{-\lambda_{mn}Kt} \qquad m, n = 1, 2, \ldots$$

$$\lambda_{mn} = \left(\frac{m\pi}{L_1}\right)^2 + \left(\frac{n\pi}{L_2}\right)^2$$

The indices m, n are independent of each other. A general solution of the heat equation with zero boundary conditions is obtained as a superposition

$$\sum_{m,n=1}^{\infty} B_{mn} \sin \frac{m\pi x}{L_1} \sin \frac{n\pi y}{L_2} e^{-\lambda_{mn}Kt} \qquad (2.5.4)$$

We can use these to solve initial-value problems for the heat equation.

Example 2.5.1. Solve the initial-value problem for the heat equation $u_t = K\nabla^2 u$ in the column $0 < x < L_1$, $0 < y < L_2$, with the boundary conditions $u(0, y; t) = 0$, $u(L_1, y; t) = 0$, $u(x, 0; t) = 0$, $u(x, L_2; t) = 0$ and the initial condition $u(x, y; 0) = 1$, for $0 < x < L_1$, $0 < y < L_2$. Find the relaxation time.

Solution. We look for the solution as a sum of separated solutions (2.5.4). The Fourier coefficients B_{mn} are obtained by setting $t = 0$ and using the initial conditions. Thus we have

$$1 = \sum_{m,n=1}^{\infty} B_{mn} \sin \frac{m\pi x}{L_1} \sin \frac{n\pi y}{L_2}$$

$$B_{mn} = \frac{4}{L_1 L_2} \int_0^{L_2} \int_0^{L_1} \sin \frac{m\pi x}{L_1} \sin \frac{n\pi y}{L_2} \, dx \, dy$$

$$= \frac{4}{\pi^2} \frac{[1 - (-1)^m][1 - (-1)^n]}{mn}$$

The solution is

$$u(x, y; t) = \frac{4}{\pi^2} \sum_{m,n=1}^{\infty} \frac{1 - (-1)^m}{m} \frac{1 - (-1)^n}{n} \sin \frac{m\pi x}{L_1} \sin \left(\frac{n\pi y}{L_2}\right) e^{-\lambda_{mn}Kt}$$

For each $t > 0$, the series for u, u_x, u_y, u_{xx}, u_{yy}, u_t converge uniformly for $0 \le x \le L_1$, $0 \le y \le L_2$, and hence u is a rigorous solution of the heat equation. The relaxation time is given by the first term of the series, when $m = 1$, $n = 1$. Thus $(\pi^2/L_1^2 + \pi^2/L_2^2)K\tau = 1$, and the relaxation time is $\tau = L_1^2 L_2^2/[K\pi^2(L_1^2 + L_2^2)]$. ●

Initial-value problems for the heat equation in a three-dimensional box can be handled similarly, by using Fourier series in three variables. For example, if we have the cube $0 < x < L$, $0 < y < L$, $0 < z < L$, then we find the separated solutions in the form $\sin(m\pi x/L) \sin(n\pi y/L) \sin(p\pi z/L) e^{-\lambda Kt}$, where (m, n, p) are independent indices and $\lambda = (m\pi/L)^2 + (n\pi/L)^2 + (p\pi/L)^2$. The initial conditions determine an expansion in a triple Fourier series and hence lead to a series solution of the initial-value problem.

Application to the Laplace Equation

We now turn to Laplace's equation $\nabla^2 u = 0$ in rectangular coordinates. To find separated solutions, we let $u(x, y, z) = f_1(x)f_2(y)f_3(z)$ and substitute in

$\nabla^2 u = 0$, with the result

$$0 = \frac{f_1''(x)}{f_1(x)} + \frac{f_2''(y)}{f_2(y)} + \frac{f_3''(z)}{f_3(z)}$$

By the methods of separation of variables, each of these must be constant, and we have the ordinary differential equations

$$f_1''(x) + \mu_1 f_1(x) = 0$$

$$f_2''(y) + \mu_2 f_2(y) = 0 \qquad (2.5.5)$$

$$f_3''(z) + \mu_3 f_3(z) = 0$$

where the separation constants (μ_1, μ_2, μ_3) must obey the relation $\mu_1 + \mu_2 + \mu_3 = 0$. To proceed further, we must know the form of the boundary conditions. The method will become clear from the following examples.

Example 2.5.2. Find the separated solutions $u(x, y)$ of Laplace's equation in the column $0 < x < L_1$, $0 < y < L_2$ satisfying the boundary conditions $u(0, y) = 0$, $u(L_1, y) = 0$.

Solution. Since u depends on (x, y), we take $\mu_3 = 0$, $\mu_1 + \mu_2 = 0$. The boundary conditions require that we solve the Sturm-Liouville problem $f_1''(x) + \mu_1 f_1(x) = 0$, $f_1(0) = 0$, $f_1(L_1) = 0$, whose solution is $f_1(x) = \sin(n\pi x/L_1)$. Thus $\mu_2 = -\mu_1 = -(n\pi/L_1)^2$, and the equation for f_2 is $f_2'' - (n\pi/L_1)^2 f_2 = 0$, whose solution is $f_2(y) = A \cosh(n\pi y/L_1) + B \sinh(n\pi y/L_1)$. We have the separated solutions $u(x, y) = \sin(n\pi x/L_1) [A \cosh(n\pi y/L_1) + B \sinh(n\pi y/L_1)]$, $n = 1, 2, \ldots$ ●

Once we have determined the separated solutions of Laplace's equation, we may solve boundary-value problems for Laplace's equation by the methods of Fourier series. The success of the method depends on considering one side at a time.

Example 2.5.3. Solve Laplace's equation in the column $0 < x < L_1$, $0 < y < L_2$ with the boundary conditions $u(0, y) = 0$, $u(L_1, y) = 0$, $u(x, 0) = 0$, $u(x, L_2) = 1$.

Solution. In Example 2.5.2 we found the separated solutions satisfying the first two boundary conditions

$$u(x, y) = \sin \frac{n\pi x}{L_1} \left(A \cosh \frac{n\pi y}{L_1} + B \sinh \frac{n\pi y}{L_1} \right)$$

The third boundary condition requires $0 = A \sin(n\pi x/L_1)$, $0 < x < L_1$; hence $A = 0$. We look for the solution as a superposition

$$u(x, y) = \sum_{n=1}^{\infty} B_n \sin \frac{n\pi x}{L_1} \sinh \frac{n\pi y}{L_1}$$

The fourth boundary condition requires that

$$1 = \sum_{n=1}^{\infty} B_n \sin \frac{n\pi x}{L_1} \sinh \frac{n\pi L_2}{L_1}$$

which is a Fourier sine series in x. But we know the Fourier sine expansion of

$$1 = \frac{2}{\pi} \sum_{n=1}^{\infty} \frac{1 - (-1)^n}{n} \sin \frac{n\pi x}{L_1}$$

and therefore

$$B_n \sinh \frac{n\pi L_2}{L_1} = \frac{2}{\pi} \frac{1 - (-1)^n}{n}$$

and the solution is

$$u(x, y) = \frac{2}{\pi} \sum_{n=1}^{\infty} \frac{[1 - (-1)^n] \sin (n\pi x/L_1) \sinh (n\pi y/L_1)}{n \sinh (n\pi L_2/L_1)} \qquad \bullet$$

We now outline the procedure for solving Laplace's equation $\nabla^2 u = 0$ in the column $0 < x < L_1$, $0 < y < L_2$ with four nonzero boundary conditions: $u(x, 0) = T_1$, $u(x, L_2) = T_2$, $u(0, y) = T_3$, $u(L_1, y) = T_4$. Following Example 2.5.3, we can obtain $u_2(x, y)$, the solution of the problem when T_1, T_3, T_4 are replaced by zero. By interchanging the roles of x and y, we can similarly obtain $u_4(x, y)$, the solution of the problem with T_1, T_2, T_3 replaced by zero.

$$u_4(x, y) = \frac{2T_4}{\pi} \sum_{n=1}^{\infty} [1 - (-1)^n] \frac{\sin (n\pi y/L_2) \sinh (n\pi x/L_2)}{n \sinh (n\pi L_1/L_2)}$$

The remainder of the solution can be obtained by the substitutions $x \to L_1 - x$ and $y \to L_2 - y$; thus

$$u_3(x, y) = \frac{2T_3}{\pi} \sum_{n=1}^{\infty} [1 - (-1)^n] \frac{\sin [n\pi(L_1 - x)/L_2] \sinh (n\pi y/L_2)}{n \sinh (n\pi L_1/L_2)}$$

with $u_1(x, y)$ obtained similarly. Adding these four functions gives a solution of Laplace's equation which satisfies all four boundary conditions; thus we have $u(x, y) = u_1(x, y) + u_2(x, y) + u_3(x, y) + u_4(x, y)$. This can be illustrated, as in Fig. 2.5.1.

In these examples the solution of Laplace's equation in the column $0 < x < L_1$, $0 < y < L_2$ was written as a sum of ordinary Fourier series. If we consider problems in a cube, we encounter double Fourier series.

Example 2.5.4. Find the separated solutions of Laplace's equation in the cube $0 < x < L$, $0 < y < L$, $0 < z < L$ satisfying the boundary conditions $u(0, y, z) = 0$, $u(L, y, z) = 0$, $u(x, 0, z) = 0$, $u(x, L, z) = 0$.

Solution. Referring to (2.5.5), we have the Sturm-Liouville problems $f_1'' + \mu_1 f_1 = 0$, $f_2'' + \mu_2 f_2 = 0$ with the boundary conditions $f_1(0) = 0$, $f_1(L_1) = 0$, $f_2(0) = 0$, $f_2(L_2) = 0$. Thus $f_1(x) = \sin (m\pi x/L_1)$, $f_2(y) = \sin (n\pi y/L_2)$, $\mu_1 = -(m\pi/L_1)^2$,

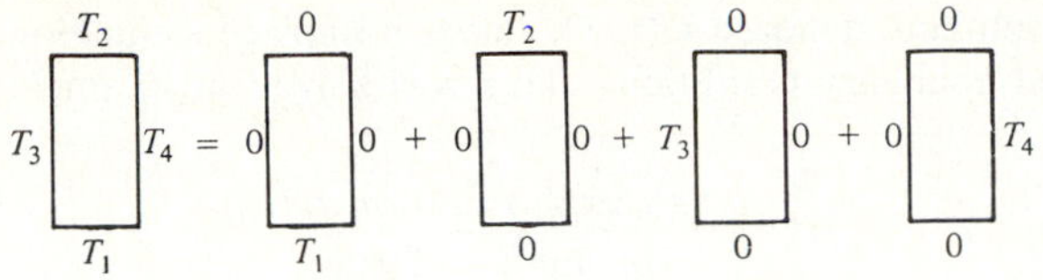

FIGURE 2.5.1

$\mu_2 = -(n\pi/L_2)^2$, and $\mu_3 = -(\mu_1 + \mu_2)$. Thus

$$f_3(z) = A \cosh \frac{\pi z}{L} \sqrt{m^2 + n^2} + B \sinh \frac{\pi z}{L} \sqrt{m^2 + n^2}$$

and the separated solutions are

$$\sin \frac{m\pi x}{L} \sin \frac{n\pi y}{L} \left(A \cosh \frac{\pi z}{L} \sqrt{m^2 + n^2} + B \sinh \frac{\pi z}{L} \sqrt{m^2 + n^2} \right) \qquad \bullet$$

These may be used to solve a boundary-value problem, as in the case of the rectangular column.

Example 2.5.5. Solve Laplace's equation $\nabla^2 u = 0$ in the cube $0 < x < L$, $0 < y < L$, $0 < z < L$ with the boundary conditions $u(0, y, z) = 0$, $u(L, y, z) = 0$, $u(x, 0, z) = 0$, $u(x, L, z) = 0$, $u(x, y, 0) = 0$, $u(x, y, L) = 1$.

Solution. Proceeding as in Example 2.5.3, we take $A = 0$ in the separated solutions found in Example 2.5.4 and obtain the solution

$$u(x, y, z) = \frac{4}{\pi^2} \sum_{m,n=1}^{\infty} \frac{[1 - (-1)^m][1 - (-1)^n] \sin \frac{m\pi x}{L} \sin \frac{n\pi y}{L} \sinh \frac{\pi z}{L} \sqrt{m^2 + n^2}}{mn \sinh \pi\sqrt{m^2 + n^2}} \qquad \bullet$$

Application to the Heat Equation (Inhomogeneous Boundary Conditions)

We now combine the above methods to solve initial-value problems for the heat equation where the boundary conditions are *inhomogeneous*. To do this, we use the five-stage method developed in Sec. 2.3. In stage 1 we obtain the steady-state solution by solving Laplace's equation, whose solution has been discussed above. In stage 2 we use this steady-state solution to transform the problem to homogeneous boundary conditions, for which the separated solutions have been obtained. In stage 3 we form a superposition of these and use the initial conditions to find the Fourier coefficients. In stage 4 we verify the solution, and in stage 5 we obtain the relaxation time.

Example 2.5.6. Solve the initial-value problem for the heat equation $u_t = K\nabla^2 u$ in the column $0 < x < L_1$, $0 < y < L_2$ with the boundary conditions $u(0, y; t) = 0$, $u(L_1, y; t) = 0$, $u(x, 0; t) = 0$, $u(x, L_2; t) = T_2$ and the initial condition $u(x, y; 0) = T_1$, where T_1, T_2 are positive constants.

Solution. The steady-state solution, denoted $U(x, y)$, satisfies Laplace's equation $\nabla^2 U = 0$ with the indicated boundary conditions. This was solved in Example 2.5.3:

$$U(x, y) = \frac{2T_2}{\pi} \sum_{m=1}^{\infty} [1 - (-1)^m] \frac{\sin (m\pi x/L_1) \sinh (m\pi y/L_1)}{m \sinh (m\pi L_2/L_1)}$$

Letting $v(x, y; t) = u(x, y; t) - U(x, y)$, we have $v_t = K\nabla^2 v$ with $v = 0$ on all four sides, while $v(x, y; 0) = T_1 - U(x, y)$. We look for $v(x, y; t)$ as a superposition of separated solutions with zero boundary conditions:

$$v(x, y; t) = \sum_{m,n} B_{mn} \sin \frac{m\pi x}{L_1} \sin \frac{n\pi y}{L_2} e^{-\lambda_{mn}Kt}$$

The Fourier coefficients B_{mn} are obtained by setting $t = 0$; thus $T_1 - U(x, y) = \sum_{m,n=1}^{\infty} B_{mn} \sin (m\pi x/L_1) \sin (n\pi y/L_2)$. To obtain the required coefficients B_{mn}, we begin with the Fourier sine expansion of the hyperbolic sine:

$$\sinh \frac{a\pi y}{L_2} = \frac{2 \sinh a\pi}{\pi} \sum_{n=1}^{\infty} \frac{n(-1)^{n+1}}{a^2 + n^2} \sin \frac{n\pi y}{L_2} \qquad 0 < y < L_2$$

Letting $a = mL_2/L_1$ and substituting in the formula for $U(x, y)$, we have the double Fourier sine series

$$U(x, y) = \frac{4T_2}{\pi^2} \sum_{m,n=1}^{\infty} \frac{[1 - (-1)^m]n(-1)^{n+1}}{m[n^2 + (mL_2/L_1)^2]} \sin \frac{m\pi x}{L_1} \sin \frac{n\pi y}{L_2}$$

Likewise, the double Fourier sine expansion of T_1 is

$$T_1 = \frac{4T_1}{\pi^2} \sum_{m,n=1}^{\infty} \frac{1 - (-1)^m}{m} \frac{1 - (-1)^n}{n} \sin \frac{m\pi x}{L_1} \sin \frac{n\pi y}{L_2}$$

From these we can obtain B_{mn} as the coefficient of $\sin (m\pi x/L_1) \sin (n\pi y/L_2)$. The solution of the original boundary-value problem is

$$u(x, y; t) = U(x, y) + \sum_{m,n=1}^{\infty} B_{mn} \sin \frac{m\pi x}{L_1} \sin \frac{n\pi y}{L_2} e^{-\lambda_{mn}Kt}$$

For each $t > 0$, the series for u, u_x, u_{xx}, u_y, u_{yy}, u_t are uniformly convergent for $0 < x < L_1$, $0 < y < L_2$; thus $u(x, y; t)$ is a rigorous solution of the boundary-value problem. The relaxation time is given by the first term of the series, $(\pi^2/L_1^2 + \pi^2/L_2^2)K\tau = 1$ if $B_{11} \neq 0$. ●

Application to the Wave Equation (Nodal Lines)*

We now turn to an example involving the wave equation. The small transverse vibrations of a tightly stretched membrane are governed by the two-dimensional wave equation. This has the form

$$u_{tt} = c^2(u_{xx} + u_{yy})$$

where $u(x, y; t)$ denotes the transverse displacement of the membrane from its

---

* See also the Appendix on "Using Mathematica."

equilibrium position and c is a positive constant. The membrane is supposed to cover the rectangle $0 < x < L_1$, $0 < y < L_2$ with the edges fixed; thus $u(x, y; t) = 0$ for $x = 0$, $x = L_1$, $y = 0$, $y = L_2$.

As a first step, we find the separated solutions of the wave equation satisfying these boundary conditions. These have the form

$$u(x, y; t) = f_1(x)f_2(y)g(t)$$

Substituting in the wave equation and separating variables, we have

$$\frac{g''(t)}{c^2 g(t)} = \frac{f_1''(x)}{f_1(x)} + \frac{f_2''(y)}{f_1(y)}$$

The left side depends on t alone, while the right side depends on (x, y); thus each is a constant, say $-\lambda$. Introducing further separation constants μ_1, μ_2, we have the ordinary differential equations

$$g''(t) + \lambda c^2 g(t) = 0$$

$$f_1''(x) + \mu_1 f_1(x) = 0$$

$$f_2''(y) + \mu_2 f_2(y) = 0$$

with $\lambda = \mu_1 + \mu_2$. The boundary conditions require $f_1(0) = 0$, $f_1(L_1) = 0$, $f_2(0) = 0$, $f_2(L_2) = 0$. These Sturm-Liouville problems have the solutions $f_1(x) = \sin(m\pi x/L_1)$, $f_2(y) = \sin(n\pi y/L_2)$, $\mu_1 = (m\pi/L_1)^2$, $\mu_2 = (n\pi/L_2)^2$. Thus $\lambda > 0$, and we can write $g(t) = A \cos \sqrt{\lambda} ct + B \sin \sqrt{\lambda} ct$; thus we have the separated solutions of the rectangular membrane

$$u_{mn}(x, y; t) = \sin \frac{m\pi x}{L_1} \sin \frac{n\pi y}{L_2} (A \cos \omega_{mn} t + B \sin \omega_{mn} t)$$

$$\omega_{mn} = c\left[\left(\frac{m\pi}{L_1}\right)^2 + \left(\frac{n\pi}{L_2}\right)^2\right]^{1/2}$$

(2.5.6)

The constants A, B can be chosen to fit various initial conditions by using the superposition principle and the methods of double Fourier series.

Example 2.5.7. Solve the initial-value problem for the vibrating membrane with the initial conditions $u(x, y; 0) = 0$, $u_t(x, y; 0) = 1$.

Solution. We look for the solution as a superposition of separated solutions.

$$u(x, y; t) = \sum_{m,n} \sin \frac{m\pi x}{L_1} \sin \frac{n\pi y}{L_2} (A_{mn} \cos \omega_{mn} t + B_{mn} \sin \omega_{mn} t)$$

Setting $t = 0$, we must have $0 = \sum_{m,n} A_{mn} \sin(m\pi x/L_1) \sin(n\pi y/L_2)$; thus $A_{mn} = 0$. Differentiating and setting $t = 0$, we must have

$$1 = \sum_{m,n} B_{mn} \omega_{mn} \sin \frac{m\pi x}{L_1} \sin \frac{n\pi y}{L_2}$$

Thus $B_{mn}\omega_{mn} = (2/\pi)^2[1 - (-1)^m][1 - (-1)^n]/(mn)$, and we have found the formal solution of the problem:

$$u(x, y; t) = \frac{4}{\pi^2} \sum_{m,n=1}^{\infty} \frac{[1 - (-1)^m][1 - (-1)^n]}{mn\omega_{mn}} \sin \frac{m\pi x}{L_1} \sin \frac{n\pi y}{L_2} \sin \omega_{mn}t \qquad \bullet$$

In contrast with the heat equation, the double Fourier series solutions obtained do not converge sufficiently fast to verify the convergence of the series for the various derivatives u_x, u_{xx}, u_y, u_{yy}, u_t, u_{tt}. Therefore we usually restrict our attention to solutions which contain only a finite number of terms. In particular, it is interesting to examine the separated solutions obtained previously. Indeed, these solutions have an important physical interpretation as *standing waves*. The profile is given by the function $\sin(m\pi x/L_1) \sin(n\pi y/L_2)$. This function undergoes a periodic oscillation with period $T = 2\pi/\omega_{mn}$, owing to the time dependence $A \cos \omega_{mn}t + B \sin \omega_{mn}t$. The membrane can be divided into various zones, depending on the sign of u; these zones are divided by curves, which are called *nodal lines*. We illustrate in Fig. 2.5.2 the nodal lines for some of the separated solutions we have just found.

We now consider in more detail the vibrating *square* membrane with $0 < x < L, 0 < y < L$. Thus we take $L_1 = L, L_2 = L$. The first 10 frequencies of the separated solutions are listed in Table 2.5.1.

We distinguish between *simple frequencies* and *multiple frequencies*. For example, $\omega_{11} = (\pi c/L)\sqrt{2}$ is a simple frequency, whereas $\omega_{12} = (\pi c/L)\sqrt{5}$ is a multiple frequency, of multiplicity 2. We may obtain solutions with a more

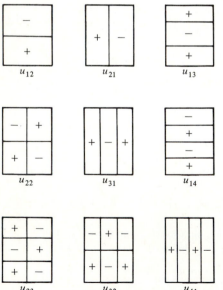

FIGURE 2.5.2
Nodal lines of a rectangular membrane.

TABLE 2.5.1

(m, n)	ω_{mn}	$u_{mn}(x, y)$
$(1, 1)$	$(\pi c/L)\sqrt{2}$	$\sin(\pi x/L)\sin(\pi y/L)$
$(2, 1)$	$(\pi c/L)\sqrt{5}$	$\sin(2\pi x/L)\sin(\pi y/L)$
$(1, 2)$	$(\pi c/L)\sqrt{5}$	$\sin(\pi x/L)\sin(2\pi y/L)$
$(2, 2)$	$(\pi c/L)\sqrt{8}$	$\sin(2\pi x/L)\sin(2\pi y/L)$
$(3, 1)$	$(\pi c/L)\sqrt{10}$	$\sin(3\pi x/L)\sin(\pi y/L)$
$(1, 3)$	$(\pi c/L)\sqrt{10}$	$\sin(\pi x/L)\sin(3\pi y/L)$
$(3, 2)$	$(\pi c/L)\sqrt{13}$	$\sin(3\pi x/L)\sin(2\pi y/L)$
$(2, 3)$	$(\pi c/L)\sqrt{13}$	$\sin(2\pi x/L)\sin(3\pi y/L)$
$(4, 1)$	$(\pi c/L)\sqrt{17}$	$\sin(4\pi x/L)\sin(\pi y/L)$
$(1, 4)$	$(\pi c/L)\sqrt{17}$	$\sin(\pi x/L)\sin(4\pi y/L)$

complex nodal structure by taking sums of solutions corresponding to a multiple frequency. For example,

$$u_{12} - u_{21} = \sin\frac{\pi x}{L}\sin\frac{2\pi y}{L} - \sin\frac{2\pi x}{L}\sin\frac{\pi y}{L}$$

$$= 2\sin\frac{\pi x}{L}\sin\frac{\pi y}{L}\left(\cos\frac{\pi y}{L} - \cos\frac{\pi x}{L}\right)$$

Thus $u_{12} - u_{21} = 0$ for $x = y$, since the factor $\cos(\pi y/L) - \cos(\pi x/L) = 0$ on that line. If we consider the multiple frequency $\omega_{13} = (\pi c/L)\sqrt{10}$ and study the difference $u_{13} - u_{31}$, we can show that this function is zero along both lines $x + y = L$ and $y = x$. These possibilities are illustrated in the following diagram. More complex diagrams may be obtained by considering higher values of (m, n).

$u_{12} - u_{21}$

$u_{13} - u_{31}$

FIGURE 2.5.3

Application to Poisson's Equation

Poisson's equation $\nabla^2 u = -\rho$ is very similar to Laplace's equation. We look for a particular solution U which does not necessarily satisfy all the boundary conditions. The function $v = u - U$ then satisfies Laplace's equation with some new boundary conditions. The solution can be found by the above method for Laplace's equation. We illustrate with a simple example.

Example 2.5.8. Find the solution of Poisson's equation $\nabla^2 u = -1$ in the rectangle $0 < x < L_1$, $0 < y < L_2$ satisfying the boundary condition that $u = 0$ on all four sides.

Solution. A particular solution depending on x alone satisfies $u_{xx} = -1$, thus $U = \frac{1}{2}x(L_1 - x)$ satisfies the equation and two of the boundary conditions. The function $v = u - U$ satisfies Laplace's equation $\nabla^2 v = 0$ with the boundary conditions that $v = 0$ when $x = 0$, $x = L_1$, and $v = -\frac{1}{2}x(L_1 - x)$ when $y = 0$, $y = L_2$. This is sought as a series of separated solutions in the form

$$v(x, y) = \sum_{n \geq 1} \sin \frac{n\pi x}{L_1} \left(A_n \cosh \frac{n\pi y}{L_1} + B_n \sinh \frac{n\pi y}{L_1} \right)$$

To satisfy the remaining boundary conditions, we must have

$$-\frac{1}{2} x(L_1 - x) = \sum_{n \geq 1} A_n \sin \frac{n\pi x}{L_1}$$

$$-\frac{1}{2} x(L_1 - x) = \sum_{n \geq 1} \sin \frac{n\pi x}{L_1} \left(A_n \cosh \frac{n\pi y}{L_2} + B_n \sinh \frac{n\pi y}{L_2} \right)$$

The Fourier series of $x(L_1 - x)$ is $(8L^2/\pi^3)\sum_{n \text{ odd}} (1/n^3) \sin (n\pi x/L_1)$. Equating coefficients gives $A_n = -[4L^2/(n^3\pi^3)]$ and

$$B_n = -\frac{4L^2}{n^3\pi^3} \frac{1 - \sinh (n\pi L_2/L_1)}{\cosh (n\pi L_2/L_1)}$$

for n odd and $B_n = 0$ for n even. This leads to

$$v(x, y) = -\frac{4L_1^2}{\pi^3} \sum_{n \text{ odd}} \frac{\sin (n\pi x/L_1)}{n^3 \sinh (n\pi L_2/L_1)} \left[\sinh \frac{n\pi y}{L_1} + \sinh \frac{n\pi(L_2 - y)}{L_1} \right]$$

Combining this with the Fourier series for U, we may write the solution of Poisson's equation as a single series in the form

$$u(x, y) = \frac{4L_1^2}{\pi^3} \sum_{n \text{ odd}} n^{-3} \sin \frac{n\pi x}{L_1} \left[2 - \frac{\sinh (n\pi y/L_1) + \sinh (n\pi L_2 y/L_1)}{\sinh (n\pi L_2/L_1)} \right] \quad \bullet$$

We now compare this with another method of solution.

Example 2.5.9. Find the solution of Poisson's equation $u_{xx} + u_{yy} = -1$ in the rectangle $0 < x < L_1$, $0 < y < L_2$ in the form of a double Fourier sine series $u(x, y) = \sum_{m,n \geq 1} A_{mn} \sin (m\pi x/L_1) \sin (n\pi y/L_2)$.

Solution. The indicated sine products form a complete set of functions in the square and satisfy the required boundary condition. Therefore it is natural to expect that we can expand the solution of Poisson's equation in such a series. For this sum we have $u_{xx} + u_{yy} = -\sum_{mn \geq 1} A_{mn}[(m\pi/L_1)^2 + (n\pi/L_2)^2] \sin (m\pi x/L_1) \sin (n\pi y/L_2)$. The Fourier series of the constant function is

$$1 = \frac{4}{\pi} \sum_{m \text{ odd}, n \text{ odd}} \frac{1}{mn} \sin \frac{m\pi x}{L_1} \sin \frac{n\pi y}{L_2}$$

The equality of the two series is guaranteed by choosing $A_{mn} = [4/(mn\pi)] \cdot [(m\pi/L_1)^2 + (n\pi/L_2)^2]$ for m and n both odd and choosing $A_{mn} = 0$ otherwise. The resulting solution is written as

$$u(x, y) = \frac{4}{\pi} \sum_{m,n \text{ odd}} \frac{1}{mn[(m\pi/L_1)^2 + (n\pi/L_2)]} \sin \frac{m\pi x}{L_1} \sin \frac{n\pi y}{L_2} \quad \bullet$$

It is interesting to compare the two forms of series solutions obtained for Poisson's equation. In the first case, the nth term of the series for $v(x, y)$ tends to zero as an exponential does, since the numerator $\simeq \exp(n\pi y/L_1)$ while the denominator $\simeq \exp(n\pi L_2/L_1)$ so that their quotient $\simeq \exp[n\pi(y - L_2)/L_1]$, an exponential rate of decrease to zero. In the second case, the (m, n)th term of the double Fourier sine series tends to zero at the rate of $1/[mn(m^2 + n^2)]$, an algebraic rate of decay; the series for the derivatives u_x, u_{xx}, u_y, u_{yy} converge at an even slower rate. Thus for the purposes of numerical approximations, the first method is far preferable to the second. With hindsight, the superiority of this method is explained by the fact that the functions which we chose in the series have some relation to the differential equation $u_{xx} + u_{yy} = -\rho$ and were not simply an arbitrary set of functions satisfying the boundary conditions, as was the case for the double Fourier sine series in the second method.

EXERCISES 2.5

1. Solve the initial-value problem for the heat equation $u_t = K\nabla^2 u$ in the column $0 < x < L_1$, $0 < y < L_2$ with the boundary conditions $u(0, y; t) = 0$, $u_x(L_1, y; t) = 0$, $u(x, 0; t) = 0$, $u_y(x, L_2; t) = 0$ and the initial condition $u(x, y; 0) = 1$. Find the relaxation time.

2. Solve the initial-value problem for the heat equation $u_t = K\nabla^2 u$ in the column $0 < x < L_1$, $0 < y < L_2$ with the boundary conditions $u(0, y; t) = 0$, $u(L_1, y; t) = 0$, $u_y(x, 0; t) = 0$, $u_y(x, L_2; t) = 0$ and the initial condition $u(x, y; 0) = 1$. Find the relaxation time.

3. Find separated solutions of Laplace's equation $\nabla^2 u = 0$ in the column $0 < x < L_1$, $0 < y < L_2$ and satisfying the boundary conditions $u_x(0, y) = 0$, $u_x(L, y) = 0$.

4. Solve Laplace's equation $\nabla^2 u = 0$ in the column $0 < x < L_1$, $0 < y < L_2$ with the boundary conditions $u_x(0, y) = 0$, $u_x(L_1, y) = 0$, $u(x, 0) = 0$, $u(x, L_2) = x$.

5. Solve Laplace's equation $\nabla^2 u = 0$ in the column $0 < x < L_1$, $0 < y < L_2$ with the boundary conditions $u_x(0, y) = 0$, $u_x(L_1, y) = 0$, $u(x, 0) = 0$, $u(x, L_2) = 1$.

6. Solve Laplace's equation $\nabla^2 u = 0$ in the column $0 < x < L_1$, $0 < y < L_2$ with the boundary conditions $u_x(0, y) = 0$, $u_x(L_1, y) = 0$, $u(x, 0) = T_1$, $u(x, L_2) = T_2$, where T_1, T_2 are constants.

7. Find separated solutions of Laplace's equation $\nabla^2 u = 0$ in the cube $0 < x < L$, $0 < y < L$, $0 < z < L$ and satisfying the boundary conditions $u(0, y, z) = 0$, $u_x(L, y, z) = 0$, $u(x, 0, z) = 0$, $u_y(x, L, z) = 0$.

8. Find the separated solutions of Laplace's equation $\nabla^2 u = 0$ in the cube $0 < x < L$, $0 < y < L$, $0 < z < L$ and satisfying the boundary conditions $u_x(0, y, z) = 0$, $u_x(L, y, z) = 0$, $u_y(x, 0, z) = 0$, $u_y(x, L, z) = 0$.

9. Solve Laplace's equation $\nabla^2 u = 0$ in the cube $0 < x < L$, $0 < y < L$, $0 < z < L$ with the boundary conditions

$$u(0, y, z) = 0 \qquad u_x(L, y, z) = 0 \qquad u(x, 0, z) = 0$$

$$u_y(x, L, z) = 0 \qquad u(x, y, 0) = 0 \qquad u(x, y, L) = 1$$

10. Solve Laplace's equation $\nabla^2 u = 0$ in the cube $0 < x < L$, $0 < y < L$, $0 < z < L$ with the boundary conditions

$$u_x(0, y, z) = 0 \qquad u_x(L, y, z) = 0 \qquad u_y(x, 0, z) = 0$$

$$u_y(x, L, z) = 0 \qquad u_z(x, y, 0) = 0 \qquad u(x, y, L) = 1$$

11. Solve the initial-value problem for the heat equation $u_t = K\nabla^2 u$ in the column $0 < x < L_1$, $0 < y < L_2$ with the boundary conditions $u_x(0, y; t) = 0$, $u_x(L, y; t) = 0$, $u(x, 0; t) = T_1$, $u(x, L_2; t) = T_2$ and the initial condition $u(x, y; 0) = T_3$, where T_1, T_2, T_3 are constants.

12. Solve the initial-value problem for the heat equation $u_t = K\nabla^2 u$ in the square column $0 < x < L$, $0 < y < L$ with the boundary conditions $u(0, y; t) = 0$, $u(L, y; t) = 0$, $u(x, 0; t) = 0$, $u(x, L_2; t) = T_1$ and the initial condition $u(x, y; 0) = 0$. Find the relaxation time.

13. Solve the initial-value problem for the vibrating membrane in the square $0 < x < L$, $0 < y < L$ with the initial conditions $u(x, y; 0) = 3 \sin(\pi x/L) \sin(2\pi x/L) + 4 \sin(3\pi x/L) \sin(5\pi y/L)$, $u_t(x, y; 0) = 0$.

14. Find the separated solutions of the wave equation $u_{tt} = c^2(u_{xx} + u_{yy})$ in the square $0 < x < L$, $0 < y < L$ with the boundary conditions $u_x(0, y; t) = 0$, $u_x(L, y; t) = 0$, $u_y(x, 0; t) = 0$, $u_y(x, L; t) = 0$ and the initial condition $u(x, y; 0) = 0$.

15. Find the first 10 frequencies of the separated solutions found in Exercise 14.

16. A vibrating membrane in the shape of an isosceles right triangle covers the region $0 < y < x < L$. Show that $u_{mn} - u_{nm}$ satisfies the wave equation with zero boundary conditions, where $m < n$ and u_{mn} is given by (2.5.6) with $L_1 = L_2$.

17. Find the first 10 frequencies of the separated solutions found in Exercise 16.

18. Solve the initial-value problem for the wave equation on the isosceles right triangle $0 < y < x < L$ with zero boundary conditions and the initial conditions $u(x, y, 0) = 0$, $u_t(x, y, 0) = 1$.

19. Consider a vibrating membrane covering the equilateral triangle $0 < y < x\sqrt{3}$, $0 < y < \sqrt{3}(L - x)$. Let $d_1 = y$, $d_2 = \frac{1}{2}(x\sqrt{3} - y)$, $d_3 = \frac{1}{2}[\sqrt{3}(L - x) - y]$ be the distance from the point (x, y) to the ith side of the triangle, $i = 1, 2, 3$. For $n = 1, 2, \ldots$, let

$$u_n(x, y; t) = \left(\sin \frac{4\pi n d_1}{L\sqrt{3}} + \sin \frac{4\pi n d_2}{L\sqrt{3}} + \sin \frac{4\pi n d_3}{L\sqrt{3}} \right) \cos \omega t$$

Show that u_n satisfies the wave equation with zero boundary conditions if ω is suitably chosen. (*Hint:* To check the boundary conditions, you may use the fact that $d_1 + d_2 + d_3 = L\sqrt{3}/2$.)

*20. Let $n = 1$ in Exercise 19. Show that if $u_1(x, y; 0) = 0$ and $d_1 > 0$, $d_2 > 0$, then $d_3 = 0$.

*21. Use Exercise 20 to show that $u_1(x, y; 0) \neq 0$ inside the equilateral triangle.

*22. Let $n = 2$ in Exercise 19. Show that $u_2(x, y; 0) = 0$ along the lines $d_1 = L\sqrt{3}/4$, $d_2 = L\sqrt{3}/4$, $d_3 = L\sqrt{3}/4$ and draw a diagram.

*23. Let $n = 3$ in Exercise 19. Plot the nodal lines along which $u_3(x, y; 0) = 0$.

24. In Example 2.5.6 compute B_{11}. Show that $B_{11} = 0$ if and only if $2T_1/L_1^2 = T_2/(L_1^2 + L_2^2)$. Show that, for a square column, this is the statement that the initial temperature is the average of the boundary temperatures.

CHAPTER
3

BOUNDARY-VALUE PROBLEMS IN CYLINDRICAL COORDINATES

INTRODUCTION

In this chapter we consider boundary-value problems in regions with circular or cylindrical boundaries. Section 3.1 is devoted to Laplace's equation in a circle, which can be solved in terms of elementary functions, that is, trigonometric Fourier series. Then we develop the properties of Bessel functions in Sec. 3.2 in order to solve more complicated problems. These include the vibrating drumhead and heat flow in a cylinder, which are treated in Secs. 3.3 and 3.4.

3.1 LAPLACE'S EQUATION IN CYLINDRICAL COORDINATES

Laplacian in Polar Coordinates

As a first step, we express the laplacian ∇^2 in cylindrical coordinates. Recall the equations of transformation between rectangular and cylindrical coordinates

$$x = \rho \cos \varphi \qquad (3.1.1a)$$

$$y = \rho \sin \varphi \qquad (3.1.1b)$$

$$z = z \qquad (3.1.1c)$$

These are polar coordinates in the xy plane. Let $u(x, y, z)$ be a smooth function and $U(\rho, \varphi, z)$ the corresponding function in the cylindrical coordinates: $U(\rho, \varphi, z) = u(\rho \cos \varphi, \rho \sin \varphi, z)$. We wish to express $u_{xx} + u_{yy}$ in terms of the partial derivatives $U_{\rho\rho}$, $U_{\varphi\varphi}$, $U_{\rho\varphi}$, U_{ρ}, U_{φ}.

We begin with the chain rule for partial derivatives:

$$u_x = \frac{\partial u}{\partial x} = \frac{\partial U}{\partial \rho} \frac{\partial \rho}{\partial x} + \frac{\partial U}{\partial \varphi} \frac{\partial \varphi}{\partial x}$$

$$u_y = \frac{\partial u}{\partial y} = \frac{\partial U}{\partial \rho} \frac{\partial \rho}{\partial y} + \frac{\partial U}{\partial \varphi} \frac{\partial \varphi}{\partial y}$$

We must therefore determine $\partial \rho/\partial x$, $\partial \varphi/\partial x$, $\partial \rho/\partial y$, $\partial \varphi/\partial y$. From (3.1.1$a$) and (3.1.1$b$) we have $\rho^2 = x^2 + y^2$, so that

$$2\rho \frac{\partial \rho}{\partial x} = 2x \qquad\qquad 2\rho \frac{\partial \rho}{\partial y} = 2y$$

$$\frac{\partial \rho}{\partial x} = \frac{x}{\rho} = \cos \varphi \qquad\qquad \frac{\partial \rho}{\partial y} = \frac{y}{\rho} = \sin \varphi$$

Differentiating both sides of (3.1.1b) with respect to x, we have

$$0 = \frac{\partial \rho}{\partial x} \sin \varphi + \rho \cos \varphi \frac{\partial \varphi}{\partial x}$$

$$= \cos \varphi \sin \varphi + \rho \cos \varphi \frac{\partial \varphi}{\partial x}$$

$$\frac{\partial \varphi}{\partial x} = -\frac{\sin \varphi}{\rho}$$

Differentiating both sides of (3.1.1a) with respect to y, we have

$$0 = \frac{\partial \rho}{\partial y} \cos \varphi - \rho \sin \varphi \frac{\partial \varphi}{\partial y}$$

$$= \sin \varphi \cos \varphi - \rho \sin \varphi \frac{\partial \varphi}{\partial y}$$

$$\frac{\partial \varphi}{\partial y} = \frac{\cos \varphi}{\rho}$$

$$\frac{\partial u}{\partial x} = \cos \varphi \frac{\partial U}{\partial \rho} - \frac{\sin \varphi}{\rho} \frac{\partial U}{\partial \varphi} \qquad\qquad (3.1.2)$$

$$\frac{\partial u}{\partial y} = \sin \varphi \frac{\partial U}{\partial \rho} + \frac{\cos \varphi}{\rho} \frac{\partial U}{\partial \varphi}$$

Having written the operators $\partial/\partial x$ and $\partial/\partial y$ in cylindrical coordinates, we apply them again to obtain

$$\frac{\partial^2 u}{\partial x^2} = \cos^2 \varphi \, \frac{\partial^2 U}{\partial \rho^2} + \frac{2 \cos \varphi \sin \varphi}{\rho^2} \frac{\partial U}{\partial \varphi} - \frac{2 \sin \varphi \cos \varphi}{\rho} \frac{\partial^2 U}{\partial \rho \, \partial \varphi}$$

$$+ \frac{\sin^2 \varphi}{\rho} \frac{\partial U}{\partial \rho} + \frac{\sin^2 \varphi}{\rho^2} \frac{\partial^2 U}{\partial \varphi^2} \tag{3.1.3}$$

$$\frac{\partial^2 u}{\partial y^2} = \sin^2 \varphi \, \frac{\partial^2 U}{\partial \rho^2} - \frac{2 \sin \varphi \cos \varphi}{\rho^2} \frac{\partial U}{\partial \varphi} + \frac{2 \sin \varphi \cos \varphi}{\rho} \frac{\partial^2 U}{\partial \rho \, \partial \varphi}$$

$$+ \frac{\cos^2 \varphi}{\rho} \frac{\partial U}{\partial \rho} + \frac{\cos^2 \varphi}{\rho^2} \frac{\partial^2 U}{\partial \varphi^2} \tag{3.1.4}$$

Adding (3.1.3) and (3.1.4) gives

$$\frac{\partial^2 u}{\partial x^2} + \frac{\partial^2 u}{\partial y^2} = \frac{\partial^2 U}{\partial \rho^2} + \frac{1}{\rho} \frac{\partial U}{\partial \rho} + \frac{1}{\rho^2} \frac{\partial^2 U}{\partial \varphi^2}$$

so that the laplacian becomes

$$\boxed{\nabla^2 u = \frac{\partial^2 U}{\partial \rho^2} + \frac{1}{\rho} \frac{\partial U}{\partial \rho} + \frac{1}{\rho^2} \frac{\partial^2 U}{\partial \varphi^2} + \frac{\partial^2 U}{\partial z^2}} \tag{3.1.5}$$

Example 3.1.1. Compute $\nabla^2[x(x^2 + y^2)^3]$.

Solution. The function $u = x(x^2 + y^2)^3$ is expressed in cylindrical coordinates as $U = \rho^7 \cos \varphi$; we have $U_\rho = 7\rho^6 \cos \varphi$, $U_{\rho\rho} = 42\rho^5 \cos \varphi$, $U_{\varphi\varphi} = -\rho^7 \cos \varphi$. Therefore the laplacian is given by $\nabla^2 u = 48\rho^5 \cos \varphi = 48x(x^2 + y^2)^2$.   ●

The correspondence $u \to U$ produces a smooth function $U(\rho, \varphi, z)$ for every smooth function $u(x, y, z)$. But many smooth functions of (ρ, φ, z) do not arise in this manner. For example, $U = \rho$ is a function of (x, y, z), but it is not smooth since the partial derivative $\partial\rho/\partial x$ is undefined at $\rho = 0$. The example $U = \varphi$ does not correspond to a smooth function of (x, y, z) since φ changes by 2π when we make a 360° rotation about the z axis and return to the same point. These theoretical difficulties need not hinder us in our work if we check that each solution we obtain in cylindrical coordinates corresponds to a smooth function of (x, y, z). For example, $U = \rho^n \cos n\varphi$ can be written as a polynomial in (x, y) if n is an integer and therefore is a smooth function; if n is not an integer, U is not a smooth function of (x, y, z). With these precautions in mind, we now formulate and solve some boundary-value problems in cylindrical coordinates. By abuse of notation, we write the solution as $u = u(\rho, \varphi, z)$, assumed to be a smooth function of (x, y, z).

Separated Solutions of Laplace's Equation

As our first application of (3.1.5), we obtain separated solutions of Laplace's equation in cylindrical coordinates independent of z. Assuming a solution of the form $u(\rho, \varphi) = f_1(\rho)f_2(\varphi)$, we substitute in the equation $\nabla^2 u = 0$, with the result

$$0 = u_{\rho\rho} + \frac{1}{\rho} u_\rho + \frac{1}{\rho^2} u_{\varphi\varphi} = f_1'' f_2 + \frac{1}{\rho} f_1' f_2 + \frac{1}{\rho^2} f_1 f_2''$$

Dividing by $f_1 f_2$ and multiplying by ρ^2, we have

$$\rho^2 \frac{f_1'' + (1/\rho)f_1'}{f_1} + \frac{f_2''}{f_2} = 0$$

The first term depends only on ρ and the second only on φ; therefore both are constant. This leads to the ordinary differential equations

$$f_2'' + \lambda f_2 = 0 \qquad f_2(-\pi) = f_2(\pi) \qquad f_2'(-\pi) = f_2'(\pi)$$

$$f_1'' + \frac{1}{\rho} f_1' - \frac{\lambda}{\rho^2} f_1 = 0$$

where λ is the separation constant; f_2 must satisfy the indicated periodic boundary conditions because u is supposed to be a smooth (single-valued) function of (x, y, z). The solution to this Sturm-Liouville problem for f_2 was obtained in Sec. 0.4, with the result

$$\lambda = m^2 \qquad f_2(\varphi) = A_m \cos m\varphi + B_m \sin m\varphi \qquad m = 0, 1, 2, \ldots$$

The equation for f_1 is a form of Euler's equidimensional equation. For $m \neq 0$, it has solutions $f_1(\rho) = \rho^m, \rho^{-m}$; for $m = 0$, the solutions are $f_1(\rho) = 1, \ln \rho$. Combining these, we have found the following separated solutions of Laplace's equation:

$$u(\rho, \varphi) = \begin{cases} \rho^m(A_m \cos m\varphi + B_m \sin m\varphi) & m = 1, 2, \ldots \\ A_0 + B_0 \ln \rho & m = 0 \\ \rho^{-m}(C_m \cos m\varphi + D_m \sin m\varphi) & m = 1, 2, \ldots \end{cases} \qquad (3.1.6)$$

Application to Boundary-Value Problems

Example 3.1.2. Find the solution of Laplace's equation in the region $R_1 < \rho < R_2$ with the boundary conditions $u(R_1, \varphi) = T_1$, $u(R_2, \varphi) = T_2$, where T_1, T_2 are constants.

Solution. Since the boundary conditions are independent of φ, we use the previous separated solutions with $m = 0$. Thus

$$u(\rho) = A_0 + B_0 \ln \rho$$

To satisfy the boundary conditions, we must have

$$T_1 = A_0 + B_0 \ln R_1 \qquad \rho = R_1$$

$$T_2 = A_0 + B_0 \ln R_2 \qquad \rho = R_2$$

Solving these simultaneous linear equations yields

$$B_0 = \frac{T_2 - T_1}{\ln (R_2/R_1)} \qquad A_0 = T_1 - (T_2 - T_1) \frac{\ln R_1}{\ln (R_2/R_1)}$$

The solution can be written in the form

$$u(\rho) = T_1 + (T_2 - T_1) \frac{\ln (\rho/R_1)}{\ln (R_2/R_1)} \qquad \bullet$$

This example shows that the average temperature is not assumed at the average radius $\rho = \frac{1}{2}(R_1 + R_2)$ but instead at the geometric mean $\rho = (R_1 R_2)^{1/2}$. Indeed, $u[(R_1 R_2)^{1/2}] = T_1 + \frac{1}{2}(T_2 - T_1) = \frac{1}{2}(T_1 + T_2)$.

In many problems of practical interest, it is required to compute the steady-state flux.

Example 3.1.3. Two concentric cylinders of radii $R_1 = 10$ cm and $R_2 = 50$ cm are maintained at temperatures $T_1 = 100°C$ and $T_2 = 0°C$. Find the steady-state flux from the outer cylinder if the conductivity is $k = 0.35$ cal/(s $\cdot$ cm $\cdot$ °C).

Solution. The flux is given by

$$-k \frac{\partial u}{\partial \rho} = \frac{-k}{R_2} \frac{T_2 - T_1}{\ln (R_2/R_1)}$$

$$= \frac{(0.35)(100)}{50 \ln 5} = 0.435 \qquad \bullet$$

We now use separation of variables to solve the boundary-value problem for Laplace's equation when the boundary values depend on φ. We have the problem

$$\nabla^2 u = u_{\rho\rho} + \frac{1}{\rho} u_\rho + \frac{1}{\rho^2} u_{\varphi\varphi} = 0 \qquad R_1 < \rho < R_2, -\pi \leq \varphi \leq \pi$$

$$u(R_1, \varphi) = G_1(\varphi) \qquad -\pi \leq \varphi \leq \pi$$

$$u(R_2, \varphi) = G_2(\varphi) \qquad -\pi \leq \varphi \leq \pi$$

where $G_1(\varphi)$ and $G_2(\varphi)$ are piecewise smooth functions which give the temperature on the inner and outer cylinders. We obtain the solution in the form

$$u(\rho, \varphi) = A_0 + B_0 \ln \rho + \sum_{m=1}^{\infty} \rho^m (A_m \cos m\varphi + B_m \sin m\varphi)$$

$$+ \sum_{m=1}^{\infty} \rho^{-m} (C_m \cos m\varphi + D_m \sin m\varphi) \qquad (3.1.7)$$

To satisfy the boundary conditions, we must have

$$G_1(\varphi) = A_0 + B_0 \ln R_1 + \sum_{m=1}^{\infty} (R_1^m A_m + R_1^{-m} C_m) \cos m\varphi$$

$$+ \sum_{m=1}^{\infty} (R_1^m B_m + R_1^{-m} D_m) \sin m\varphi$$

$$G_2(\varphi) = A_0 + B_0 \ln R_2 + \sum_{m=1}^{\infty} (R_2^m A_m + R_2^{-m} C_m) \cos m\varphi$$

$$+ \sum_{m=1}^{\infty} (R_2^m B_m + R_2^{-m} D_m) \sin m\varphi$$

Using the orthogonality of the functions $\{1, \cos m\varphi, \sin m\varphi\}$, we can obtain the coefficients by the Fourier formulas:

$$\frac{1}{2\pi} \int_{-\pi}^{\pi} G_1(\varphi) \, d\varphi = A_0 + B_0 \ln R_1$$

$$\frac{1}{2\pi} \int_{-\pi}^{\pi} G_2(\varphi) \, d\varphi = A_0 + B_0 \ln R_2$$

$$\frac{1}{\pi} \int_{-\pi}^{\pi} G_1(\varphi) \cos m\varphi \, d\varphi = R_1^m A_m + R_1^{-m} C_m \qquad m = 1, 2, \ldots$$

$$\frac{1}{\pi} \int_{-\pi}^{\pi} G_2(\varphi) \cos m\varphi \, d\varphi = R_2^m A_m + R_2^{-m} C_m \qquad m = 1, 2, \ldots$$

$$\frac{1}{\pi} \int_{-\pi}^{\pi} G_1(\varphi) \sin m\varphi \, d\varphi = R_1^m B_m + R_1^{-m} D_m \qquad m = 1, 2, \ldots$$

$$\frac{1}{\pi} \int_{-\pi}^{\pi} G_2(\varphi) \sin m\varphi \, d\varphi = R_2^m B_m + R_2^{-m} D_m \qquad m = 1, 2, \ldots$$

These three systems of simultaneous equations can be solved to obtain the coefficients A_m, B_m, C_m, D_m.

Example 3.1.4. Solve Laplace's equation in the cylinder $1 < \rho < 2$ with the boundary conditions $u(1, \varphi) \equiv 0$ and $u(2, \varphi) = 1$ if $0 < \varphi < \pi$ and $u(2, \varphi) = -1$ if $-\pi < \varphi < 0$.

Solution. In this case the six equations become $A_0 + B_0 \ln R_1 = 0$, $A_0 + B_0 \ln R_2 = 0$, $A_m + C_m = 0$, $2^m A_m + 2^{-m} C_m = 0$, $B_m + D_m = 0$, $2^m B_m + 2^{-m} D_m = 2[1 - (-1)^m]/(m\pi)$. This gives $A_0 = 0$, $B_0 = 0$, $A_m = 0$, $C_m = 0$, $B_m = -D_m = 2[1 - (-1)^m]/[m\pi(2^m - 2^{-m})]$. The solution is

$$u(\rho, \varphi) = \frac{2}{\pi} \sum_{m=1}^{\infty} \frac{\rho^m - \rho^{-m}}{2^m - 2^{-m}} \frac{1 - (-1)^m}{m} \sin m\varphi \qquad 1 < \rho < 2, -\pi \leq \varphi \leq \pi \quad \bullet$$

It is not difficult to show that the formal solution (3.1.7) of Laplace's equation is indeed a smooth function. To simplify the writing, we consider the case $R_1 = 0$, when we solve the problem in the interior of the cylinder $0 \leq \rho \leq R = R_2$. The formal solution is

$$u(\rho, \varphi) = A_0 + \sum_{m=1}^{\infty} \rho^m (A_m \cos m\varphi + B_m \sin m\varphi) \qquad 0 < \rho < R \quad (3.1.8)$$

where $R^m A_m$ and $R^m B_m$ are the Fourier coefficients of the boundary function $G(\varphi)$, $-\pi < \varphi < \pi$. This function is piecewise smooth and therefore bounded by a constant M. This means that we must have

$$|R^m A_m| \leq 2M \qquad |R^m B_m| \leq 2M \qquad m = 1, 2, \ldots$$

Therefore the terms of the series (3.1.8) are no larger than $4M(\rho/R)^m$. But this is the general term of a convergent series for $\rho < R$. This also shows that the series (3.1.8) is uniformly convergent for $0 \leq \rho \leq \rho_0$, where ρ_0 is any number less than R. Furthermore, each term in the series $\rho^m \cos m\varphi$ and $\rho^m \sin m\varphi$ is a smooth function of (x, y). The partial derivatives are easily found to be

$$\frac{\partial}{\partial x} (\rho^m \cos m\varphi) = m\rho^{m-1} \cos (m - 1)\varphi$$

$$\frac{\partial}{\partial x} (\rho^m \sin m\varphi) = m\rho^{m-1} \sin (m - 1)\varphi$$

The general term of the series for $\partial u/\partial x$ is no larger than $(4Mm/R)(\rho/R)^{m-1}$. Therefore this series is also uniformly convergent for $0 \leq \rho \leq \rho_0$, where ρ_0 is any number less than R. Continuing in this way, we see that the series for u_{xx}, u_y, u_{yy} are uniformly convergent, and therefore we can differentiate term by term and verify that u satisfies Laplace's equation.

Uniqueness of Solutions

The solution of Laplace's equation in a bounded region is *unique*. To be specific, suppose that we have two solutions u_1, u_2 of Laplace's equation $\nabla^2 u = 0$ in the region $0 \leq \rho < R$, with the same boundary values $u_1(R, \varphi) = u_2(R, \varphi)$ for $-\pi \leq \varphi \leq \pi$. The function $u = u_1 - u_2$ satisfies Laplace's equation $\nabla^2 u = 0$ in the region $0 \leq \rho < R$, with boundary value $u(R, \varphi) = 0$. We may apply Green's theorem $\oint (M \, dx + N \, dy) = \iint (N_x - M_y) \, dx \, dy$ where the line integral is taken over the circle $\rho = R$ and the double integral is taken over the disk $\rho < R$, with $M = -uu_y$, $N = uu_x$. The line integral is zero since $u = 0$ on the circle, while the integrand in the double integral is $N_x - M_y = (uu_x)_x + (uu_y)_y = uu_{xx} + (u_x)^2 + uu_{yy} + (u_y)^2 = u(u_{xx} + u_{yy}) + (u_x^2 + u_y^2)$. The first term in parentheses is zero since $\nabla^2 u = 0$, and we are left with $0 = \iint (u_x^2 + u_y^2) \, dx \, dy$. Both u_x^2 and u_y^2 are nonnegative, and their integrals are zero; hence they must both be zero. This means that $u(x, y)$ must be a constant. But $u = 0$ on the circle, which proves that $u_1 = u_2$, the desired uniqueness result.

Exterior Problems

The separated solutions of Laplace's equation can be used to solve boundary-value problems in the *exterior* of a cylinder. Suppose that we wish to determine the solution $u(\rho, \varphi)$ of Laplace's equation $\nabla^2 u = 0$ for $\rho > R$ and satisfying the boundary condition $u(R, \varphi) = G(\varphi)$, a given piecewise smooth function. We require in addition that the solution be bounded: $|u(\rho, \varphi)| \leq M$ for some constant M; otherwise, we may have nonuniqueness. As an example, the function $u_1(\rho, \varphi) = 1 + (\rho/R) \cos \varphi - (R/\rho) \cos \varphi$ satisfies Laplace's equation in the exterior of the cylinder $\rho > R$ and $u_1(R, \varphi) = 1$. The function $u_2(\rho, \varphi) \equiv 1$ also satisfies Laplace's equation with the same boundary values.

Example 3.1.5. Find the bounded solution of Laplace's equation $\nabla^2 u = 0$ in the exterior of the cylinder $\rho > R$ and satisfying the boundary conditions $u(R, \varphi) = 1$ if $0 < \varphi < \pi$, $u(R, \varphi) = -1$ if $-\pi < \varphi < 0$.

Solution. To ensure boundedness, we take a sum of separated solutions of the form

$$u(\rho, \varphi) = A_0 + \sum_{n=1}^{\infty} \rho^{-n}(A_n \cos n\varphi + B_n \sin n\varphi)$$

To satisfy the boundary conditions, we must have $A_n = 0$, $R^{-n}B_n = [2/(n\pi)] \cdot [1 - (-1)^n]$. The solution is

$$u(\rho, \varphi) = \frac{2}{\pi} \sum_{n=1}^{\infty} [1 - (-1)^n] \frac{R^n}{n\rho^n} \sin n\varphi \qquad \bullet$$

Wedge Domains

Often we encounter boundary-value problems for Laplace's equation in a *wedge domain,* of the form $0 < \rho < R$, $0 < \varphi < \alpha$ where $\alpha < 2\pi$. In this case the separated solutions are still of the form $\rho^n(A \cos n\varphi + B \sin n\varphi)$, but n is no longer necessarily an integer. The allowed values of n will depend on the wedge opening α and the nature of the boundary conditions which are imposed at $\varphi = 0$, $\varphi = \alpha$.

Example 3.1.6. Find the solution of Laplace's equation in the wedge domain $0 < \rho < 1$, $0 < \varphi < \alpha$ and satisfying the boundary conditions $u(\rho, 0) = 0$, $u(\rho, \alpha) = 0$ for $0 < \rho < 1$ and $u(1, \varphi) = 1$ for $0 < \varphi < \alpha$.

Solution. For the separated solutions $\rho^n(A \cos n\varphi + B \sin n\varphi)$, the boundary conditions at $\varphi = 0$ and $\varphi = \alpha$ require that $A = 0$, $\sin n\alpha = 0$. Therefore $n = m\pi/\alpha$ for $m = 1, 2, 3, \ldots$. To satisfy the boundary condition at $\rho = 1$, we try a sum of separated solutions: $u(\rho, \varphi) = \sum_{m=1}^{\infty} B_m \rho^{m\pi/\alpha} \sin (m\pi\varphi/\alpha)$. We must have $1 = \sum_{m=1}^{\infty} B_m \sin (m\pi\varphi/\alpha)$, $0 < \varphi < \alpha$; therefore $B_m = [2/(m\pi)][1 - (-1)^m]$. The solution is $u(\rho, \varphi) = (2/\pi) \sum_{m=1}^{\infty} [1 - (-1)^m] m^{-1} \rho^{m\pi/\alpha} \sin (m\pi\varphi/\alpha)$.   ●

Neumann's Problem

In all these problems we have solved Laplace's equation with *Dirichlet* boundary conditions, where $u(R, \varphi)$ is given. We can also solve problems with *Neumann* boundary conditions where $\partial u/\partial \rho = G(\varphi)$ is a given piecewise smooth function. This problem features nonuniqueness: If $u(\rho, \varphi)$ is a solution, then $u(\rho, \varphi) + C$ is also a solution for any constant C. To ensure a unique solution, we may require that $u = 0$ when $\rho = 0$. In addition, $G(\varphi)$ must satisfy the condition $\int_{-\pi}^{\pi} G(\varphi) \, d\varphi = 0$. This is not too surprising since the solutions of Laplace's equation represent steady-state temperature distributions and $\partial u/\partial \rho = G(\varphi)$ is proportional to the flux across the boundary of the cylinder. It is natural to expect that, in steady state, the total flux across the boundary is zero.

To solve Neumann problems for Laplace's equation in the cylinder $0 \leq \rho \leq R$, we try a sum of separated solutions of the form

$$u(\rho, \varphi) = \sum_{n=1}^{\infty} \rho^n(A_n \cos n\varphi + B_n \sin n\varphi)$$

The boundary conditions require that $nR^{n-1}A_n$ and $nR^{n-1}B_n$ be the Fourier coefficients of the piecewise smooth function $G(\varphi)$, $-\pi < \varphi < \pi$.

Example 3.1.7. Find the solution of Laplace's equation in the cylinder $0 < \rho < R$ and satisfying the Neumann boundary condition $\partial u/\partial \rho = 1$ for $0 < \varphi < \pi$ and $\partial u/\partial \rho = -1$ for $-\pi < \varphi < 0$.

Solution. The solution is sought in the series form $u(\rho, \varphi) = \sum_{n=1}^{\infty} \rho^n(A_n \cos n\varphi + B_n \sin n\varphi)$. The boundary conditions require that $A_n = 0$ and $nR^{n-1}B_n = [2/(n\pi)][1 - (-1)^n]$. The solution is

$$u(\rho, \varphi) = \frac{2}{\pi} \sum_{n=1}^{\infty} \frac{R}{n^2} [1 - (-1)^n] \left(\frac{\rho}{R}\right)^n \sin n\varphi$$ ●

Explicit Representation by Poisson's Formula

We have obtained the Fourier representation of the solution of Laplace's equation in the cylinder $0 \le \rho < R$ by means of formula (3.1.8). We now show that this can be converted to an explicit representation, the *Poisson integral formula*. To do this, we recall the formulas for the Fourier coefficients, written with the integration variable θ:

$$A_0 = \frac{1}{2\pi} \int_{-\pi}^{\pi} G(\theta)\, d\theta$$

$$R^m A_m = \frac{1}{\pi} \int_{-\pi}^{\pi} G(\theta) \cos m\theta\, d\theta \qquad m = 1, 2, \ldots$$

$$R^m B_m = \frac{1}{\pi} \int_{-\pi}^{\pi} G(\theta) \sin m\theta\, d\theta \qquad m = 1, 2, \ldots$$

We rewrite the expression $A_m \cos m\varphi + B_m \sin m\varphi$ in the form

$$\pi R^m (A_m \cos m\varphi + B_m \sin m\varphi)$$

$$= \int_{-\pi}^{\pi} G(\theta)(\cos m\theta \cos m\varphi + \sin m\theta \sin m\varphi)\, d\theta$$

$$= \int_{-\pi}^{\pi} G(\theta) \cos m(\theta - \varphi)\, d\theta \qquad m = 1, 2, \ldots$$

Substituting this in (3.1.8), we have formally

$$u(\rho, \varphi) = \frac{1}{\pi} \int_{-\pi}^{\pi} G(\theta) \left[\frac{1}{2} + \sum_{m=1}^{\infty} \left(\frac{\rho}{R}\right)^m \cos m(\theta - \varphi) \right] d\theta \qquad 0 \le \rho < R$$

The inner sum was evaluated in Exercise 1.5.6, with the result

$$\frac{1}{2} + \sum_{m=1}^{\infty} \left(\frac{\rho}{R}\right)^m \cos m(\theta - \varphi) = \frac{R^2 - \rho^2}{2[R^2 + \rho^2 - 2\rho R \cos(\theta - \varphi)]}$$

This gives the Poisson integral formula

$$\boxed{u(\rho, \varphi) = \frac{1}{2\pi} \int_{-\pi}^{\pi} \frac{R^2 - \rho^2}{R^2 + \rho^2 - 2R\rho \cos(\theta - \varphi)} G(\theta)\, d\theta \\ 0 \le \rho < R} \qquad (3.1.9)$$

The Poisson integral formula has some important theoretical consequences for solutions of Laplace's equation. We list these facts in the form of a proposition.

Proposition. Let $u(\rho, \varphi)$ be a solution of Laplace's equation in the cylinder $0 \le \rho < R$, represented by the Poisson integral formula (3.1.9), where $G(\theta)$ is a given continuous function for $-\pi \le \theta \le \pi$. Then

1. $$u(\rho, \varphi)\bigg|_{\rho=0} = \frac{1}{2\pi} \int_{-\pi}^{\pi} G(\theta)\, d\theta \qquad \text{average of } G$$

2.
$$u(\rho, \varphi) \le \max_{-\pi \le \theta \le \pi} G(\theta)$$

3. If $u(\rho_0, \varphi_0) = \max_{-\pi \le \theta \le \pi} G(\theta)$ for some $\rho_0 < R$, $-\pi \le \varphi_0 \le \pi$, then $u(\rho, \varphi)$ is a constant for all $0 \le \rho \le R$, $-\pi \le \varphi \le \pi$.

Proof. Taking $\rho = 0$ in the Poisson integral formula gives property 1. To prove 2 and 3, let $M = \max_{-\pi \le \theta \le \pi} G(\theta)$; integrating the uniformly convergent Fourier series $\frac{1}{2} + \sum_{n=1}^{\infty} (\rho/R)^n \cos n(\theta - \varphi)$ term by term for $-\pi \le \theta \le \pi$, we see that the total integral of the Poisson kernel is 1. Thus for any (ρ, φ) we have

$$2\pi[M - u(\rho, \varphi)] = \int_{-\pi}^{\pi} [M - G(\theta)] \frac{R^2 - \rho^2}{R^2 + \rho^2 - 2R\rho \cos(\theta - \varphi)} \, d\theta$$

The integrand is a nonnegative continuous function for $-\pi \le \theta \le \pi$. Therefore $M - u(\rho, \varphi) \ge 0$, and we have proved 2. If this is zero for some (ρ_0, φ_0), then the integrand must be zero for all θ; thus $G(\theta) = M$ for all θ, $-\pi \le \theta \le \pi$. Referring to Poisson's formula, we see that $u(\rho, \varphi) = M$ for all $0 \le \rho < R$, $-\pi \le \varphi \le \pi$. ∎

EXERCISES 3.1

In Exercises 1 to 5 use formula (3.1.5) to compute the indicated quantities.

1. $\nabla^2(\rho^4 \cos 2\varphi)$
2. $\nabla^2(\rho^2 \cos 2\varphi)$
3. $\nabla^2(\rho^n)$, $n = 1, 2, \ldots$
4. $\nabla^2(\rho^n \cos m\varphi)$, $m, n = 1, 2, \ldots$
5. $\nabla^2(e^\rho \cos \varphi)$

6. Which of the functions in Exercises 1 to 5 corresponds to a smooth function of (x, y, z)?
7. Show that formula (3.1.5) can be written in the form

$$\nabla^2 u = \frac{1}{\rho} \frac{\partial}{\partial \rho} \left(\rho \frac{\partial U}{\partial \rho} \right) + \frac{1}{\rho^2} \frac{\partial^2 U}{\partial \varphi^2} + \frac{\partial^2 U}{\partial z^2}$$

8. Let $u = f(x^2 + y^2)^{1/2}$, where f is a smooth function. Show that $\nabla^2 u = f'' + (1/\rho)f' = (1/\rho)(\rho f')'$.
9. Use Exercise 8 to find the general solution of $\nabla^2 f(\rho) = 0$.
10. Use Exercise 8 to find the general solution of $\nabla^2 f(\rho) = -1$.
11. Find the solution of $\nabla^2 f(\rho) = 0$ satisfying the boundary conditions $f(1) = 3$, $f(2) = 5$.
12. Find the solution of $\nabla^2 f(\rho) = -1$ satisfying the boundary conditions $f(1) = 0$, $f(2) = 1$.
13. Find the solution $u(\rho, \varphi)$ of Laplace's equation in the cylinder $0 \le \rho < R$ satisfying the boundary conditions $u(R, \varphi) = 1 + \cos 2\varphi + 3 \sin 3\varphi$, $-\pi \le \varphi \le \pi$.
14. Find the solution $u(\rho, \varphi)$ of Laplace's equation in the cylindrical region $1 < \rho < 2$ and satisfying the boundary conditions $u(1, \varphi) = \cos 2\varphi$, $u(2, \varphi) = 1$ for $-\pi \le \varphi \le \pi$.

15. Find the solution $u(\rho, \varphi)$ of Laplace's equation in the cylindrical region $1 < \rho < 2$ and satisfying the boundary conditions $u(1, \varphi) \equiv 0$, $u(2, \varphi) = 0$ for $-\pi < \varphi < 0$ and $u(2, \varphi) = 1$ for $0 < \varphi < \pi$.

16. Find the bounded solution of Laplace's equation in the exterior of the cylinder $\rho > R$ and satisfying the boundary condition $u(R, \varphi) = 3 + 4 \cos 2\varphi + 5 \sin 3\varphi$, $-\pi \le \varphi \le \pi$.

17. Let $u(\rho, \varphi)$ be the bounded solution of Laplace's equation in the exterior of the cylinder $\rho > R$, with the boundary condition $u(R, \varphi) = G(\varphi)$, a given piecewise smooth function. Show formally that

$$u(\rho, \varphi) \rightarrow \frac{1}{2\pi} \int_{-\pi}^{\pi} G(\varphi) \, d\varphi \qquad \text{when } \rho \rightarrow \infty$$

18. Find the separated solutions of Laplace's equation in the wedge domain $0 < \rho < 1$, $0 < \varphi < \pi/2$ and satisfying the boundary conditions $u(\rho, 0) = 0$, $u(\rho, \pi/2) = 0$.

19. Use the separated solutions found in Exercise 18 to solve Laplace's equation in the wedge domain $0 < \rho < 1$, $0 < \varphi < \pi/2$ with the boundary conditions $u(\rho, 0) = 0$, $u(\rho, \pi/2) = 0$ for $0 < \rho < 1$ and $u(1, \varphi) = 1$ for $0 < \varphi < \pi/2$.

20. Find separated solutions of Laplace's equation in the wedge domain $0 < \rho < 1$, $0 < \varphi < \pi$ and satisfying the boundary conditions $u_\varphi(\rho, 0) = 0$, $u_\varphi(\rho, \pi) = 0$ for $0 < \rho < 1$.

21. Use the separated solutions found in Exercise 20 to find the solution of Laplace's equation in the wedge domain $0 < \rho < 1$, $0 < \varphi < \pi$ and satisfying the boundary conditions $u_\varphi(\rho, 0) = 0$, $u_\varphi(\rho, \pi) = 0$ for $0 < \rho < 1$ and $u(1, \varphi) = \varphi(\pi - \varphi)$ for $0 < \varphi < \pi$.

22. Let

$$u(\rho) = T_1 + (T_2 - T_1) \frac{\ln \rho - \ln R_1}{\ln R_2 - \ln R_1}$$

be the solution of Laplace's equation in the cylindrical region $R_1 < \rho < R_2$.
(a) Show that $u(\rho) \rightarrow T_2$ if $R_1 \rightarrow 0$ and ρ is a fixed number with $0 < \rho \le R_2$.
(b) Show that $u(\rho) \rightarrow T_1$ if $R_2 \rightarrow \infty$ and ρ is a fixed number with $\rho \ge R_1$.

*23. Let $u(\rho, \varphi)$ be the solution of Laplace's equation in the cylinder $0 \le \rho \le R$ with the boundary conditions $u(R, \varphi) = T$ if $0 < \varphi < \pi$ and $u(R, \varphi) = 0$ if $-\pi < \varphi < 0$.
(a) Show that $u(\rho, 0) = \frac{1}{2}T$ for $0 \le \rho < R$.
(b) Show that $u(\rho, \pi/2) = \frac{1}{2}T + (2T/\pi) \tan^{-1} (\rho/R)$ for $0 \le \rho < R$.

*24. Let $u(\rho, \varphi)$ be a solution of Laplace's equation in the cylinder $0 \le \rho < R$ and represented by the Poisson integral formula (3.1.9) with $0 \le G(\varphi)$.
(a) Show that

$$\frac{R - \rho}{R + \rho} u(0, \varphi) \le u(\rho, \varphi) \le \frac{R + \rho}{R - \rho} u(0, \varphi) \qquad 0 \le \rho \le R$$

(b) Use this inequality to prove that

$$\left| \frac{1}{u} \frac{\partial u}{\partial \rho} \right|_{\rho=0} \le \frac{2}{R}$$

3.2   BESSEL FUNCTIONS†

To treat more general boundary-value problems in cylindrical coordinates, we need to study the properties of Bessel functions. The reader may be familiar with many properties of Bessel functions from previous work in ordinary differential equations. In this case much of this section can be read quickly and used for later reference when we study the applications to boundary-value problems. Our treatment, which requires no previous knowledge of Bessel functions, is especially designed for the applications to boundary-value problems in cylindrical coordinates in this chapter and for applications to spherical coordinates in Chap. 4. For applications to boundary-value problems, the important facts about Bessel functions are contained in Proposition 3.2.4 and the examples which follow.

Bessel's Equation

Bessel functions originate as solutions of the following equation containing three parameters (d, λ, μ):

$$\boxed{y'' + (d - 1)\frac{y'}{x} + \left(\lambda - \frac{\mu}{x^2}\right)y = 0} \tag{3.2.1}$$

The parameter $d \geq 1$ is the *dimension,* the parameter λ is the *eigenvalue,* and the parameter $\mu \geq 0$ is the *angular index.* For solutions in cylindrical coordinates we need $d = 2$, while for problems in spherical coordinates we need $d = 3$. For solutions with circular symmetry in cylindrical coordinates, we need $\mu = 0$, while for nonsymmetric solutions in cylindrical coordinates we need $\mu = 1, 4, 9, \ldots$. The case $d = 1$, $\mu = 0$ is the equation $y'' + \lambda y = 0$, which has already been treated in detail.

The Bessel equation (3.2.1) is an equation of Sturm-Liouville type, since it can be written in the self-adjoint form

$$(x^{d-1}y')' + (\lambda x^{d-1} - \mu x^{d-3})y = 0$$

We recognize the weight function $\rho(x) = x^{d-1}$, while $s(x) = x^{d-1}$, $q(x) = \mu x^{d-3}$.

The Power Series Solution of Bessel's Equation

The Bessel equation becomes singular at $x = 0$; therefore we cannot expect two linearly independent solutions which remain bounded when $x \to 0$. There is always one solution which remains bounded when $x \to 0$. To find it, we follow the method of Frobenius and look for the solution as a power series

$$y = x^\gamma \sum_{n=0}^{\infty} a_n x^n = \sum_{n=0}^{\infty} a_n x^{n+\gamma} \tag{3.2.2}$$

---

† Certain aspects of Bessel functions are discussed in the Appendix on "Using Mathematica."

where $(\gamma, a_0, a_1, \ldots)$ are to be determined. The power series (3.2.2) will converge for all x. To show this, we substitute the series (3.2.2) into (3.2.1) and rewrite the result as a single power series. Thus

$$y' = \sum_{n=0}^{\infty} (n + \gamma)a_n x^{n+\gamma-1} \tag{3.2.2'}$$

$$y'' = \sum_{n=0}^{\infty} (n + \gamma)(n + \gamma - 1)a_n x^{n+\gamma-2} \tag{3.2.2''}$$

$$y'' + \frac{d-1}{x}y' - \frac{\mu}{x^2}y = \sum_{n=0}^{\infty} [(n+\gamma)(n+\gamma-1) + (d-1)(n+\gamma) - \mu]a_n x^{n+\gamma-2}$$

$$= \sum_{n=0}^{\infty} [(n+\gamma)(n+\gamma+d-2) - \mu]a_n x^{n+\gamma-2}$$

In order for this to be equal to the series for $-\lambda y$, the two series must agree, term by term. The series for λy begins with the power x^{γ}, while the above series begins with $x^{\gamma-2}$. Therefore we must have

$$[\gamma(\gamma + d - 2) - \mu]a_0 = 0 \qquad n = 0 \tag{3.2.3}$$

$$[(1 + \gamma)(\gamma + d - 1) - \mu]a_1 = 0 \qquad n = 1 \tag{3.2.4}$$

$$[(n + \gamma)(n + \gamma + d - 2) - \mu]a_n + \lambda a_{n-2} = 0 \qquad n = 2, 3, \ldots \tag{3.2.5}$$

We obtain the Frobenius solution by taking

$$a_0 \neq 0 \qquad a_1 = 0$$

$$\gamma = 1 - \frac{d}{2} + \left[\mu + \left(\frac{d}{2} - 1\right)^2\right]^{1/2} \tag{3.2.6}$$

The exponent γ, which is *positive*, is the largest root of the indicial equation $\gamma(\gamma + d - 2) - \mu = 0$, from (3.2.3). To determine a_n, $n \geq 2$, we use the indicial equation to write

$$(n + \gamma)(n + \gamma + d - 2) - \mu = n^2 + n(\gamma + d - 2) + n\gamma + \gamma(\gamma + d - 2) - \mu$$

$$= n(n + 2\gamma + d - 2)$$

Thus (3.2.5) becomes

$$n(n + 2\gamma + d - 2)a_n + \lambda a_{n-2} = 0 \qquad n = 2, 3, \ldots$$

$a_1 = 0$ requires $a_3 = 0 = a_5 = \cdots$, while

$$a_2 = \frac{-\lambda}{2(d + 2\gamma)}a_0$$

$$a_4 = \frac{-\lambda}{2(d + 2\gamma)}\frac{-\lambda}{4(d + 2\gamma + 2)}a_0$$

$$a_{2n} = \frac{(-\lambda)^n}{2(d + 2\gamma)4(d + 2\gamma + 2) \cdots 2n(d + 2\gamma + 2n - 2)} a_0 \qquad n = 1, 2, \ldots$$

Hence we have obtained the desired function

$$y(x) = a_0 x^\gamma \left[1 + \sum_{n=1}^\infty \frac{(-\lambda)^n x^{2n}}{2(d + 2\gamma)4(d + 2\gamma + 2) \cdots 2n(d + 2\gamma + 2n - 2)} \right] \qquad (3.2.7)$$

We may check convergence of series (3.2.7) by the ratio test. The ratio of two consecutive terms is $a_{2n+2}x^{2n+2}/(a_{2n}x^{2n}) = -\lambda x^2/[(2n + 2)(d + 2\gamma + 2n)]$. For any x this tends to zero when $n \to \infty$. Therefore series (3.2.7) converges for all x. By a similar use of the ratio test, it may be shown that the differentiated series (3.2.2′), (3.2.2″) converge for all x. This convergence is uniform on all finite intervals, and therefore we may differentiate the series term by term and verify that (3.2.7) is a solution of Bessel's equation.

In case $d = 2$ and $\mu = m^2$ ($m = 0, 1, 2, \ldots$), there is a standard choice of a_0, which we adopt. From (3.2.6) we see that $\gamma = m$ when $d = 2$. Therefore the formula for a_{2n} simplifies to

$$a_{2n} = \frac{(-\lambda)^n a_0}{2(2 + 2m)4(4 + 2m) \cdots 2n(2n + 2m)} \qquad n = 1, 2, \ldots$$

$$= \frac{(-\lambda)^n a_0}{2^{2n} n!(1 + m) \cdots (n + m)}$$

We follow established usage and choose $a_0 = 1/(m! 2^m)$, $\lambda = 1$. This leads to the definition

$$J_m(x) = \frac{x^m}{2^m m!} \left[1 + \sum_{n=1}^\infty \frac{(-1)^n x^{2n}}{2^{2n} n!(1 + m) \cdots (n + m)} \right] \qquad (3.2.8a)$$

This may also be written in the form

$$J_m(x) = \sum_{n=0}^\infty \frac{(-1)^n x^{2n+m}}{2^{m+2n}(m + n)! n!} \qquad (3.2.8b)$$

If m is not an integer, we *define* $m! = \int_0^\infty t^m e^{-t} dt$, a convergent improper integral for $m > -1$.† Integration by parts shows that

$$(m + 1)! = \int_0^\infty t^{m+1} e^{-t} dt$$

$$= -\int_0^\infty t^{m+1} de^{-t}$$

$$= (m + 1) \int_0^\infty t^m e^{-t} dt = (m + 1)m!$$

---

† This is often denoted $\Gamma(m + 1) = \int_0^\infty t^m e^{-t} dt$, the *gamma function*.

Therefore the fundamental property of factorials is preserved: $(m + 1)! = (m + 1)m!$ for $m > -1$. We now *define* the Bessel function $J_m(x)$ for arbitrary $m > -1$ by formula (3.2.8*a*). The series converges for all x and is a solution of the Bessel equation (3.2.1) with $d = 2$, $\lambda = 1$. Formula (3.2.8*b*) is also valid for arbitrary $m > -1$, since the factorials have the property $(m + n)! = (m + n) \cdots (m + 1)m!$ for $n = 0, 1, 2, \ldots$ and arbitrary $m > -1$.

Example 3.2.1. Find the power series solution of the Bessel equation with $d = 2$, $\mu = 0$, $\lambda > 0$.

Solution. In this case we have $\gamma = 0$ and

$$a_{2n} = \frac{(-\lambda)^n}{2^{2n}(n!)^2} a_0$$

The solution is

$$y = \sum_{n=0}^{\infty} \frac{(-\lambda)^n a_0}{2^{2n}(n!)^2} x^{2n}$$

$$= a_0 \sum_{n=0}^{\infty} \frac{(-1)^n (x\sqrt{\lambda})^{2n}}{2^{2n}(n!)^2}$$

$$= a_0 J_0(x\sqrt{\lambda}) \qquad \bullet$$

Example 3.2.2. Find the power series solution of the Bessel equation with $d = 3$, $\mu = 0$, $\lambda > 0$.

Solution. From (3.2.6) we have $\gamma = -\frac{1}{2} + \frac{1}{2} = 0$, and

$$a_{2n} = \frac{(-\lambda)^n a_0}{2 \cdot 3 \cdot 4 \cdot 5 \cdots 2n(2n + 1)}$$

The solution is

$$y = \sum_{n=0}^{\infty} \frac{(-\lambda)^n a_0 x^{2n}}{(2n + 1)!}$$

$$= \frac{a_0}{x\sqrt{\lambda}} \sum_{n=0}^{\infty} \frac{(-1)^n (x\sqrt{\lambda})^{2n+1}}{(2n + 1)!}$$

$$= a_0 \frac{\sin x\sqrt{\lambda}}{x\sqrt{\lambda}}$$

In this example the solution of Bessel's equation is an elementary function. $\bullet$

In some problems we encounter Bessel's equation with $\lambda < 0$. To treat such problems, we define the *modified Bessel function* by

$$I_m(x) = i^{-m} J_m(ix) = \frac{x^m}{2^m m!} \left[1 + \sum_{n=1}^{\infty} \frac{x^{2n}}{2^{2n} n! (1 + m) \cdots (n + m)} \right]$$

where i is the imaginary unit, $i^2 = -1$; $I_m(x)$ is a real-valued function.

Example 3.2.3. Find the power series solution of Bessel's equation with $d = 2$, $\mu = 0$, $\lambda = -c^2 < 0$.

Solution. We have $\gamma = 0$, and

$$a_{2n} = \frac{c^{2n}a_0}{2^{2n}(n!)^2}x^{2n}$$

The solution is

$$y = \sum_{n=0}^{\infty} \frac{c^{2n}a_0}{2^{2n}(n!)^2}x^{2n} = a_0 I_0(cx)$$

●

We now show that the power series solution (3.2.7) with $\lambda > 0$ can be expressed in terms of the standard Bessel function $J_m(x)$. To do this, we write the denominator of (3.2.7) as

$$2(d + 2\gamma)4(d + 2\gamma + 2) \cdots 2n(d + 2\gamma + 2n - 2)$$
$$= 2[2 + 2(\gamma + \tfrac{1}{2}d - 1)] \cdots 2n[2n + 2(\gamma + \tfrac{1}{2}d - 1)]$$

If $d = 2$, then $\gamma = |\mu|$ and the denominator will reduce to $2(2 + 2m)$ $\cdots 2n(2n + 2m)$, where $m = |\mu|$. In the general case, the denominator has this form if we choose $m = \gamma + \tfrac{1}{2}d - 1$. We summarize this as follows.

Proposition 3.2.0. The power series solution (3.2.7) with $\lambda > 0$ can be expressed in terms of the standard Bessel function by the relation

$$\frac{y(x)}{x^\gamma} = a_0 \frac{J_m(x\sqrt{\lambda})}{x^m} \qquad m = \gamma + \tfrac{1}{2}d - 1$$

Example 3.2.4. Find the power series solution of Bessel's equation with $d = 3$, $\mu = p(p + 1)$ for an integer p.

Solution. In this case we have $\gamma = -\tfrac{1}{2} + [p(p + 1) + \tfrac{1}{4}]^{1/2} = p$ and thus $m = p + \tfrac{3}{2} - 1 = p + \tfrac{1}{2}$. Therefore $y(x) = a_0(x^p/x^{p+1/2})J_{p+1/2}(x\sqrt{\lambda}) = a_0 J_{p+1/2}(x\sqrt{\lambda})/x^{1/2}$.

●

These results are used in Sec. 4.2 in connection with the *spherical Bessel functions*, which arise from separation of variables in spherical coordinates.

Integral Representation of Bessel Functions

In many problems the power series is not the most efficient representation of Bessel functions. For example, if we wish to determine the asymptotic behavior of $J_m(t)$ when $t \to \infty$, the power series provides no useful information. For these purposes we will prove the following integral formulas:

$$J_m(t) = \frac{i^{-m}}{2\pi} \int_{-\pi}^{\pi} e^{it\cos\theta}e^{-im\theta}\, d\theta \qquad m = 0, 1, 2, \ldots \qquad (3.2.9)$$

In other words, $i^m J_m(t)$ is the mth Fourier coefficient of the complex function $\theta \to e^{it \cos \theta}$. Since we know that J_m is a real function, the imaginary part of this integral is zero, and we can obtain an equivalent real form. For example, when $m = 0$, we can write

$$J_0(t) = \frac{1}{2\pi} \int_{-\pi}^{\pi} \cos (t \cos \theta) \, d\theta$$

To prove (3.2.9), we expand $e^{it \cos \theta}$ in a power series:

$$e^{it \cos \theta} = \sum_{n=0}^{\infty} \frac{(it \cos \theta)^n}{n!}$$

For a fixed t, this series converges uniformly for $-\pi \le \theta \le \pi$. When we multiply by $e^{-im\theta}$, we still have uniform convergence, and we can therefore integrate term by term, with the result

$$\int_{-\pi}^{\pi} e^{it \cos \theta} e^{-im\theta} \, d\theta = \sum_{n=0}^{\infty} \frac{(it)^n}{n!} \int_{-\pi}^{\pi} \cos^n \theta \, e^{-im\theta} \, d\theta$$

This integral was worked out in Sec. 1.5, where we found a nonzero value only for $m + n$ even, $0 \le m + n \le 2n$, in particular $n \ge m$. For fixed m, the nonzero coefficients are obtained when $n = m, m + 2, m + 4, \ldots$. Introducing a new summation variable j through the equation $n = m + 2j$, we have

$$\frac{1}{2\pi} \int_{-\pi}^{\pi} e^{it \cos \theta} e^{-im\theta} \, d\theta = \sum_{j=0}^{\infty} \frac{(it)^{m+2j}}{(m + 2j)!} \frac{1}{2\pi} \int_{-\pi}^{\pi} (\cos \theta)^{m+2j} e^{-im\theta} \, d\theta$$

$$= \sum_{j=0}^{\infty} \frac{(it)^{m+2j}}{(m + 2j)!} \frac{1}{2^{m+2j}} \binom{m + 2j}{j}$$

$$= \sum_{j=0}^{\infty} \frac{(it)^{m+2j}}{2^{m+2j} j! (m + j)!}$$

$$= i^m t^m \sum_{j=0}^{\infty} \frac{(-t^2)^j}{2^{m+2j} j! (m + j)!}$$

$$= i^m J_m(t)$$

This completes the proof of (3.2.9).

Example 3.2.5. Show that $|J_m(t)| \le 1$ for $m = 0, 1, 2, \ldots$.

Solution. The function $e^{it \cos \theta} e^{-im\theta}$ has absolute value 1. Therefore the integral (3.2.9) has absolute value no greater than 1. ●

We now prove a differentiation formula. Beginning with the integral representation (3.2.9), we have

$$2\pi i^m J'_m(t) = \int_{-\pi}^{\pi} e^{it \cos \theta} e^{-im\theta} i \cos \theta \, d\theta$$

$$= \frac{i}{2} \int_{-\pi}^{\pi} e^{it \cos \theta} (e^{-i(m-1)\theta} + e^{-i(m+1)\theta}) \, d\theta \qquad m = 0, 1, 2, \ldots$$

If $m \geq 1$, we may use (3.2.9) to rewrite this as

$$2\pi i^m J'_m(t) = \frac{i}{2} [2\pi i^{m-1} J_{m-1}(t) + 2\pi i^{m+1} J_{m+1}(t)]$$

Thus we have proved the *differentiation formula*

$$\boxed{J'_m(t) = \tfrac{1}{2}[J_{m-1}(t) - J_{m+1}(t)] \qquad m = 1, 2, \ldots} \qquad (3.2.10)$$

If $m = 0$, we use (3.2.9) to write

$$2\pi (J'_0 + J_1)(t) = \int_{-\pi}^{\pi} (i \cos \theta \, e^{it \cos \theta} - i e^{i\theta} e^{it \cos \theta}) \, d\theta$$

$$= \int_{-\pi}^{\pi} \sin \theta \, e^{it \cos \theta} \, d\theta$$

$$= \int_{-\pi}^{\pi} \sin \theta \cos (t \cos \theta) \, d\theta$$

$$= 0$$

where we have used the fact that the integral is real and the final integrand is an odd function of θ, $-\pi < \theta < \pi$. Thus

$$\boxed{J'_0(t) = -J_1(t)} \qquad (3.2.11)$$

We now use integration by parts to find another useful formula, known as the recurrence formula. From (3.2.9) we have

$$2\pi i^m J_m(t) = \int_{-\pi}^{\pi} e^{it \cos \theta} d \frac{e^{-im\theta}}{-im} \qquad m = 1, 2, \ldots$$

$$= e^{it \cos \theta} \frac{e^{-im\theta}}{-im} \Big|_{-\pi}^{\pi} - \frac{1}{m} \int_{-\pi}^{\pi} e^{it \cos \theta} t \sin \theta \, e^{-im\theta} \, d\theta$$

$$= -\frac{t}{m} \int_{-\pi}^{\pi} e^{it \cos \theta} \frac{e^{-i(m-1)\theta} - e^{-i(m+1)\theta}}{2i} \, d\theta$$

$$= \frac{it}{2m} [2\pi i^{m-1} J_{m-1}(t) - 2\pi i^{m+1} J_{m+1}(t)]$$

In the second line we have used periodicity to discard the first term. Therefore we have the *recurrence formula*

$$J_m(t) = \frac{t}{2m}[J_{m-1}(t) + J_{m+1}(t)] \qquad m = 1, 2, \ldots \qquad (3.2.12)$$

This formula allows us to compute J_{m+1} in terms of J_m and J_{m-1}. Combining this with (3.2.10) and (3.2.11), adding, and subtracting we obtain the *differentiation formulas*

$$J_m'(t) + \frac{m}{t}J_m(t) = J_{m-1}(t) \qquad m = 1, 2, \ldots$$

$$J_m'(t) - \frac{m}{t}J_m(t) = -J_{m+1}(t) \qquad m = 0, 1, 2, \ldots$$

By using the integrating factors $t^{\pm m}$, these can be rewritten in the form

$$\frac{d}{dt}[t^m J_m(t)] = t^m J_{m-1}(t) \qquad m = 1, 2, \ldots$$

$$\frac{d}{dt}[t^{-m}J_m(t)] = -t^{-m}J_{m+1}(t) \qquad m = 0, 1, 2, \ldots \qquad (3.2.13)$$

These formulas can be used to reduce certain integrals which occur in the normalization of the Bessel functions. For example, with $m = 1$ we have

$$\frac{d}{dt}(tJ_1) = tJ_0$$

Integrating this for $0 \le t \le R$, we have

$$RJ_1(R) = \int_0^R tJ_0(t)\, dt$$

The Second Solution of Bessel's Equation

Since Bessel's equation becomes singular at $x = 0$, we cannot expect two linearly independent solutions in the form of power series. Let $y_1(x) = \sum_{n=0}^{\infty} a_n x^{n+\gamma}$ be the solution found above. To find the second solution $y_2(x)$, we use the method of reduction of order: Let $v = y_2/y_1$ and find a first-order equation satisfied by v. Thus

$$y_2' = vy_1' + v'y_1$$

$$y_2'' = vy_1'' + 2v'y_1' + v''y_1$$

Assuming y_2 is a solution, we must have

$$0 = y_2'' + \frac{d-1}{x}y_2' + \left(\lambda - \frac{m^2}{x^2}\right)y_2$$

$$= v\left[y_1'' + \frac{d-1}{x}y' + \left(\lambda - \frac{m^2}{x^2}\right)y_1\right] + 2v'y_1' + v''y_1 + \frac{d-1}{x}(v'y_1)$$

We require that v be a solution of the equation

$$v''y_1 + v'\left(2y_1' + \frac{d-1}{x}y_1\right) = 0$$

This is a first-order linear equation, which may be solved with the integrating factor $y_1 x^{d-1}$. Thus we obtain a solution by writing

$$(y_1^2 x^{d-1} v')' = 0$$

$$y_1^2 x^{d-1} v' = c \neq 0$$

$$v(x) = \int_x^1 \frac{c}{y_1^2 x^{d-1}}\, dx$$

The second constant of integration gives a multiple of y_1 and hence is omitted. The integration begins at $x > 0$ since the integrand becomes infinite when $x \to 0$.

To see this, recall that $y_1(x) \sim x^\gamma$, $x \to 0$. Therefore $y_1^2 x^{d-1} \sim x^{2\gamma+d-1}$. From (3.2.6)

$$2\gamma + d - 1 = 2\left\{1 - \frac{d}{2} + \left[\left(\frac{d}{2} - 1\right)^2 + m^2\right]^{1/2}\right\} + d - 1$$

$$= 1 + 2\left[\left(\frac{d}{2} - 1\right)^2 + m^2\right]^{1/2}$$

This is greater than or equal to 1, and hence the integral for $v(x)$ diverges when $x \to 0$.

To study this more precisely, we consider separate cases. If $d = 2$ and $m = 2$, then $\gamma = 0$ and we have

$$v(x) \sim c \int_x^1 \frac{dx}{x} = -c \log x$$

$$y_2(x) = v(x)y_1(x) \sim -c \log x \qquad x \to 0$$

If $d \neq 2$ or $m \neq 0$, then $2\gamma + d - 1 > 1$ and

$$v(x) \sim c\, \frac{x^{2-2\gamma-d}}{2 - 2\gamma - d} \qquad x \to 0$$

But $y_1(x) \sim x^\gamma$, $x \to 0$. Therefore

$$y_2(x) \sim cx^{-(\gamma+d-2)} \qquad x \to 0$$

But $\gamma + d - 2 = d/2 - 1 + [m^2 + (d/2 - 1)^2]^{1/2}$, which is positive if $d > 2$ or $d = 2$, $m \neq 0$.

To summarize, we have found a second solution of Bessel's equation

$$\boxed{y_2(x) = y_1(x) \int_x^1 \frac{dx'}{y_1(x')^2(x')^{d-1}}} \tag{3.2.14}$$

and y_2 becomes infinite when $x \to 0$.

Zeros of Bessel Functions: Asymptotic Behavior

We first discuss the equation $J_0(x) = 0$. For this purpose we write the series for J_0 in the form

$$J_0(x) = 1 - \frac{x^2}{4} + \sum_{p=1}^{\infty} \left\{ \frac{x^{4p}}{2^{4p}[(2p)!]^2} - \frac{x^{4p+2}}{2^{4p+2}[(2p+1)!]^2} \right\}$$

$$= 1 - \frac{x^2}{4} + \sum_{p=1}^{\infty} \frac{x^{4p}}{2^{4p}[(2p)!]^2} \left[1 - \frac{x^2}{4(2p+1)^2} \right]$$

If $0 \le x \le 2$, all the terms in the summation are positive since

$$x^2 \le 4 < 4(2p+1)^2 \qquad p = 1, 2, \ldots$$

Therefore we have the inequality

$$J_0(x) > 1 - \frac{x^2}{4} \qquad 0 \le x \le 2$$

In particular, $J_0(2) > 0$.

But we may write the series in the form

$$J_0(x) = 1 - \frac{x^2}{4} + \frac{x^4}{64} - \sum_{p=1}^{\infty} \left\{ \frac{x^{4p+2}}{2^{4p+2}[(2p+1)!]^2} - \frac{x^{4p+4}}{2^{4p+4}[(2p+2)!]^2} \right\}$$

$$= 1 - \frac{x^2}{4} + \frac{x^4}{64} - \sum_{p=1}^{\infty} \frac{x^{4p+2}}{2^{4p+2}[(2p+1)!]^2} \left(1 - \frac{x^2}{4(2p+2)^2} \right)$$

If $0 \le x \le 3$, all the terms in the summation are positive, since

$$x^2 \le 9 < 4(2p+2)^2 \qquad p = 1, 2, \ldots$$

Therefore we have the inequality

$$J_0(x) < 1 - \frac{x^2}{4} + \frac{x^4}{64} \qquad 0 \le x \le 3$$

In particular, $J_0(2\sqrt{2}) < 1 - \frac{8}{4} + \frac{64}{64} = 0$.

Now we can apply the intermediate-value theorem. J_0 is a continuous function with $J_0(2) > 0 > J_0(2\sqrt{2})$. Therefore $J_0(x) = 0$ for some x with $2 < x < 2\sqrt{2}$. But $J_0(x) > 0$ for $0 < x < 2$. Therefore we have proved that the smallest solution x_1 of the equation $J_0(x) = 0$ satisfies $2 < x_1 < 2\sqrt{2} = 2.828$.

The zeros of J_0 have been computed numerically. The first five are listed in the following table:

n	x_n
1	2.405
2	5.520
3	8.654
4	11.792
5	14.931

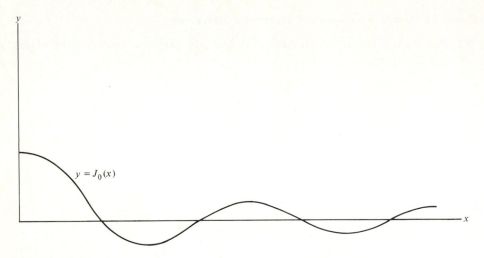

FIGURE 3.2.1
Graph of the Bessel function $y = J_0(x)$.

By using this information, it is possible to sketch a graph of the function $y = J_0(x)$. (See Fig. 3.2.1.)

We now show that the solutions of Bessel's equations with $\lambda > 0$ have *infinitely many* zeros. The power series representation is an inefficient tool for this purpose. Instead we make a *phase plane analysis*. When x is large, we expect that the terms $(d - 1)y'/x$ and $m^2 y/x^2$ will be of little importance and that the solutions will resemble the solutions of $y'' + \lambda y = 0$, which are of the form $A \cos x\sqrt{\lambda} + B \sin x\sqrt{\lambda}$. Since we expect this oscillatory behavior when $x \to \infty$, it is natural to take polar coordinates to analyze the behavior of $y(x)$, $y'(x)$ when $x \to \infty$.

The first step in the analysis is to transform the Bessel equation to remove the first-derivative term $[(d - 1)/x]y'$. We do this by the transformation

$$z(x) = x^{(d-1)/2}y(x)$$

We have

$$z'(x) = x^{(d-1)/2}y'(x) + \tfrac{1}{2}(d - 1)x^{(d-3)/2}y(x)$$

$$z''(x) = x^{(d-1)/2}y''(x) + (d - 1)x^{(d-3)/2}y'(x) + \tfrac{1}{4}(d - 1)(d - 3)x^{(d-5)/2}y(x)$$

$$= x^{(d-1)/2}\left[y''(x) + \frac{d - 1}{x} y'(x) + \frac{1}{4}\frac{(d - 1)(d - 3)}{x^2} y(x) \right]$$

We use the Bessel equation and the definition $z(x) = x^{(d-1)/2}y(x)$ to rewrite the expression for $z''(x)$ as $z(x)[-\lambda + m^2/x^2 + \tfrac{1}{4}(d - 1)(d - 3)/x^2]$. Therefore $z(x)$ satisfies the equation

$$z''(x) + \left[\lambda - \frac{1}{4}\frac{(d - 1)(d - 3)}{x^2} - \frac{m^2}{x^2} \right] z(x) = 0 \qquad (3.2.15)$$

Since the solution of Bessel's equation depends on the product $x\sqrt{\lambda}$, it is no loss of generality to take $\lambda = 1$. To further simplify the notation, we write $C = -[\frac{1}{4}(d-1)(d-3) + m^2]$. With this simplification, $z(x)$ satisfies the equation $z''(x) + z(x)(1 + C/x^2) = 0$.

The next step is to take polar coordinates in the plane of (z, z'):

$$z(x) = R(x) \cos \theta(x)$$

$$z'(x) = -R(x) \sin \theta(x)$$

(3.2.16)

[The minus sign in the definition of $z'(x)$ is for technical convenience and has no other significance.] If we try to use these equations directly and define $\theta(x)$ by the formula $\theta(x) = \tan^{-1}[z'(x)/z(x)]$, there are some theoretical difficulties. Instead we use the equation for $z(x)$ to determine a system of equations which $(R(x), \theta(x))$ must satisfy, and we use them as the definition of $(R(x), \theta(x))$. To do this, we differentiate (3.2.16) and obtain

$$z'(x) = R'(x) \cos \theta(x) - R(x)\theta'(x) \sin \theta(x)$$

$$= -R(x) \sin \theta(x)$$

$$z''(x) = -R'(x) \sin \theta(x) - R(x)\theta'(x) \cos \theta(x)$$

$$= \left(-1 - \frac{C}{x^2}\right) R(x) \cos \theta(x)$$

If we solve these equations for $R(x)$, $\theta(x)$, we find

$$\theta'(x) = 1 - \frac{C \cos^2 \theta(x)}{x^2}$$

(3.2.17)

$$R'(x) = -\frac{C}{x^2} R(x) \sin \theta(x) \cos \theta(x)$$

These equations will be used to define $(R(x), \theta(x))$. To do this, we choose an initial point $x_0 > 0$ and solve the equations with initial values R_0, θ_0 chosen so that $R_0 \cos \theta_0 = z(x_0)$, $R_0 \sin \theta_0 = -z'(x_0)$. It is a straightforward but tedious exercise to show that $R(x) \cos \theta(x)$ satisfies Bessel's equation with the initial conditions $z(x_0) = R_0 \cos \theta_0$, $z'(x_0) = -R_0 \sin \theta_0$. With this done, we have the desired polar coordinate representation of $z(x)$.

Equation (3.2.17) is a first-order nonlinear ordinary differential equation for the function $\theta(x)$. The right side is bounded and smooth for $x \geq x_0 > 0$, and therefore the solution exists for all $x \geq x_0$. Having thus determined $\theta(x)$, we can determine $R(x)$ by a single integration:

$$R(x) = R_0 \exp\left[-C \int_{x_0}^x \frac{\sin \theta(t) \cos \theta(t)}{t^2} dt\right] \qquad x \geq x_0$$

Now $z(x_0)$ and $z'(x_0)$ are not both zero; therefore $R_0 \neq 0$, and thus $R(x) \neq 0$ for all x. Therefore the zeros of $z(x)$ occur precisely when $\cos \theta(x) = 0$,

$\theta(x) = (n = \frac{1}{2})\pi$. To study these, we integrate equation (3.2.17) for $x_0 \le t \le x$

$$\theta(x) - \theta(x_0) = \int_{x_0}^{x} \left[1 - \frac{C \cos^2 \theta(t)}{t^2} \right] dt \qquad x \ge x_0$$

$$= x - x_0 - C \int_{x_0}^{x} \frac{\cos^2 \theta(t)}{t^2} dt$$

$$= x - x_0 + O(1) \qquad x \to \infty$$

since the integral is no larger than $\int_{x_0}^{x} dt/t^2 = 1/x_0 - 1/x \le 1/x_0$. This computation shows that $\theta(x)$ tends to ∞ when x tends to ∞. The zeros of $z(x)$ are obtained when $\theta(x) = (n - \frac{1}{2})\pi$. Denoting these $x_1, x_2, \ldots$, we have $\theta(x_n) = (n - \frac{1}{2})\pi$. Integrating equation (3.2.17) for $x_n \le t \le x_{n+1}$, we have

$$\pi = \theta(x_{n+1}) - \theta(x_n)$$

$$= \int_{x_n}^{x_{n+1}} \left[1 - C \frac{\cos^2 \theta(t)}{t^2} \right] dt$$

$$= x_{n+1} - x_n - C \int_{x_n}^{x_{n+1}} \frac{\cos^2 \theta(t)}{t^2} dt$$

This integral is no larger than

$$\int_{x_n}^{x_{n+1}} \frac{dt}{t^2} = \frac{1}{x_n} - \frac{1}{x_{n+1}}$$

which tends to zero when $n \to \infty$. Therefore we have shown that

$$\boxed{\lim_{n \to \infty} (x_{n+1} - x_n) = \pi}$$

By specializing to the case $d = 2$, $\lambda = 1$, we can state a result which summarizes the computations.

Proposition 3.2.1. The equation $J_m(x) = 0$ has infinitely many positive solutions $\{x_n\}$, $n = 1, 2, \ldots$. They satisfy

$$\lim_{n \to \infty} x_n = \infty$$

$$\lim_{n \to \infty} (x_{n+1} - x_n) = \pi$$

In many problems it is important to have information about the zeros of $\cos \beta \, x J'_m(x) + \sin \beta \, J_m(x)$, where $0 \le \beta \le \pi/2$.

Proposition 3.2.2. For any $0 \le \beta \le \pi/2$, $\cos \beta \, x J'_m(x) + \sin \beta \, J_m(x) = 0$ has infinitely many positive solutions $\{x_n\}$, $n = 1, 2, \ldots$. They satisfy

$$\lim_{n \to \infty} x_n = \infty$$

$$\lim_{n \to \infty} (x_{n+1} - x_n) = \pi$$

(3.2.18)

Proof. We use the phase plane representation $\sqrt{x}J_m(x) = R(x)\cos\theta(x)$, $(\sqrt{x}J_m)'(x) = -R(x)\sin\theta(x)$. The equation $\cos\beta\, xJ'_m(x) + \sin\beta\, J_m(x) = 0$ is equivalent to $0 = x[(\sqrt{x}J_m)'(x) - (1/2\sqrt{x})J_m(x)] + \tan\beta\,[\sqrt{x}J_m(x)] = x\{-R(x)\sin\theta(x) - [1/(2x)]R(x)\cos\theta(x)\} + \tan\beta\, R(x)\cos\theta(x)$ or to the equation $(\tan\beta - \frac{1}{2})\cos\theta(x) = x\sin\theta(x)$, which is written as

$$\tan\theta(x) = \frac{\tan\beta - \frac{1}{2}}{x}$$

From the graph of the tangent function we see that this equation has a unique solution x_n satisfying $(n - \frac{1}{2})\pi < \theta(x_n) < (n + \frac{1}{2})\pi$. To analyze the asymptotic behavior of x_n, we use the addition formula for $\tan\theta$; thus

$$\tan[\theta(x_n) - n\pi] = \frac{\tan\theta(x_n) - \tan n\pi}{1 + \tan\theta(x_n)\tan n\pi} = \frac{\tan\beta - \frac{1}{2}}{x_n}$$

$$= O\left(\frac{1}{n}\right) \qquad n \to \infty$$

where we have used the fact that $\theta(x) = x + O(1)$, $x \to \infty$. Referring again to the graph of the tangent function, since $|\theta(x_n) - n\pi| \le \pi/2$, we must have $\theta(x_n) - n\pi = O(1/n)$, $n \to \infty$. But $\theta(x) = x + C + O(1/x)$. Therefore $\theta(x_{n+1}) - \theta(x_n) = (x_{n+1} - x_n) + O(1/n)$. But $\theta(x_{n+1}) - \theta(x_n) = \pi + O(1/n)$. Combining these, we have proved (3.2.18). ∎

The phase plane analysis can now be used to obtain an asymptotic formula for solutions of Bessel's equation when $x \to \infty$. The formula contains two constants, the *asymptotic amplitude* and the *asymptotic phase*. They depend on the specific solution as well as the parameters (d, λ, m). [In Chap. 6 we make an exact determination of these constants for the power series solutions $J_m(x)$ of the Bessel equation with $d = 2$.]

Proposition 3.2.3. Let $y(x)$ be a solution of Bessel's equation (3.2.1) with $\lambda > 0$. There exist constants C_1 and C_2 such that

$$\boxed{x^{(d-1)/2}y(x) = e^{C_1}\cos(x\sqrt{\lambda} + C_2) + O\left(\frac{1}{x}\right) \qquad x \to \infty} \qquad (3.2.19)$$

Proof. By changing x to $x\sqrt{\lambda}$, we can suppose that $\lambda = 1$. We have already shown above that $y(x) = x^{(1-d)/2}R(x)\cos\theta(x)$, where $R(x)$, $\theta(x)$ are determined from (3.2.17). We have

$$\theta(x) - \theta(x_0) = x - x_0 + C\int_{x_0}^x \frac{\cos^2\theta(u)}{u^2}\,du$$

$$= x - x_0 + C\int_{x_0}^\infty \frac{\cos^2\theta(u)}{u^2}\,du - C\int_x^\infty \frac{\cos^2\theta(u)}{u^2}$$

$$\theta(x) = x + C_2 + O\left(\frac{1}{x}\right) \qquad x \to \infty \qquad (3.2.20)$$

where we have written $C_2 = \theta(x_0) - x_0 + C \int_{x_0}^{\infty} [\cos^2 \theta(u)]/u^2 \, du$ and observed that the second integral is $O(1/x)$ when $x \to \infty$. In a similar manner we may analyze $R(x)$. Thus

$$\log R(x) - \log R(x_0) = -C \int_{x_0}^{x} \frac{\sin \theta(u) \cos \theta(u)}{u_2} \, du$$

When we write this integral as the difference of two improper integrals, the first is constant and the second is $O(1/x)$, $x \to \infty$. Therefore we have $\log R(x) = C_1 + O(1/x)$, $x \to \infty$. Combining these, we have

$$R(x) = \exp\left[C_1 + O\left(\frac{1}{x}\right) \right] = e^{C_1}\left[1 + O\left(\frac{1}{x}\right) \right] \qquad x \to \infty$$

$$\cos \theta(x) = \cos\left[x + C_2 + O\left(\frac{1}{x}\right) \right] = \cos(x + C_2) + O\left(\frac{1}{x}\right) \qquad x \to \infty$$

Multiplying these two expressions gives the desired result. ∎

Fourier-Bessel Series

In many problems it is important to expand a given function in a series of the form $\sum_{n=1}^{\infty} A_n J_m(xx_n)$, where m is a fixed positive number and $\{x_n\}$ are determined from a suitable boundary condition. The boundary condition might be $J_m(x_n) = 0$, $J'_m(x_n) = 0$, or some linear combination of these that equals zero. To study series of this type, first we derive the orthogonality properties of the functions $J_m(xx_n)$.

Proposition 3.2.4. Let $\{x_n\}$ be the nonnegative solutions of the equation $\cos \beta \, x_n J'_m(x_n) + \sin \beta \, J_m(x_n) = 0$, where $m \geq 0$ and $0 \leq \beta \leq \pi/2$. Then

$$\boxed{\int_0^1 J_m(xx_{n_1}) J_m(xx_{n_2}) x \, dx = 0 \qquad n_1 \neq n_2} \tag{3.2.21}$$

$$\boxed{\int_0^1 J_m(xx_n)^2 x \, dx = \begin{cases} \frac{1}{2} J_{m+1}(x_n)^2 & \beta = \pi/2 \\[2mm] \dfrac{(x_n^2 - m^2 + \tan^2 \beta) J_m(x_n)^2}{2x_n^2} & 0 \leq \beta < \pi/2 \end{cases}} \tag{3.2.22}$$

Proof. If $y(x) = J_m(xx_n)$, then $y'(x) = x_n J'_m(xx_n)$. In this notation the equation for x_n becomes $\cos \beta \, y'(1) + \sin \beta \, y(1) = 0$. Since the Bessel equation is an equation of the Sturm-Liouville type with $s(x) = x = \rho(x)$, $\lambda = x_n^2$, we can apply the orthogonality properties of general Sturm-Liouville systems. The basic condition $s(y_1 y_2' - y_1' y_2)|_0^1 = 0$ is satisfied since $s(0) = 0$, while the boundary condition implies that $y_1(1)y_2'(1) - y_1'(1)y_2(1) = 0$. Thus we have proved (3.2.21).

To compute the integrals (3.2.22), we write the Bessel equation in the form $(xy')' + (\lambda x - m^2/x)y = 0$, where $\lambda = x_n^2$. Multiplying by xy', we obtain $[(xy')^2]' + (\lambda x^2 - m^2)(y^2)' = 0$. Integrating this equation for $0 < x < 1$ and integrating the second term by parts, we have

$$y'(1)^2 + (\lambda - m^2)y(1)^2 - 2\lambda \int_0^1 xy(x)^2 \, dx = 0 \qquad (3.2.23)$$

If $\beta = \pi/2$, the boundary condition is $y(1) = 0$, which gives

$$2\lambda \int_0^1 xy(x)^2 \, dx = y'(1)^2 = x_n^2 J_m'(x_n)^2 = x_n^2 J_{m+1}(x_n)^2$$

To handle the case $0 \le \beta < \pi/2$, solve for $y'(1)$ in the form $y'(1) = -\tan \beta \, y(1)$. Substituting this in (3.2.23), we have

$$2\lambda \int_0^1 xy(x)^2 \, dx = y(1)^2(\tan^2 \beta + \lambda - m^2)$$

Noting that $y(1) = J_m(x_n)$, we have the required form. ■

These orthogonality relations permit us to compute the coefficients in the expansion of a piecewise smooth function $f(x)$, $0 < x < 1$, in a series of the form

$$f(x) = \sum_{n=1}^{\infty} A_n J_m(xx_n) \qquad 0 < x < 1 \qquad (3.2.24)$$

where $\{x_n\}$ are defined as the nonnegative solutions of the equation $\cos \beta \, xJ_m'(x) + \sin \beta \, J_m(x) = 0$. This is called a *Fourier-Bessel expansion*. To obtain $\{A_n\}$, we proceed formally; we multiply the equation by $J_m(xx_n)$ and integrate with respect to the weight $x \, dx$ from 0 to 1. This gives the formula

$$\int_0^1 f(x)J_m(xx_n)x \, dx = A_n \int_0^1 J_m(xx_n)^2 x \, dx \qquad n = 1, 2, \ldots \quad (3.2.25)$$

We state without proof a theorem concerning this expansion.†

Theorem 3.2.1. Let $m \ge 0$, $0 \le \beta \le \pi/2$, and let $\{x_n: n \ge 1\}$ be the nonnegative solutions of $\cos \beta \, x_n J_m'(x_n) + \sin \beta \, J_m(x_n) = 0$. If $f(x)$, $0 < x < 1$, is a piecewise smooth function, define $\{A_n: n \ge 1\}$ by (3.2.25). Then the series $\sum_{n=1}^{\infty} A_n J_m(xx_n)$ converges for each x, $0 < x < 1$, and the sum is $\frac{1}{2}[f(x + 0) + f(x - 0)]$.

It is important to note under what conditions we may have $x_1 = 0$. Since $y(x) = J_m(xx_1)$ must be nonzero, this implies that $0 \ne J_m(0)$, that is, $m = 0$. We must also have the boundary condition $\sin \beta \, J_m(0) = 0$, which requires $\beta = 0$. Conversely, the function $y = J_0(xx_1)$ is a solution of the Bessel equation satisfying the boundary condition $\cos \beta \, x_1 J_0'(x_1) + \sin \beta \, J_0(x_1) = 0$ if $x_1 = 0$. We record this as a proposition.

---

† See H. F. Weinberger, *A First Course in Partial Differential Equations*, Ginn-Blaisdell, Boston, 1965, pp. 176–178.

Proposition 3.2.5. If $m > 0$ or $\beta \neq 0$, then $x_n > 0$ for all $n \geq 1$. If $m = 0$ and $\beta = 0$, then $x_1 = 0$ and $x_n > 0$ for all $n \geq 2$.

Example 3.2.6. Compute the Fourier-Bessel expansion of the function $f(x) = 1$, $0 < x < 1$, where $m = 0$, $\beta = \pi/2$.

Solution. We have $1 = \sum_{n=1} A_n J_0(x x_n)$, where $J_0(x_n) = 0$ and

$$\int_0^1 x J_0(x x_n) \, dx = A_n \int_0^1 J_0(x x_n)^2 x \, dx \qquad n = 1, 2, \ldots$$

To compute the first integral, we use (3.2.13) with $m = 1$. With the substitution $t = x x_n$, we have

$$\int_0^1 x J_0(x x_n) \, dx = \frac{1}{x_n^2} \int_0^{x_n} t J_0(t) \, dt = \frac{1}{x_n} J_1(x_n)$$

The integral $\int_0^1 J_0(x x_n)^2 x \, dx$ was already shown to be $\frac{1}{2} J_1(x_n)^2$, by (3.2.22). The required expansion is

$$1 = 2 \sum_{n=1}^{\infty} \frac{J_0(x x_n)}{x_n J_1(x_n)} \qquad 0 < x < 1 \qquad \bullet$$

Example 3.2.7. Compute the Fourier-Bessel expansion of the function $f(x) = 1 - x^2$, $0 < x < 1$, where $m = 0$, $\beta = \pi/2$.

Solution. We have $1 - x^2 = \sum_{n=1}^{\infty} A_n J_0(x x_n)$ where $J_0(x_n) = 0$ and

$$\int_0^1 (1 - x^2) J_0(x x_n) x \, dx = A_n \int_0^1 J_0(x x_n)^2 x \, dx$$

To compute the first integral, let $t = x x_n$ and use formulas (3.2.13) in the forms $(d/dt)(t J_1) = t J_0$, $(d/dt)(J_0) = -J_1$. Thus

$$\int_0^1 (1 - x^2) J_0(x x_n) x \, dx = \frac{1}{x_n^4} \int_0^{x_n} (x_n^2 - t^2) t J_0(t) \, dt$$

$$= \frac{2}{x_n^4} \int_0^{x_n} t^2 J_1(t) \, dt$$

$$= \frac{4}{x_n^4} \int_0^{x_n} t J_0(t) \, dt$$

$$= \frac{4}{x_n^3} J_1(x_n)$$

from Example 3.2.6 and integration by parts. The second integral $\int_0^1 J_0(x x_n)^2 x \, dx$ was shown to be $\frac{1}{2} J_1(x_n)^2$. The required expansion is therefore

$$1 - x^2 = 8 \sum_{n=1}^{\infty} \frac{J_0(x x_n)}{x_n^3 J_1(x_n)} \qquad 0 < x < 1 \qquad \bullet$$

These examples suggest a general method for computing the Fourier-Bessel expansion of certain polynomials. Let $P_0(x) = \frac{1}{2}$ and let $P_{2n}(x)$ be the polynomial of degree $2n$ which satisfies

$$(xP'_{2n})' = -xP_{2n-2} \qquad P_{2n}(1) = 0 \qquad n = 1, 2, 3, \ldots$$

For example, $P_2(x) = (1 - x^2)/8$, $P_4(x) = \frac{1}{128}(3 - 4x^2 + x^4)$. To find the Fourier-Bessel expansion, we write $\lambda = x_n^2$, $y(x) = J_0(xx_n)$, $y'(x) = x_n J'_0(xx_n)$, $(xy')' + \lambda(xy) = 0$. Multiplying this Bessel equation by $P_{2n}(x)$ and integrating, we have

$$\lambda \int_0^1 P_{2n}(x) xy(x)\, dx = \int_0^1 P_{2n}(x)(xy')'(x)\, dx$$

$$= \int_0^1 P'_{2n}(x)(xy')(x)\, dx$$

$$= -\int_0^1 [xP'_{2n}(x)]' y(x)\, dx$$

$$= \int_0^1 P_{2n-2}(x) xy(x)\, dx$$

Therefore the Fourier-Bessel coefficients of P_{2n} are obtained from those of P_{2n-2} upon division by $\lambda = x_n^2$. For example, beginning with the expansion

$$P_0(x) = \frac{1}{2} = \sum_{n=1}^{\infty} \frac{J_0(xx_n)}{x_n J_1(x_n)} \qquad 0 < x < 1$$

we have

$$P_2(x) = \frac{1 - x^2}{8} = \sum_{n=1}^{\infty} \frac{J_0(xx_n)}{x_n^3 J_1(x_n)} \qquad 0 < x < 1$$

$$P_4(x) = \frac{1}{128}(3 - 4x^2 + x^4) = \sum_{n=1}^{\infty} \frac{J_0(xx_n)}{x_n^5 J_1(x_n)} \qquad 0 < x < 1$$

In some problems it is necessary to find the Fourier-Bessel expansion of discontinuous functions. The following example gives a typical case.*

Example 3.2.8. Let $f(x)$, $0 < x < 1$, be defined by $f(x) = 1$ for $0 < x < a$ and $f(x) = 0$ for $a < x < 1$. Find the Fourier-Bessel expansion of $f(x)$, $0 < x < 1$, with $m = 0$, $\beta = \pi/2$.

Solution. The desired expansion is of the form $f(x) = \sum_{n=1}^{\infty} A_n J_0(xx_n)$, where $J_0(x_n) = 0$. The coefficients are determined by orthogonality by $A_n \int_0^1 J_0(xx_n)^2 x\, dx = \int_0^1 f(x) J_0(xx_n) x\, dx = \int_0^a J_0(xx_n) x\, dx$. The first integral was evaluated in Proposition 3.2.4 as $\frac{1}{2} J_1(x_n)^2$. To evaluate the second integral, we make the substitution

---

* See also Appendix on "Using Mathematica."

$t = xx_n$ and find that

$$\int_0^a J_0(xx_n)x\ dx = \frac{1}{x_n^2} \int_0^{ax_n} tJ_0(t)$$

$$= \frac{1}{x_n^2} tJ_1(t) \Big|_{t=0}^{t=ax_n}$$

$$= \frac{a}{x_n} J_1(ax_n)$$

Therefore $A_n = (2a/x_n)J_1(ax_n)/J_1(x_n)^2$, and the expansion is

$$f(x) = \sum_{n=1}^{\infty} \frac{2aJ_1(ax_n)}{x_nJ_1(x_n)^2} J_0(xx_n) \qquad 0 < x < 1, x \neq a \qquad \bullet$$

In all these examples of Fourier-Bessel expansions, we have taken $m = 0$, $\beta = \pi/2$. This is well suited to problems with radial symmetry, where the boundary conditions do not involve the derivative. In the following examples, we give Fourier-Bessel expansions with $m > 0$ or $\beta < \pi/2$.

Example 3.2.9. For $m > 0$, $\beta = \pi/2$, compute the Fourier-Bessel expansion of the function $f(x) = x^m$, $0 < x < 1$.

Solution. The Fourier coefficients $\{A_n\}$ are given by $A_n \int_0^1 J_m(xx_n)^2x\ dx = \int_0^1 x^{m+1} J_m(xx_n)\ dx$. The first integral is given by (3.2.22): $\int_0^1 J_m(xx_n)^2x\ dx = \frac{1}{2}J_{m+1}(x_n)^2$. To compute the second integral, we use the first differentiation formula (3.2.13) with m replaced by $m + 1$:

$$\frac{d}{dt} [t^{m+1}J_{m+1}(t)] = t^{m+1}J_m(t)$$

Thus
$$\int_0^1 x^{m+1}J_m(xx_n)\ dx = \frac{1}{x_n^{m+2}} \int_0^{x_n} t^{m+1}J_m(t)\ dt$$

$$= \frac{1}{x_n^{m+2}} x_n^{m+1} J_{m+1}(x_n)$$

$$= \frac{1}{x_n} J_{m+1}(x_n)$$

Hence $A_n = 2/[x_nJ_{m+1}(x_n)]$, and we have the expansion

$$x^m = 2 \sum_{n=1}^{\infty} \frac{J_m(xx_n)}{x_nJ_{m+1}(x_n)} \qquad x_n = x_n^{(m)}, 0 < x < 1 \qquad \bullet$$

Example 3.2.10. For $m = 0$, $0 < \beta < \pi/2$, compute the Fourier-Bessel expansion of the functions $f(x) = 1$ and $f(x) = 1 - x^2$.

Solution. The first expansion is of the form $1 = \sum_{n=1}^{\infty} A_nJ_0(xx_n)$, where $\cos \beta\ x_nJ_0'(x_n) + \sin \beta\ J_0(x_n) = 0$. From the orthogonality relations we must have $A_n \int_0^1 J_0(xx_n)^2\ dx = \int_0^1 xJ_0(xx_n)\ dx$. From (3.2.22) the first integral is

$(x_n^2 + \tan^2 \beta)J_0(x_n)^2/(2x_n^2)$. To compute the second integral, we use the differentiation formula (3.2.13) with $m = 1$: $(d/dt)(tJ_1) = tJ_0$. Thus $\int_0^1 xJ_0(xx_n) \, dx = (1/x_n^2) \int_0^{x_n} tJ_0(t) \, dt = J_1(x_n)/x_n$. From the boundary condition we have $x_nJ_1(x_n) = -x_nJ_0'(x_n) = \tan \beta \, J_0(x_n)$ and the expansion

$$1 = 2 \tan \beta \sum_{n=1}^{\infty} \frac{J_0(xx_n)}{(x_n^2 + \tan^2 \beta)J_0(x_n)} \qquad 0 < x < 1$$

The expansion of $f(x) = 1 - x^2$ is handled similarly by using integration by parts to reduce the integral $\int_0^1 (1 - x^2)J_0(xx_n)x \, dx$ to the integral which has already been computed. The result is

$$1 - x^2 = 8 \tan \beta \sum_{n=1}^{\infty} \frac{J_0(xx_n)}{x_n^2(x_n^2 + \tan^2 \beta)J_0(x_n)} \qquad 0 < x < 1 \qquad \bullet$$

EXERCISES 3.2

1. Show that $(-\frac{1}{2})! = \int_0^{\infty} t^{-1/2}e^{-t} \, dt = 2 \int_0^{\infty} e^{-x^2} \, dx = \sqrt{\pi}$. (You may assume it is known that $\int_{-\infty}^{\infty} e^{-x^2} \, dx = \sqrt{\pi}$.)

2. Show that $(\frac{1}{2})! = \frac{1}{2}\sqrt{\pi}$, $(\frac{3}{2})! = \frac{3}{4}\sqrt{\pi}$.

3. Show that $(n + \frac{1}{2})! = \{[(2n + 1)!]/(2^{2n+1}n!)\}\sqrt{\pi}$, $n = 0, 1, 2, \ldots$.

4. Show that $J_{1/2}(x) = \sqrt{2/(\pi x)} \sin x$.

5. Show that $J_{-1/2}(x) = \sqrt{2/(\pi x)} \cos x$.

6. Use the ratio test to show that series (3.2.2′) converges for all x.

7. Use the ratio test to show that series (3.2.2″) converges for all x.

8. Write the first four nonzero terms in the power series expansions of $J_0(x)$ and $J_1(x)$.

9. Use Example 3.2.5 and equation (3.2.10) to show that $|J_m'(t)| \leq 1$ for $m = 1, 2, \ldots$.

10. Let $m > 0$, not necessarily an integer. Prove the differentiation formula $J_m'(x) = \frac{1}{2}[J_{m-1}(x) - J_{m+1}(x)]$ directly from the power series definition (3.2.8b).

11. Let $m > 0$, not necessarily an integer. Prove the recurrence formula $J_{m+1}(x) = (2m/x)J_m(x) - J_{m-1}(x)$ directly from the power series definition (3.2.8b).

12. Let $m > 0$, not necessarily an integer. Use Exercises 10 and 11 to verify the formula $xJ_m'(x) = mJ_m(x) - xJ_{m+1}(x)$.

13. Let $m > 0$, not necessarily an integer. Use Exercises 10 and 11 to verify the formula $xJ_m'(x) = xJ_{m-1}(x) - mJ_m(x)$.

14. Let $m > 0$, not necessarily an integer. Prove the differentiation formulas

$$\frac{d}{dx} x^m J_m(x) = x^m J_{m-1}(x) \qquad \text{and} \qquad \frac{d}{dx} x^{-m}J_m(x) = -x^{-m}J_{m+1}(x)$$

15. Let x_n be a solution of the equation $J_m(x_n) = 0$, $m > 0$. Use Exercises 12 and 13 to show that $J_m'(x_n) = J_{m-1}(x_n) = -J_{m+1}(x_n)$.

16. Use Exercises 4 and 14 to show that

$$J_{3/2}(x) = \sqrt{\frac{2}{\pi x}} \left(\frac{\sin x}{x} - \cos x \right)$$

17. Let $\theta(x)$, $R(x)$ be functions which satisfy the equations (3.2.17). Show that $z(x) = R(x) \cos \theta(x)$ satisfies the equation $z''(x) + (1 + C/x^2)z(x) = 0$.

18. Show that $\int_{-\pi}^{\pi} e^{it \cos \theta} e^{-im\theta} \, d\theta = \int_{-\pi}^{\pi} e^{it \cos \theta} e^{im\theta} \, d\theta$ for $m = 0, 1, 2, \ldots$.

19. Use Exercise 18 and the integral representation of $J_m(t)$ to show that $\int_{-\pi}^{\pi} e^{it \cos \theta} e^{im\theta} \, d\theta = 2\pi i^{|m|} J_{|m|}(t)$ for $m = 0, \pm 1, \pm 2, \ldots$.

20. Use Exercise 19 and the properties of complex Fourier series to show that $e^{it \cos \theta} = \sum_{-\infty}^{\infty} i^{|m|} J_{|m|}(t) e^{im\theta}$.

21. Use Exercise 20 to show that

$$\cos (t \cos \theta) = J_0(t) - 2J_2(t) \cos 2\theta + 2J_4(t) \cos 4\theta \cdots$$

$$-\sin (t \cos \theta) = -2J_1(t) \cos \theta + 2J_3(t) \cos 3\theta - 2J_5(t) \cos 5\theta \cdots$$

22. Use Exercise 21 and Parseval's theorem to show that

$$1 = J_0(t)^2 + 2 \sum_{m=1}^{\infty} J_m(t)^2$$

23. Show that

$$(-1)^m J_{2m}(t) = \frac{1}{\pi} \int_0^{\pi} \cos (t \cos \theta) \cos 2m\theta \, d\theta$$

$$(-1)^m J_{2m+1}(t) = \frac{1}{\pi} \int_0^{\pi} \sin (t \cos \theta) \cos (2m + 1)\theta \, d\theta$$

24. Show that

$$1 = J_0(t) + 2J_2(t) + 2J_4(t) + \cdots$$

25. Show that

$$\cos t = J_0(t) - 2J_2(t) + 2J_4(t) \cdots \qquad \text{and} \qquad \sin t = 2J_1(t) - 2J_3(t) + 2J_5(t)$$

26. Show that $\lim_{m \to \infty} J_m(t) = 0$. (*Hint:* Apply the riemannian lemma from Sec. 1.2.)

27. Show that for each $p = 1, 2, \ldots$, $\lim_{m \to \infty} m^p J_m(t) = 0$. (*Hint:* Integrate by parts in Exercise 23.)

28. Let $f(x) = (1 - x^2)^2$, $0 < x < 1$. Find the coefficients $\{A_n\}$ in the Fourier-Bessel expansion $f(x) = \sum_{n=1}^{\infty} A_n J_0(xx_n)$, where $J_0(x_n) = 0$. [*Hint:* Write $f(x)$ as a linear combination of $P_2(x)$ and $P_4(x)$.]

29. Find $P_6(x)$, the solution of $(xP_6')' = -xP_4$, $P_6(1) = 0$.

30. Let $f(x) = (1 - x^2)^3$, $0 < x < 1$. Find the coefficients $\{A_n\}$ in the Fourier-Bessel expansion $f(x) = \sum_{n=1}^{\infty} A_n J_0(xx_n)$, where $J_0(x_n) = 0$.

31. Compute $P_8(x)$, $P_{10}(x)$.

32. Let $f(\rho), = \rho$, $0 < \rho < 1$. Find the coefficients $\{A_n\}$ in the Fourier-Bessel expansion $f(\rho) = \sum_{n=1}^{\infty} A_n J_1(\rho x_n)$, where $J_1(x_n) = 0$.

*33. Obtain the sum of the series $F_3 = \sum_{n=1}^{\infty} J_1(\rho x)/[x_n^3 J_2(x_n)]$, where $J_1(x_n) = 0$. [*Hint:* Use Exercise 32 to show that $\nabla^2(F_3 \cos \varphi) = -(\rho/2) \cos \varphi$.]

*34. Obtain the sum of the series $F_5 = \sum_{n=1}^{\infty} J_1(\rho x_n)/[x_n^5 J_2(x_n)]$, where $J_1(x_n) = 0$.

*35. Let $y(x)$ be a solution of the Bessel equation (3.2.1). Define a new function by $z(x) = x^{-s}y(x^r)$ where r and s are constants with $r > 0$. Show that $z(x)$ satisfies the differential equation

$$z''(x) + \frac{2s + 1 + r(d - 2)}{x} z'(x) + \left[\lambda r^2 x^{(s+2)(r-1)} - \frac{r^2 m^2 + s^2 + rs(d - 2)}{x^2} \right] z(x) = 0$$

[*Hint:* Let $t = x^r$, $y(t) = t^{s/r} z(t^{1/r})$ and express $y'(t)$, $y''(t)$ in terms of z, z', z''.]

3.3 THE VIBRATING DRUMHEAD*

Wave Equation in Polar Coordinates

As a first application of Bessel functions, we study the small transverse vibrations of a circular membrane whose perimeter is fixed. This gives a mathematical model of a drumhead and is closely related to the rectangular membrane, treated in Sec. 2.5.

To be specific, suppose that the drumhead occupies a disk $x^2 + y^2 \leq R^2$. Let $u(x, y; t)$ be the displacement of the point (x, y) at time t. By an argument entirely similar to the derivation of the one-dimensional wave equation and used previously in Sec. 2.5, we see that $u(x, y; t)$ satisfies the partial differential equation

$$u_{tt} = c^2(u_{xx} + u_{yy}) \tag{3.3.1}$$

where c is a positive constant, expressible in terms of the mass, area, and tension of the drumhead. The wave equation (3.3.1) is second-order in time, and therefore it is natural to specify two initial conditions

$$u(x, y; 0) = u_1(x, y) \qquad x^2 + y^2 < R^2$$
$$u_t(x, y; 0) = u_2(x, y) \qquad x^2 + y^2 < R^2 \tag{3.3.2}$$

These correspond to the initial position and velocity of the drumhead. Finally we have the boundary condition

$$u(x, y; t) = 0 \qquad x^2 + y^2 = R^2 \tag{3.3.3}$$

This means that the perimeter of the drumhead is fixed during the motion. The equation and boundary conditions are both homogeneous; therefore we can immediately proceed to look for separated solutions of the wave equation. To do this, we take polar coordinates $x = \rho \cos \varphi$, $y = \rho \sin \varphi$. By abuse of notation, we let $u(\rho, \varphi; t)$ denote the displacement in polar coordinates. Thus the wave equation takes the form

$$u_{tt} = c^2 \left(u_{\rho\rho} + \frac{1}{\rho} u_\rho + \frac{1}{\rho^2} u_{\varphi\varphi} \right) \tag{3.3.4}$$

Separated Solutions

We look for separated solutions in the form

$$u(\rho, \varphi; t) = f_1(\rho) f_2(\varphi) g(t) \tag{3.3.5}$$

---

* See also the Appendix on "Using Mathematica."

Substituting into (3.3.4) and dividing by c^2u, we have

$$\frac{g''(t)}{c^2g(t)} = \frac{f_1''(\rho) + (1/\rho)f_1'(\rho)}{f_1(\rho)} + \frac{f_2''(\varphi)}{\rho^2 f_2(\varphi)} \tag{3.3.6}$$

The left member depends only on t, while the right member depends only on (ρ, φ). Therefore each side equals the same constant, which we call $-\lambda$. This is rewritten in the form

$$g''(t) + \lambda c^2 g(t) = 0 \tag{3.3.7}$$

The remaining equations take the form

$$\rho^2 \left[\lambda + \frac{f_1''(\rho) + (1/\rho)f_1'(\rho)}{f_1(\rho)} \right] = -\frac{f_2''(\varphi)}{f_2(\varphi)}$$

The right member depends only on φ, and the left member depends only on ρ; therefore each equals the same constant μ. We rewrite this in the form

$$f_2''(\varphi) + \mu f_2(\varphi) = 0 \tag{3.3.8}$$

$$f_1''(\rho) + \frac{1}{\rho} f_1'(\rho) + \left(\lambda - \frac{\mu}{\rho^2} \right) f_1(\rho) = 0 \tag{3.3.9}$$

Equations (3.3.7) to (3.3.9) are the ordinary differential equations whose solutions describe the vibrating membrane.

First we treat (3.3.8). The membrane occupies the disk whose equation in polar coordinates is $0 \le \rho < R$, $-\pi \le \varphi \le \pi$. Therefore f_2 must be a smooth periodic function

$$f_2(-\pi) = f_2(\pi)$$

$$f_2'(-\pi) = f_2'(\pi)$$

The general solution of (3.3.8) can be analyzed according to three cases. This Sturm-Liouville problem, which has been solved previously, will be reviewed for completeness. We have

Case 1: $\mu > 0$, $f_2(\varphi) = A \cos \varphi\sqrt{\mu} + B \sin \varphi\sqrt{\mu}$

Case 2: $\mu = 0$, $f_2(\varphi) = A + B\varphi$

Case 3: $\mu < 0$, $f_2(\varphi) = A \cosh \varphi\sqrt{-\mu} + B \sinh \varphi\sqrt{-\mu}$

In case 1 the boundary conditions require that

$$A \cos(-\pi\sqrt{\mu}) + B \sin(-\pi\sqrt{\mu}) = A \cos \pi\sqrt{\mu} + B \sin \pi\sqrt{\mu}$$

$$-A\sqrt{\mu} \sin(-\pi\sqrt{\mu}) + B\sqrt{\mu} \cos(-\pi\sqrt{\mu})$$

$$= -A\sqrt{\mu} \sin \pi\sqrt{\mu} + B\sqrt{\mu} \cos \pi\sqrt{\mu}$$

The first equation states that $B \sin \pi\sqrt{\mu} = 0$, and the second equation states that $A\sqrt{\mu} \sin \pi\sqrt{\mu} = 0$. These are clearly satisfied if $\sqrt{\mu} = 1, 2, \ldots$. If $\sqrt{\mu}$ is not an integer, then we must have $A = B = 0$, yielding the trivial solution $f_2(\varphi) = 0$. Thus we have found the general solution in case 1

$$f_2(\varphi) = A \cos m\varphi + B \sin m\varphi \qquad m = 1, 2, 3, \ldots \qquad (3.3.10)$$

In case 2 the boundary conditions require that $B = 0$; hence we obtain the additional solution $f_2(\varphi) = A$. This can be included in (3.3.10) if we let $m = 0$. In case 3 the boundary conditions cannot be satisfied. (This is left as an exercise.)
Returning to (3.3.9) with $\mu = m^2$ $(m = 0, 1, 2, \ldots)$, we have

$$f_1''(\rho) + \frac{1}{\rho} f_1'(\rho) + \left(\lambda - \frac{m^2}{\rho^2}\right) f_1(\rho) = 0 \qquad 0 \leq \rho \leq R$$

$$f_1(R) = 0$$

From the discussion of Bessel functions in Sec. 3.2, the solution is

$$f_1(\rho) = J_m(\rho\sqrt{\lambda})$$

where λ is chosen so that $J_m(R\sqrt{\lambda}) = 0$. Thus

$$R\sqrt{\lambda} = x_n^{(m)}$$

where $x_n^{(m)}$ are the positive zeros of the Bessel function J_m. Finally the time dependence is obtained by solving (3.3.7):

$$g(t) = \tilde{A} \cos ct\sqrt{\lambda} + \tilde{B} \sin ct\sqrt{\lambda} \qquad \lambda = \frac{(x_n^{(m)})^2}{R^2}$$

Therefore we have found separated solutions of the form

$$u(\rho, \varphi; t) = J_m\left(\frac{\rho x_n^{(m)}}{R}\right) (A \cos m\varphi + B \sin m\varphi)\left(\tilde{A} \cos \frac{ctx_n^{(m)}}{R} + \tilde{B} \sin \frac{ctx_n^{(m)}}{R}\right)$$

$$(3.3.11)$$

Solution of Initial-Value Problems

By taking a superposition of separated solutions, we may satisfy the initial conditions (3.3.2). These solutions become quite complicated; hence it is instructive to briefly study the separated solutions for small values of m, n. At time $t = 0$, the drumhead can be divided into zones, depending on whether u is positive or negative. The curves which divide the zones are called *nodal lines*. For concreteness we take $\tilde{A} = 1$, $\tilde{B} = 0$, $A = 1$, $B = 0$. Thus, for $n = 1$, we have the diagrams of Fig. 3.3.1. For $n = 2$ the Bessel function has an interior zero, and the drumhead appears as in Fig. 3.3.2.

For larger values of (m, n) the diagrams become successively more complex. As time progresses, each of these profiles is multiplied by

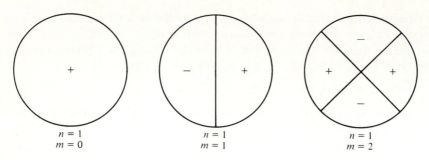

$n = 1$ $n = 1$ $n = 1$
$m = 0$ $m = 1$ $m = 2$

FIGURE 3.3.1
Nodal lines of a vibrating drumhead for $n = 1$.

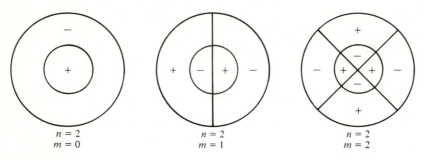

$n = 2$ $n = 2$ $n = 2$
$m = 0$ $m = 1$ $m = 2$

FIGURE 3.3.2
Nodal lines of a vibrating drumhead for $n = 2$.

$\cos (ctx_n^{(m)}/R)$, a periodic function. But when we form a superposition of these separated solutions, the resulting solution is *no longer periodic*. Indeed, the numbers $x_n^{(m)}$ are not multiples of a fixed fundamental frequency, as was the case for the vibrating string, where we had separated solutions of the form

$$\sin \frac{n\pi x}{L} \left(\tilde{A} \cos \frac{n\pi ct}{L} + \tilde{B} \sin \frac{n\pi ct}{L} \right)$$

We now use the separated solutions (3.3.11) to solve the initial-value problem (3.3.2).

Example 3.3.1. Find the solution of the vibrating membrane problem in the case where $u_2(\rho, \varphi) = 0$.

Solution. In this case we use the separated solutions (3.3.11), which satisfy the second initial condition, that is, $\tilde{B} = 0$. We write the solution as a formal sum

$$u(\rho, \varphi; t) = \sum_{m,n} J_m \left(\frac{\rho x_n^{(m)}}{R} \right) (A_{mn} \cos m\varphi + B_{mn} \sin m\varphi) \cos \frac{ctx_n^{(m)}}{R}$$

The first initial condition requires that

$$u_1(\rho, \varphi) = u(\rho, \varphi; 0)$$

$$= \sum_{m,n} J_m \left(\frac{\rho x_n^{(m)}}{R} \right) (A_{mn} \cos m\varphi + B_{mn} \sin m\varphi)$$

$$= \sum_{m=0}^{\infty} \cos m\varphi \sum_{n=1}^{\infty} A_{mn} J_m \left(\frac{\rho x_n^{(m)}}{R} \right) + \sum_{m=1}^{\infty} \sin m\varphi \sum_{n=1}^{\infty} B_{mn} J_m \left(\frac{\rho x_n^{(m)}}{R} \right)$$

This is a Fourier series in φ, whose coefficients are Fourier-Bessel series with $\beta = \pi/2$, $m = 0, 1, 2, \ldots$. Therefore, to solve the problem, we must develop the function $u_1(\rho, \varphi)$ in a series of this type and identify the coefficients A_{mn}, B_{mn}. The next example illustrates a specific case. $\bullet$

Example 3.3.2. Find the solution of the vibrating membrane problem for the case where $u_2(\rho, \varphi) = 0$, $u_1(\rho, \varphi) = R^2 - \rho^2$, $0 < \rho < R$.

Solution. From Sec. 3.2 we have the Fourier-Bessel expansion

$$1 - x^2 = 8 \sum_{n=1}^{\infty} \frac{J_0(xx_n^{(0)})}{(x_n^{(0)})^3 J_1(x_n^{(0)})} \qquad 0 < x < 1$$

Making the substitution $x = \rho/R$, we have the required expansion

$$u_1(\rho, \varphi) = R^2 - \rho^2 = 8R^2 \sum_{n=1}^{\infty} \frac{J_0(\rho x_n^{(0)}/R)}{(x_n^{(0)})^3 J_1(x_n^{(0)})}$$

We see that $B_{mn} = 0$ for all m, n and $A_{mn} = 0$ for $m > 0$, while $A_{0n} = 8R^2/[(x_n^{(0)})^3 J_1(x_n^{(0)})]$. The solution of the problem is

$$u(\rho, \varphi; t) = 8R^2 \sum_{n=1}^{\infty} \frac{J_0(\rho x_n^{(0)}/R)}{(x_n^{(0)})^3 J_1(x_n^{(0)})} \cos \frac{ctx_n^{(0)}}{R} \qquad \bullet$$

Example 3.3.3. Find the solution of the vibrating membrane problem for the case where $u_2(\rho, \varphi) = 0$, $u_1(\rho, \varphi) = J_3(\rho x_1^{(3)}/R) \cos 3\varphi$.

Solution. In this case the initial data are already written as a Fourier-Bessel series as in the previous example, with $A_{31} = 1$ and all other coefficients zero. Therefore the solution of the problem is

$$u(\rho, \varphi; t) = J_3 \left(\frac{\rho x_1^{(3)}}{R} \right) \cos 3\varphi \cos \frac{ctx_1^{(3)}}{R} \qquad \bullet$$

EXERCISES 3.3

1. Suppose that the drumhead is under the influence of gravity. The wave equation takes the form

$$u_{tt} = c^2(u_{xx} + u_{yy}) - g \qquad x^2 + y^2 \le R^2$$

Find the steady-state solution of the form $u = U(\rho)$ which satisfies the boundary condition $U(R) = 0$.

2. Let $f(\varphi) = A \cosh \varphi\sqrt{-\mu} + B \sinh \varphi\sqrt{-\mu}$, $\mu < 0$. Show that if f satisfies the boundary conditions $f(0) = f(2\pi)$, $f'(0) = f'(2\pi)$, then $A = B = 0$.

3. Sketch the drumhead profiles for $n = 3, 4$.

4. Find the solution of the vibrating membrane problem where $u_1(\rho, \varphi) = F(\rho)$, $u_2(\rho, \varphi) = 0$.

5. Find the solution of the vibrating membrane problem where $u_1(\rho, \varphi) = 0$ and $u_2(\rho, \varphi) = F_2(\rho)$.

6. Find the solution of the vibrating membrane problem where $u_1(\rho, \varphi) = 0$, $u_2(\rho, \varphi) = 1$, $0 < \rho < R$.

7. Find the solution of the vibrating membrane problem where $u_1(\rho, \varphi) = 0$, $u_2(\rho, \varphi) = R^2 - \rho^2$, $0 < \rho < R$.

8. Find the solution of the vibrating membrane problem where $u_1(\rho, \varphi) = 0$, $u_2(\rho, \varphi) = J_3(\rho x_1^{(3)}/R) \cos 3\varphi$.

9. Consider a membrane in the shape of a half-circle $0 \leq \varphi \leq \pi$, $0 \leq \rho \leq R$. Show that the separated solutions of the wave equation with zero boundary conditions have the form

$$u(\rho, \varphi; t) = \left(A \cos \frac{ctx_n^{(m)}}{R} + B \sin \frac{ctx_n^{(m)}}{R} \right) J_m \left(\frac{\rho x_n^{(m)}}{R} \right) \sin m\varphi$$

where $m = 1, 2, \ldots$, and $x_n^{(m)}$ are the positive zeros of the Bessel function J_m.

3.4   HEAT FLOW IN A CYLINDER

In this section we study initial-value problems for the heat equation in cylindrical coordinates. To solve these problems, we combine the five-stage method of Chap. 2 with the separated solutions, which are expressed in terms of Bessel functions.

Separated Solutions

To begin, we look for separated solutions of the heat equation in cylindrical coordinates, independent of z. We have

$$u(\rho, \varphi; t) = f_1(\rho)f_2(\varphi)g(t) \tag{3.4.1}$$

$$u_t = K\nabla^2 u = K \left(u_{\rho\rho} + \frac{1}{\rho} u_\rho + \frac{1}{\rho^2} u_{\varphi\varphi} \right) \tag{3.4.2}$$

Substituting (3.4.1) into (3.4.2) and dividing by Ku, we have

$$\frac{g'(t)}{Kg(t)} = \frac{f_1'' + (1/\rho)f_1'}{f_1} + \frac{(1/\rho^2)f_2''}{f_2}$$

The right side depends on (ρ, φ), whereas the left side depends on t. Therefore each is a constant, which we call $-\lambda$. Thus

$$g'(t) + \lambda Kg(t) = 0 \tag{3.4.3}$$

Likewise, the ratio f_2''/f_2 is a constant, which we call $-\mu$. Thus we have the equations

$$f_2'' + \mu f_2 = 0 \tag{3.4.4}$$

$$f_1'' + \frac{1}{\rho} f_1' + \left(\lambda - \frac{\mu}{\rho^2}\right) f_1 = 0 \tag{3.4.5}$$

Equation (3.4.3) is solved by $g(t) = e^{-\lambda K t}$. Equation (3.4.4) with the periodic boundary conditions $f_2(-\pi) = f_2(\pi)$, $f_2'(-\pi) = f_2'(\pi)$ is solved by $f_2(\varphi) = A \cos m\varphi + B \sin m\varphi$, $m = 0, 1, 2, \ldots$. Equation (3.4.5) is a form of Bessel's equation with $d = 2$. The power series solution is $J_m(\rho\sqrt{\lambda})$, and the second solution may be obtained from (3.2.14). Thus we have obtained the separated solutions of the heat equation in cylindrical coordinates:

$$u(\rho, \varphi; t) = J_m(\rho\sqrt{\lambda})(A \cos m\varphi + B \sin m\varphi)e^{-\lambda K t} \tag{3.4.6}$$

The separation constant λ is obtained from the boundary conditions of the problem.

Initial-Value Problems in a Cylinder

Example 3.4.1. Find the solution of the heat equation in the infinite cylinder $0 \le \rho < R$ satisfying the boundary condition $u(R, \varphi; t) = 0$ and the initial condition $u(\rho, \varphi; 0) = f(\rho, \varphi)$.

Solution. The steady-state solution is zero. The solution must be finite at $\rho = 0$, so we do not take the second solution of Bessel's equation. The required separated solutions are $J_m(\rho\sqrt{\lambda})(A \cos m\varphi + B \sin m\varphi)e^{-\lambda K t}$. The boundary condition requires that $J_m(R\sqrt{\lambda}) = 0$, thus $R\sqrt{\lambda} = x_n^{(m)}$, a positive zero of the Bessel function J_m. The solution takes the form

$$u(\rho, \varphi; t) = \sum_{m,n} J_m\left(\frac{\rho x_n^{(m)}}{r}\right)(A_{mn} \cos m\varphi + B_{mn} \sin m\varphi) \exp\left[\frac{-(x_n^{(m)})^2 K t}{R^2}\right]$$

To satisfy the initial condition, we must have

$$f(\rho, \varphi) = \sum_{m,n} J_m\left(\frac{\rho x_n^{(m)}}{R}\right)(A_{mn} \cos m\varphi + B_{mn} \sin m\varphi)$$

This is a Fourier series in $(\cos m\varphi, \sin m\varphi)$, whose coefficients are Fourier-Bessel expansions with $\beta = \pi/2$, $m = 0, 1, 2, \ldots$. The problem is completely solved once we have an expanded $f(\rho, \varphi)$ in a series of this type. ●

The next example gives a specific case.

Example 3.4.2. Find the solution of the heat equation in the infinite cylinder $0 \le \rho < R$ satisfying the boundary condition $u(R, \varphi; t) = 0$ and the initial condition $u(\rho, \varphi; 0) = 1$. Find the relaxation time.

Solution. We use the method of the previous example with $f(\rho, \varphi) = 1$. In this case, we use the Fourier-Bessel expansion

$$1 = 2 \sum_{n=1}^{\infty} \frac{J_0(xx_n)}{x_n J_1(x_n)} = 2 \sum_{n=1}^{\infty} \frac{J_0(\rho x_n/R)}{x_n J_1(x_n)} \qquad 0 < \rho < R$$

where $J_0(x_n) = 0$ and we have made the substitution $x = \rho/R$. Therefore the solution of the initial-value problem is

$$u(\rho, \varphi; t) = 2 \sum_{n=1}^{\infty} \frac{J_0(\rho x_n/R)}{x_n J_1(x_n)} \exp\left(\frac{-x_n^2 Kt}{R^2}\right)$$

For each $t > 0$, the series for u, u_ρ, $u_{\rho\rho}$, u_t converge uniformly; hence u is a rigorous solution. When $t \to \infty$, the solution tends to zero, the steady-state solution. The relaxation time can be computed from the first term of the series.

$$-\frac{1}{\tau} = \lim_{t \to \infty} \frac{1}{t} \ln |u(\rho, \varphi; t)| = -\frac{x_1^2 K}{R^2}$$

$$\tau = \frac{R^2}{Kx_1^2} = 0.1729 \frac{R^2}{K} \qquad \bullet$$

In the next example we add an inhomogeneous boundary condition and an inhomogeneous term to the heat equation, representing a heat source. To solve problems of this type, first we find the steady-state solution, according to the five-stage method of Chap. 2.

Example 3.4.3. Find the solution of the following heat equation in the infinite cylinder $0 \le \rho < R$:

$$u_t = K\nabla^2 u + \sigma \qquad t > 0, 0 \le \rho < R, -\pi \le \varphi \le \pi$$

$$u(R, \varphi; t) = T_1 \qquad t > 0, -\pi \le \varphi \le \pi$$

$$u(\rho, \varphi; 0) = T_2 \qquad 0 < \rho < R, -\pi \le \varphi \le \pi$$

where σ, T_1, T_2 are positive constants.

Solution

Stage 1. To find the steady-state solution, we try $U = U(\rho)$, independent of φ (since the boundary condition is independent of φ). We obtain the ordinary differential equation $K[U'' + (1/\rho)U'] + \sigma = 0$ with the boundary condition $U(R) = 0$. The general solution of the equation is $U = -\sigma\rho^2/(4K) + A + B \log \rho$. We must have $U(R) = 0$, $U(0)$ finite; hence $B = 0$, $A = T_1 + \sigma R^2/(4K)$. The steady-state solution is

$$U(\rho) = T_1 + \frac{\sigma}{4K}(R^2 - \rho^2)$$

Stage 2. We use the steady-state solution to transform to a homogeneous equation with homogeneous boundary conditions. Thus, letting $v = u - U$, we have $v_t = K\nabla^2 v$, $v(R, \varphi; t) = 0$, $v(\rho, \varphi; 0) = T_2 - U(\rho)$.

Stage 3. We look for v as a sum of separated solutions of the homogeneous equation satisfying the homogeneous boundary conditions. These were found in Example 3.4.1:

$$v(\rho, \varphi; t) = \sum_{m,n} J_m \left(\frac{\rho x_n^{(m)}}{R} \right) (A_{mn} \cos m\varphi + B_{mn} \sin m\varphi) \exp \left[-\frac{(x_n^{(m)})^2 Kt}{R^2} \right]$$

To satisfy the initial condition, we must have

$$T_2 - U(\rho) = v(\rho, \varphi; 0) = \sum_{m,n} J_m \left(\frac{\rho x_n^{(m)}}{R} \right) (A_{mn} \cos m\varphi + B_{mn} \sin m\varphi)$$

To obtain the required expansion of $T_2 - U(\rho)$, recall the Fourier-Bessel expansions from Sec. 3.2:

$$1 = 2 \sum_{n=1}^{\infty} \frac{J_0(x x_n)}{x_n J_1(x_n)} \qquad 0 < x < 1, \, J_0(x_n) = 0$$

$$1 - x^2 = 8 \sum_{n=1}^{\infty} \frac{J_0(x x_n)}{x_n^3 J_1(x_n)} \qquad 0 < x < 1, \, J_0(x_n) = 0$$

Letting $x = \rho/R$, we have

$$T_2 - U(\rho) = (T_2 - T_1) - \frac{\sigma R^2}{4K} \left(1 - \frac{\rho^2}{R^2} \right)$$

$$= \sum_{n=1}^{\infty} A_n J_0 \left(\frac{\rho x_n}{R} \right) \qquad 0 < \rho < R$$

where
$$A_n = \frac{2(T_2 - T_1)/x_n - 2\sigma R^2/(K x_n^3)}{J_1(x_n)} \qquad n = 1, 2, \ldots$$

The solution of the original problem is therefore

$$u(\rho, \varphi; t) = U(\rho) + \sum_{n=1}^{\infty} A_n J_0 \left(\frac{\rho x_n}{R} \right) \exp \left(-\frac{x_n^2 Kt}{R^2} \right) \qquad J_0(x_n) = 0$$

Stages 4 and 5. For each $t > 0$, the series for $u, u_\rho, u_{\rho\rho}, u_t$ converge uniformly for $0 \le \rho \le R$; hence u satisfies the heat equation. If $A_1 \ne 0$, the relaxation time is given by the first term of the series: $\tau = R^2/(x_1^2 K)$. ●

The next example illustrates the possibility of boundary conditions which involve radiation of heat to the exterior of the cylinder.

Example 3.4.4. Find the solution of the following heat equation in the infinite cylinder $0 \le \rho < R$:

$$u_t = K\nabla^2 u + \sigma \qquad t > 0, 0 \le \rho < R$$

$$-k u_\rho|_{\rho=R} = h(u - T_1)|_{\rho=R}$$

$$u(\rho, \varphi; 0) = T_2$$

Here $\sigma, K, k, h, T_1, T_2$ are positive constants.

Solution

Stage 1. We look for the steady-state solution in the form $U = U(\rho)$. We obtain the ordinary differential equation

$$K \left(U'' + \frac{1}{\rho} U' \right) + \sigma = 0$$

with the boundary condition $-kU'(R) = h[U(R) - T_1]$. The general solution of this ordinary differential equation is $U(\rho) = -[\sigma\rho^2/(4K)] + A + B \log \rho$. The condition $U(0)$ finite requires $B = 0$, while the boundary condition requires $k\sigma R/(2K) = h\{A - [\sigma R^2/(4K)] - T_1\}$, $A = T_1 + \sigma Rk/(2hK) + \sigma R^2/(4K)$. The steady-state solution is

$$U(\rho) = T_1 + \frac{\sigma Rk}{2hK} + \frac{\sigma}{4K} (R^2 - \rho^2)$$

Stage 2. We introduce the function $v = u - U$, which satisfies the homogeneous equation $v_t = K\nabla^2 v$ with the homogeneous boundary condition $-kv_\rho|_{\rho=R} = hv|_{\rho=R}$.

Stage 3. The separated solutions of the heat equation are $J_m(\rho\sqrt{\lambda})(A_{mn} \cos m\varphi + B_{mn} \sin m\varphi)e^{-\lambda Kt}$. To satisfy the new boundary condition, we must have $k\sqrt{\lambda}J'_m(R\sqrt{\lambda}) + hJ_m(R\sqrt{\lambda}) = 0$. We saw in Sec. 3.2 that this equation has infinitely many solutions $R\sqrt{\lambda} = x_n^{(m)}$, $n = 1, 2, \ldots$. To satisfy the initial condition, we take a sum of these separated solutions. Since the initial condition is independent of φ, we may write

$$v = v(\rho; t) = \sum_{n=1}^{\infty} A_n J_0 \left(\frac{\rho x_n}{R} \right) \exp \left(\frac{x_n^2 Kt}{R^2} \right)$$

To satisfy the initial condition, we must have

$$T_2 - T_1 - \frac{\sigma Rk}{2hK} - \frac{\sigma}{4K} (R^2 - \rho^2) = \sum_{n=1}^{\infty} A_n J_0 \left(\frac{\rho x_n}{R} \right)$$

The Fourier coefficients $\{A_n\}$ are obtained from the Fourier-Bessel expansion of Example 3.2.10 with $\tan \beta = hR/k$:

$$1 = 2 \sum_{n=1}^{\infty} \frac{J_0(xx_n)}{x_n\{1 + [x_n k/(hR)]^2\}J_1(x_n)} \qquad 0 < x < 1$$

$$1 - x^2 = 8 \sum_{n=1}^{\infty} \frac{J_0(xx_n)}{x_n^3\{1 + [x_n k/(hR)]^2\}J_1(x_n)} \qquad 0 < x < 1$$

Making the substitution $x = \rho/R$, we have

$$A_n = \frac{2}{x_n} \left[1 + \left(\frac{x_n k}{hR} \right)^2 \right]^{-1} \left(T_2 - T_1 - \frac{\sigma Rk}{2hK} - \frac{\sigma R^2}{Kx_n^2} \right)$$

The solution of the original problem is written in the form

$$u(\rho; t) = U(\rho) + \sum_{n=1}^{\infty} A_n J_0 \left(\frac{\rho x_n}{R} \right) \exp \left(-\frac{x_n^2 Kt}{R^2} \right)$$

Stage 4. For $t > 0$, the series for u, u_ρ, $u_{\rho\rho}$, and u_t converge uniformly for $0 \le \rho \le R$. Hence u is a rigorous solution of the heat equation.

Stage 5. When $t \to \infty$, $u(\rho; t)$ tends to $U(\rho)$, the steady-state solution. The relaxation time is $\tau = R^2/(x_1^2 K)$, where $(k/R)x_1 J_0'(x_1) + hJ_0(x_1) = 0$, $A_1 \ne 0$. ●

Initial-Value Problems between Two Cylinders

As our final application, we consider heat flow between two cylinders, of inner radius R_1 and outer radius R_2. To solve problems of this type, we need both solutions of Bessel's equation. We consider the following initial-boundary-value problem:

$$u_t = K\nabla^2 u \qquad t > 0, R_1 < \rho < R_2, -\pi \le \varphi \le \pi$$

$$u(R_1, \varphi; t) = T_1 \qquad t > 0, -\pi \le \varphi \le \pi$$

$$u(R_2, \varphi; t) = T_2 \qquad t > 0, -\pi \le \varphi \le \pi$$

$$u(\rho, \varphi; 0) = f(\rho, \varphi) \qquad R_1 < \rho < R_2, -\pi \le \varphi \le \pi$$

STAGE 1. The steady-state equation is $\nabla^2 u = 0$, with the above boundary conditions. Since these are independent of φ, we look for the steady-state solution in the form $U = U(\rho)$. The general solution of $\nabla^2 u = 0$ is $U = A + B \ln \rho$. The boundary conditions require $A + B \ln R_1 = T_1$, $A + B \ln R_2 = T_2$; thus

$$U(\rho) = T_1 + (T_2 - T_1) \frac{\ln \rho - \ln R_1}{\ln R_2 - \ln R_1}$$

STAGE 2. We use the steady-state solution to transform to homogeneous boundary conditions. Thus, letting $v = u - U$, we have $v_t = K\nabla^2 v$, $v(R_1, \varphi; t) = 0$, $v(R_2, \varphi; t) = 0$, $v(\rho, \varphi; 0) = f(\rho, \varphi) - U(\rho)$.

STAGE 3. The separated solutions of the heat equation in cylindrical coordinates are $f_1(\rho)(A \cos m\varphi + B \sin m\varphi)e^{-\lambda Kt}$, where $m = 0, 1, 2, \ldots$ and f_1 is a solution of Bessel's equation with $d = 2$ and satisfying the boundary conditions $f_1(R_1) = 0$, $f_1(R_2) = 0$. Since $R_1 > 0$, this is a *regular Sturm-Liouville problem*. The theory of Sec. 0.4 guarantees an infinite number of eigenvalues $\lambda_n^{(m)}$, with $\sqrt{\lambda_n^{(m)}} = n\pi/(R_2 - R_1) + O(1/n)$, $n \to \infty$, and normalized eigenfunctions $\Phi_n^{(m)}$ with

$$\sqrt{\rho}\Phi_n^{(m)}(\rho) = \left(\frac{2}{R_2 - R_1}\right)^{1/2} \sin \frac{n\pi(\rho - R_1)}{R_2 - R_1} + O\left(\frac{1}{n}\right) \qquad n \to \infty$$

The solution is written in the form

$$v(\rho, \varphi; t) = \sum_{m,n} \Phi_n^{(m)}(\rho)(A_{mn} \cos m\varphi + B_{mn} \sin m\varphi)e^{-\lambda_n^{(m)}Kt}$$

The eigenfunctions $\Phi_n^{(m)}$ are orthogonal with respect to the weight $\rho \, d\rho$ for different values of n. Therefore we can obtain the Fourier coefficients by the formulas

$$\int_{-\pi}^{\pi} \int_{R_1}^{R_2} [f(\rho, \varphi) - U(\rho)]\Phi_n^{(m)}(\rho) \cos m\varphi \, \rho \, d\rho \, d\varphi$$

$$= A_{mn} \int_{-\pi}^{\pi} \cos^2 m\varphi \, d\varphi \int_{R_1}^{R_2} \Phi_n^{(m)}(\rho)^2 \rho \, d\rho$$

$$\int_{-\pi}^{\pi} \int_{R_1}^{R_2} [f(\rho, \varphi) - U(\rho)]\Phi_n^{(m)}(\rho) \sin m\varphi \, \rho \, d\rho \, d\varphi$$

$$= B_{mn} \int_{-\pi}^{\pi} \sin^2 m\varphi \, d\varphi \int_{R_1}^{R_2} \Phi_n^{(m)}(\rho)^2 \rho \, d\rho$$

STAGE 4. The formal solution to our problem is

$$u(\rho, \varphi; t) = U(\rho) + \sum_{m,n} \Phi_n^{(m)}(\rho)(A_{mn} \cos m\varphi + B_{mn} \sin m\varphi)e^{-\lambda_n^{(m)}Kt}$$

where the Fourier coefficients A_{mn}, B_{mn} are obtained in stage 3. To verify that this series converges and represents a solution of the heat equation, we consider the special case of radially symmetric solutions, where $f(\rho, \varphi) = f(\rho)$ independent of φ. In this case the solution assumes the simpler form

$$u(\rho; t) = U(\rho) + \sum_{1}^{\infty} A_n \Phi_n(\rho)e^{-\lambda_n Kt}$$

where $\lambda_n = \lambda_n^{(0)}$, $\Phi_n = \Phi_n^{(0)}$, $A_n = A_{0n}$. Suppose $|f(\rho)| \leq M$, a constant. The Fourier coefficients can be estimated by

$$2\pi|A_n| = \left| \int_{R_1}^{R_2} [f(\rho) - U(\rho)]\Phi_n(\rho)\rho \, d\rho \right|$$

$$\leq M_2 \int_{R_1}^{R_2} \rho|\Phi_n(\rho)| \, d\rho$$

$$\leq M_2 \left(\frac{R_2^2 - R_1^2}{2} \right)^{1/2} \qquad M_2 = M + T_1 + T_2$$

where we have used the Schwarz inequality for integrals and the normalization $\int_{R_1}^{R_2} \Phi_n(\rho)^2 \rho \, d\rho = 1$. Therefore the terms of the series $\sum_{n=1}^{\infty} A_n \Phi_n(\rho)e^{-\lambda_n Kt}$ are bounded by the terms of the series $M_3 \sum_{n=1}^{\infty} (e^{-at})^n$, $a = \pi K/(R_2 - R_1)$, $M_3 = M_2(1 + R_2/R_1)^{1/2}$. Likewise, the series for u_ρ, $u_{\rho\rho}$, and u_t are bounded by convergent numerical series and hence are uniformly convergent.

STAGE 5. When $t \to \infty$, the solution $u(\rho, \varphi; t)$ tends to the steady-state solution $U(\rho)$. To compute the relaxation time, we again restrict our attention to radially symmetric solutions, where $A_{mn} = 0$ for $m \neq 0$ and $B_{mn} = 0$. In this case the eigenvalues $\{\lambda_n\}$ have been tabulated numerically for various values of the ratio R_1/R_2. The following table lists some representative values:†

R_1/R_2	$R_2\sqrt{\lambda_1}$	$R_2\sqrt{\lambda_2}$	$R_2\sqrt{\lambda_3}$
0.8	12.56	25.13	37.70
0.6	4.70	9.42	14.13
0.4	2.07	4.18	6.27
0.2	0.76	0.69	2.35
0.1	0.33	0.53	1.04

For example, if $R_1 = 3$, $R_2 = 5$, then we can read from the second line of the table that $\lambda_1 \approx 0.88$, $\lambda_2 \approx 3.53$, $\lambda_3 = 8.01$. The relaxation time can be obtained from the first term of the series for $u(\rho; t)$; thus $\lambda_1 K\tau = 1$. For example, if $R_1 = 3$, $R_2 = 5$, then $\tau = 1/(0.88K) = 1.14/K$ to two decimal places.

EXERCISES 3.4

1. Find the solution of the heat equation $u_t = K\nabla^2 u$ in the infinite cylinder $0 \le \rho < R$ that will satisfy the boundary condition $u(R, \varphi; t) = 0$ and the initial condition $u(\rho, \varphi; 0) = R^2 - \rho^2$.

2. Find the solution of the heat equation $u_t = K\nabla^2 u$ in the infinite cylinder $0 \le \rho < R$ that will satisfy the boundary condition $u(R, \varphi; t) = 1$ and the initial condition $u(\rho, \varphi; 0) = 0$. Find the relaxation time.

3. Find the solution of the heat equation $u_t = K\nabla^2 u$ in the infinite cylinder $0 \le \rho < R$ satisfying the boundary condition $u(R, \varphi; t) = 1 + \frac{1}{2} \cos \varphi$ and the initial condition $u(\rho, \varphi; 0) = 0$. (*Hint:* Use Example 3.2.9.)

4. Find the solution of the heat equation $u_t = K\nabla^2 + \sigma$ in the infinite cylinder $0 \le \rho < R$ satisfying the boundary condition $u(R, \varphi; t) = T_1$ and the initial condition $u(\rho, \varphi; 0) = T_2(1 - \rho^2/R^2)$.

5. It is required to solve the heat equation $u_t = K\nabla^2 u$ in the infinite half-cylinder $0 \le \rho < R$, $0 < \varphi < \pi$. Find the separated solutions of the heat equation satisfying the boundary conditions $u(\rho, 0; t) = 0$, $u(\rho, \pi; t) = 0$, $u(R, \varphi; t) = 0$.

*6. Find the solution of the heat equation $u_t = K\nabla^2 u$ in the infinite half-cylinder $0 \le \rho < R$, $0 < \varphi < \pi$, satisfying the boundary conditions $u(\rho, 0; t) = 0$, $u(\rho, \pi; t) = 0$, $u(R, \varphi; t) = 0$ and the initial conditions $u(\rho, \varphi; 0) = f(\rho)$.

7. Consider heat flow in the region $R_1 < \rho < R_2$ with $R_1 = 3$ cm, $R_2 = 15$ cm. The boundary conditions are $T_1 = 0°C$, $T_2 = 100°C$. Find the steady-state solution and the relaxation time.

*8. Consider heat flow in the cylinder $0 \le \rho < 2$, where the surface $\rho = 2$ is maintained at $100°C$. At $t = 0$, we have $u = 0$ for $0 \le \rho < 1$ and $u = 50°C$ for $1 \le \rho < 2$. Find the solution $u = u(\rho; t)$ for all $t > 0$, $0 \le \rho < 2$. (*Hint:* Use the method of Example 3.2.8.)

---

† Milton Abramowitz and Irene A. Stegun, *Handbook of Mathematical Functions*, Dover, New York, 1972, p. 415.

9. Consider heat flow in the cylinder $0 \leq \rho < 2$, where the surface $\rho = 2$ is *insulated*, $\partial u/\partial \rho = 0$ at $\rho = 2$. Find the separated solutions of the heat equation which satisfy this boundary condition. Solve the problem in the case where $u(\rho; 0) = 100°C$. (*Hint:* $x_1 = 0$ from Proposition 3.2.5.)

10. Find all the separated solutions $u(\rho, \varphi, z)$ of Laplace's equation $\nabla^2 u = 0$ in the finite cylinder $0 \leq \rho < R$, $0 < z < L$ and satisfying the boundary condition $u(R, \varphi, z) = 0$ for $-\pi \leq \varphi \leq \pi$, $0 < z < L$.

11. Find the solution $u(\rho, \varphi, z)$ of Laplace's equation $\nabla^2 u = 0$ in the finite cylinder $0 \leq \rho < R$, $0 < z < L$ and satisfying the boundary condition $u(R, \varphi, z) = 0$ for $-\pi \leq \varphi \leq \pi$, $0 < z < L$; $u(\rho, \varphi, 0) = 0$ for $0 \leq \rho < R$, $-\pi \leq \varphi \leq \pi$; $u(\rho, \varphi, L) = 1$ for $0 \leq \rho < R$, $-\pi \leq \varphi \leq \pi$.

12. Find the solution of Laplace's equation in the finite cylinder $0 \leq \rho < R$, $0 < z < L$ and satisfying the boundary conditions $u(R, \varphi, z) = 0$ for $-\pi \leq \varphi \leq \pi$, $0 < z < L$; $u(\rho, \varphi, 0) = T_1$ for $0 \leq \rho < R$, $-\pi \leq \varphi \leq \pi$; $u(\rho, \varphi, L) = T_2$ for $0 \leq \rho < R$, $-\pi \leq \varphi \leq \pi$.

13. Find all the separated solutions $u(\rho, \varphi, z)$ of Laplace's equation $\nabla^2 u = 0$ in the finite cylinder $0 \leq \rho < R$, $0 < z < L$ and satisfying the boundary conditions $u(\rho, \varphi, 0) = 0 = u(\rho, \varphi, L)$ for $0 \leq \rho < R$, $-\pi \leq \varphi \leq \pi$.

14. Find the solution $u(\rho, \varphi, z)$ of Laplace's equation $\nabla^2 u = 0$ in the finite cylinder $0 \leq \rho < R$, $0 < z < L$ and satisfying the boundary conditions $u(\rho, \varphi, 0) = 0 = u(\rho, \varphi, L)$ for $0 \leq \rho < R$, $-\pi \leq \varphi < \pi$; $u(R, \varphi, z) = 1$ for $-\pi \leq \varphi \leq \pi$, $0 < z < L$.

15. Find the solution $u(\rho, \varphi, z)$ of Laplace's equation $\nabla^2 u = 0$ in the finite cylinder $0 \leq \rho < R$, $0 < z < L$ and satisfying the boundary conditions $u(\rho, \varphi, 0) = 0 = u(\rho, \varphi, L)$ for $0 \leq \rho < R$, $-\pi \leq \varphi < \pi$; $u(R, \varphi, z) = z$ for $-\pi \leq \varphi \leq \pi$, $0 < z < L$.

16. Find the solution $u(\rho, \varphi; t)$ of the heat equation $u_t = K\nabla^2$ in the infinite half-cylinder $0 \leq \rho < R$, $0 < \varphi < \pi$ satisfying the boundary conditions $u(\rho, 0; t) = 0 = u(\rho, \pi; t) = 0$ for $0 \leq \rho < R$, $t > 0$, $u(R, \varphi; t) = 0$, and the initial condition $u(\rho, \varphi; 0) = \rho \sin \varphi$.

17. Find the solution $u(\rho, \varphi; t)$ of the heat equation $u_t = K\nabla^2 u$ in the infinite cylinder $0 \leq \rho < R$ satisfying the boundary condition $u(R, \varphi; t) = 0$ for $-\pi \leq \varphi \leq \pi$, $t > 0$ and the initial condition $u(\rho, \varphi; 0) = \rho^2 \cos 2\varphi$ for $0 \leq \rho < R$, $-\pi \leq \varphi \leq \pi$.

18. Find the solution of the heat equation $u_t = K\nabla^2 u$ in the infinite half-cylinder $0 \leq \rho < R$ and satisfying the boundary condition $-ku_\rho|_{\rho=R} = h(u - T_1)|_{\rho=R}$ and the initial condition $u(\rho, \varphi; 0) = T_2(1 - \rho^2/R^2)$, where K, h, k, T_1, T_2 are positive constants.

CHAPTER 4

BOUNDARY-VALUE PROBLEMS IN SPHERICAL COORDINATES

Introduction

In this chapter we consider boundary-value problems in regions with spherical boundaries. Section 4.1 treats some problems which can be solved in terms of elementary functions, i.e., Laplace's equation in a sphere and the heat equation with spherically symmetric boundary temperature. In Sec. 4.2 we develop the properties of Legendre functions and spherical Bessel functions which are applied in Sec. 4.3 to the solution of more general boundary-value problems in a sphere.

4.1 SPHERICALLY SYMMETRIC SOLUTIONS

Laplacian in Spherical Coordinates

Recall that spherical coordinates are defined by the formulas

$$x = r \sin \theta \cos \varphi$$

$$y = r \sin \theta \sin \varphi \qquad (4.1.1)$$

$$z = r \cos \theta$$

The polar angle θ measures the colatitude, i.e., the angle which the vector (x, y, z) makes with the positive z axis. The azimuthal angle φ measures the longitude, i.e., the angle which vector $(x, y, 0)$ makes with the positive x axis

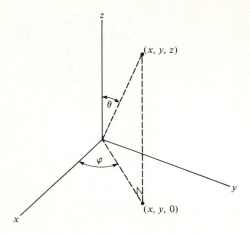

FIGURE 4.1.1
Spherical coordinates.

in the xy plane. The radial distance r measures the length of vector (x, y, z). By convention, $0 \le \theta \le \pi$ and $-\pi \le \varphi \le \pi$, where $\theta = 0$ and $\theta = \pi$ correspond to the positive (respectively, negative) z axis and the variable φ is undefined. This is illustrated in Fig. 4.1.1.

Let $u(x, y, z)$ be a smooth function and $\nabla^2 u = u_{xx} + u_{yy} + u_{zz}$, the laplacian of u. A new function $U(r, \theta, \varphi)$ is defined by the formula $U(r, \theta, \varphi) = u(r \sin \theta \cos \varphi, r \sin \theta \sin \varphi, r \cos \theta)$. We wish to compute $\nabla^2 u$ in terms of U_{rr}, U_r, $U_{\theta\theta}$, etc. To simplify the exposition, we write instead u_{rr}, u_r, u_θ for the indicated partial derivatives. This simplification will cause no confusion.

To compute $\nabla^2 u$, first we recall the result from Sec. 3.1 for the laplacian in cylindrical coordinates

$$u_{xx} + u_{yy} = u_{\rho\rho} + \frac{1}{\rho} u_\rho + \frac{1}{\rho^2} u_{\varphi\varphi} \tag{4.1.2}$$

The coordinates (ρ, φ) obey the relations

$$x = \rho \cos \varphi \qquad y = \rho \sin \varphi$$

From the formulas for z and x, we have

$$z = r \cos \theta \qquad \rho = r \sin \theta$$

Therefore (z, ρ) *are obtained from* (r, θ) *in exactly the same way as* (x, y) *are obtained from* (ρ, φ). Thus we must have

$$u_{zz} + u_{\rho\rho} = u_{rr} + \frac{1}{r} u_r + \frac{1}{r^2} u_{\theta\theta} \tag{4.1.3}$$

Adding (4.1.2) and (4.1.3), we have

$$u_{xx} + u_{yy} + u_{zz} = u_{rr} + \frac{1}{r} u_r + \frac{1}{r^2} u_{\theta\theta} + \frac{1}{\rho} u_\rho + \frac{1}{\rho^2} u_{\varphi\varphi} \tag{4.1.4}$$

It remains to compute u_ρ. From the chain rule for partial derivatives, we have

$$u_\rho = u_r \frac{\partial r}{\partial \rho} + u_\theta \frac{\partial \theta}{\partial \rho} + u_\varphi \frac{\partial \varphi}{\partial \rho}$$

The transformation from cylindrical to spherical coordinates gives us $r = (\rho^2 + z^2)^{1/2}$, $\theta = \tan^{-1}(\rho/z)$, $\varphi = \varphi$. From this it follows that $\partial r/\partial \rho = \rho/r$, $\partial \theta/\partial \rho = (\cos\theta)/r$, $\partial \varphi/\partial \rho = 0$. Substituting these in the equation for u_ρ, we have

$$u_\rho = u_r \frac{\rho}{r} + u_\theta \frac{\cos\theta}{r}$$

Substituting this into (4.1.4) and rearranging yield

$$\boxed{\nabla^2 u = u_{rr} + \frac{2}{r} u_r + \frac{1}{r^2}(u_{\theta\theta} + \cot\theta\, u_\theta + \csc^2\theta\, u_{\varphi\varphi})} \qquad (4.1.5)$$

This is the required formula.

Example 4.1.1. Compute $\nabla^2(x^2 + y^2 + z^2)^{5/2}$.

Solution. If we were to use the definition of ∇^2, this would be a tedious computation. Noting that the function is expressed in spherical coordinates as r^5, we have

$$\nabla^2(r^5) = 20r^3 + \frac{2}{r} 5r^4 = 30r^3 \qquad \bullet$$

Time-Periodic Heat Flow: Applications to Geophysics

As a first application of spherical coordinates, we reconsider the geophysics problem from Sec. 2.1. Now we assume that the earth is a perfect sphere of radius a and, as before, that the surface temperature is a periodic function of time. Thus we have the boundary-value problem

$$\begin{aligned} u_t &= K\nabla^2 u && t > 0, 0 \le x^2 + y^2 + z^2 < a^2 \\ u(x, y, z; t) &= f(t) && t > 0, x^2 + y^2 + z^2 = a^2 \end{aligned} \qquad (4.1.6)$$

Taking spherical coordinates, we look for the solution in the form $u = u(r; t)$. Thus we must have

$$u_t = K\left(u_{rr} + \frac{2}{r} u_r\right) \qquad t > 0, 0 \le r < a$$

$$u(a; t) = f(t) \qquad t > 0$$

This can be reduced to the heat equation in one dimension by introducing the new function

$$w(r; t) = ru(r; t)$$

Thus $w_t = ru_t$, $w_r = ru_r + u$, $w_{rr} = ru_{rr} + 2u_r$. Multiplying the heat equation by r and making these substitutions, we have

$$w_t = Kw_{rr} \qquad t > 0, 0 \le r < a$$

$$w(a; t) = af(t) \qquad\qquad (4.1.6a)$$

$$w(0; t) = 0$$

The final boundary condition comes from the fact that the temperature at the center of the sphere must be finite for all time. Mathematically *we have reduced the problem of spherically symmetric heat flow to heat flow in a slab with zero temperature on one face.*† This problem can be solved by looking for complex separated solutions

$$w(r; t) = e^{i\omega t}e^{\gamma r}$$

Substituting in the heat equation (4.1.6a), we find that $\gamma^2 = i\omega/K$; hence $\gamma = \pm\sqrt{\omega/(2K)}\,(1 + i)$, which yields the complex separated solutions

$$\left.\begin{array}{c} e^{-cr}e^{i(\omega t - cr)} \\[4pt] e^{cr}e^{i(\omega t + cr)} \end{array}\right\} \qquad c = \sqrt{\frac{\omega}{2K}}$$

To match the boundary conditions, we assume that $f(t)$ has been expanded as a Fourier series.

$$f(t) = A_0 + \sum_{n=1}^{\infty} \left(A_n \cos\frac{2\pi nt}{T} + B_n \sin\frac{2\pi nt}{T} \right)$$

$$= \sum_{-\infty}^{\infty} \alpha_n e^{2\pi int/T} \qquad\qquad (4.1.7)$$

Therefore it suffices to solve the problem with $f(t) = e^{i\omega t}$ and to take the real and imaginary parts. For this we try a linear combination of complex separated solutions

$$w(r; t) = C_1 e^{-cr}e^{i(\omega t - cr)} + C_2 e^{cr}e^{i(\omega t + cr)}$$

The boundary conditions at $r = 0$ and $r = a$ require that

$$0 = C_1 e^{i\omega t} + C_2 e^{i\omega t}$$

$$ae^{i\omega t} = C_1 e^{-ca}e^{i(\omega t - ca)} + C_2 e^{ca}e^{i(\omega t + ca)}$$

---

† This general principle applies to all the applications discussed in this section.

Thus $C_1 + C_2 = 0$, $a = C_1 e^{-ca} e^{-ica} + C_2 e^{ca} e^{ica}$. Solving these two equations simultaneously and simplifying, we have

$$w(r; t) = a e^{i\omega t} \frac{e^{c(1+i)r} - e^{-c(1+i)r}}{e^{c(1+i)a} - e^{-c(1+i)a}}$$

The original problem is therefore solved by

$$u(r; t) = \frac{a}{r} \sum_{-\infty}^{\infty} \alpha_n e^{i\omega_n t} \frac{e^{c_n(1+i)r} - e^{-c_n(1+i)r}}{e^{c_n(1+i)a} - e^{-c_n(1+i)a}}$$

where we have set $\omega_n = 2\pi n/T$, $c_n = \sqrt{\pi n/(KT)}$.

Example 4.1.2. Find the solution of the heat equation $u_t = K\nabla^2 u$ for $-\infty < t < \infty$, in the sphere $0 < r < a$, and satisfying the boundary condition $u(a; t) = A_0 + A_1 \cos \omega t$. Find $u(0; t)$.

Solution. We take $\alpha_0 = A_0$, $\alpha_1 = \alpha_{-1} = \frac{1}{2}A_1$. The solution is

$$u(r; t) = A_0 + \frac{aA_1}{r} \operatorname{Re} \left(e^{i\omega t} \frac{e^{c(1+i)r} - e^{-c(1+i)r}}{e^{c(1+i)a} - e^{-c(1+i)a}} \right) \qquad (4.1.8)$$

where $c = \sqrt{\omega/(2K)}$. To find $u(0; t)$, we may appeal to L'Hospital's rule to take the limit when $r \to 0$. Thus

$$u(0; t) = \lim_{r \to 0} u(r; t) = A_0 + aA_1 \operatorname{Re} \left[e^{i\omega t} \frac{2c(1 + i)}{e^{c(1+i)a} - e^{-c(1+i)a}} \right] \qquad \bullet$$

It is instructive to compare the solution of Example 4.1.2 with the solution for the flat earth in Chap. 2. To do this, we let $z = a - r$ and consider the limit when $a \to \infty$ with z fixed. This corresponds to observing the temperature at a shallow depth. Removing a factor of $e^{c(1+i)a}$ from the denominator and removing the factor $e^{c(1+i)r}$ from the numerator, we have

$$u(r; t) = A_0 + \frac{aA_1}{r} \operatorname{Re} \left(e^{c(r-a)} e^{i[\omega t + c(r-a)]} \frac{1 - e^{-2cr(1+i)}}{1 - e^{-2ca(1+i)}} \right)$$

If $a \to \infty$, $r \to \infty$ with $z = a - r$ fixed, the final fraction tends to 1 and

$$\lim_{a \to \infty} u(r; t) = A_0 + A_1 \operatorname{Re} \left(e^{-cz} e^{i(\omega t - cz)} \right) = A_0 + A_1 e^{-cz} \cos (\omega t - cz)$$

This is exactly the solution which we found in Sec. 2.1 for the flat earth. Hence we see that, for shallow depths, the earth's surface can be assumed flat. For example, if $a = 4000$ mi and $z = 1$ mi, then $1 - a/r = 0.00026$, less than 0.1 of 1 percent error in the solution.

Heat Flow in a Sphere

As a second application of spherical coordinates, we consider the following initial-value problem for heat flow in a sphere:

$$u_t = K\nabla^2 u \qquad t > 0, 0 \le r < a$$

$$u(a; t) = T_1 \qquad t > 0 \tag{4.1.9}$$

$$u(r; 0) = \varphi(r) \qquad 0 \le r < a$$

where $\varphi(r)$ is a given piecewise smooth function, defined for $0 \le r < a$. This problem is solved by the five-stage method introduced in Chap. 2. The steady-state solution which satisfies the heat equation and boundary condition is $U = T_1$. Subtracting this, we have the transformed problem with T_1 replaced by zero and $\varphi(r)$ replaced by $\varphi(r) - T_1$. Introducing the function $w = r(u - U)$, we see that the problem reduces to

$$w_t = Kw_{rr} \qquad t > 0, 0 \le r < a$$

$$w(a; t) = 0 \qquad t > 0$$

$$w(0; t) = 0 \qquad t > 0 \tag{4.1.9a}$$

$$w(r; 0) = r[\varphi(r) - T_1] \qquad 0 \le r < a$$

This is a one-dimensional problem for heat flow in a slab, for which we know the separated solutions

$$w(r; t) = \left(\sin \frac{n\pi r}{a}\right) e^{-(n\pi/a)^2 Kt} \qquad n = 1, 2, \ldots$$

Therefore, by superposition, we have solved the original problem in the form

$$\boxed{u(r; t) = T_1 + \frac{1}{r} \sum_{n=1}^{\infty} B_n \left(\sin \frac{n\pi r}{a}\right) e^{-(n\pi/a)^2 Kt}} \tag{4.1.10}$$

The coefficients B_n can be obtained by setting $t = 0$ and using the formulas from Fourier series. Thus,

$$\varphi(r) = T_1 + \frac{1}{r} \sum_{n=1}^{\infty} B_n \sin \frac{n\pi r}{a} \qquad 0 < r < a$$

$$\tag{4.1.11}$$

$$B_n = \frac{2}{a} \int_0^a r[\varphi(r) - T_1] \sin \frac{n\pi r}{a} \, dr \qquad n = 1, 2, \ldots$$

If $\varphi(r)$ is piecewise smooth, the series converge and u satisfies the heat equation and initial and boundary conditions. The relaxation time can be found by taking the first term of the series; thus $\tau = a^2/(\pi^2 K)$ if $B_1 \neq 0$.

The temperature at the center of the sphere is found by noting that $\lim_{r \to 0} (1/r) \sin (n\pi r/a) = n\pi/a$. Thus,

$$u(0; t) = \lim_{r \to 0} u(r; t) = T_1 + \sum_{n=1}^{\infty} \frac{n\pi}{a} B_n e^{-(n\pi/a)^2 Kt}$$

In using this to compute the temperature at the center for small times, we must be careful in estimating the sum of the series. Indeed, we expect on physical grounds (and it can be proved mathematically) that $u(0; t)$ is no larger than T_1 and the maximum of φ, no smaller than T_1 and the minimum of φ. This should be reflected in practical computations based on the series solution we have just found.

Example 4.1.3. Let $\varphi(r) = T_2$, a constant. Find the solution $u(r; t)$. For the numerical values $K = 0.03$, $a = 0.5$, $T_1 = 0$, $T_2 = 100$, find the relaxation time and estimate $u(0; t)$ for $t = 5$, $t = 1$, $t = 0.1$.

Solution. In this case the formula (4.1.11) gives

$$B_n = \frac{2(T_2 - T_1)}{a} \int_0^a r \sin \frac{n\pi r}{a} \, dr$$

These integrals were computed in Example 1.1.1, with the result

$$\frac{2}{a} \int_0^a r \sin \frac{n\pi r}{a} \, dr = \frac{(-1)^{n+1}}{n} \frac{2a}{\pi}$$

Therefore, from (4.1.10), we have the solution

$$u(r; t) = T_1 + \frac{2a(T_2 - T_1)}{\pi r} \sum_{n=1}^{\infty} \frac{(-1)^{n+1}}{n} \left(\sin \frac{n\pi r}{a} \right) e^{-(n\pi/a)^2 Kt}$$

At the center we have

$$u(0; t) = T_1 + 2(T_2 - T_1) \sum_{n=1}^{\infty} (-1)^{n+1} e^{-(n\pi/a)^2 Kt}$$

For $K = 0.03$, $a = 0.5$, we have $\pi^2 K/a^2 = 1.18$, to two decimals. The relaxation time is $\tau = 1/1.18 = 0.85$, to two decimals. For $t = 5$, the first term of the series is 0.0027, and the remaining terms are less than 10^{-11}. This leads to the estimate $u(0; t) = 0.54$. For $t = 1$, the first two terms of the series are $0.3073 - 0.0089$, and the remaining terms are less than 10^{-4}. This leads to the estimate $u(0; 1) = 59.68$, to two decimal places. For $t = 0.1$, the first five terms of the series are $0.8887 - 0.6237 + 0.3458 - 0.1514 + 0.0523 = 0.5117$. If we use this to estimate the temperature, we have $u(0; 0.1) = 102.34$, to two decimals, a physically unrealistic result. To obtain a more realistic result, we may average the fourth and fifth partial sums of the series. This leads to the estimate $u(0; 0.1) = 97.11$, to two decimals. ●

As our next application, we consider the problem of a sphere which exchanges heat with the outside according to Newton's law of cooling; i.e., the heat flux across the boundary is proportional to the difference between the surface temperature and the outside temperature. In addition, we assume that heat is produced at a constant rate σ. Problems of this type occur when we consider apples which are placed in a refrigerator.

Mathematically we have the problem

$$u_t = K\nabla^2 u + \sigma \qquad\qquad 0 \le r < a, t > 0$$

$$-k\frac{\partial u}{\partial r}(a; t) = h[u(a; t) - T_1] \qquad t > 0 \qquad\qquad (4.1.12)$$

$$u(r; 0) = \varphi(r) \qquad\qquad 0 \le r < a$$

where $h > 0$, $k > 0$, and $\varphi(r)$ is a function which is unspecified. The initial and boundary conditions are independent of (θ, φ), and therefore we may expect the solution in the form $u = u(r; t)$. Letting $w(r; t) = ru(r; t)$, we have $w_t = ru_t$, $w_r = ru_r + u$, $w_{rr} = ru_{rr} + 2u_r = r\nabla^2 u$. We multiply the equations (4.1.12) by r. In terms of $w(r; t)$, the problem becomes

$$w_t = Kw_{rr} + \sigma r \qquad\qquad 0 \le r < a, t > 0$$

$$-k\left(w_r - \frac{w}{r}\right)(a; t) = h[w(a; t) - aT_1] \qquad t > 0 \qquad\qquad (4.1.13)$$

$$w(0; t) = 0 \qquad\qquad t > 0$$

$$w(r; 0) = r\varphi(r) \qquad\qquad 0 \le r < a$$

This boundary-value problem for w is a one-dimensional problem which can be solved by the five-stage method of Chap. 2. Note that we have the additional boundary condition at $r = 0$.

Stage 1. The steady-state equation is $KW_{rr} + \sigma r = 0$ with the two boundary conditions at $r = 0$, $r = a$. The general solution of this ordinary differential equation is

$$W(r) = -\frac{\sigma r^3}{6K} + A + Br$$

where A, B are arbitrary constants. The boundary condition $W(0) = 0$ requires that $A = 0$. To analyze the boundary condition at $r = a$, we write

$$W'(r) = -\frac{\sigma r^2}{2K} + B$$

$$W'(r) - \frac{W(r)}{r} = B - \frac{\sigma r^2}{2K} - B + \frac{\sigma r^2}{6K}$$

$$= -\frac{\sigma r^2}{3K}$$

$$h[W(r) - rT_1] = h\left(Br - \frac{\sigma r^3}{6K} - rT_1\right)$$

Therefore B is determined by the equation

$$-k\left(-\frac{\sigma a^2}{3K}\right) = h\left(Ba - \frac{\sigma a^3}{6K} - aT_1\right)$$

The solution is

$$W(r) = \frac{r\sigma}{6K}(a^2 - r^2) + r\left(T_1 + \frac{\sigma ak}{3hK}\right) \qquad (4.1.14)$$

We can use this to compute the total flux out of the sphere in steady state. In terms of the original temperature function u, we have

$$-k\frac{\partial u}{\partial r}\bigg|_{r=a} = h(u - T_1)|_{r=a} = \frac{h(w - rT_1)}{r}\bigg|_{r=a} = \frac{\sigma a}{3}\frac{k}{K}$$

Multiplying this by the surface area $4\pi a^2$, we have the total flux $= (\sigma 4\pi a^3/3) \cdot (k/K)$. This agrees with physical intuition, since the only way that heat can flow across the boundary in steady state is from the source term σ.

Stage 2. We use the steady-state solution to transform the problem. Letting $v(r; t) = w(r; t) - W(r)$, we have the equation for v:

$$v_t = Kv_{rr} \qquad\qquad 0 \le r < a, t > 0$$

$$v(0; t) = 0 \qquad\qquad t > 0$$

$$-kv_r(a; t) = hv(a; t) - \frac{k}{a}v(a; t) \qquad t > 0$$

$$v(r; 0) = r\varphi(r) - W(r) \qquad\qquad 0 \le r < a$$

Stage 3. The separated solutions of the problem for v can be easily obtained. Writing $v(r; t) = f(r)g(t)$, we have the equations $g' + \lambda Kg = 0$, $f'' + \lambda f = 0$ with the boundary conditions $f(0) = 0$, $f'(a) = (1/a - h/k)f(a)$. We consider separately three cases.

Case 1: $1/a < h/k$
Case 2: $1/a = h/k$
Case 3: $1/a > h/k$

In case 1 the boundary conditions are of the form $f(0) = 0$, $f'(a) = -\alpha f(a)$, $\alpha > 0$. This boundary condition satisfies the positivity condition for Sturm-Liouville problems. Therefore the eigenvalues λ_n are all positive. From the differential equation, they are of the form $f_n(r) = \sin r\sqrt{\lambda_n}$. The eigenvalues $\{\lambda_n\}$ are solutions of the transcendental equation $a\sqrt{\lambda_n} \cot a\sqrt{\lambda_n} = 1 - ha/k$. They may be obtained graphically when we have the numerical values of a, h, k. For large n they have the asymptotic behavior $a\sqrt{\lambda_n} = n\pi + O(1)$, $n \to \infty$.

In case 2 the boundary conditions are $f(0) = 0$, $f'(a) = 0$. This Sturm-Liouville problem is solved by the eigenfunctions $f_n(r) = \sin [(n - \frac{1}{2})\pi r/a]$ and the eigenvalues $\lambda_n = (n - \frac{1}{2})^2\pi^2/a^2$, all of which are positive.

In case 3 the boundary conditions are $f(0) = 0$, $f'(a) = \alpha f(a)$ with $\alpha = 1/a - h/k > 0$. This eigenvalue problem was analyzed in Sec. 2.2, where we considered the three cases $\alpha a < 1$, $\alpha a = 1$, $\alpha a > 1$. In our present case, $\alpha a = 1 - ha/k < 1$; according to the results of Sec. 2.2, all the eigenvalues are positive. They are obtained as solutions of the transcendental equation $a\sqrt{\lambda} \cot a\sqrt{\lambda} = 1 - ha/k$. The eigenfunctions are $f_n(r) = \sin r\sqrt{\lambda_n}$.

In all three cases we can write the superposition of separated solutions as

$$v(r; t) = \sum_{n=1}^{\infty} A_n \sin (r\sqrt{\lambda_n})e^{-\lambda_n Kt} \qquad (4.1.15)$$

Since the boundary conditions are separable, the eigenfunctions must be orthogonal. Thus

$$\int_0^a \sin r\sqrt{\lambda_n} \sin r\sqrt{\lambda_m} \, dr = 0 \qquad n \neq m$$

The normalization can be computed as the integral

$$\int_0^a \sin^2 r\sqrt{\lambda_n} \, dr = \frac{1}{2} \int_0^a (1 - \cos 2r\sqrt{\lambda_n}) \, dr$$

$$= \frac{1}{2} \left(a - \frac{\sin 2a\sqrt{\lambda_n}}{2\sqrt{\lambda_n}} \right)$$

The Fourier coefficients A_n can be obtained by using these relations. Thus by setting $t = 0$ in (4.1.15), multiplying by $\sin r\sqrt{\lambda_n}$, and integrating, we have

$$\int_0^a [r\varphi(r) - W(r)] \sin r\sqrt{\lambda_n} \, dr = A_n \int_0^a \sin^2 r\sqrt{\lambda_n} \, dr \qquad (4.1.16)$$

Example 4.1.4. Find the Fourier coefficients in the case where $\varphi(r) = T_2$, a constant.

Solution. In this case we must compute the integral

$$\int_0^a [rT_2 - W(r)] \sin r\sqrt{\lambda_n} \, dr$$

$$= \int_0^a \left[r\left(T_2 - T_1 - \frac{\sigma ak}{3hK} \right) - \frac{r\sigma}{6K} (a^2 - r^2) \right] \sin r\sqrt{\lambda_n} \, dr$$

We use the integrals

$$\int_0^a r \sin r\sqrt{\lambda} \, dr = \frac{\sin a\sqrt{\lambda}}{\lambda} - \frac{a \cos a\sqrt{\lambda}}{\sqrt{\lambda}}$$

$$\int_0^a r(a^2 - r^2) \sin r\sqrt{\lambda} \, dr = \frac{6 \sin a\sqrt{\lambda}}{\lambda^2} - \frac{6a \cos a\sqrt{\lambda}}{\lambda^{3/2}} - \frac{2a^2 \sin a\sqrt{\lambda}}{\lambda}$$

We substitute these in the above formulas and use the relation $a\sqrt{\lambda} \cot a\sqrt{\lambda} = 1 - ha/k$, with the result

$$\int_0^a [rT_2 - W(r)] \sin r\sqrt{\lambda}\, dr = \left(T_2 - T_1 - \frac{\sigma ak}{3Kh}\right) \frac{\sin a\sqrt{\lambda}}{\lambda} \frac{ha}{k}$$

$$- \frac{\sigma}{K} \frac{\sin a\sqrt{\lambda}}{\lambda^2} \left\{\frac{ha}{k} - \frac{1}{3}\left[1 - \left(\frac{ha}{k}\right)^2\right]\right\} \quad \bullet$$

Stage 4. We have obtained the formal solution of the problem (4.1.12) as

$$u(r; t) = U(r) + \frac{1}{r} \sum_{n=1}^\infty A_n \sin (r\sqrt{\lambda_n}) e^{-\lambda_n Kt} \qquad (4.1.17)$$

where the steady-state solution $U(r) = [\sigma/(6K)](a^2 - r^2) + T_1 + \sigma ak/(3Kh)$; the eigenvalues $\{\lambda_n\}$ are determined from the transcendental equation $a\sqrt{\lambda_n} \cot a\sqrt{\lambda_n} = 1 - ha/k$, and the Fourier coefficients A_n are obtained from (4.1.16). We have $A_n = O(1)$ and $a\sqrt{\lambda_n} = n\pi + O(1)$ when $n \to \infty$. Therefore, for each $t > 0$, the series (4.1.17) and the differentiated series for u_r, u_{rr}, and u_t all converge uniformly for $0 \le r \le a$. Thus $u(r; t)$ is a rigorous solution of the heat equation.

Stage 5. When $t \to \infty$, the solution $u(r; t)$ tends to the steady-state solution $U(r)$. We use the method from Chap. 2 to estimate the rate of approach; thus,

$$\frac{1}{r} \sum_{n=1}^\infty A_n \sin (r\sqrt{\lambda_n}) e^{-\lambda_n Kt} = O(e^{-\lambda_1 Kt}) \qquad t \to \infty$$

Therefore $u(r; t) - U(r) = O(e^{-\lambda_1 Kt})$, $t \to \infty$. Finally we compute the relaxation time by noting that

$$u(r; t) - U(r) = A_1 \frac{\sin r\sqrt{\lambda_1}}{r} e^{-\lambda_1 Kt} + O(e^{-\lambda_2 Kt}) \qquad t \to \infty$$

If $A_1 \ne 0$, the relaxation time is given by

$$\tau = \frac{1}{\lambda_1 K} \qquad (4.1.18)$$

To obtain numerical estimates of the relaxation time, we may appeal to the graphs in Fig. 4.1.2.

Example 4.1.5. If $K = 0.30$, $a = 0.50$, $h = 0.08$, $k = 0.303$, find the relaxation time.

Solution. We have $1 - ha/k = 0.868$. The smallest solution of the equation $x \cot x = 0.868$ is $x = 0.62$; thus $\sqrt{\lambda_1} = 0.62/0.50 = 1.24$. The relaxation time is $\tau = 1/(0.30)(1.24)^2$, or about 2 s. $\quad \bullet$

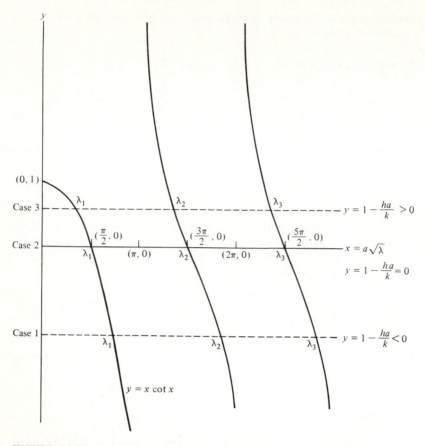

FIGURE 4.1.2
Determination of eigenvalues λ_n as solutions of the transcendental equation $a\sqrt{\lambda}\cot(a\sqrt{\lambda}) = 1 - ha/k$. Case 1: $1 - ha/k < 0$. Case 2: $1 - ha/k = 0$. Case 3: $1 - ha/k > 0$.

The heat equation can be used to study the cooling of apples that are placed into a refrigerator. Assuming a perfect sphere of radius $a = 2$ in, diffusivity $K = 0.720$ in^2/h, and Biot modulus $ha/k = 1.0$, we have, for the first eigenvalue $a\sqrt{\lambda_1} = \pi/2$, $\sqrt{\lambda_1} = 0.7854$ and the relaxation time $\tau = 1/(\lambda_1 K) = 1/[(0.720)(0.7854)^2]$, about 2 h. From this we expect that within 10 h the apple will be within 1 percent of the ambient refrigerator temperature, relative to its initial temperature.

The Three-Dimensional Wave Equation

We close this section by remarking that these techniques can also be used to find spherically symmetric solutions of the three-dimensional wave equation

$$u_{tt} = c^2\nabla^2 u = c^2\left(u_{rr} + \frac{2}{r}u_r\right)$$

which is satisfied by the scalar potential function of electromagnetic theory or by the small pressure variations of a gas. To obtain solutions, we let $w(r; t) = ru(r; t)$ and obtain the equation for w

$$w_{tt} = c^2 w_{rr}$$

This is the one-dimensional wave equation which was encountered in the discussion of the vibrating string in Sec. 2.4, where we found the separated solutions $w(r; t) = (A \sin kr + B \cos kr)(C \cos kct + D \sin kct)$. The function w must satisfy the additional boundary condition that $w(0; t) = 0$. This gives the solutions of the wave equation in the form

$$u(r; t) = \frac{1}{r} \sin kr(C \cos kct + D \sin kct)$$

These may be used to solve initial-value problems for the three-dimensional wave equation.

EXERCISES 4.1

In Exercises 1 to 7, use formula (4.1.5) to compute the indicated expressions.

1. $\nabla^2(r^3)$ 2. $\nabla^2(r^2 \sin^3 \theta)$ 3. $\nabla^2(r)$ 4. $\nabla^2(\theta)$
5. $\nabla^2(r^2 \sin^2 \theta \cos 2\varphi)$
6. $\nabla^2(e^{3r})$
7. $\nabla^2(r^n)$, $n = 1, 2, \ldots$

8. Show that $\nabla^2[f(r)] = (1/r)(rf)''$, where $f(r)$ is a smooth function of one variable.
9. Show that the general solution of the equation $\nabla^2[f(r)] = 0$ is $f(r) = A + B/r$ for arbitrary constants A, B.
10. Show that the general solution of the equation $\nabla^2[f(r)] = -1$ is $f(r) = A + B/r - r^2/6$ for arbitrary constants A, B.
11. Solve the equation $\nabla^2[f(r)] = -1$ with the boundary condition $f(a) = 0$ and $f(0)$ finite.
12. Solve the equation $\nabla^2[f(r)] = -r^2$ with the boundary condition $f(a) = 0$ and $f(0)$ finite.
13. Solve the equation $\nabla^2[f(r)] = -r^4$ with the boundary condition $f(a) = 0$ and $f(0)$ finite.
14. Find the solution $u(r; t)$ of the heat equation $u_t = K\nabla^2 u$ for $-\infty < t < \infty$, in the sphere $0 \leq r < a$, and satisfying the boundary condition $u(a; t) = 3 \cos 2t$.
15. Find the solution $u(r; t)$ of the heat equation $u_t = K\nabla^2 u$ for $-\infty < t < \infty$, in the sphere $0 \leq r < a$, and satisfying the boundary condition $u(a; t) = A_1 \cos \omega(t - t_0)$, where A_1, ω, t_0 are positive constants.
16. Find the solution $u(r; t)$ of the heat equation $u_t = K\nabla^2 u$ for $-\infty < t < \infty$, in the sphere $0 \leq r < a$, and satisfying the boundary condition $u(a; t) = 2 \cos 3t + 5 \cos \pi t$.
17. Find the solution $u(r; t)$ of the heat equation $u_t = K\nabla^2 u + \sigma$ in the sphere $0 \leq r < a$ and satisfying the boundary condition $u(a; t) = T_1$ and the initial condition $u(r; 0) = T_2$. Use the five-stage method, and find the relaxation time.

18. Find the solution $u(r; t)$ of the heat equation $u_t = K\nabla^2 u$ in the sphere $0 \le r < 2a$ and satisfying the boundary conditions $u(2a; t) = T_1$ and the initial conditions $u(r; 0) = T_2$ for $0 \le r < a$ and $u(r; 0) = 0$ for $a \le r < 2a$.

19. Suppose that a ball 0.5 m in diameter, initially at 100°C, is placed into a refrigerator which instantaneously cools the surface to 0°C. Find the approximate temperature at the center at $t = 5$ min if the ball is made of (a) iron ($K = 0.15$ cgs unit), (b) concrete ($K = 0.005$ cgs unit).

20. Consider the heat equation $u_t = K\nabla^2 u$ in the sphere $0 \le r < a$ with the boundary condition $\partial u / \partial r = 0$ at $r = a$. Find the separated solutions of the form $u(r; t) = f(r)g(t)$.

21. Find the solution of the heat equation $u_t = K\nabla^2 u$ in the sphere $0 \le r < a$ with the boundary condition $\partial u / \partial r = 0$ at $r = a$ and the initial condition $u(r; 0) = T_1$ for $0 \le r < \frac{1}{2}a$ and $r(r; 0) = 0$ for $\frac{1}{2}a \le r < a$.

22. Find the solution of the heat equation $u_t = K\nabla^2 u$ in the sphere $0 \le r < a$, with the boundary condition $-k \, \partial u / \partial r = (k/a)(u - T_1)$ at $r = a$ and the initial condition $u(r; 0) = T_2$, $0 \le r < a$.

23. A green pea of radius $a = 0.25$ in, diffusivity $K = 0.75$ in²/h, and Biot modulus $ha/k = 1.0$ is removed from a freezer and allowed to thaw. Find the relaxation time.

24. This problem is designed to show that $a\sqrt{\lambda_n} = (n - \frac{1}{2})\pi + O(1/n)$, $n \to \infty$, for the solutions of the equation $a\sqrt{\lambda_n} \cot a\sqrt{\lambda_n} = b$, $b = 1 - ah/k$.

    (a) From the graph in Fig. 4.1.2 show that $(n - 1)\pi < a\sqrt{\lambda_n} < n\pi$, $n = 1, 2, 3, \ldots$.

    (b) Defining new variables ϵ and y by $a\sqrt{\lambda_n} = (n - \frac{1}{2})\pi + \epsilon$ and $y = 1/[(n - \frac{1}{2})\pi]$, show that $by = (1 + \epsilon y) \tan \epsilon$ and $-\pi/2 < \epsilon < \pi/2$.

    (c) Use the method of implicit differentiation to show that the solution $\epsilon(y)$ with $\epsilon(0) = 0$ has $\epsilon'(0) = b$.

    (d) Conclude that $a\sqrt{\lambda_n} = (n - \frac{1}{2})\pi + b/n + O(1/n^2)$, $n \to \infty$.

25. Let $u = u(r; t)$ be a solution of the three-dimensional wave equation $u_{tt} = c^2\nabla^2 u$, and let $w(r; t) = ru(r; t)$. Show that w is a solution of the one-dimensional wave equation $w_{tt} = c^2 w_{rr}$.

26. Solve the following initial-wave problem for the three-dimensional wave equation $u_{tt} = c^2\nabla^2 u$ in a sphere of radius a: $u(a; t) = 0$, $u(r; 0) = 0$, $u_t(r; 0) = (A/r) \sin (n\pi r/a)$, where $A > 0$ and $n = 1, 2, \ldots$.

4.2 LEGENDRE POLYNOMIALS, SPHERICAL HARMONICS, AND SPHERICAL BESSEL FUNCTIONS

Separated Solutions of the Heat Equation in Spherical Coordinates

Having obtained the form of $\nabla^2 u$, we can obtain general solutions of the heat equation in spherical coordinates. A fundamental first step is the construction of separated solutions:

$$u(r, \theta, \varphi; t) = f_1(r)f_2(\theta)f_3(\varphi)g(t) \tag{4.2.1}$$

Substituting this into the heat equation in spherical coordinates, we obtain

$$\frac{1}{f_1(r)} \left[f_1''(r) + \frac{2}{r} f_1'(r) \right] + \frac{1}{r^2 f_2(\theta)} [f_2''(\theta) + \cot\theta\, f_2'(\theta)] + \frac{\csc^2\theta\, f_3''(\varphi)}{r^2 f_3(\varphi)} - \frac{g'(t)}{Kg(t)}$$
$$= 0 \quad (4.2.2)$$

The first three terms are independent of t, whereas the final term is a function of t alone and therefore a constant, to be called $-\lambda$. Thus

$$\boxed{g'(t) + \lambda K g(t) = 0} \quad (4.2.3)$$

Multiplying (4.2.2) by r^2, we see that the second and third terms are independent of r, while the remainder depends on r. Therefore we introduce a new separation constant $-\mu$. This produces the equation

$$\frac{1}{f_2(\theta)} [f_2''(\theta) + \cot\theta\, f_2'(\theta)] + (\csc^2\theta) \frac{f_3''(\varphi)}{f_3(\varphi)} = -\mu \quad (4.2.4)$$

Multiplying (4.2.4) by $\sin^2\theta$, we see that the term $f_3''(\varphi)/f_3(\varphi)$ is independent of θ, whereas the remainder of (4.2.4) depends on θ. Introducing a new separation constant $-\nu$, we have the equations

$$\boxed{f_3''(\varphi) + \nu f_3(\varphi) = 0} \quad (4.2.5)$$

and

$$\boxed{f_2''(\theta) + \cot\theta\, f_2'(\theta) + (\mu - \nu\csc^2\theta) f_2(\theta) = 0} \quad (4.2.6)$$

Finally, the remaining part of (4.2.2) depends on r alone. So we have the equation

$$\boxed{f_1''(r) + \frac{2}{r} f_1'(r) + \left(\lambda - \frac{\mu}{r^2}\right) f_1(r) = 0} \quad (4.2.7)$$

Therefore we have reduced the heat equation to four ordinary differential equations: (4.2.3), (4.2.5), (4.2.6), (4.2.7). These involve the three separation constants (λ, μ, ν), whose values are determined below.

The solution of (4.2.3) is straightforward:

$$g(t) = e^{-\lambda K t}$$

This is identical to the form obtained previously in rectangular and cylindrical coordinates.

We are now ready to solve the angular equations (4.2.5) and (4.2.6). The product $f_2(\theta)f_3(\varphi)$ is called a *spherical harmonic*. It is our goal in this section to obtain a complete system of spherical harmonics.

Equation (4.2.5) is straightforward. Indeed, from the physical meaning of φ, we have the periodic boundary conditions $f_3(-\pi) = f_3(\pi)$, $f_3'(-\pi) = f_3'(\pi)$. Therefore,

$$f_3(\varphi) = A \cos m\varphi + B \sin m\varphi \qquad m = 0, 1, 2, \ldots$$

$$\nu = m^2$$

Equation (4.2.6) is the *associated Legendre equation*. To obtain solutions of this, we let $z = \cos \theta$. Therefore,

$$\frac{df}{d\theta} = -\sin \theta \frac{df}{dz} \quad \text{and} \quad \frac{d^2f}{d\theta^2} = \sin^2 \theta \frac{d^2f}{dz^2} - \cos \theta \frac{df}{dz}$$

Thus (4.2.6) becomes

$$\boxed{(1 - z^2) \frac{d^2f_2}{dz^2} - 2z \frac{df_2}{dz} + \left(\mu - \frac{m^2}{1 - z^2} \right) f_2 = 0} \qquad (4.2.8)$$

This equation is of the Sturm-Liouville type, with $s(z) = 1 - z^2$, $\rho(z) = 1$, $q(z) = m^2/(1 - z^2)$; $z = 0$ is an ordinary point for this equation, and therefore we can expect a power series solution, of the form

$$f(z) = \sum_{n=0}^{\infty} a_n z^n$$

Legendre Polynomials†

We consider first the case where $m = 0$, which is called the *Legendre equation:*

$$\boxed{(1 - z^2)f'' - 2zf' + \mu f = 0} \qquad (4.2.9)$$

Assuming a power series solution $f(z) = \sum_{n=0}^{\infty} a_n z^n$, we have

$$f'(z) = \sum_{n=0}^{\infty} na_n z^{n-1} \qquad f''(z) = \sum_{n=0}^{\infty} n(n - 1)a_n z^{n-2}$$

Substituting these in the Legendre equation, we have

$$0 = (1 - z^2) \sum_{n=0}^{\infty} n(n - 1)a_n z^{n-2} - 2z \sum_{n=0}^{\infty} na_n z^{n-1} + \mu \sum_{n=0}^{\infty} a_n z^n$$

$$= \sum_{n=0}^{\infty} n(n - 1)a_n z^{n-2} + \sum_{n=0}^{\infty} [\mu - 2n - n(n - 1)]a_n z^n$$

---

† A number of properties of Legendre polynomials are treated in the Appendix on "Using Mathematica."

The first sum is the same as $\sum_{n=0}^{\infty} (n + 2)(n + 1)a_{n+2}z^n$. Therefore,

$$0 = \sum_{n=0}^{\infty} (n + 2)(n + 1)a_{n+2}z^n + \sum_{n=0}^{\infty} [\mu - n(n + 1)]a_n z^n$$

Since a power series is zero if and only if all coefficients are zero, we must have

$$(n + 2)(n + 1)a_{n+2} + [\mu - n(n + 1)]a_n = 0 \qquad n = 0, 1, 2, \ldots$$

This yields the *recurrence relation*

$$a_{n+2} = \frac{n(n + 1) - \mu}{(n + 2)(n + 1)} a_n \qquad n = 0, 1, 2, \ldots \qquad (4.2.10)$$

Constants a_0, a_1 may be chosen arbitrarily, and for any value of μ we obtain two linearly independent power series solutions of the Legendre equation. If μ is of the form $l(l + 1)$ for some integer l, then the recurrence relation (4.2.10) shows that $a_n = 0$ for $n = l + 2, l + 4, \ldots$. Therefore we may obtain a polynomial solution in the following way: If l is even, choose $a_1 = 0$, $a_0 \neq 0$. Then from the recurrence relation $a_n = 0$ for $n = 3, 5, \ldots$. Combining this with the fact that $l + 2$ is even, we see that $a_n = 0$ for all $n \geq 1 + l$; in other words, $f(z)$ is a polynomial of degree l. In case l is odd, we choose $a_0 = 0$, $a_1 \neq 0$ and conclude in the same way that $a_n = 0$ for $n \geq l + 1$. In either case we have shown that *if $\mu = l(l + 1)$ for an integer l, then a_0, a_1 can be chosen so that $f(z)$ is a polynomial of degree l.* This polynomial is denoted $P_l(z)$ and is called the *Legendre polynomial* of degree l.

To uniquely define $P_l(z)$, we require a standard choice of a_0, a_1. Although by no means canonical, we follow established conventions and choose these so that a_l, the coefficient of z^l, is $(2l)!/[2^l(l!)^2]$. We can show that this implies the easily remembered fact that $P_l(1) = 1$ (see Exercise 8).

Example 4.2.1. Compute $P_2(z)$.

Solution. We take $l = 2$, $\mu = 2 \cdot 3 = 6$, and $a_2 = 4!/[2^2(2!)^2] = \frac{3}{2}$. Taking $n = 0$ in (4.2.10) yields $a_2 = -\frac{6}{2}a_0$; hence $a_0 = -\frac{1}{2}$. Thus $P_2(z) = \frac{3}{2}z^2 - \frac{1}{2}$. ●

We list the first few Legendre polynomials.

l	μ	$P_l(z)$
0	0	1
1	2	z
2	6	$(3z^2 - 1)/2$
3	12	$(5z^3 - 3z)/2$
4	20	$(35z^4 - 30z^2 + 3)/8$

The Legendre polynomials satisfy an important orthogonality relation, a special case of the Sturm-Liouville theory discussed in Sec. 0.4:

$$\int_{-1}^{1} P_{l_1}(z)P_{l_2}(z)\ dz = 0 \qquad l_1 \neq l_2$$

(4.2.11)

It is also important to recognize the orthogonality relation in spherical coordinates. Making the substitution $z = \cos\theta$, $dz = -\sin\theta\ d\theta$, we are led to

$$\int_{0}^{\pi} P_{l_1}(\cos\theta)P_{l_2}(\cos\theta)\sin\theta\ d\theta = 0 \qquad l_1 \neq l_2$$

(4.2.12)

The graphs in Fig. 4.2.1 give the Legendre polynomials for $l = 0, 1, 2, 3, 4$.

We now use the orthogonality of Legendre polynomials to obtain an important representation of $P_l(z)$, known as *Rodrigues' formula*. To do this, we note that any polynomial Q of degree l can be written as a finite sum $\Sigma_{k=0}^{l} c_k P_k(z)$. Indeed, $P_l(z)$ is a polynomial of degree l, and thus by choosing the coefficient c_l properly, we can arrange that $Q(z) - c_l P_l(z)$ is a polynomial of degree $l - 1$. Continuing this iteratively, we arrive at the desired representation. We now apply this observation to $Q(z) = (d/dz)^l(z^2 - 1)^l$, a polynomial of degree l. Thus

$$\left(\frac{d}{dz}\right)^l (z^2 - 1)^l = \sum_{k=0}^{l} c_k P_k(z)$$

(4.2.13)

By orthogonality,

$$\int_{-1}^{1} P_k(z)\left(\frac{d}{dz}\right)^l (z^2 - 1)^l\ dz = c_k \int_{-1}^{1} P_k(z)^2\ dz \qquad 0 \leq k \leq l$$

(4.2.14)

The left side can be integrated by parts l times. The endpoint terms vanish, and if $k < l$, $(d/dz)^l P_k(z) = 0$. Therefore $c_k = 0$ for $k < l$. To compute c_l, recall that the coefficient of z^l in $P_l(z)$ was taken to be $(2l)!/[2^l(l!)^2]$. But the coefficient of z^l in $(d/dz)^l(z^2 - 1)^l$ is $2l(2l - 1) \cdots (l + 1) = (2l)!/(l!)$. Comparing the two, we see that $c_l(2l)!/[2^l(l!)^2] = (2l)!/(l!)$; thus $c_l = 2^l l!$ Hence we have proved *Rodrigues' formula*

$$P_l(z) = \frac{1}{2^l l!} \left(\frac{d}{dz}\right)^l (z^2 - 1)^l$$

(4.2.15)

This can now be used to compute the integrals which appear on the right side of (4.2.14). For this purpose we again integrate by parts l times and use $(d/dz)^{2l}(z^2 - 1)^l = (2l)!$ to write

$$(2^l l!)^2 \int_{-1}^{1} P_l(z)^2\ dz = \int_{-1}^{1} \left[\left(\frac{d}{dz}\right)^l (z^2 - 1)^l\right]^2 dz$$

$$= (-1)^l \int_{-1}^{1} (z^2 - 1)^l \left(\frac{d}{dz}\right)^{2l} (z^2 - 1)^l\ dz$$

$$= (2l)! \int_{-1}^{1} (1 - z^2)^l\ dz$$

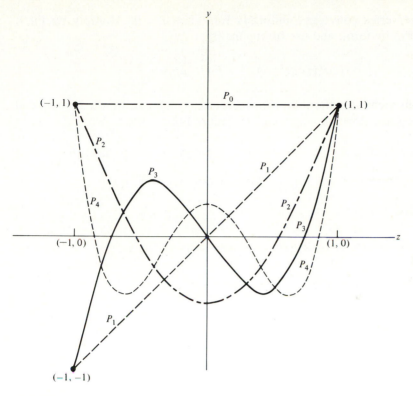

FIGURE 4.2.1
Graphs of the Legendre polynomials $y = P_l(z)$ for $l = 0, 1, 2, 3, 4$ and $-1 \leq z \leq 1$.

To do the final integral, denote its value by d_l. Then

$$d_l - d_{l-1} = -\int_{-1}^{1} z^2(1 - z^2)^{l-1} \, dz = \frac{1}{2l} \int_{-1}^{1} z d(1 - z^2)^l = -\frac{d_l}{2l}$$

Thus we have the recurrence relation $d_l/d_{l-1} = 2l/(2l + 1)$ and the initial value $d_0 = 2$. Iterating these, we have

$$d_l = \frac{2l(2l - 2) \cdots 4\cdot2}{(2l + 1)(2l - 1) \cdots 5\cdot3} \, 2 = \frac{2^{2l+1}(l!)^2}{(2l + 1)(2l)!}$$

Substituting this in the previous equation, we find that

$$\boxed{\int_{-1}^{1} P_l(z)^2 \, dz = \frac{2}{2l + 1}} \tag{4.2.16}$$

Series Expansions

These integrals can be used to compute the coefficients in the expansion of a function in a series of Legendre polynomials. Suppose that

$$f(z) = \sum_{l=0}^{\infty} A_l P_l(z)$$

and that the series converges uniformly for $-1 \le z \le 1$. Multiply by $P_l(z)$, integrate term by term, and use orthogonality:

$$\int_{-1}^{1} f(z)P_l(z)\, dz = A_l \int_{-1}^{1} P_l(z)^2\, dz = \frac{2A_l}{2l + 1}$$

This formula motivates the following theorem, which can be used to expand an "arbitrary function" in a series of Legendre polynomials.

Theorem 4.2.1. Let $f(z)$, $-1 < z < 1$, be a piecewise smooth function. Let

$$A_l = \tfrac{1}{2}(2l + 1) \int_{-1}^{1} f(z)P_l(z)\, dz \qquad l = 0, 1, 2, \dots \tag{4.2.17}$$

Then

$$\sum_{l=0}^{\infty} A_l P_l(z) = \tfrac{1}{2}[f(z + 0) + f(z - 0)] \qquad -1 < z < 1 \tag{4.2.18}$$

At $z = 1$ (respectively, $z = -1$), the series converges to $f(1 - 0)$ [respectively, $f(-1 + 0)$].

The proof of this theorem is not given. We now give an example of the computation of the coefficients.

Example 4.2.2. Let

$$f(z) = \begin{cases} 1 & a < z < b, \ -1 \le a < b \le 1 \\ 0 & \text{otherwise} \end{cases}$$

Find the expansion of $f(z)$ in a series of Legendre polynomials.

Solution. To find the coefficients A_l, $l \ge 1$, we write Legendre's equation in the form $[(1 - z^2)f']' + l(l + 1)f = 0$. Integrating this equation for $a < z < b$, we find that

$$(1 - z^2)P_l'(z)\Big|_{a}^{b} + l(l + 1) \int_{a}^{b} P_l(z)\, dz = 0$$

Therefore,

$$2A_l = \frac{2l + 1}{l(l + 1)}[(1 - a^2)P_l'(a) - (1 - b^2)P_l'(b)] \qquad l \ge 1$$

The coefficient A_0 is obtained from (4.2.17) as $A_0 = \tfrac{1}{2}(b - a)$.   ●

On the interval $0 < z < 1$, it is possible to expand a piecewise smooth function $f(z)$ in a Legendre series of the form $\sum_{n=0}^{\infty} A_{2n+1} P_{2n+1}(z)$. To do this, we extend f as an *odd* function, defined for $-1 < z < 1$, by setting $f(-z) = -f(z)$. Then the product $f(z)P_l(z)$ is an odd function for $l = 0, 2, 4, \dots$ and an even function for $l = 1, 3, 5, \ldots$. Therefore we have

$$A_l = \tfrac{1}{2}(2l + 1) \int_{-1}^{1} f(z)P_l(z)\, dz$$

$$= \begin{cases} 0 & l = 0, 2, 4, \ldots \\ (2l + 1) \int_{0}^{1} f(z)P_l(z)\, dz & l = 1, 3, 5, \ldots \end{cases}$$

Example 4.2.3. Expand $f(z) = 1$ in a series of the form $\sum_{n=0}^{\infty} A_{2n+1} P_{2n+1}(z)$.

Solution. We have

$$\int_{0}^{1} P_l(z)\, dz = \frac{1}{l(l + 1)}\, P_l'(0)$$

Therefore,

$$1 = \sum_{n=0}^{\infty} \frac{4n + 3}{(2n + 1)(2n + 2)}\, P_{2n+1}'(0) P_{2n+1}(z) \qquad 0 < z < 1$$

$$= \tfrac{3}{2}P_1(z) - \tfrac{7}{8}P_3(z) + \tfrac{11}{16}P_5(z) \cdots \qquad\qquad \bullet$$

Associated Legendre Functions

Returning to the theory, we now consider the associated Legendre equation (4.2.6) for $m \neq 0$. To solve this equation, we write the equation for $\nu = 0$:

$$f''(\theta) + \cot \theta\, f'(\theta) + l(l + 1)f(\theta) = 0$$

Differentiating with respect to θ yields

$$f'''(\theta) + \cot \theta\, f''(\theta) + \left[l(l + 1) - \frac{1}{\sin^2 \theta} \right] f'(\theta) = 0$$

Comparing this with (4.2.6), we see that we have obtained a solution for the value $\mu = l(l + 1)$, $m = 1$. This is called the *associated Legendre function* $P_{l,1}(\cos \theta)$.

To obtain associated Legendre functions for higher values of m, we can use the formula

$$P_{l,m}(\cos \theta) = \sin^m \theta\, P_l^{(m)}(\cos \theta) \qquad\qquad (4.2.19)$$

To prove (4.2.19), we differentiate the Legendre equation (4.2.9) m times with respect to z. Thus

$$0 = (1 - z^2)f'' - 2zf' + l(l + 1)f \qquad\qquad f(z) = P_l(z)$$

$$0 = (1 - z^2)f^{(m+2)} - 2mzf^{(m+1)} - m(m + 1)f^{(m)}$$
$$\quad - 2(zf^{(m+1)} + mf^{(m)}) + l(l + 1)f^{(m)}$$

$$= (1 - z^2)f^{(m+2)} - (2m + 2)zf^{(m+1)} + [l(l + 1) - m(m + 1)]f^{(m)}$$

Now let $g(z) = (1 - z^2)^{m/2} f^{(m)}(z)$. We have

$$g' = (1 - z^2)^{m/2} f^{(m+1)} - mz(1 - z^2)^{m/2-1} f^{(m)}$$

$$(1 - z^2)g' = (1 - z^2)^{1+m/2} f^{(m+1)} - mz(1 - z^2)^{m/2} f(m)$$

$$[(1 - z^2)g']' = (1 - z^2)^{1+m/2} f^{(m+2)} - (m + 2)z(1 - z^2)^{m/2} f^{(m+1)}$$

$$- mz(1 - z^2)^{m/2} f^{(m+1)} - mf^{(m)}(1 - z^2)^{m/2-1}[1 - (m + 1)z^2]$$

We use the equation connecting $f^{(m+2)}$, $f^{(m+1)}$, $f^{(m)}$ and the definition of g to write

$$[(1 - z^2)g']' = (1 - z^2)^{m/2}[m(m + 1) - l(l + 1)]f^{(m)}$$

$$- m[1 - (m + 1)z^2]f^{(m)}(1 - z^2)^{m/2-1}$$

$$= [m(m + 1) - l(l + 1)]g - \frac{m[1 - (m + 1)z^2]g}{1 - z^2}$$

$$= \left[-l(l + 1) + \frac{m^2}{1 - z^2}\right] g$$

Therefore g satisfies the associated Legendre equation (4.2.8) for the value $\mu = l(l + 1)$, which was to be proved.

Since $P_l(z)$ is a polynomial of degree l, this procedure yields a nonzero function $g(z)$, provided that $m \leq l$. This can be written in the form

$$\boxed{P_{l,m}(z) = (1 - z^2)^{m/2} \left(\frac{d}{dz}\right)^m P_l(z)}$$

The associated Legendre functions $P_{l,m}$ for $l = 0, 1, 2, 3$ are listed in Table 4.2.1.

The associated Legendre functions satisfy the following orthogonality relation:

$$\boxed{\int_{-1}^{1} P_{l_1,m}(z)P_{l_2,m}(z) \, dz = 0 \qquad l_1 \neq l_2} \tag{4.2.20}$$

TABLE 4.2.1

l	m	$P_{l,m} (\cos \theta)$
0	0	1
1	0	$\cos \theta$
1	1	$\sin \theta$
2	0	$\frac{1}{2}(3 \cos^2 \theta - 1)$
2	1	$3 \cos \theta \sin \theta$
2	2	$3 \sin^2 \theta$
3	0	$\frac{1}{2}(5 \cos^3 \theta - 3 \cos \theta)$
3	1	$\frac{1}{2}(15 \cos^2 \theta - 3) \sin \theta$
3	2	$15 \cos \theta \sin^2 \theta$
3	3	$15 \sin^3 \theta$

This is a special case of the Sturm-Liouville theory where choice $s(z) = 1 - z^2$, $q(z) = m^2/(1 - z^2)$, $\rho(z) = 1$. Although $q(z)$ becomes infinite when $z \to \pm 1$, $P_{l,m}(z)$ tends to zero and the quotient $P_{l,m}(z)/(1 - z^2)$ remains bounded. Therefore the proof of orthogonality given in Sec. 0.4 remains valid here.

To obtain the normalization coefficients, we write

$$\int_{-1}^{1} P_{l,m}(z)^2 \, dz = \int_{-1}^{1} [(1 - z^2)^{m/2} D^m P_l(z)]^2 \, dz$$

$$= \int_{-1}^{1} (1 - z^2)^m D^m P_l(z) D^m P_l(z) \, dz$$

$$= (-1)^m \int_{-1}^{1} P_l(z) D^m (1 - z^2)^m D^m P_l(z) \, dz$$

where we have used the definition of $P_{l,m}$ and integrated by parts m times. The next step is to write the coefficient of $P_l(z)$ as a multiple of $P_l(z)$ plus lower-order terms. To do this, we recall that $P_l(z) = a_l z^l$ plus lower-order terms. Therefore $D^m P_l(z) = [l!a_l/(l - m)!]z^{l-m}$ plus lower-order terms, and $(1 - z^2)^m D^m P_l(z) = (-1)^m[l!a_l/(l - m)!]z^{l+m}$ plus lower-order terms, $D^m(1 - z^2)^m D^m P_l(z) = (-1)^m \times [(l + m)!/(l - m)!]a_l z^l$ plus lower-order terms $= (-1)^m[(l + m)!/(l - m)!]P_l(z)$ plus lower-order terms. These lower-order terms can be written as linear combinations of the Legendre polynomials $P_n(z)$ with $n \leq l - 1$. By orthogonality all the resulting integrals are zero. But the integral of $P_l(z)^2$ was shown to be $2/(2l + 1)$. Therefore we have proved that

$$\boxed{\int_{-1}^{1} P_{l,m}(z)^2 \, dz = \frac{(l + m)!}{(l - m)!} \frac{2}{2l + 1} \qquad 0 \leq m \leq l} \qquad (4.2.21)$$

Spherical Bessel Functions

We now turn to the analysis of the radial equation (4.2.7). This is a form of the general Bessel equation with the parameters $d = 3$, $\mu = l(l + 1)$. According to the theory from Sec. 3.2, this equation has a power series solution of the form $\sum_{n=0}^{\infty} a_n r^{\gamma+n}$, where $\gamma = -\frac{1}{2} + l + \frac{1}{2} = l$. This function is the *spherical Bessel function* of order l:

$$f_1(r) = j_l(r\sqrt{\lambda}) = r^l \left(a_0 + \sum_{n=1}^{\infty} a_n r^n \right)$$

This function is expressed in terms of the standard Bessel functions $J_m(x)$ by the relation

$$\frac{f_1(r)}{r^\gamma} = a_0 \frac{J_m(r)}{r^m}$$

where we have used the theory of Sec. 3.2 (especially Example 3.2.4) with $\gamma = l$, $m = \gamma + \frac{1}{2} = l + \frac{1}{2}$. Therefore $f_1(r) = a_0 r^{\gamma-m} J_m(r) = a_0 r^{-1/2} J_{l+1/2}(r)$. The constant a_0 is determined by requiring that $j_0(r) = (\sin r)/r$, which leads to the choice $a_0 = \sqrt{\pi/2}$ and the definition

$$j_l(r) = \left(\frac{\pi}{2r}\right)^{1/2} J_{l+1/2}(r) \tag{4.2.22}$$

Hence the spherical Bessel functions are expressed in terms of ordinary Bessel functions J_m.

The orthogonality and normalization of the spherical Bessel functions are a special case of the general theory of Bessel functions, treated in Chap. 2. To discuss these, we assume a boundary condition of the form

$$\cos \beta \, f'(R) + \sin \beta \, f(R) = 0 \qquad 0 \le \beta \le \frac{\pi}{2} \tag{4.2.23}$$

Then we can summarize the orthogonality properties as follows:

Proposition. Let $j_l(r\sqrt{\lambda})$ and $j_l(r\sqrt{\lambda'})$ be two solutions of the spherical Bessel equation satisfying the boundary conditions (4.2.23). Then we have

$$\int_0^R j_l(r\sqrt{\lambda})j_l(r\sqrt{\lambda'})r^2 \, dr = 0 \qquad 0 \le \beta \le \frac{\pi}{2}, \lambda \ne \lambda' \tag{4.2.24}$$

$$\int_0^R j_l(r\sqrt{\lambda})^2 r^2 \, dr = \begin{cases} \dfrac{\pi}{4} R^2 J'_{l+1/2}(R\sqrt{\lambda})^2 & \beta = \dfrac{\pi}{2} \\[2mm] \pi R^2 \left(\lambda + \tan^2 \beta - \dfrac{m^2}{R^2} \right) J_{l+1/2}(R\sqrt{\lambda})^2 & 0 \le \beta < \dfrac{\pi}{2} \end{cases}$$

$$\tag{4.2.25}$$

If $\lambda = 0$, the radial equation (4.2.7) is

$$f_1'' + \frac{2}{r} f_1' - \frac{l(l+1)}{r^2} f_1 = 0$$

This is a form of Euler's equidimensional equation. We obtain the general solution by trying $f_1(r) = r^\gamma$, which leads to the quadratic equation $\gamma(\gamma + 1) - l(l + 1) = 0$. The solutions of these equations are $\gamma = l$, $\gamma = -(l + 1)$.

Summarizing the results of this section, we have found the separated solutions of the heat equation in spherical coordinates

$$u(r, \theta, \varphi) = j_l(r\sqrt{\lambda})P_{l,m}(\cos \theta)(A \cos m\varphi + B \sin m\varphi)e^{-\lambda K t} \qquad \lambda > 0$$

In the case where $\lambda = 0$, we have the separated solutions of Laplace's equation

$$u(r, \theta, \varphi) = r^l P_{l,m}(\cos \theta)(A \cos m\varphi + B \sin m\varphi) \qquad \lambda = 0$$

Both give solutions valid throughout all space. In some problems we may need solutions which are valid outside of some sphere. In this case the analysis of the radial equation allows us to choose the second solution of Bessel's equation, which becomes infinite when $r \to 0$. This function is written $f_1(r) = n_l(r\sqrt{\lambda})$, $\lambda > 0$. Similarly when $\lambda = 0$, we may choose $f_1(r) = r^{-(l+1)}$. Thus we have the separated solutions, valid for $r \neq 0$,

$$u(r, \theta, \varphi) = \begin{cases} n_l(r\sqrt{\lambda})P_{l,m}(\cos\theta)(A\cos m\varphi + B\sin m\varphi)e^{-\lambda Kt} & \lambda > 0 \\ r^{-(l+1)}P_{l,m}(\cos\theta)(A\cos m\varphi + B\sin m\varphi) & \lambda = 0 \end{cases}$$

We list below the spherical Bessel functions for $l = 0, 1, 2$.

l	$j_l(r)$
0	$\dfrac{1}{r}\sin r$
1	$\dfrac{1}{r^2}\sin r - \dfrac{1}{r}\cos r$
2	$\left(\dfrac{3}{r^3} - \dfrac{1}{r}\right)\sin r - \dfrac{3}{r^2}\cos r$

EXERCISES 4.2

1. Compute $P_1(0)$, $P_2(0)$, $P_3(0)$, $P_4(0)$.
2. Compute $P_1'(0)$, $P_2'(0)$, $P_3'(0)$, $P_4'(0)$.
3. Write the Legendre polynomials $P_5(z)$, $P_6(z)$.
4. Show that $P_l(z)$ is an even function if $l = 0, 2, 4, \ldots$.
5. Show that $P_l(z)$ is an odd function if $l = 1, 3, 5, \ldots$.
6. Show that $P_l'(z)$ is an odd function if $l = 0, 2, 4, \ldots$.
7. Show that $P_l'(z)$ is an even function if $l = 1, 3, 5, \ldots$.
8. Use Rodrigues' formula to show that $P_l(1) = 1$ for $l = 0, 1, 2, 3, 4, \ldots$.
9. Use Exercises 4, 5, and 8 to show that $P_l(-1) = (-1)^l$ for $l = 0, 1, 2, 3, 4, \ldots$.
10. It is known that $P_l(z)$ has exactly l zeros on the interval $-1 \le z \le 1$. Find these zeros for $l = 0, 1, 2, 3, 4$.
11. Let $f(z)$, $-1 < z < 1$, be a piecewise smooth function, $a_l = (2l + 1)/2 \int_{-1}^{1} f(z)P_l(z)\,dz$. (a) If f is odd, show that $a_{2n} = 0$, $n = 0, 1, 2, \ldots$. (b) If f is even, show that $a_{2n+1} = 0$.
12. Let $f(z) = 0$ for $-1 < z < 0$ and $f(z) = 1$ for $0 < z < 1$. Find the expansion of $f(z)$ in a series of Legendre polynomials.
13. Let $f(z) = -1$ for $-1 < z < 0$ and $f(z) = 1$ for $0 < z < 1$. Find the expansion of $f(z)$ in a series of Legendre polynomials.
14. Let $f(z) = 1$ if $-\frac{1}{2} < z < \frac{1}{2}$ and $f(z) = 0$ otherwise. Find the first four terms in the expansion of $f(z)$ in a series of Legendre polynomials.
15. Write the associated Legendre functions $P_{41}(z)$, $P_{42}(z)$, $P_{43}(z)$, $P_{44}(z)$.
16. Show that $P_{ll}(\cos\theta) = (\sin\theta)^l$ for $l = 0, 1, 2, 3, \ldots$, $0 \le \theta \le \pi$.

17. Use Rodrigues' formula to show that

$$P'_{2n+1}(0) = (-1)^n \frac{(2n+2)!}{n!(n+1)!2^{2n+1}} \qquad P'_{2n}(0) = 0$$

for $n = 0, 1, 2, \ldots$.

18. Let $f(z)$, $-1 < z < 1$, be a function with n continuous derivatives. Use Rodrigues' formula and integration by parts to show that the coefficients in the Legendre expansion of $f(z)$ can be written in the form

$$A_l = \frac{(-1)^l}{2^l l!} \int_{-1}^{1} (z^2 - 1)^l f^{(l)}(z) \, dz \qquad 0 \leq l \leq n$$

19. Use Exercise 18 to find the Legendre expansion of z^2, z^3, z^4.

*20. Derive the generating function of Legendre polynomials: If $-1 < t < 1$, $-1 \leq z \leq 1$, then

$$(1 - 2tz + t^2)^{-1/2} = \sum_{l=0}^{\infty} t^l P_l(z)$$

Use the following steps:
(a) Write the binomial series for $(1 - \alpha)^{-1/2}$, $-1 < \alpha < 1$.
(b) Let $\alpha = 2zt - t^2$ and use the binomial theorem to expand α^n.
(c) Rearrange the resulting double series to identify the coefficient of t^l as $P_l(z)$.

4.3 LAPLACE'S EQUATION IN SPHERICAL COORDINATES

Separated Solutions of Laplace's Equation

In this section we consider boundary-value problems for Laplace's equation

$$0 = \nabla^2 u = u_{rr} + \frac{2}{r} u_r + \frac{1}{r^2} (u_{\theta\theta} + \cot \theta \, u_\theta + \csc^2 \theta \, u_{\varphi\varphi})$$

From the results in Sec. 4.2 we have the following separated solutions of Laplace's equation in spherical coordinates:

$$u(r, \theta, \varphi) = (A_1 r^l + A_2 r^{-(l+1)}) P_{l,m}(\cos \theta)(A_3 \cos m\varphi + A_4 \sin m\varphi)$$
$$l = 0, 1, 2, \ldots ; \; m = 0, 1, \ldots, l \tag{4.3.1}$$

The constants A_1, A_2, A_3, A_4 can be specialized to solve various boundary-value problems.

Boundary-Value Problems in a Sphere

As a first application, we find the solution of $\nabla^2 u = 0$ inside the sphere $r \leq a$ satisfying the boundary condition $u(a, \theta, \varphi) = G(\theta)$, where $G(\theta)$ is a given piecewise smooth function. To do this, we note that since we are looking for solutions inside the sphere, we need only consider separated solutions with

$A_2 = 0$; otherwise the solution would not be defined at $r = 0$. Furthermore the boundary condition is independent of φ; therefore we only need the separated solutions for $m = 0$. This leads to the choice

$$u(r, \theta) = \sum_{l=0}^{\infty} B_l r^l P_l(\cos \theta)$$

The boundary condition requires that $u(a, \theta) = \sum_{l=0}^{\infty} B_l a^l P_l(\cos \theta)$. By the orthogonality of Legendre polynomials we must have

$$\int_0^{\pi} u(a, \theta) P_l(\cos \theta) \sin \theta \, d\theta = B_l a^l \int_0^{\pi} P_l(\cos \theta)^2 \sin \theta \, d\theta = B_l a^l \frac{2}{2l + 1}$$

Therefore the coefficients B_l must be taken to be

$$B_l = \frac{l + \frac{1}{2}}{a^l} \int_0^{\pi} G(\theta) P_l(\cos \theta) \sin \theta \, d\theta$$

Example 4.3.1. Find the solution $u(r, \theta)$ of Laplace's equation in the sphere $r \leq a$ satisfying $u(a, \theta) = 1$ if $0 < \theta < \pi/2$, $u(a, \theta) = 0$ if $\pi/2 < \theta < \pi$. Show that $u(r, \pi/2) = \frac{1}{2}$ for $0 \leq r \leq a$.

Solution. In this case we can apply the above method with

$$B_l = \frac{l + \frac{1}{2}}{a^l} \int_0^{\pi/2} P_l(\cos \theta) \sin \theta \, d\theta \qquad l = 0, 1, 2, \ldots$$

This integral was evaluated in Sec. 4.2 with the result

$$\int_0^{\pi/2} P_l(\cos \theta) \sin \theta \, d\theta = \frac{P_l'(0)}{l(l + 1)} \qquad l = 1, 2, \ldots$$

Therefore $B_0 = \frac{1}{2}$, $B_l = [(l + \frac{1}{2})/l(l + 1)] P_l'(0)$. The solution to the problem is

$$u(r, \theta) = \frac{1}{2} + \sum_{l=1}^{\infty} \left(\frac{r}{a}\right)^l P_l'(0) \frac{l + \frac{1}{2}}{l(l + 1)} P_l(\cos \theta)$$

and $P_l(0) = 0$ if l is odd, while $P_l'(0) = 0$ if l is even. This shows that $u(r, \pi/2) = \frac{1}{2}$. ●

In some problems we can avoid the computation of the integrals in the definition of B_l, when the boundary function $G(\theta)$ can be written as a sum of Legendre polynomials.

Example 4.3.2. Find the solution $u(r, \theta)$ of Laplace's equation in the sphere $r \leq a$ satisfying $u(a, \theta) = 1 + 3 \cos \theta + 3 \cos^2 \theta$.

Solution. The boundary condition can be written in the form

$$u(a, \theta) = G(\theta) = 2 + 3 \cos \theta + 3 \cos^2 \theta - 1$$

$$= 2P_0(\cos \theta) + 3P_1(\cos \theta) + 2P_2(\cos \theta)$$

Therefore the required solution to Laplace's equation is

$$u(r, \theta) = 2 \left(\frac{r}{a}\right)^2 P_2(\cos \theta) + 3 \frac{r}{a} P_1(\cos \theta) + 2P_0(\cos \theta) \qquad \bullet$$

If the boundary conditions depend on (θ, φ), then we must use the separated solutions with $m \neq 0$. Consider, e.g., the boundary-value problem for Laplace's equation inside the sphere $r \leq a$ with $u(a, \theta, \varphi) = G(\theta, \varphi)$ a given function. We look for solutions in the form

$$u(r, \theta, \varphi) = \sum_{l=0}^{\infty} \sum_{m=0}^{l} r^l P_{lm}(\cos \theta)(A_{lm} \cos m\varphi + B_{lm} \sin m\varphi)$$

Example 4.3.3. Find the solution $u(r, \theta, \varphi)$ of Laplace's equation in the sphere $r \leq a$ satisfying

$$u(a, \theta, \varphi) = \sin \theta \cos \varphi + \cos \theta \sin \theta \sin \varphi$$

Solution. From the table of associated Legendre functions in Sec. 4.2, we have $P_{11} = \sin \theta$, $P_{21} = 3 \sin \theta \cos \theta$. Thus $u(a, \theta, \varphi) = P_{11}(\cos \theta) \cos \varphi + \frac{1}{3}P_{21}(\cos \theta) \cdot \cos 2\varphi$. Comparing this with the separated solutions in (4.3.1), we must have

$$u(r, \theta, \varphi) = rP_{11}(\cos \theta) \cos \varphi + \frac{r^2}{3} P_{21}(\cos \theta) \sin \varphi \qquad \bullet$$

If $u(a, \theta, \varphi)$ is not already expressed as a sum of spherical harmonics, then we must compute integrals in order to solve the problem. Using orthogonality of the associated Legendre functions, we have

$$\int_0^{2\pi} \int_0^{\pi} u(a, \theta, \varphi)P_{lm}(\cos \theta) \cos m\varphi \sin \theta \, d\theta \, d\varphi$$

$$= A_{lm} \int_0^{2\pi} \int_0^{\pi} P_{lm}(\cos \theta)^2 \sin \theta \cos^2 m\varphi \, d\theta \, d\varphi$$

$$\int_0^{2\pi} \int_0^{\pi} u(a, \theta, \varphi)P_{lm}(\cos \theta) \sin m\varphi \sin \theta \, d\theta \, d\varphi$$

$$= B_{lm} \int_0^{2\pi} \int_0^{\pi} P_{lm}(\cos \theta)^2 \sin \theta \sin^2 m\varphi \, d\theta \, d\varphi$$

Boundary-Value Problems Exterior to a Sphere

We can also use the separated solutions (4.3.1) to solve Laplace's equation in the exterior of a sphere, $r \geq a$. In this case we impose the requirement that $u(r, \theta, \varphi) \to 0$ when $r \to \infty$. This ensures the uniqueness of the solution. For example, the function $u_1(r) = 1 - a/r$ satisfies Laplace's equation for $r \geq a$ and is zero on the sphere $r = a$. The function $u_2(r) \equiv 0$ satisfies these same conditions. With this in mind we choose $A_1 = 0$ in the separated solutions (4.3.1) and get the general exterior solution

$$u(r, \theta, \varphi) = \sum_{l=0}^{\infty} \sum_{m=0}^{l} r^{-(l+1)} P_{lm}(\cos \theta)(A_{lm} \cos m\varphi + B_{lm} \sin m\varphi)$$

Example 4.3.4. Find the solution of Laplace's equation outside the sphere $r = a$ and satisfying $u(a, \theta) = 1$ if $0 < \theta < \pi/2$ and $u(a, \theta) = 0$ if $\pi/2 < \theta < \pi$.

Solution. Since the boundary condition is independent of φ, we may take the separated solutions with $m = 0$; thus

$$u(r, \theta) = \sum_{l=0}^{\infty} A_l r^{-(l+1)} P_l(\cos \theta)$$

The Legendre expansion of $u(a, \theta)$ was obtained in Example 4.3.1.

$$u(a, \theta) = \frac{1}{2} + \sum_{l=1}^{\infty} \frac{l + \frac{1}{2}}{l(l + 1)} P_l'(0) P_l(\cos \theta)$$

Therefore the required solution is

$$u(r, \theta) = \frac{1}{2} \frac{a}{r} + \sum_{l=1}^{\infty} \left(\frac{a}{r}\right)^{l+1} \frac{l + \frac{1}{2}}{l(l + 1)} P_l'(0) P_l(\cos \theta) \qquad \bullet$$

We now consider a more general boundary condition for Laplace's equation in the sphere $r \leq a$:

$$\cos \alpha \frac{\partial u}{\partial r} (a, \theta, \varphi) + \sin \alpha \, u(a, \theta, \varphi) = G(\theta, \varphi)$$

Here α is a constant which may assume values between 0 and $\pi/2$; $\alpha = \pi/2$ corresponds to the *Dirichlet problem* which has already been solved; $\alpha = 0$ corresponds to the *Neumann problem* where $\partial u/\partial r$ is specified on the boundary. The intermediate values of α correspond to the *mixed problem*. Physically this occurs when we have free radiation of heat according to Newton's law of cooling.

To solve the problem, we begin with the series of separated solutions

$$u(r, \theta, \varphi) = \sum_{l=0}^{\infty} \sum_{m=0}^{l} r^l P_{lm}(\cos \theta)(A_{lm} \cos m\varphi + B_{lm} \sin m\varphi)$$

The boundary condition requires that

$$G(\theta, \varphi) = \sum_{l=0}^{\infty} \sum_{m=0}^{l} (l \cos \alpha \, a^{l-1} + \sin \alpha \, a^l)$$
$$\cdot P_{lm}(\cos \theta)(A_{lm} \cos m\varphi + B_{lm} \sin m\varphi)$$

To obtain A_{lm}, B_{lm}, we expand $G(\theta, \varphi)$ in a series of spherical harmonics and equate coefficients. If

$$G(\theta, \varphi) = \sum_{l=0}^{\infty} \sum_{m=0}^{l} P_{lm}(\cos \theta)(\overline{A}_{lm} \cos m\varphi + \overline{B}_{lm} \sin m\varphi)$$

then we must have

$$\overline{A}_{lm} = (l \cos \alpha \, a^{l-1} + \sin \alpha \, a^l)A_{lm} \qquad l = 0, 1, 2, \ldots ; 0 \le m \le l$$

$$\overline{B}_{lm} = (l \cos \alpha \, a^{l-1} + \sin \alpha \, a^l)B_{lm} \qquad l = 0, 1, 2, \ldots ; 0 \le m \le l$$

The coefficients of A_{lm}, B_{lm} are nonzero unless both $\alpha = 0$ and $l = 0$. Thus we may solve for A_{lm} and B_{lm} and obtain the formal solution of the problem. If $\alpha = 0$, then we see that the coefficient of A_{00} is zero. In fact, we have the additional condition $\int_0^{2\pi} \int_0^{\pi} G(\theta, \varphi) \sin \theta \, d\theta \, d\varphi = 0$. The value of A_{00} is undetermined; to uniquely specify the solution, we take $A_{00} = 0$. We summarize these computations in the following proposition:

Proposition. If $0 < \alpha \le \pi/2$, the solution of Laplace's equation in the sphere $r < a$ with the boundary condition $\cos \alpha \, \partial u/\partial r + \sin \alpha \, u = G$ is uniquely determined by the above procedure. If $\alpha = 0$, then G must satisfy the additional condition $\int_0^{2\pi} \int_0^{\pi} G(\theta, \varphi) \sin \theta \, d\theta \, d\varphi = 0$. In this case $u(r, \theta, \varphi)$ is uniquely determined by requiring $u(0, \theta, \varphi) = 0$.

Example 4.3.5. Find the solution of $\nabla^2 u = 0$ in the sphere $r < a$ and satisfying the boundary condition $\partial u/\partial r + u = \cos \theta$.

Solution. According to the above procedure, we have $\alpha = \pi/4$ and $G(\theta, \varphi) = (1/\sqrt{2}) \cos \theta$. Assuming a solution of the form $u = \sum_{l=0}^{\infty} A_l r^l P_l(\cos \theta)$, we must have $\cos \theta = \sum_{l=0}^{\infty} (l + a)a^{l-1}A_l P_l(\cos \theta)$; thus $A_l = 0$ for $l \ne 1$ and $1 = (l + a)A_1$. Thus the solution is $u(r, \theta) = (r \cos \theta)/(1 + a)$.   ●

The formal solutions which we have obtained may be established as rigorous solutions with little difficulty. To be specific, we consider the Laplace equation in the sphere $0 \le r < a$ with the boundary condition $u(a, 0) = G(\theta)$, a given piecewise smooth function for $0 < \theta < \pi$. The formal solution to the problem was found to be

$$u(r, \theta) = \sum_{l=0}^{\infty} A_l \left(\frac{r}{a}\right)^l P_l(\cos \theta) \qquad 0 \le r \le a \qquad (4.3.2)$$

where $A_l = (l + \tfrac{1}{2}) \int_0^{\pi} G(\theta)P_l(\cos \theta) \sin \theta \, d\theta \qquad l = 0, 1, 2, \ldots$

The Legendre polynomials satisfy $|P_l| \le 1$, $|P_l'| \le l^2$, $|P_l''| \le l^4$. To show that $u_r = \sum_{l=0}^{\infty} A_l l(r/a)^{l-1}P_l(\cos \theta)$, we note that the terms of this series are no larger than $M(2l + 1)l(r/a)^{l-1}$, where M is the maximum of $G(\theta)$, $0 \le \theta \le \pi$.

If $r_0 < a$, these terms are no larger than $M(2l + 1)/[(a + r_0)/(2a)]^{l-1}$ whenever $0 \leq r \leq \frac{1}{2}(a + r_0)$. By the Weierstrass M test, the series for u_r is uniformly convergent for $0 \leq r \leq \frac{1}{2}(a + r_0)$. In a similar fashion it may be shown that the series for u_{rr}, u_θ, $u_{\theta\theta}$ are also uniformly convergent. Thus $u(r, \theta)$ defined by (4.3.2) is, in fact, a rigorous solution of Laplace's equation in the sphere $0 \leq r < a$.

EXERCISES 4.3

1. Find the solution of Laplace's equation $\nabla^2 u = 0$ in the sphere $0 \leq r < a$ and satisfying the boundary condition $u(a, \theta) = 3 + 4 \cos \theta + 2 \cos^2 \theta$.

2. Find the solution of Laplace's equation $\nabla^2 u = 0$ in the sphere $0 \leq r < a$ and satisfying the boundary condition $u(a, \theta) = 1 + \cos 2\theta$.

*3. Find the solution of Laplace's equation $\nabla^2 u = 0$ in the sphere $0 \leq r < a$ and satisfying the boundary condition $u(a, \theta) = (5 - 4 \cos \theta)^{-1/2}$. *Hint:* Use the generating function for Legendre polynomials

$$(1 - 2xt + t^2)^{-1/2} = \sum_{n=0}^{\infty} t^n P_n(x) \qquad -1 < t < 1, \; -1 \leq x \leq 1$$

4. Find the solution of Laplace's equation $\nabla^2 u = 0$ in the sphere $0 \leq r < a$ and satisfying the boundary conditions $u(a, \theta) = 1$ if $0 < \theta < \pi/2$ and $u(a; \theta) = 0$ otherwise.

5. Find the solution of Laplace's equation $\nabla^2 u = 0$ in the sphere $0 \leq r < a$ and satisfying the boundary conditions $u(a, \theta) = 1$ if $0 < \theta < \pi/2$ and $u(a, \theta) = -1$ if $\pi/2 < \theta < \pi$. Find $u(r, \pi/2)$.

6. Show that the functions $r^{2n+1} P_{2n+1}(\cos \theta)$, $n = 0, 1, 2, 3, \ldots$, satisfy Laplace's equation in the hemisphere $0 \leq r < a$, $0 < \theta < \pi/2$, with the boundary condition $u(r, \pi/2) = 0$, $0 \leq r < a$.

7. Find the solution of Laplace's equation $\nabla^2 u = 0$ in the hemisphere $0 \leq r < a$, $0 < \theta < \pi/2$ and satisfying the boundary conditions $u(r, \pi/2) = 0$, $0 \leq r < a$, and $u(a, \theta) = 4 \cos \theta + 2 \cos^3 \theta$, $0 < \theta < \pi/2$.

8. Find the solution of Laplace's equation $\nabla^2 u = 0$ in the hemisphere $0 \leq r < a$, $0 < \theta < \pi/2$ and satisfying the boundary conditions $u(r, \pi/2) = 0$ for $0 < r < a$ and $u(a, \theta) = 1$ for $0 < \theta < \pi/2$.

9. Find the solution of Laplace's equation $\nabla^2 u = 0$ outside the sphere $r \geq a$ and satisfying the boundary condition $u(a, \theta) = 1 + 2 \cos \theta + \cos^4 \theta$, $0 < \theta < \pi$.

10. Find the solution of Laplace's equation $\nabla^2 u = 0$ outside the sphere $r \geq a$ and satisfying the boundary conditions $u(a, \theta) = 1$ if $0 < \theta < \pi/2$, $u(a, \theta) = -1$ if $\pi/2 < \theta < \pi$.

11. Find the solution of Laplace's equation $\nabla^2 u = 0$ outside the sphere $r \geq a$ and satisfying the boundary condition $(\partial u/\partial r)(a, \theta) = \cos \theta + 3 \cos^3 \theta$, $0 < \theta < \pi$.

12. Find the solution of Laplace's equation $\nabla^2 u = 0$ outside the sphere $r \geq a$ and satisfying the boundary condition $(\partial u/\partial r)(a, \theta) + u(a, \theta) = 3 \cos \theta$, $0 < \theta < \pi$.

13. Find the solution of Laplace's equation $\nabla^2 u = 0$ in the sphere $0 \leq r < a$ and satisfying the boundary condition $u(a, \theta, \varphi) = \sin \theta \cos \varphi + \sin^2 \theta \sin 2\varphi$.

14. Find the solution of Laplace's equation $\nabla^2 u = 0$ outside the sphere $r > a$ and satisfying the boundary condition $u(a, \theta, \varphi) = \sin \theta \cos \varphi + \sin^2 \theta \sin 2\varphi$.

15. Let $u(r, \theta, \varphi)$ be the solution of Laplace's equation $\nabla^2 u = 0$ in the sphere $0 \leq r < a$ with the boundary condition $u(a, \theta, \varphi) = G(\theta, \varphi)$ for $0 < \theta < \pi, 0 < \varphi < 2\pi$. Show formally that $u(0, \theta, \varphi) = [1/(4\pi)] \int_0^{2\pi} \int_0^\pi G(\theta, \varphi) \sin \theta \, d\theta \, d\varphi$.

*16. Let $u(r, \theta, \varphi)$ be the solution of Laplace's equation $\nabla^2 u = 0$ outside the sphere $r \geq a$ with the boundary condition $u(a, \theta, \varphi) = G(\theta, \varphi)$ for $0 < \theta < \pi, 0 < \varphi < 2\pi$. Show formally that when $r \to \infty$,

$$u(r, \theta, \varphi) = \frac{1}{4\pi r} \int_0^{2\pi} \int_0^\pi G(\theta, \varphi) \sin \theta \, d\theta \, d\varphi + O\left(\frac{1}{r^2}\right)$$

FOURIER TRANSFORMS

Introduction

In previous chapters we obtained the solution of initial- and boundary-value problems in terms of Fourier *series* of separated solutions. In this chapter we consider certain problems which require a continuous superposition of separated solutions, leading to the notion of a Fourier *integral* representation. In many cases this can be rewritten to give an explicit representation as a "convolution" of the initial and boundary data with a standard function, the *fundamental solution* of the problem.

In Sec. 5.1 we develop the properties of the Fourier transform in its own right. The following sections reveal the application to problems involving the heat equation, Laplace's equation, the wave equation, and the telegraph equation.

5.1 DEFINITIONS AND EXAMPLES

Passage from Fourier Series to Fourier Integrals

To motivate the definition of the Fourier transform, we look for a suitable generalization of the Fourier series to infinite intervals. We begin with the form of a (complex) Fourier series on $-L < x < L$:

$$f(x) = \sum_{-\infty}^{\infty} \alpha_n e^{in\pi x/L} \tag{5.1.1}$$

$$\alpha_n = \frac{1}{2L} \int_{-L}^{L} f(x)e^{-in\pi x/L} \, dx \tag{5.1.2}$$

We assume that $f(x)$ is a real-valued piecewise smooth function on $(-\infty, \infty)$ and that all the relevant integrals are absolutely convergent. Define a discrete variable μ_n by

$$\mu_n = \frac{n\pi}{L} \qquad n = 0, \pm 1, \ldots$$

$$\Delta \mu_n = \frac{\pi}{L} = \mu_{n+1} - \mu_n$$

Formulas (5.1.1) and (5.1.2) can be written in the form

$$f(x) = \sum_{-\infty}^{\infty} F_L(\mu_n)e^{i\mu_n x} \, \Delta \mu_n \tag{5.1.3}$$

$$F_L(\mu) = \frac{1}{2\pi} \int_{-L}^{L} f(x)e^{-i\mu x} \, dx \qquad \alpha_n = F_L(\mu_n) \, \Delta \mu_n \tag{5.1.4}$$

In this form we take the limit $L \to \infty$; (5.1.3) is an approximating sum for an (improper) reimannian integral. Thus we have

$$\boxed{f(x) = \int_{-\infty}^{\infty} F(\mu)e^{i\mu x} \, d\mu} \tag{5.1.5}$$

$$\boxed{F(\mu) = \frac{1}{2\pi} \int_{-\infty}^{\infty} f(x)e^{-i\mu x} \, dx} \tag{5.1.6}$$

The functions $f(x)$, $F(\mu)$ are called *Fourier transform pairs*. Defined by (5.1.6), $F(\mu)$ is the *Fourier transform* of $f(x)$. Formula (5.1.5) is the *Fourier inversion formula*.

In addition to these formulas, we derive the form of Parseval's equality. To obtain this, we recall from (5.1.1) and (5.1.4) that

$$\int_{-L}^{L} |f(x)|^2 \, dx = 2L \sum_{-\infty}^{\infty} |\alpha_n|^2 = 2\pi \sum_{-\infty}^{\infty} |F_L(\mu_n)|^2 \, \Delta \mu_n$$

Taking the limit as $L \to \infty$, we have the *Parseval theorem for Fourier transforms*

$$\boxed{\int_{-\infty}^{\infty} |f(x)|^2 \, dx = 2\pi \int_{-\infty}^{\infty} |F(\mu)|^2 \, d\mu} \tag{5.1.7}$$

The above derivations, although intuitively attractive, must be interpreted correctly. In particular, the integral (5.1.5) must be understood as the limit of the integral on $(-L, L)$ when $L \to \infty$, a so-called *Cauchy principal value*. This is clarified in the next subsection where we give the mathematically rigorous definitions and theorems.

Definition and Properties of the Fourier Transform

The Fourier transform of a complex-valued function $f(x)$, $-\infty < x < \infty$, is defined by the integral (5.1.6). The following theorem gives precise conditions for the existence of the integral and for the "Fourier representation" formula (5.1.5).

Convergence theorem for Fourier transforms.† Let $f(x)$, $-\infty < x < \infty$, be piecewise smooth on each finite interval with $\int_{-\infty}^{\infty} |f(x)|\, dx$ convergent. Define the Fourier transform $F(\mu)$ by (5.1.6). Then for each x

$$\lim_{L \to \infty} \int_{-L}^{L} F(\mu)e^{i\mu x}\, d\mu = \tfrac{1}{2}f(x + 0) + \tfrac{1}{2}f(x - 0)$$

This theorem is proved at the end of the section. The following theorem gives some basic properties of the Fourier transform.

Theorem 5.1.1 (a) *Fourier transform preserves linearity;* i.e., if F_1 is the Fourier transform of f_1 and F_2 is the Fourier transform of f_2, then $a_1 F_1 + a_2 F_2$ is the Fourier transform of $a_1 f_1 + a_2 f_2$ for any constants a_1, a_2.

(b) *Fourier transform exchanges multiplication with convolution;* i.e., if F_1 is the Fourier transform of f_1 and F_2 is the Fourier transform of f_2, then $(2\pi)F_1 F_2$ is the Fourier transform of $f_1 * f_2$, where the *convolution* of two functions is defined by the integral

$$(f_1 * f_2)(x) = \int_{-\infty}^{\infty} f_1(z)f_2(x - z)\, dz$$

(c) *Fourier transform exchanges differentiation and multiplication;* i.e., if $F(\mu)$ is the Fourier transform of $f(x)$, then $i\mu F(\mu)$ is the Fourier transform of $f'(x)$ and $F'(\mu)$ is the Fourier transform of $-ixf(x)$.

(d) *Fourier transform interchanges translation and phase factor;* i.e., if $F(\mu)$ is the Fourier transform of $f(x)$, then $e^{ia\mu}F(\mu)$ is the Fourier transform of $f(x - a)$ and $F(\mu + b)$ is the Fourier transform of $e^{ibx}f(x)$.

(e) *Fourier transform preserves the mean square integral:* i.e., if F is the Fourier transform of f, then Parseval's theorem holds in the form $2\pi \int_{-\infty}^{\infty} |F(\mu)|^2\, d\mu = \int_{-\infty}^{\infty} |f(x)|^2\, dx$.

Proof. The proofs of properties (a) and (d) are left to the exercises. Property (e) is Parseval's theorem, noted earlier. We outline the proofs of properties (b) and (c).

---

† Some authors call this result the inversion theorem for Fourier transforms. Our terminology is designed to illustrate the close connection with the theory of Fourier series developed in Sec. 1.2.

For (*b*), we write the product $F_1(\mu)F_2(\mu)$ in the form of a double integral:

$$(2\pi)^2 F_1(\mu)F_2(\mu) = \int_{-\infty}^{\infty} e^{-i\mu x} f_1(x) \, dx \int_{-\infty}^{\infty} e^{-i\mu y} f_2(y) \, dy$$

$$= \int_{-\infty}^{\infty} \int_{-\infty}^{\infty} e^{-i\mu(x+y)} f_1(x)f_2(y) \, dx \, dy$$

The integration extends over the entire *xy* plane. By making the change of variable $x + y = z$, the (*x*, *z*) integration still extends over the entire *xz* plane, and we have

$$(2\pi)^2 F_1(\mu)F_2(\mu) = \int_{-\infty}^{\infty} \int_{-\infty}^{\infty} e^{-i\mu z} f_1(x)f_2(z - x) \, dx \, dz$$

$$= \int_{-\infty}^{\infty} e^{-i\mu z} \left[\int_{-\infty}^{\infty} f_1(x)f_2(z - x) \, dx \right] dz$$

as required.

For the proof of (*c*), we have

$$2\pi i\mu F(\mu) = \int_{-\infty}^{\infty} f(x) i\mu e^{-i\mu x} \, dx$$

$$= -\int_{-\infty}^{\infty} f(x) \left(\frac{d}{dx}\right) e^{-i\mu x} \, dx$$

If we integrate by parts and discard the term at the limits, we obtain

$$2\pi i\mu F(\mu) = \int_{-\infty}^{\infty} f'(x) e^{-i\mu x} \, dx$$

Similarly,

$$2\pi F'(\mu) = \int_{-\infty}^{\infty} f(x)(-ixe^{-i\mu x}) \, dx$$

as required. Note that in this proof we have assumed that $f(x)$ tends to zero at infinity and that we can differentiate the Fourier integral under the integral sign. Both of these are valid if f' and F' exist and are integrable. ∎

We now give some examples of computation of Fourier transforms.

Examples of Fourier Transforms

Example 5.1.1. Find the Fourier transform of the "square wave"

$$f(x) = \begin{cases} 0 & x < a \\ 1 & a \leq x \leq b \\ 0 & x > b \end{cases}$$

Solution. From (5.1.6) we have

$$F(\mu) = \frac{1}{2\pi} \int_{-\infty}^{\infty} f(x)e^{-i\mu x} \, dx = \frac{1}{2\pi} \int_{a}^{b} e^{-i\mu x} \, dx$$

$$= \begin{cases} \dfrac{e^{-i\mu b} - e^{-i\mu a}}{-2\pi i\mu} & \mu \neq 0 \\[3mm] \dfrac{b - a}{2\pi} & \mu = 0 \end{cases}$$

The convergence theorem states that

$$\lim_{L \to \infty} \int_{-L}^{L} \frac{(e^{-i\mu b} - e^{-i\mu a})e^{i\mu x}}{-2\pi i\mu} \, d\mu = \begin{cases} 0 & x < a \\ \frac{1}{2} & x = a \\ 1 & a < x < b \\ \frac{1}{2} & x = b \\ 0 & x > b \end{cases}$$

Parseval's theorem states that

$$\frac{1}{2\pi} \int_{-\infty}^{\infty} \frac{|e^{-i\mu b} - e^{-i\mu a}|^2}{\mu^2} \, d\mu = b - a \qquad \bullet$$

Example 5.1.2. Find the Fourier transform of the *normal density function*

$$f(x) = \frac{\exp\left[-(x - m)^2/(2\sigma^2)\right]}{\sqrt{2\pi\sigma^2}}$$

Solution. The graph of $y = f(x)$ is a bell-shaped curve with unit area, as will be shown. To compute the Fourier transform $F(\mu)$, we make the change of variable $z = (x - m)/\sigma$ and expand $e^{-i\mu\sigma z}$ in a Taylor series. Thus

$$2\pi F(\mu) = \int_{-\infty}^{\infty} \frac{\exp\left[-(x - m)^2/(2\sigma^2)\right]}{\sqrt{2\pi\sigma^2}} e^{-i\mu x} \, dx$$

$$= \int_{-\infty}^{\infty} \frac{e^{-z^2/2}}{\sqrt{2\pi}} e^{-i\mu(m+\sigma z)} \, dz$$

$$= e^{-i\mu m} \sum_{n=0}^{\infty} \int_{-\infty}^{\infty} \frac{e^{-z^2/2}}{\sqrt{2\pi}} \frac{(-i\mu\sigma z)^n}{n!} \, dz$$

This interchange of summation and integration can be fully justified and is done at the end of the example.

To evaluate each of these integrals, we use mathematical induction. Let

$$I_n = \int_{-\infty}^{\infty} z^n \frac{e^{-z^2/2}}{\sqrt{2\pi}} \, dz \qquad n = 0, 1, 2, \ldots$$

Thus, $\qquad\qquad 2\pi F(\mu) = e^{-i\mu m} \displaystyle\sum_{n=0}^{\infty} \frac{(-i\mu\sigma)^n}{n!} I_n$

For $n = 0$, we have a classical calculation using polar coordinates. Thus,

$$I_0^2 = \left(\int_{-\infty}^{\infty} \frac{e^{-z^2/2}}{\sqrt{2\pi}} \, dz \right)^2$$

$$= \int_{-\infty}^{\infty} \int_{-\infty}^{\infty} \frac{e^{-z_1^2/2} e^{-z_2^2/2}}{2\pi} \, dz_1 \, dz_2$$

$$= \int_0^{2\pi} \int_0^{\infty} \frac{e^{-r^2/2}}{2\pi} r \, dr \, d\theta$$

$$= 1$$

Hence $I_0 = 1$. To compute I_n for $n > 0$, we integrate by parts:

$$\sqrt{2\pi} I_n = -\int_{-\infty}^{\infty} z^{n-1} \, d(e^{-z^2/2})$$

$$= (n - 1) \int_{-\infty}^{\infty} z^{n-2} e^{-z^2/2} \, dz$$

$$= (n - 1)\sqrt{2\pi} I_{n-2}$$

Thus we have the recurrence formula $I_n = (n - 1)I_{n-2}$. Finally we notice that $I_n = 0$ whenever n is odd, since $z^n e^{-z^2/2}$ is an odd function in that case. Putting these facts together, we have

$$I_{2n} = (2n - 1)I_{2n-2}$$

$$= (2n - 1)(2n - 3)I_{2n-4}$$

$$\vdots$$

$$= (2n - 1)(2n - 3) \cdots 3 \cdot 1$$

To obtain a more compact form, we write

$$(2n - 1)(2n - 3) \cdots 3 \cdot 1 = \frac{(2n)(2n - 1)(2n - 2) \cdots 3 \cdot 2 \cdot 1}{(2n)(2n - 2) \cdots 4 \cdot 2}$$

$$= \frac{(2n)!}{2^n n!}$$

Finally we have

$$2\pi F(\mu) = e^{-im\mu} \sum_{n=0}^{\infty} \frac{(-i\mu\sigma)^{2n}}{(2n)!} \frac{(2n)!}{2^n n!}$$

$$= e^{-im\mu} \sum_{n=0}^{\infty} \frac{1}{n!} \left(\frac{-\mu^2 \sigma^2}{2} \right)^n$$

$$= e^{-im\mu} e^{-\mu^2 \sigma^2/2}$$

Thus we have the Fourier pair

$$\boxed{f(x) = \frac{\exp[-(x - m)^2/(2\sigma^2)]}{(2\pi\sigma^2)^{1/2}} \qquad F(\mu) = \frac{\exp(-im\mu - \mu^2\sigma^2/2)}{2\pi}}$$

The graph of $y = f(x)$ is a bell-shaped curve centered on $x = m$. The parameter σ measures the spread of the graph about its midpoint. Analytically, we have

$$\sigma^2 = \int_{-\infty}^{\infty} (x - m)^2 f(x) \, dx$$

which suggests the name *variance parameter* for σ^2. Meanwhile the function $F(\mu)$ also has a bell-shaped curve. The parameter σ^2 now appears in the numerator of the exponent; this means that the spread of this graph about its midpoint is *inversely proportional* to σ^2. The larger σ^2 is, the more closely concentrated the values of $F(\mu)$ will be about $\mu = 0$.

To justify the interchange of summation and integration, we consider the Taylor approximation of the complex exponential:

$$e^{itx} = \sum_{k=0}^{N} \frac{(itx)^k}{k!} + R_N(tx)$$

Integrating this finite sum with respect to the normal density, we have

$$\int_{-\infty}^{\infty} e^{-x^2/2} e^{itx} \, dx = \sum_{k=0}^{N} \frac{1}{k!} \int_{-\infty}^{\infty} e^{-x^2/2} (itx)^k \, dx + \int_{-\infty}^{\infty} e^{-x^2/2} R_N(tx) \, dx$$

We must show that the final integral tends to zero when $N \uparrow \infty$. To do this, we represent the Taylor remainder in integral form by

$$R_N(u) = \frac{i^{N+1}}{N!} \int_0^u (u - s)^N e^{is} \, ds$$

The integrand is bounded by $(u - s)^N$, and therefore the integral is less than or equal to $u^{N+1}/(N + 1)$, leading to the bound

$$|R_N(u)| \le \frac{u^{N+1}}{(N + 1)!}$$

Replacing N by $2N - 1$ and u by tx, we have

$$\left| \int_{-\infty}^{\infty} e^{-x^2/2} R_{2N-1}(tx) \, dx \right| \le \int_{-\infty}^{\infty} e^{-x^2/2} \frac{|tx|^{2N}}{(2N)!} \, dx$$

$$= |t|^{2N} \frac{\sqrt{2\pi} 2^N}{N!}$$

which tends to zero when $N \uparrow \infty$. This completes the proof. ●

Computation of certain Fourier transforms may be facilitated by the elementary integral formulas

$$\int_0^{\infty} x^n e^{-cx} \, dx = \frac{n!}{c^{n+1}}$$

where c may be a complex number with positive real part and $n > -1$. (When n is not a positive integer, $n!$ must be interpreted by the gamma function.)

Example 5.1.3. Let $f(x) = x^2 e^{-x}$ for $x > 0$ and $f(x) = 0$ for $x < 0$. Find the Fourier transform $F(\mu)$.

Solution. We have

$$2\pi F(\mu) = \int_0^\infty f(x)e^{-i\mu x}\,dx = \int_0^\infty x^2 e^{-x} e^{-i\mu x}\,dx$$

$$= \int_0^\infty x^2 e^{-x(1+i\mu)}\,dx = \frac{2}{(1 + i\mu)^3}$$

The Fourier transform is given by $F(\mu) = (1/\pi)/(1 + i\mu)^3$.   ●

Example 5.1.4. Let $f(x) = \sin x\, e^{-x}$ for $x > 0$ and $f(x) = 0$ for $x < 0$. Find the Fourier transform $F(\mu)$.

Solution. We have

$$2\pi F(\mu) = \int_0^\infty \sin x\, e^{-x} e^{-i\mu x}\,dx$$

$$= (2i)^{-1} \int_0^\infty (e^{-x(1+i\mu-i)} - e^{-x(1+i\mu+i)})\,dx$$

$$= (2i)^{-1} \left(\frac{1}{1 + i\mu - i} - \frac{1}{1 + i\mu + i} \right)$$

$$= \frac{1}{(1 + i\mu)^2 + 1}$$

The Fourier transform is $F(\mu) = [1/(2\pi)][1/(1 + i\mu)^2 + 1]$.   ●

Example 5.1.5. Let $f(x) = e^{-x}$ for $x > 0$ and $f(x) = e^{2x}$ for $x < 0$. Find the Fourier transform $F(\mu)$.

Solution. We have

$$2\pi F(\mu) = \int_0^\infty e^{-x} e^{-i\mu x}\,dx + \int_{-\infty}^0 e^{2x} e^{-i\mu x}\,dx$$

$$= \int_0^\infty e^{-x(1+i\mu)}\,dx + \int_{-\infty}^0 e^{x(2-i\mu)}\,dx$$

$$= \frac{1}{1 + i\mu} + \frac{1}{2 - i\mu}$$

The Fourier transform is $F(\mu) = [3/(2\pi)]/(2 + i\mu + \mu^2)$.   ●

Example 5.1.6. Find the Fourier transform of $f(x) = e^{-|x|}$.

Solution. We have

$$2\pi F(\mu) = \int_0^\infty e^{-x} e^{-i\mu x}\,dx + \int_{-\infty}^0 e^{x} e^{-i\mu x}\,dx$$

$$= \frac{1}{1 + i\mu} + \frac{1}{1 - i\mu} = \frac{2}{1 + \mu^2}$$

The Fourier transform is $F(\mu) = (1/\pi)/(1 + \mu^2)$.   ●

Often new Fourier transforms can be found from old ones by interchanging the roles of x and μ.

Example 5.1.7. Find the Fourier transform of $f(x) = 1/(1 + x^2)$.

Solution. We have shown in Example 5.1.6 that the Fourier transform of $e^{-|x|}$ is $(1/\pi)(1 + \mu^2)$. Applying the convergence theorem for Fourier transforms, we have

$$\pi e^{-|x|} = \lim_{L \to \infty} \int_{-L}^{L} e^{i\mu x} \frac{1}{1 + \mu^2} d\mu$$

an absolutely convergent improper integral. Changing the roles of x and μ, we see that the Fourier transform of $f(x) = 1/(1 + x^2)$ is $F(\mu) = (1/2)e^{-|\mu|}$. ●

Fourier Sine and Cosine Transforms

We now discuss the Fourier cosine transform and the Fourier sine transform. These result when we specialize the Fourier transform to functions defined only for $x > 0$.

In detail, let $f(x)$ be defined for $x > 0$. We extend f to negative x by defining $f(-x) = f(x)$. Taking the Fourier transform of the resulting even function, we have

$$2\pi F(\mu) = \int_{-\infty}^{\infty} f(x)e^{-i\mu x} \, dx$$

$$= \int_{0}^{\infty} f(x)e^{-i\mu x} \, dx + \int_{-\infty}^{0} f(-x)e^{-i\mu x} \, dx$$

$$= \int_{0}^{\infty} f(x)(e^{-i\mu x} + e^{i\mu x}) \, dx$$

$$= 2 \int_{0}^{\infty} f(x) \cos \mu x \, dx \qquad \cos \theta = \tfrac{1}{2}(e^{i\theta} + e^{-i\theta})$$

Therefore the Fourier transform is also an even function. Hence from (5.1.5)

$$f(x) = \int_{-\infty}^{\infty} F(\mu)e^{i\mu x}$$

$$= \int_{0}^{\infty} F(\mu)e^{i\mu x} \, d\mu + \int_{-\infty}^{0} F(-\mu)e^{i\mu x} \, d\mu$$

$$= \int_{0}^{\infty} F(\mu)(e^{i\mu x} + e^{-i\mu x}) \, d\mu$$

$$= 2 \int_{0}^{\infty} F(\mu) \cos \mu x \, d\mu$$

Writing $F_c(\mu) = 2F(\mu)$, we have the *Fourier cosine formulas*

$$f(x) = \int_0^\infty F_c(\mu) \cos \mu x \, d\mu \qquad F_c(\mu) = \frac{2}{\pi} \int_0^\infty f(x) \cos \mu x \, dx \qquad (5.1.8)$$

where $F_c(\mu)$ is the *Fourier cosine transform* of $f(x)$, $-\infty < x < \infty$.

To obtain the Fourier sine transform, we begin with a function f, defined for $x > 0$, and extend it as an odd function: $f(-x) = -f(x)$. Following the same steps as for the cosine formulas, we obtain the *Fourier sine formulas*

$$f(x) = \int_0^\infty F_s(\mu) \sin \mu x \, d\mu$$

$$F_s(\mu) = \frac{2}{\pi} \int_0^\infty f(x) \sin \mu x \, dx$$

$$(5.1.9)$$

And $F_s(\mu)$ is the *Fourier sine transform* of $f(x)$, $-\infty < x < \infty$.

Example 5.1.8. Let $f(x) = e^{-ax}$ for $x > 0$ where $a > 0$. Find the Fourier cosine transform $F_c(\mu)$ and the Fourier sine transform $F_s(\mu)$.

Solution. We have

$$\int_0^\infty e^{-ax} e^{i\mu x} \, dx = \int_0^\infty e^{-x(a-i\mu)} \, dx = \frac{1}{a - i\mu} = \frac{a + i\mu}{a^2 + \mu^2}$$

Taking the real and imaginary parts, we have the integrals

$$\int_0^\infty e^{-ax} \cos \mu x \, dx = \frac{a}{a^2 + \mu^2} \qquad \int_0^\infty e^{-ax} \sin \mu x \, dx = \frac{\mu}{a^2 + \mu^2}$$

Therefore, $\qquad F_c(\mu) = \dfrac{2a/\pi}{a^2 + \mu^2} \qquad F_s(\mu) = \dfrac{2\mu/\pi}{a^2 + \mu^2}$ ●

Fourier Transforms in Several Variables

The preceding formulas and theorems for the one-dimensional Fourier transform can be extended to functions in several variables. For concreteness we illustrate the theory for the case of three variables (x_1, x_2, x_3).

Let $f(x_1, x_2, x_3)$ be a complex-valued function which is piecewise continuous in each variable separately and is absolutely integrable: $\iiint |f(x_1, x_2, x_3)| \, dx_1 \, dx_2 \, dx_3 < \infty$. The three-dimensional Fourier transform is defined by the improper integral

$$F(\mu_1, \mu_2, \mu_3) = \left(\frac{1}{2\pi}\right)^3 \int\int\int_{-\infty}^{\infty} e^{-i(\mu_1 x_1 + \mu_2 x_2 + \mu_3 x_3)} f(x_1, x_2, x_3) \, dx_1 \, dx_2 \, dx_3$$

So defined, the Fourier transform has the properties of linearity, differentiation, convolution, and translation, which were described in detail for the one-dimensional case in Theorem 5.1.1. Parseval's theorem for the three-dimensional Fourier transform is written

$$(2\pi)^3 \int_{-\infty}^{\infty} \int \int |F(\mu_1, \mu_2, \mu_3)|^2 \, d\mu_1 \, d\mu_2 \, d\mu_3 = \int_{-\infty}^{\infty} \int \int |f(x_1, x_2, x_3)|^2 \, dx_1 \, dx_2 \, dx_3$$

The original function $f(x_1, x_2, x_3)$ can be represented as the integral

$$f(x_1, x_2, x_3) = \int_{-\infty}^{\infty} \int \int e^{i(\mu_1 x_1 + \mu_2 x_2 + \mu_3 x_3)} F(\mu_1, \mu_2, \mu_3) \, d\mu_1 \, d\mu_2 \, d\mu_3$$

which is valid if, e.g., the Fourier transform is absolutely integrable.

Examples of Fourier transforms in three variables can be easily obtained from the one-dimensional case by separation of variables: if f_i, $i = 1, 2, 3$, are functions with Fourier transforms F_i, then the Fourier transform of the function $f_1(x_1)f_2(x_2)f_3(x_3)$ is the function $F_1(\mu_1)F_2(\mu_2)F_3(\mu_3)$. For example, the Fourier transform of the normal density function $f(x_1, x_2, x_3) = (2\pi)^{-3/2} \exp \left[-\frac{1}{2}(x_1^2 + x_2^2 + x_3^2)\right]$ is the function $F(\mu_1, \mu_2, \mu_3) = (2\pi)^{-3} \exp \left[-\frac{1}{2}(\mu_1^2 + \mu_2^2 + \mu_3^2)\right]$.

We now consider an example which arises in the study of the three-dimensional wave equation in Sec. 5.4.

Example 5.1.9. Find the three-dimensional Fourier transform of the function

$$f(x_1, x_2, x_3) = \begin{cases} 1 & R_1^2 < x_1^2 + x_2^2 + x_3^2 < R_2^2 \\ 0 & \text{otherwise} \end{cases}$$

Solution. The Fourier transform is given by the integral

$$(2\pi)^3 F(\mu_1, \mu_2, \mu_3) = \int \int \int_{R_1 < |x| < R_2} e^{-i(\mu_1 x_1 + \mu_2 x_2 + \mu_3 x_3)} \, dx_1 \, dx_2 \, dx_3$$

To evaluate this integral, we take a system of spherical polar coordinates (r, θ, ϕ) in (x_1, x_2, x_3)-space so that the north pole points along the vector (μ_1, μ_2, μ_3). Then $\mu_1 x_1 + \mu_2 x_2 + \mu_3 x_3 = |\mu| r \cos \theta$, $dx_1 \, dx_2 \, dx_3 = r^2 \sin \theta \, dr \, d\theta \, d\phi$, and the integral is

$$\int_{-\pi}^{\pi} \int_0^{\pi} \int_{R_1}^{R_2} e^{-i|\mu| r \cos \theta} r^2 \sin \theta \, dr \, d\theta \, d\phi$$

$$= \frac{4\pi}{|\mu|} \int_{R_1}^{R_2} r \sin (r|\mu|) \, dr$$

$$= \frac{4\pi}{|\mu|} \left(\frac{R_1 \cos (R_1 \mu) - R_2 \cos (R_2 \mu)}{|\mu|} + \frac{\sin (R_2 |\mu|) - \sin (R_1 |\mu|)}{|\mu|^2} \right) \quad \bullet$$

The above example arises in the theory of the three-dimensional wave equation in the following way. If we divide both sides by $R_2 - R_1$, then the (volume) integral over the shell $R_1 < r < R_2$ tends to the (surface) integral over the sphere $r = R_1$ if R_2 tends to R_1. The limit of the right side may be expressed in terms of ordinary differentiations; thus we have

$$\iint\limits_{|x|=R} e^{-i(\mu_1 x_1 + \mu_2 x_2 + \mu_3 x_3)} \, dS$$

$$= \int_{-\pi}^{\pi} \int_0^{\pi} e^{-i|\mu|R \cos \theta} R^2 \sin \theta \, d\theta \, d\phi$$

$$= \frac{4\pi R \sin (|\mu|R)}{|\mu|}$$

This can be expressed in terms of "mean values" over the sphere as

$$\frac{1}{4\pi R^2} \iint\limits_{|x|=R} e^{-i(\mu_1 x_1 + \mu_2 x_2 + \mu_3 x_3)} \, dS = \frac{\sin(R|\mu|)}{R|\mu|}$$

This formula is applied in Sec. 5.3 to obtain an explicit representation of the solution of the three-dimensional wave equation.

The Uncertainty Principle

In the case of a normal density function, we have $f(x) = \text{(const.)} \exp(-\frac{1}{2}x^2/\sigma^2)$ with Fourier transform $F(\mu) = \text{(const.)} \exp(-\frac{1}{2}\mu^2 \sigma^2)$. If σ is small, the function $f(x)$ is highly peaked about $x = 0$, while $F(\mu)$ is relatively flat; if σ is large, the function $F(\mu)$ is highly peaked about $\mu = 0$ while $f(x)$ is relatively flat. To make this precise in general, we define the *dispersion about zero* of a complex-valued function $f(x)$ by the formula

$$D_0(f) = \frac{\displaystyle\int_{-\infty}^{\infty} x^2 |f(x)|^2 \, dx}{\displaystyle\int_{-\infty}^{\infty} |f(x)|^2 \, dx}$$

This gives a quantitative measure of the "spread" of the function about $x = 0$. In case $f(x) = \exp(-\frac{1}{2}x^2/\sigma^2)$, we have $D_0(f) = \sigma^2/2$, $D_0(F) = 1/(2\sigma^2)$, and $D_0(f)D_0(F) = \frac{1}{4}$. In the general case we have the following inequality.

> **Uncertainty principle.** Let $f(x)$ be a piecewise smooth function with finite $D_0(f)$ and with Fourier transform $F(\mu)$ having a finite $D_0(F)$. Then $D_0(f)D_0(F) \geq \frac{1}{4}$. If equality holds, then $F(\mu) = \text{(const.)}e^{-A\mu^2/2}$ and $f(x) = \text{(const.)}e^{-A^{-1}x^2/2}$ for some $A > 0$.

> **Remark.** The reason for the name *uncertainty principle* comes from the interpretation that we cannot localize both $f(x)$ and $F(\mu)$ in their respective spaces. If we make $f(x)$ peaked near $x = 0$, then $D_0(f)$ will be small; the uncertainty principle then states that $D_0(F)$ will be correspondingly large, indicating a lack of localization.

Proof. Under the given hypotheses, both $F'(\mu)$ and $\mu F(\mu)$ exist and have finite square integrals. Applying the Schwarz inequality yields

$$\left| \int_{-\infty}^{\infty} F'(\mu)\mu\overline{F(\mu)} \, d\mu \right|^2 \leq \int_{-\infty}^{\infty} |F'(\mu)|^2 \, d\mu \int_{-\infty}^{\infty} |\mu F(\mu)|^2 \, d\mu$$

Equality holds if and only if $F'(\mu)$ and $\mu F(\mu)$ are proportional to each other. But we may write the real part of twice the integral on the left side in the form

$$2 \, \mathrm{Re}(\textstyle\int \mu \overline{F} F' \, d\mu) = \int \mu (\overline{F} F' + F \overline{F}') \, d\mu = \int \mu (F\overline{F})' \, d\mu = \int \mu (|F|^2)' \, d\mu$$

Integrating the latter by parts gives

$$2 \, \mathrm{Re}(\textstyle\int \mu \overline{F} F' \, d\mu) = -\int |F|^2$$

We insert this in the above Schwarz inequality and use Parseval's theorem twice, to write $\int |F|^2 = (2\pi)^{-1}\int |f|^2$ and $\int |F'|^2 = (2\pi)^{-1}\int x^2 |f(x)|^2 \, dx$ [since the Fourier transform of $xf(x)$ is $iF'(\mu)$]. Making these substitutions gives the desired result in the form $\frac{1}{4}\int |f|^2 \int |F|^2 \leq \int |xf|^2 \int |\mu F|^2$.

In the case of equality, we have $\mathrm{Im}(\int \mu \overline{F} F') = 0$ and the proportionality $F'(\mu) = -A\mu F(\mu)$ for some (complex) number A; the unique solution of this differential equation is of the form $F(\mu) = (\text{const.})e^{-A\mu^2/2}$. This function has a finite $D_0(F)$ if and only if the complex number A has a strictly positive real part: $\mathrm{Re}\, A > 0$. It remains to show that $\mathrm{Im}\, A = 0$. For this purpose, let $A = \alpha + i\beta$ and write $F(\mu) = (\text{const.})e^{-(\alpha+i\beta)\mu^2/2}$, $f(x) = (\text{const.})e^{-x^2/[2(\alpha+i\beta)]}$. Then $|F(\mu)|^2 = e^{-\alpha\mu^2}$ and $|f(x)|^2 = e^{-\alpha x^2/(\alpha^2+\beta^2)}$ so that $D_0(F) = 1/(2\alpha)$, $D_0(f) = (\alpha^2 + \beta^2)/(2\alpha)$, and $D_0(f)D_0(F) = (\alpha^2 + \beta^2)/(4\alpha^2)$. If β is nonzero, then $D_0(f)D_0(F) > \frac{1}{4}$, which contradicts equality. Thus $\beta = 0$, and the proof is complete. ∎

Proof of Convergence

We close this section by giving the proof of the convergence theorem for Fourier transforms. We are given a function $f(x)$, $-\infty < x < \infty$, which is piecewise smooth on each finite interval and for which $\int_{-\infty}^{\infty} |f(x)| \, dx$ is convergent. Defining the Fourier transform by (5.1.6), we make the following transformations:

$$2\pi \int_{-L}^{L} F(\mu)e^{i\mu x} \, d\mu = \int_{-L}^{L} \left[\int_{-\infty}^{\infty} e^{-i\mu\xi} f(\xi) \, d\xi \right] e^{i\mu x} \, d\mu$$

$$= \int_{-\infty}^{\infty} \left(\int_{-L}^{L} e^{i\mu(x-\xi)} \, d\mu \right) f(\xi) \, d\xi$$

$$= 2 \int_{-\infty}^{\infty} \frac{\sin L(x - \xi)}{x - \xi} f(\xi) \, d\xi$$

$$= 2 \int_{-\infty}^{\infty} \frac{\sin L\eta}{\eta} f(x + \eta) \, d\eta$$

where we have changed the order of integration and made the substitution $\eta = \xi - x$. By the riemannian lemma of Chap. 1, it follows that the integral $\int_{|\eta| \geq \delta} [(\sin L\eta)/\eta]f(x + \eta) \, d\eta$ tends to zero when L tends to infinity, for any positive number δ. It remains to analyze the integral for $-\delta \le \eta \le \delta$. Another use of the riemannian lemma shows that each of the integrals

$$2 \int_0^\delta \frac{\sin L\eta}{\eta} [f(x + \eta) - f(x + 0)] \, d\eta$$

and

$$2 \int_{-\delta}^0 \frac{\sin L\eta}{\eta} [f(x + \eta) - f(x - 0)] \, d\eta$$

tends to zero when L tends to infinity. Therefore,

$$\lim_{L \to \infty} 2 \int_{-\delta}^\delta \frac{\sin L\eta}{\eta} f(x + \eta) \, d\eta = [f(x + 0) + f(x - 0)] \lim_{L \to \infty} \int_{-\delta}^\delta \frac{\sin L\eta}{\eta} \, d\eta$$

$$= \pi[f(x + 0) + f(x - 0)]$$

Dividing by 2π, we have the desired result. ∎

The unrestricted improper integral in (5.1.5) doesn't exist in general. (This is analogous to the situation for complex Fourier series; see Exercise 8 in Sec. 1.5). An example is given in Exercise 28.

EXERCISES 5.1

In Exercises 1 to 10, find the Fourier transforms of the indicated functions.

1. $f(x) = \begin{cases} 1 & -2 < x < 2 \\ 0 & \text{otherwise} \end{cases}$

2. $f(x) = \begin{cases} -4 & -1 < x < 0 \\ 4 & 0 < x < 1 \\ 0 & \text{otherwise} \end{cases}$

3. $f(x) = \begin{cases} e^{-3x} & x > 0 \\ e^{2x} & x < 0 \end{cases}$

4. $f(x) = xe^{-|x|}$

5. $f(x) = \cos x \, e^{-|x|}$

6. $f(x) = \cos^2 x \, e^{-|x|}$

7. $f(x) = 2x/(1 + x^2)^2$

8. $f(x) = \exp[-(x^2 + 3x)/2]$

9. $f(x) = \cos x \, e^{-x^2/2}$

10. $f(x) = xe^{-x^2/2}$

11. Suppose that $f(x)$, $-\infty < x < \infty$, is continuous and piecewise smooth on every finite interval and both $\int_{-\infty}^\infty |f(x)| \, dx$ and $\int_{-\infty}^\infty |f'(x)| \, dx$ are absolutely convergent. Using integration by parts, show that the Fourier transform of f' is $i\mu F(\mu)$.

12. Apply Exercise 11 to Exercises 4 to 10 to obtain additional examples of Fourier transforms.

13. Let $f(x) = xe^{-x}$ for $x > 0$. Find the Fourier cosine transform $F_c(\mu)$ and the Fourier sine transform $F_s(\mu)$.

14. Complete the derivation of the Fourier sine formulas (5.1.9).

15. Let $a > 0$. Show that the Fourier transform of $f(ax)$ is $F(\mu/a)/a$.

16. Show that the Fourier transform of $f(x - a)$ is $e^{-ia\mu}F(\mu)$.

Use Exercises 15 and 16 to find the Fourier transforms of the following functions.

17. $f(x) = 1/[1 + (x - 3)^2]$

18. $f(x) = e^{-(x-2)^2/2}$

19. $f(x) = e^{-3|x-2|}$

20. $f(x) = \begin{cases} \sin 2x e^{-2x} & x > 0 \\ 0 & x < 0 \end{cases}$ **21.** $f(x) = \begin{cases} e^{-2x} & x > 0 \\ 0 & x < 0 \end{cases}$

22. Compute $D_0(f)D_0(F)$ for the following functions.
   (a) $f(x) = (\text{const.})xe^{-x^2/2}$ $F(\mu) = (\text{const.})\mu e^{-\mu^2/2}$
   (b) $f(x) = (\text{const.})x^2 e^{-x^2/2}$ $F(\mu) = (\text{const.})(1 - \mu^2)e^{-\mu^2/2}$
   (c) $f(x) = (\text{const.})e^{-x-1^2/2}$ $F(\mu) = (\text{const.})e^{-i\mu}e^{-\mu^2/2}$

23. If $f(x)$ is a complex-valued function with Fourier transform $F(\mu)$, let $f_{a,m}(x) = e^{-imx}f(x + a)$, where a, m are real. Show that the Fourier transform of $f_{a,m}$ is $e^{-ia(\mu+m)}F(\mu + m) = e^{-iam}F_{m,a}(\mu)$.

24. The *dispersion about a* of a complex-valued function is defined by $D_a(f) = \int_{-\infty}^{\infty} (x - a)^2 |f(x)|^2 \, dx / \int_{-\infty}^{\infty} |f(x)|^2 \, dx$. Show $D_a(f) = D_0(f_{a,m})$ and $D_m(F) = D_0(F_{m,a})$ for any (a, m).

25. Combine Exercises 23 and 24 with the uncertainty principle to obtain the general inequality $D_a(f)D_m(F) \geq \frac{1}{4}$ with equality iff $f(x) = (\text{const.})e^{-A^{-1}(x-a)^2/2}e^{imx}$, $F(\mu) = (\text{const.})e^{-A(\mu-m)^2/2}e^{-ia\mu}$ for suitable complex constants.

26. Prove Theorem 5.1.1a.

27. Prove Theorem 5.1.1d.

28. Let $f(x) = e^{-x}$ for $x > 0$ and $f(x) = -e^x$ for $x < 0$. Show that the unrestricted improper integral $\int_{-\infty}^{\infty} F(\mu)e^{i\mu x} \, d\mu$ doesn't exist for all x. [*Hint:* Examine separately the integrals $\int_0^{\infty} F(\mu)e^{i\mu x} \, d\mu$ and $\int_{-\infty}^{0} F(\mu)e^{i\mu x} \, d\mu$ at $x = 0$.]

29. (*Fourier transforms in real form.*) Suppose that $f(x)$, $-\infty < x < \infty$, is a real-valued function with $\int_{-\infty}^{\infty} |f(x)| \, dx < \infty$ and is piecewise smooth on each finite interval. Define the real-valued functions $A(\mu) = (1/\pi) \int_{-\infty}^{\infty} f(x) \cos \mu x \, dx$, and $B(\mu) = (1/\pi) \int_{-\infty}^{\infty} f(x) \sin \mu x \, dx$. Show that we have for each x, $-\infty < x < \infty$,

$$\lim_{L \to \infty} \int_0^L [A(\mu) \cos \mu x + B(\mu) \sin \mu x] \, dx = \tfrac{1}{2}f(x + 0) + \tfrac{1}{2}f(x - 0)$$

5.2   SOLUTION OF THE HEAT EQUATION FOR AN INFINITE ROD

We will now see how the Fourier transform applies to the heat equation. For this purpose we consider the following initial-value problem on the entire axis:

$$u_t = Ku_{xx} \qquad t > 0, \ -\infty < x < \infty$$
$$u(x; 0) = f(x) \qquad -\infty < x < \infty \tag{5.2.1}$$

We solve this problem by two different methods.

First Method (Fourier Series Passage to the Limit)

We consider the entire axis as the limit when $L \to \infty$ of the interval $(-L, L)$ with the periodic boundary conditions $u(L, t) = u(-L, t)$, $u'(L, t) = u'(-L, t)$. Equation (5.2.1) with periodic boundary conditions can be solved by separation of variables. In complex form, the separated solutions are

$$e^{in\pi x/L}e^{-(n\pi/L)^2 Kt} \qquad n = 0, \pm 1, \pm 2, \ldots$$

Using the formula for complex Fourier series, we look for the solution as a superposition

$$u_L(x; t) = \sum_{-\infty}^{\infty} \alpha_n e^{in\pi x/L} e^{-(n\pi/L)^2 Kt}$$

The initial data and the coefficients are related by the Fourier series formulas

$$f(x) = \sum_{-\infty}^{\infty} \alpha_n e^{in\pi x/L}$$

$$\alpha_n = \frac{1}{2L} \int_{-L}^{L} f(x)e^{-in\pi x/L} \, dx$$

To take the limit when $L \to \infty$, we follow the method of Sec. 5.1. We let

$$F_L(\mu) = \frac{1}{2\pi} \int_{-L}^{L} f(x)e^{-i\mu x} \, dx$$

$$\mu_n = \frac{n\pi}{L}$$

$$\Delta\mu_n = \frac{\pi}{L} = \mu_{n+1} - \mu_n$$

Thus the previous formulas can be rewritten as

$$\alpha_n = F_L(\mu_n) \, \Delta\mu_n$$

$$u_L(x; t) = \sum_{-\infty}^{\infty} F_L(\mu_n)e^{i\mu_n x}e^{-\mu_n^2 Kt} \, \Delta\mu_n$$

Taking the limit as $L \to \infty$, we get the formulas

$$\boxed{u(x; t) = \int_{-\infty}^{\infty} F(\mu)e^{i\mu x}e^{-\mu^2 Kt} \, d\mu} \qquad (5.2.2a)$$

$$\boxed{F(\mu) = \frac{1}{2\pi} \int_{-\infty}^{\infty} f(x)e^{-i\mu x} \, dx} \qquad (5.2.2b)$$

This is the *Fourier representation* of the solution of the initial-value problem (5.2.1). To find $u(x; t)$, we first find $F(\mu)$, the Fourier transform of $f(x)$. Then we substitute into the first equation and perform the indicated integration.

Example 5.2.1. Solve (5.2.1) in the case where $f(x) = e^{-x^2/2}$, $-\infty < x < \infty$.

Solution. The integral (5.2.2*b*) can be done explicitly, as in Example 5.1.2, with the result

$$F(\mu) = \frac{e^{-\mu^2/2}}{\sqrt{2\pi}}$$

Substituting this into (5.2.2*a*) and using the same example, we have

$$u(x; t) = \frac{e^{-x^2/(2+4Kt)}}{\sqrt{1 + 2Kt}}$$

●

Second Method (Direct Solution by Fourier Transform)

In this method we use the Fourier transform to convert the partial differential equation to an ordinary differential equation. Let $U(\mu; t)$ be the Fourier transform of $u(x; t)$. Thus from (5.1.5) and (5.1.6) we have

$$u(x; t) = \int_{-\infty}^{\infty} U(\mu; t)e^{i\mu x} \, d\mu$$

$$U(\mu; t) = \frac{1}{2\pi} \int_{-\infty}^{\infty} u(x; t)e^{-i\mu x} \, dx$$

(5.2.3)

Assuming for the moment that the derivatives can be taken under the integral, we have

$$u_t(x; t) = \int_{-\infty}^{\infty} U_t(\mu; t)e^{i\mu x} \, d\mu$$

$$u_x(x; t) = \int_{-\infty}^{\infty} U(\mu; t)i\mu e^{i\mu x} \, d\mu$$

$$u_{xx}(x; t) = \int_{-\infty}^{\infty} U(\mu; t)(i\mu)^2 e^{i\mu x} \, d\mu$$

In order that u satisfy the heat equation, we must have

$$0 = u_t - Ku_{xx} = \int_{-\infty}^{\infty} [U_t(\mu; t) + K\mu^2 U(\mu; t)]e^{i\mu x} \, d\mu$$

Therefore we ask that U be a solution of the ordinary differential equation

$$U_t + K\mu^2 U = 0$$

The initial condition is determined by setting $t = 0$ in (5.2.3). Thus $U(\mu; 0)$ must be the Fourier transform of f:

$$U(\mu; 0) = F(\mu)$$

Therefore we must have

$$U(\mu; t) = F(\mu)e^{-\mu^2 Kt}$$

Substituting into (5.2.3) gives formula (5.2.2), as in the first method.

Verification of the Solution

We now prove that (5.2.2) is a rigorous solution of the initial-value problem (5.2.1). We assume explicitly that the initial data $f(x)$ is piecewise smooth on each finite interval with $\int_{-\infty}^{\infty} |f(x)|\, dx < \infty$ and with a Fourier transform $F(\mu)$ which satisfies $\int_{-\infty}^{\infty} \mu^2 F(\mu)\, d\mu < \infty$.

Theorem 5.2.1. Let $f(x)$ be an integrable piecewise smooth function with Fourier transform $F(\mu)$.

(a) The integral (5.2.2b) defines a solution of the heat equation $u_t = Ku_{xx}$ for $t > 0$, $-\infty < x < \infty$.

(b) If, in addition, $\int_{-\infty}^{\infty} \mu^2 |F(\mu)|\, d\mu < \infty$, then $\lim_{t \downarrow 0} u(x; t) = f(x)$, $-\infty < x < \infty$.

Proof. (a) We first prove that $u(x; t)$ satisfies the heat equation. For this purpose, it suffices to prove that

$$u_x(x; t) = \int_{-\infty}^{\infty} F(\mu)i\mu e^{i\mu x} e^{-\mu^2 Kt}\, d\mu \tag{5.2.4a}$$

$$u_{xx}(x; t) = \int_{-\infty}^{\infty} F(\mu)(i\mu)^2 e^{i\mu x} e^{-\mu^2 Kt}\, d\mu \tag{5.2.4b}$$

$$u_t(x; t) = \int_{-\infty}^{\infty} F(\mu)e^{i\mu x}(-\mu^2 K e^{-\mu^2 Kt})\, d\mu \tag{5.2.4c}$$

Once we have proved these formulas, it is apparent that $u_t = Ku_{xx}$. To prove the first formula, we write

$$\frac{u(x + h; t) - u(x; t)}{h} - \int_{-\infty}^{\infty} F(\mu)i\mu e^{i\mu x}e^{-\mu^2 Kt}\, d\mu$$

$$= \int_{-\infty}^{\infty} F(\mu)\left(\frac{e^{i\mu(x+h)} - e^{i\mu x} - i\mu h e^{i\mu x}}{h}\right)e^{-\mu^2 Kt}\, d\mu$$

To estimate this term, we notice that, for any real θ,

$$\left|e^{i\theta} - 1 - i\theta\right|^2 = (1 - \cos \theta)^2 + (\sin \theta - \theta)^2$$

$$\leq \left(\frac{\theta^2}{2}\right)^2 + \left(\frac{\theta^2}{2}\right)^2$$

Applying this with $\theta = \mu h$, we have

$$\left|\frac{u(x + h; t) - u(x; t)}{h} - \int_{-\infty}^{\infty} F(\mu)i\mu e^{i\mu x}e^{-\mu^2 Kt}\, d\mu\right| \leq h \int_{-\infty}^{\infty} |F(\mu)|\mu^2 e^{-\mu^2 Kt}\, d\mu$$

The integral on the right is absolutely convergent. Taking the limit as $h \to 0$, we have proved (5.2.4a). It is left as an exercise to prove (5.2.4b) and (5.2.4c) by the same method. This completed, we have proved that the Fourier representation (5.2.2) gives a rigorous solution of the heat equation. It is a remarkable fact that u is *differentiable to any order*, in spite of the fact that f is only assumed *piecewise* smooth.

(b) To prove that the initial condition is satisfied, we write

$$u(x; t) - f(x) = \int_{-\infty}^{\infty} (e^{-\mu^2 Kt} - 1) e^{i\mu x} F(\mu) \, d\mu$$

Now the inequality $|e^{-\theta} - 1| \le \theta$ may be applied to the integrand, with the result

$$|u(x; t) - f(x)| \le \int_{-\infty}^{\infty} \mu^2 Kt |F(\mu)| \, d\mu$$

The right side tends to zero when $t \downarrow 0$, which completes the proof.   ∎

Explicit Representation by Gauss-Weierstrass Kernel

We now show how to obtain an explicit representation of the solution involving only one integration and not involving the Fourier transform. To do this, we note that (5.2.2b) represents the Fourier transform of $u(x; t)$ as a product of the Fourier transform of f with the elementary function $e^{-\mu^2 Kt}$. But this is the Fourier transform of the normal density function which we encountered in Example 5.1.2, where $\sigma^2 = 2Kt$:

$$\int_{-\infty}^{\infty} e^{i\mu(x-\xi)} e^{-\mu^2 Kt} \, d\mu = 2\pi \frac{e^{-(x-\xi)^2/(4Kt)}}{\sqrt{4\pi Kt}}$$

Appealing to the convolution theorem, Theorem 5.1.2b, we have the *explicit representation*

$$u(x; t) = \int_{-\infty}^{\infty} f(\xi) \frac{e^{-(x-\xi)^2(4Kt)}}{\sqrt{4\pi Kt}} \, d\xi \tag{5.2.5}$$

We will give an independent proof below that (5.2.5) gives a solution of the initial-value problem (5.2.1). For each ξ the function $(t, x) \to e^{-(x-\xi)^2/(4Kt)}/\sqrt{4\pi Kt}$ is a solution of the heat equation, the *fundamental solution*. Thus we see that (5.2.5) gives a representation of the solution as a continuous superposition of fundamental solutions. This may be contrasted with the Fourier representation (5.2.2b) which represents the solution as a continuous superposition of separated solutions.

This form of the solution is preferable to the Fourier representation (5.2.2b) for several reasons. In addition to its computational directness, it is theoretically more attractive, especially since the formula makes sense for many functions $f(x)$ which do not possess a Fourier transform, e.g., any bounded continuous function. In addition, it doesn't require any smoothness conditions, which were required in dealing with the Fourier transform inversion theorem. We state the properties of this representation as Theorem 5.2.2.

Theorem 5.2.2. Suppose that $f(x)$, $-\infty < x < \infty$, is piecewise continuous and bounded on the entire axis: $|f(x)| \le A$ for $-\infty < x < \infty$. Then the Gauss-Weierstrass integral (5.2.5) defines a solution of the heat equation $u_t = Ku_{xx}$ for $t > 0$, $-\infty < x < \infty$, and $\lim_{t \downarrow 0} u(x; t) = \frac{1}{2}f(x + 0) + \frac{1}{2}f(x - 0)$.

Proof. It suffices to prove that

$$u_x(x; t) = \int_{-\infty}^{\infty} \frac{\partial}{\partial x} \left(\frac{e^{-(x-z)^2/(4Kt)}}{\sqrt{4\pi Kt}} \right) f(z) \, dz$$

$$u_{xx}(x; t) = \int_{-\infty}^{\infty} \frac{\partial^2}{\partial x^2} \left(\frac{e^{-(x-z)^2/4Kt}}{\sqrt{4\pi Kt}} \right) f(z) \, dz$$

$$u_t(x; t) = \int_{-\infty}^{\infty} \frac{\partial}{\partial t} \left(\frac{e^{-(x-z)^2/(4Kt)}}{\sqrt{4\pi Kt}} \right) f(z) \, dz$$

since $(t, x) \to e^{-(x-z)^2/(4Kt)}/\sqrt{4\pi Kt}$ may be verified to be a solution of the heat equation for each z. To prove the first formula, we introduce a time variable $s = 2Kt$ and examine the difference

$$u(x + h; s/(2K)) - u(x; s/(2K)) = \int_{-\infty}^{\infty} \frac{(e^{-(x+h-z)^2/(2s)} - e^{-(x-z)^2/(2s)}) f(z) \, dz}{\sqrt{2\pi s}}$$

$$= \int_{-\infty}^{\infty} (e^{-(y+h)^2/(2s)} - e^{-y^2/(2s)}) \left(\frac{f(x - y)}{\sqrt{2\pi s}} \right) dy$$

which is the difference for the translated function $f_x(y) \doteq f(x - y)$ at $x = 0$. Therefore it suffices to prove the formula at $x = 0$ for the new function. If the integration were over a finite interval, then one could apply Proposition 0.5.4 for the differentiation of an integral depending on a parameter. To handle the general case, we write

$$\frac{u(h; s/(2K)) - u(0; s/(2K))}{h} - \int_{-\infty}^{\infty} \frac{y}{s} \frac{e^{-y^2/(2s)} f(y) \, dy}{\sqrt{2\pi s}}$$

$$= \left\{ \int_{|y| \leq M} + \int_{|y| > M} \right\} \left(e^{-(y+h)^2/(2s)} - e^{-y^2/(2s)} - \frac{y}{s} e^{-y^2/(2s)} \right) \frac{f(y) \, dy}{\sqrt{2\pi s}}$$

The integral over $|y| \leq M$ tends to zero when $h \downarrow 0$, by Proposition 0.5.4. To handle the integral over $|y| > M$, we use the mean value theorem to write

$$e^{-(y+h)^2/(2s)} - e^{-y^2/(2s)} = -h((y + \theta)/s)e^{-(y+\theta)^2/(2s)}$$

for some θ with $|\theta| < |h|$; now expand and estimate the exponent according to

$$(y + \theta)^2 = y^2 + 2y\theta + \theta^2$$

$$\geq y^2 - (\tfrac{1}{2}y^2 + 2\theta^2) + \theta^2$$

$$= \tfrac{1}{2}y^2 - \theta^2$$

so that for $|h| \leq 1$ we can estimate the first integrand above by using $(2/s)(1 + |y|)e^{-y^2/(4s)}$. Now for any preassigned $\epsilon > 0$, choose $M = M(\epsilon)$ so that $\int_{-\infty}^{\infty} [(2/s)(1 + |y|)e^{-y^2/(4s)} + (|y|/s)e^{-y^2/(2s)}] f(y) \, dy/\sqrt{2\pi s} < \epsilon$. Taking the limit when $h \to 0$, we see that

$$\limsup_{h \to 0} \left| \frac{u(h; t) - u(0; t)}{h} - \int_{-\infty}^{\infty} \frac{y}{s} e^{-y^2/(2s)} \frac{f(y)}{\sqrt{2\pi s}} \, dy \right| < \epsilon$$

This holds for any $\epsilon > 0$; hence the indicated limsup is zero, and we have proved the first differentiation formula. The proofs of the formulas for u_{xx} and u_t are carried out in a similar manner.

To prove the stated form of the initial conditions, we first make the change of variable $z = (x - \xi)/\sqrt{2Kt}$ which leads to

$$u(x; t) = \int_{-\infty}^{\infty} f(x + z\sqrt{2Kt}) \frac{e^{-z^2/2}}{\sqrt{2\pi}} dz$$

For any preassigned $\epsilon > 0$, there is a $\delta > 0$ such that

$$|f(x') - f(x + 0)| < \epsilon \qquad x < x' < x + \delta \qquad (5.2.6)$$
$$|f(x') - f(x - 0)| < \epsilon \qquad x - \delta < x' < x$$

Now we write

$$\int_0^{\infty} f(x + z\sqrt{2Kt}) \frac{e^{-z^2/2}}{\sqrt{2\pi}} dz - \tfrac{1}{2}f(x + 0)$$

$$= \left\{ \int_0^{\delta/\sqrt{2Kt}} + \int_{\delta/\sqrt{2Kt}}^{\infty} \right\} [f(x + z\sqrt{2Kt}) - f(x + 0)] \frac{e^{-z^2/2}}{\sqrt{2\pi}} dz$$

In the first integral we use (5.2.6) and obtain the upper bound

$$\epsilon \int_0^{\delta/\sqrt{2Kt}} \frac{e^{-z^2/2} dz}{\sqrt{2\pi}} \leq \epsilon$$

To estimate the second integral, we note that $|f(x') - f(x + 0)| \leq 2A$ and that the second integral is bounded above by $2A \int_{\delta/\sqrt{2Kt}}^{\infty} e^{-z^2/2} dz/\sqrt{2\pi}$, which tends to zero when $t \downarrow 0$. Therefore

$$\limsup_{t \downarrow 0} \left| \int_0^{\infty} [f(x + z\sqrt{2Kt}) - f(x + 0)] \frac{e^{-z^2/2} dz}{\sqrt{2\pi}} \right| \leq \epsilon$$

Since this holds for every ϵ, we conclude that

$$\lim_{t \downarrow 0} \int_0^{\infty} f(x + z\sqrt{2Kt}) \frac{e^{-z^2/2} dz}{\sqrt{2\pi}} = \tfrac{1}{2}f(x + 0)$$

In the same fashion it is shown that

$$\lim_{t \downarrow 0} \int_{-\infty}^{0} f(x + z\sqrt{2Kt}) \frac{e^{-z^2/2} dz}{\sqrt{2\pi}} = \tfrac{1}{2}f(x - 0)$$

Adding these two statements gives the required result. ∎

Examples

In case the initial temperature $f(x)$, $-\infty < x < \infty$, is piecewise constant, the solution $u(x; t)$ of the heat equation can be written in terms of the *normal distribution function* Φ, defined by

$$\Phi(c) = \int_{-\infty}^{c} e^{-z^2/2} \frac{dz}{\sqrt{2\pi}}$$

This is a continuous increasing function with $\Phi(-\infty) = 0$, $\Phi(0) = \frac{1}{2}$, $\Phi(\infty) = 1$. To apply this function to the heat equation, we proceed as follows: If $a < b$, we write

$$\int_a^b \frac{e^{-(x-\xi)^2/(4Kt)}}{\sqrt{4\pi Kt}}\, d\xi = \int_{(x-b)/\sqrt{2Kt}}^{(x-a)/\sqrt{2Kt}} e^{-z^2/2}\, \frac{dz}{\sqrt{2\pi}} \qquad \frac{x-\xi}{\sqrt{2Kt}} = z$$

$$= \int_{-\infty}^{(x-a)/\sqrt{2Kt}} e^{-z^2/2}\, \frac{dz}{\sqrt{2\pi}} - \int_{-\infty}^{(x-b)/\sqrt{2Kt}} e^{-z^2/2}\, \frac{dz}{\sqrt{2\pi}}$$

$$= \Phi\left(\frac{x-a}{\sqrt{2Kt}}\right) - \Phi\left(\frac{x-b}{\sqrt{2Kt}}\right)$$

Example 5.2.2. Solve (5.2.1) in the case where

$$f(x) = \begin{cases} 0 & x < a \\ T & a \le x \le b \\ 0 & x > b \end{cases}$$

Solution. Formula (5.2.5) reduces to

$$u(x; t) = T \int_a^b \frac{\exp\left[-(x-\xi)^2/(4Kt)\right]}{\sqrt{4\pi Kt}}\, d\xi$$

$$= T \int_{(x-b)/\sqrt{2Kt}}^{(x-a)/\sqrt{2Kt}} e^{-z^2/2}\, \frac{dz}{\sqrt{2\pi}} \qquad x - \xi = z\sqrt{2Kt}$$

$$= T\Phi\left(\frac{x-a}{\sqrt{2Kt}}\right) - T\Phi\left(\frac{x-b}{\sqrt{2Kt}}\right)$$

Using the properties $\Phi(-\infty) = 0$, $\Phi(0) = \frac{1}{2}$, $\Phi(\infty) = 1$, we can verify directly the following limits:

$$\lim_{t \downarrow 0} u(x; t) = \begin{cases} 0 & x < a \\ \dfrac{T}{2} & x = a \\ T & a < x < b \\ \dfrac{T}{2} & x = b \\ 0 & x > b \end{cases}$$

By using tabulated values of $\Phi(a)$, it is possible to obtain accurate graphs of the temperature function $u(x; t)$ for various values of t. The graphs in Figure 5.2.1 assume that $K = \frac{1}{2}$, $a = -1$, $b = 1$. The values of t are $t = 0$, $t = 0.01$, $t = 1$, $t = 100$. ●

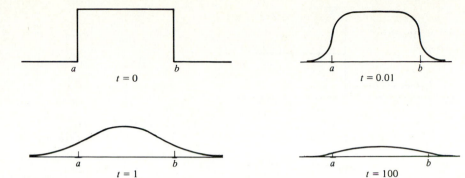

FIGURE 5.2.1
Solution of the heat equation at four different times.

Example 5.2.3. Two materials of the same conductivity are initially at different temperatures T_1, T_2. Find the temperature at all later times when heat is allowed to flow freely between the two materials. Discuss the approach to a steady state.

Solution. Suppose that the first material occupies the negative axis $x < 0$ and the second material occupies the positive axis $x > 0$. Letting $u(x; t)$ denote the temperature at the point x at the time t, we have the initial-value problem

$$u_t = Ku_{xx} \qquad t > 0, \ -\infty < x < \infty$$

$$u(x; 0) = T_1 \qquad x < 0$$

$$u(x; 0) = T_2 \qquad x > 0$$

The solution to this problem is given in (5.2.5), by the formula

$$u(x; t) = T_1 \int_{-\infty}^{0} \frac{\exp\left[-(x - \xi)^2/(4Kt)\right]}{\sqrt{4\pi Kt}} \, d\xi + T_2 \int_{0}^{\infty} \frac{\exp\left[-(x - \xi)^2/(4Kt)\right]}{\sqrt{4\pi Kt}} \, d\xi$$

$$= T_1 \int_{x/\sqrt{2Kt}}^{\infty} \frac{e^{-z^2/2}}{\sqrt{2\pi}} \, dz + T_2 \int_{-\infty}^{x/\sqrt{2Kt}} \frac{e^{-z^2/2}}{\sqrt{2\pi}} \, dz.$$

$$= T_1 \left[1 - \Phi\left(\frac{x}{\sqrt{2Kt}}\right)\right] + T_2 \, \Phi\left(\frac{x}{\sqrt{2Kt}}\right)$$

The graphs in Figure 5.2.2 depict $u(x; t)$ for three values of t when $K = 200$, $T_1 = 25$, $T_2 = 100$.

To study the approach to steady state, recall that $\lim_{x \to 0} \Phi(x) = \Phi(0) = \frac{1}{2}$. Therefore $\lim_{t \to \infty} u(x; t) = \frac{1}{2}T_1 + \frac{1}{2}T_2$ for any x. To proceed further, we write

$$u(x; t) - (\tfrac{1}{2}T_1 + \tfrac{1}{2}T_2) = T_1 \left[\frac{1}{2} - \Phi\left(\frac{x}{\sqrt{2Kt}}\right)\right] + T_2 \left[\Phi\left(\frac{x}{\sqrt{2Kt}}\right) - \frac{1}{2}\right]$$

$$= (T_2 - T_1) \left[\Phi\left(\frac{x}{\sqrt{2Kt}}\right) - \frac{1}{2}\right]$$

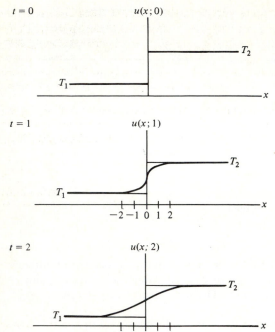

FIGURE 5.2.2
Solution of the heat equation at three different times.

When $t \to \infty$, we have $\Phi(x/\sqrt{2Kt}) - \frac{1}{2} = (x/\sqrt{2Kt})/\sqrt{2\pi} + O(1/t)$; therefore $t^{-1} \ln |u(x; t) - (\frac{1}{2}T_1 + \frac{1}{2}T_2)| = O(t^{-1} \ln t)$, which tends to zero. Therefore the relaxation time, as defined in Chap. 2, is infinite. To obtain a concrete numerical estimate of the time necessary to attain steady state, we define τ^* as the solution of

$$|u(x; \tau^*) - (\tfrac{1}{2}T_1 + \tfrac{1}{2}T_2)| = 0.1|T_1 - T_2|$$

To solve this equation, we must solve the equation $\Phi(s) - \frac{1}{2} = 0.1$ and set $s = x/\sqrt{2K\tau^*}$. From tables of the normal distribution function, we have $s = 0.25$, and thus $\tau^* = (2K)^{-1}(x/0.25)^2$. For example, if $K = 200$ cm²/s and $x = 10$ cm, then $\tau^* = 4$ s.   ●

Solutions on a Half-Line; Method of Images

Now we turn to heat flow in a semi-infinite region. Specifically, we consider the problem

$$u_t = Ku_{xx} \qquad t > 0, x > 0$$

$$u(0; t) = 0 \qquad t > 0$$

$$u(x; 0) = f(x) \qquad x > 0$$

To solve this problem, we use the *method of images*. This amounts to extending the given initial function $f(x)$ to the entire real axis in an appropriate fashion and using the explicit representation (5.2.5). (In physical terms we are creating a fictional temperature distribution on the negative axis.) Specifically we set

$$\tilde{f}(x) = \begin{cases} f(x) & x > 0 \\ -f(-x) & x < 0 \end{cases}$$

where $\tilde{f}(x)$ is an odd function, defined for $-\infty < x < \infty$. We substitute in (5.2.5) and obtain

$$u(x; t) = \int_{-\infty}^{\infty} \frac{\exp\left[-(x - \xi)^2/(4Kt)\right]}{\sqrt{4\pi Kt}} \tilde{f}(\xi) \, d\xi$$

$$= \left\{ \int_{-\infty}^{0} + \int_{0}^{\infty} \right\} \frac{\exp\left[-(x - \xi)^2/(4Kt)\right]}{\sqrt{4\pi Kt}} \tilde{f}(\xi) \, d\xi$$

In the first integral we change ξ to $-\xi$ and use the oddness of $\tilde{f}$. Thus

$$\int_{-\infty}^{0} \frac{\exp\left[-(x - \xi)^2/(4Kt)\right]}{\sqrt{4\pi Kt}} \tilde{f}(\xi) \, d\xi = -\int_{0}^{\infty} \frac{\exp\left[-(x + \xi)^2/(4Kt)\right]}{\sqrt{4\pi Kt}} f(\xi) \, d\xi$$

Putting the two together, we have

$$u(x; t) = \int_{0}^{\infty} \frac{\exp\left[-(x - \xi)^2/(4Kt)\right] - \exp\left[-(x + \xi)^2/(4Kt\right]}{\sqrt{4\pi Kt}} f(\xi) \, d\xi \qquad (5.2.7)$$

This is the required form of the solution.

The method of images also can be applied to solve the initial-boundary-value problem

$$u_t = Ku_{xx} \qquad t > 0, x > 0$$

$$u_x(0; t) = 0 \qquad t > 0$$

$$u(x; 0) = f(x) \qquad x > 0$$

Notice that we have changed only the boundary condition and that otherwise the problem is identical to the previous one. To treat this case, we extend the initial function by the rule

$$\tilde{f}(x) = \begin{cases} f(x) & x > 0 \\ f(-x) & x < 0 \end{cases}$$

In other words, $\tilde{f}$ is an even function. The reader is invited to follow the steps of the preceding problem to show that in this case we have the solution formula

$$u(x; t) = \int_{0}^{\infty} \frac{\exp\left[-(x - \xi)^2/(4Kt)\right] + \exp\left[-(x + \xi)^2/(4Kt)\right]}{\sqrt{4\pi Kt}} f(\xi) \, d\xi \qquad (5.2.8)$$

Hermite Polynomials

The normal density function $e^{-x^2/2}$ is proportional to its Fourier transform. This is the first of an infinite sequence of functions with that property. To define these, we introduce the generating function

$$e^{tx-t^2/2} = \sum_{k\geq 0} \frac{t^k}{k!} H_k(x) = H_0(x) + tH_1(x) + \tfrac{1}{2}t^2 H_2(x) + \cdots$$

This power series converges for all t, real or complex; the coefficients $H_k(x)$ are the *Hermite polynomials*. The first few are as follows:

$$H_0(x) = 1$$

$$H_1(x) = x$$

$$H_2(x) = x^2 - 1$$

$$H_3(x) = x^3 - 3x$$

$$H_4(x) = x^4 - 6x^2 + 3$$

The generating function is a Taylor series in the variable t, hence the coefficients can be obtained by successive differentiation as

$$H_k(x) = \left(\frac{d}{dt}\right)^k (e^{tx-t^2/2})\big|_{t=0}$$

The Hermite polynomials obey a system of orthogonality relations as follows:

$$\int_{-\infty}^{\infty} H_k(x)H_j(x)e^{-x^2/2}\,dx = \begin{cases} k!\sqrt{2\pi} & k = j \\ 0 & k \neq j \end{cases}$$

To prove this, multiply the generating relations for t and s:

$$e^{-x^2/2}e^{tx-t^2/2}e^{sx-s^2/2} = \sum_{j,k\geq 0} \frac{t^k}{k!}\frac{s^j}{j!} H_k(x)H_j(x)e^{-x^2/2}$$

If we integrate the right side term by term, we obtain

$$\sum_{j,k\geq 0} \frac{t^k}{k!}\frac{s^j}{j!} \int_{-\infty}^{\infty} H_k(x)H_j(x)e^{-x^2/2}\,dx = \int_{-\infty}^{\infty} e^{-x^2/2}e^{tx-t^2/2}e^{sx-s^2/2}\,dx$$

$$e^{st}\int_{-\infty}^{\infty} e^{-[x^2-2x(s+t)+(s+t)^2]/2}\,dx$$

$$= \sqrt{2\pi}\, e^{st}$$

$$= \sqrt{2\pi} \sum_{k\geq 0} \frac{(st)^k}{k!}$$

The equality of these two power series implies the equality of the corresponding coefficients, hence the orthogonality relations as stated. In particular, the functions $H_k(x)/[\sqrt{k!}(2\pi)^{1/4}]$ are orthonormal.

We now show the invariance of the Hermite polynomials under Fourier transformation, in the statement

$$(2\pi)^{-1/2} \int_{-\infty}^{\infty} H_k(x\sqrt{2})e^{-x^2/2}e^{-i\mu x}\, dx = (-i)^k H_k(\mu\sqrt{2})e^{-\mu^2/2}$$
$$k = 0, 1, 2, \ldots$$

For example, the function $(2x^2 - 1)e^{-x^2/2}$ is negatively proportional to its Fourier transform. In general, the kth function emerges with the multiplicative factor $i^k\sqrt{2\pi}$.

This general statement can also be proved by using the generating function. To do this, first we write

$$\sum_{k\geq 0} \frac{t^k}{k!} H_k(x\sqrt{2})e^{-x^2/2}e^{-i\mu x} = e^{tx\sqrt{2}-t^2/2-i\mu x-x^2/2}$$

$$= e^{-[x^2-2x(t\sqrt{2}-i\mu)]/2}e^{-t^2/2}$$

$$= e^{-[x-(t\sqrt{2}-i\mu)]^2/2}e^{\frac{1}{2}t^2-i\mu t\sqrt{2}\frac{1}{2}\mu^2}$$

Integrating term by term, we find that

$$\sum_{k\geq 0} \frac{t^k}{k!} \int_{-\infty}^{\infty} H_k(x\sqrt{2})e^{-x^2/2}e^{-i\mu x}\, dx$$

$$= \int_{-\infty}^{\infty} e^{-[x-(t\sqrt{2}-i\mu)]^2/2}\, dx\; e^{t^2/2-i\mu t\sqrt{2}-\mu^2/2}$$

$$= \sqrt{2\pi}\; e^{t^2/2-i\mu t\sqrt{2}-\mu^2/2}$$

$$= \sqrt{2\pi}\; e^{-\mu^2/2} \sum_{k\geq 0} \frac{(-it)^k}{k!} H_k(\mu\sqrt{2})$$

which gives the required identification. Here we have used the generating function with the *imaginary* parameter *it* instead of *t*.

This observation allows us to use Hermite polynomials to compute the Fourier transform. To do this, we define the *normalized Hermite functions* by $H_k(x) = H_k(x\sqrt{2})e^{-x^2/2}/\sqrt{k!}$; then we have for the Fourier transform $\hat{H}_k(\mu) = \sqrt{2\pi}i^k H_k(\mu)$. If a function f is written as a finite sum $f(x) = \Sigma a_k H_k(x)$, then the Fourier transform is $F(\mu) = \sqrt{2\pi}\,\Sigma i^k a_k H_k(\mu)$.

Hermite polynomials can also be used to solve the heat equation in one dimension with *polynomial* initial conditions. To see this, first note that the function $e^{\alpha x}e^{\alpha^2 Kt}$ is a solution of the heat equation $u_t = Ku_{xx}$ for each fixed α. As a function of α, it has derivatives of all orders. Since the heat equation is a linear homogeneous equation, it stands to reason that the Taylor coefficients at $\alpha = 0$ also should be solutions of the heat equation. To make the required connection, we write

$$e^{\alpha x + \alpha^2 Kt} = e^{(i\alpha\sqrt{2Kt}(x/i\sqrt{2Kt}) - (i\alpha\sqrt{2Kt})^2/2}$$

$$= \sum_{k \geq 0} \frac{(i\alpha\sqrt{2Kt})^k}{k!} H_k\left(\frac{x}{i\sqrt{2Kt}}\right)$$

$$= \sum_{k \geq 0} \frac{\alpha^k}{k!} (i\sqrt{2Kt})^k H_k\left(\frac{x}{i\sqrt{2Kt}}\right)$$

The coefficient of $\alpha^k/(k!)$ is a real-valued solution of the heat equation, denoted $H_k(x; t)$. In detail we have for the first few

k	$H_k(x; t)$
0	1
1	x
2	$x^2 + 2Kt$
3	$x^3 + 6Ktx$
4	$x^4 + 12x^2Kt + 12K^2t^2$

Example 5.2.4. Solve the heat equation $u_t = Ku_{xx}$ for $t > 0$, $-\infty < x < \infty$, with the initial condition $u(x; 0) = 3x^4 + 6x + 2$.

Solution. Any finite sum of the form $\Sigma a_k H_k(x; t)$ is a solution of the heat equation with initial condition $\Sigma a_k x^k$. To satisfy the stated initial conditions, we set $a_0 = 2$, $a_1 = 6$, $a_4 = 3$, and $a_k = 0$ for all other values of k. The solution is $u(x; t) = 2H_0(x; t) + 6H_1(x; t) + 3H_4(x; t) = 2 + 6x + 3(x^4 + 12x^2Kt + 12K^2t^2)$.   ●

EXERCISES 5.2

1. Show that the solution formula (5.2.5) can be written in the form

$$u(x; t) = \int_{-\infty}^{\infty} f(x - z\sqrt{2Kt}) \frac{e^{-z^2/2}}{\sqrt{2\pi}} \, dz$$

   by making the substitution $x - \xi = z\sqrt{2Kt}$.

2. Use the result of Exercise 1 to prove directly that $u(x; t)$ is a solution of the equation. You may assume for this purpose that f', f'' exist.

3. Use the result of Exercise 1 to prove that $|u(x; t) - f(x)| \leq 2M\sqrt{Kt/\pi}$. You may assume for this purpose that f' exists and $|f'(x)| \leq M$ everywhere. *Hint:* Write $u(x; t) - f(x)$ as a definite integral and apply the mean-value theorem to $f(x - z\sqrt{2Kt}) - f(x)$.

4. Show that for any real number θ,

$$|\cos \theta - 1| \leq \frac{\theta^2}{2} \qquad |\sin \theta - \theta| \leq \frac{\theta^2}{2}$$

   Hint: $1 - \cos \theta = \int_0^\theta \int_0^{\theta_1} \cos \varphi \, d\varphi \, d\theta_1 \qquad \sin \theta - \theta = -\int_0^\theta \int_0^{\theta_1} \sin \varphi \, d\varphi \, d\theta_1$

5. Complete the details of the proof of formulas (5.2.4b) and (5.2.4c).
6. Apply the method of images to solve the initial-value problem

$$u_t = Ku_{xx} \qquad t > 0, \, x > 0$$

$$u(0; t) = 0 \qquad t > 0$$

$$u(x; 0) = \begin{cases} 1 & 0 \le x \le L_1 \\ 0 & x > L_1 \end{cases}$$

Show that $u(x; t) = O(t^{-3/2})$ when $t \to \infty$. Hint: $|e^{-a} - e^{-b}| \le |a - b|$ if $a, b \ge 0$.
7. Apply the method of images to solve the initial-value problem

$$u_t = Ku_{xx} \qquad t > 0, \, x > 0$$

$$u_x(0; t) = 0 \qquad t > 0$$

$$u(x; 0) = \begin{cases} 1 & 0 \le x \le L_1 \\ 0 & x > L_1 \end{cases}$$

Show that $u(x; t) = O(t^{-1/2})$ when $t \to \infty$.
8. Consider the following initial-value problem for a heat equation with transport term:

$$u_t = Ku_{xx} + vu_x \qquad t > 0, \, -\infty < x < \infty$$

$$u(x; 0) = f(x)$$

where K is the conductivity and v is a nonzero constant which represents the mean velocity of the diffusing substance.
(a) Use the method of Fourier transforms to solve the problem in the form

$$u(x; t) = \int_{-\infty}^{\infty} e^{i\mu x} e^{-\mu^2 K t} e^{i v \mu t} F(\mu) \, d\mu$$

(b) Show that this can be rewritten in the form

$$u(x; t) = \int_{-\infty}^{\infty} \frac{\exp\left[-(x - y + vt)^2/(4Kt)\right]}{\sqrt{4\pi Kt}} f(y) \, dy$$

(c) Show that $u(x; t) = O(e^{-v^2/(4K)})$, $t \to \infty$. You may suppose for this purpose that $f(y) = 0$ for $|y| \ge 5.3 \times 10^6$.
9. Show that (5.2.7) can be written in the form

$$u(x; t) = \int_0^{\infty} F_s(\mu) e^{-\mu^2 K t} \sin \mu x \, d\mu$$

where $F_s(\mu)$ is the Fourier sine transform of f.
10. Show that (5.2.8) can be written in the form

$$u(x; t) = \int_0^{\infty} F_c(\mu) e^{-\mu^2 K t} \cos \mu x \, d\mu$$

where $F_c(\mu)$ is the Fourier cosine transform of f.
11. Solve the heat equation $u_t = Ku_{xx}$ with the initial conditions $u(x; 0) = T_1$ if $x < 0$ and $u(x; 0) = 0$ if $x > 0$. Show that the level curves $u(x; t) = C$ are half-parabolas

passing through $(0, 0)$ in the xt plane. Plot these level curves if $K = \frac{1}{2}$, $T_1 = 100$ for the values $C = 10$, $C = 30$, $C = 50$.

12. Two materials of the same conductivity K and temperatures T_1, T_2 are brought together at $t = 0$. Find the time τ^* such that $|u(x; \tau^*) - (\frac{1}{2}T_1 + \frac{1}{2}T_2)| < 0.2|T_1 - T_2|$, where $u(x; t)$ is the solution of the heat equation.

13. Solve the heat equation $u_t = Ku_{xx}$ with the initial conditions $u(x; 0) = T_1$ if $-L < x < 0$, $u(x; 0) = T_2$ if $0 < x < L$, and $u(x; 0) = 0$ if $|x| > L$. What is $\lim_{t \to \infty} u(x; t)$?

14. Verify directly that $u(x; t)$, defined by (5.2.7), satisfies the boundary condition $u(0; t) = 0$.

15. Verify directly that $u(x; t)$, defined by (5.2.8), satisfies the boundary condition $u_x(0; t) = 0$.

Hermite polynomials

16. Use the generating function for Hermite polynomials to prove the equations $H'_k(x) = kH_{k-1}(x)$, $k = 1, 2, \ldots$.

17. Use the generating function for Hermite polynomials to prove the equations $H_{k+1}(x) = xH_k(x) - kH_{k-1}(x)$, $k = 1, 2, \ldots$.

18. Combine the results of Exercises 16 and 17 to prove the following differential equations satisfied by the Hermite polynomials: $H''_k(x) - xH'_k(x) = kH_k(x)$, $k = 0, 1, 2, \ldots$.

19. Consider the partial differential equation $u_t = u_{xx} - xu_x$.
    (a) Show that separated solutions may be obtained in the form $u(x; t) = e^{-kt}H_k(x)$, $k = 0, 1, 2, \ldots$.
    (b) Show directly or by using the generating function that for any complex number α, the function $\exp(\alpha xe^{-t} - \frac{1}{2}\alpha^2 e^{-2t})$ is a solution.
    (c) Find a Fourier representation of the solution of the equation $u_t = u_{xx} - xu_x$ for $t > 0$, $-\infty < x < \infty$, with $u(x; 0) = f(x) = \int_{-\infty}^{\infty} F(\mu)e^{i\mu x} d\mu$. (Hint: Take $\alpha = i\mu$ in (b) and form a continuous superposition with respect to a suitable function of μ.)
    (d) Find an explicit representation of $u(x; t)$ by writing $F(\mu) = (2\pi)^{-1} \int f(x)e^{-i\mu x} dx$, changing the order of integration, and using the known Fourier transform of the normal density.
    [sol: $u(x; t) = (\sqrt{2\pi}(1 - e^{-2t}))^1 \int_{-\infty}^{\infty} \exp[-\frac{1}{2}(\xi - xe^{-t})^2/(1 - e^{-2t})]f(\xi) d\xi]$

20. Use Exercises 16 and 17 to show that the functions $\psi_k(x) = e^{-x^2/4}H_k(x)$ satisfy the following "ladder equations" $\psi'_k(x) + \frac{1}{2}x\psi_k(x) = k\psi_{k-1}(x)$ for $k = 1, 2, \ldots$ and $\psi'_k(x) - \frac{1}{2}x\psi_k(x) = -\psi_{k+1}(x)$ for $k = 0, 1, 2, \ldots$.

21. Show that the functions $\psi_k(x) = e^{-x^2/4}H_k(x)$ satisfy the differential equations $\psi''_k(x) - \frac{1}{4}x^2\psi_k(x) = -(k + \frac{1}{2})\psi_k$, $k = 0, 1, 2, \ldots$.

22. Carry through the computations of Exercise 19a for the partial differential equation $u_t = u_{xx} - \frac{1}{4}x^2u$.

23. Solve the heat equation $u_t = Ku_{xx}$ for $t > 0$, $-\infty < x < \infty$, with the initial conditions $u(x; 0) = 2 + 3x + 2x^3 + 6x^4$.

24. Use the definition of the Hermite polynomials to prove:

    (a) $\displaystyle\sum_{k \geq 0} \frac{t^k}{k!} e^{-x^2/2} H_k(x) = e^{-(x-t)^2/2}$

(b) $e^{-x^2/2}H_k(x) = (-1)^k \left(\dfrac{d}{dx}\right)^k (e^{-x^2/2})$ $k = 0, 1, 2, \ldots$

25. Show that the functions $H_k(x)H_l(y)e^{-(k+l)t}$ satisfy the partial differential equation $u_t = u_{xx} + u_{yy} - (xu_x + yu_y)$. Generalize to three variables.

26. Show that the Fourier transform of $H_k(x)e^{-x^2/2}$ is proportional to $(i\mu)^k e^{-\mu^2/2}$.

27. (*Generalization.*) If $a \neq 0, 1, -1$, show that the Fourier transform of $H_k(ax)e^{-x^2/2}$ is of the form $(\text{const.})H_k(\alpha\mu)e^{-\mu^2/2}$, where α is real if $|a| > 1$ and α is imaginary if $|a| < 1$. Show that $\alpha = a$ if and only if $a^2 = 2$.

28. Show that the Hermite polynomials are given explicitly by

$$H_{2n}(x) = \sum_{0 \le k \le n} \frac{(-1)^k (2n)! x^{2n-2k}}{2^k k! (2n-2k)!}$$

$$H_{2n+1}(x) = \sum_{0 \le k \le n} \frac{(-1)^k (2n+1)! x^{2n-2k+1}}{2^k k! (2n-2k+1)!}$$

29. Use the result of Exercise 28 to show that the Hermite polynomials are bounded in the form

$$H_{2n}(x) \le \frac{(2n)! 2^n (1 + |x|)^{2n}}{n!}$$

$$H_{2n+1}(x) \le \frac{(2n+1)! 2^{n+1} (1 + |x|)^{2n+1}}{n!}$$

The following exercise illustrates the application of the methods of this section to a problem involving spherically symmetric solutions of the three-dimensional heat equation. (Suggested by Professor Leon Karp)

30. Let $u(r; t)$ be the solution of the three-dimensional heat equation $u_t = K(u_{rr} + (2/r)u_r)$ defined for $r > R$, $t > 0$, and satisfying the boundary condition $u(R; t) = 0$ for $t > 0$ and the initial condition $u(r; 0) = T$ for $R_1 < r < R_2$ and $u(r; 0) = 0$ if either $R < r \le R_1$ or $r \ge R_2$. Here R, R_1, and R_2 are constants with $0 < R < R_1 < R_2$ and $T > 0$. The problem is to show that when $t \to \infty$ we have the asymptotic behavior

$$u(r; t) = \frac{A(r-R)T}{(4\pi Kt)^{3/2} r} \left(1 + O\left(\frac{1}{t}\right)\right)$$

where the constant $A = 4\pi[(R_2^3 - R_1^3)/3 - R(R_2^2 - R_1^2)/2]$

Some hints: (a) Define the new function $w(r; t) = ru(r; t)$ and check that w satisfies the one-dimensional heat equation $w_t = Kw_{rr}$ with the same boundary condition and the initial condition that $w(r; 0) = rT$ for $R_1 < r < R_2$ and $w(r; 0) = 0$ otherwise.

(b) Use the method of images to obtain the representation

$$w(x + R; t) = \frac{1}{(4\pi Kt)^{1/2}} \int_0^\infty \left[e^{-\frac{(x-y)^2}{4Kt}} - e^{-\frac{(x+y)^2}{4Kt}} \right] w(y + R; 0) \, dy$$

$$= \frac{1}{(4\pi Kt)^{1/2}} \int_{R_1-R}^{R_2-R} \left[\frac{e^{-(x-y)^2}}{4Kt} - \frac{e^{-(x+y)^2}}{4Kt} \right] (y + R)T \, dy$$

(c) Use the Taylor expansion of the exponential function $e^{-a} = 1 - a + O(a^2)$ ($a \to 0$) to obtain the indicated result.

5.3 SOLUTIONS OF THE WAVE EQUATION AND LAPLACE'S EQUATION

One-Dimensional Wave Equation and d'Alembert's Formula

The Fourier transform also can be applied to solve initial-value problems for the wave equation. The simplest problem of this type is

$$y_{tt} = c^2 y_{xx} \qquad t > 0, \ -\infty < x < \infty$$

$$y(x; 0) = f_1(x) \qquad -\infty < x < \infty \tag{5.3.1}$$

$$y_t(x; 0) = f_2(x) \qquad -\infty < x < \infty$$

We are considering a pure initial-value problem, the *Cauchy problem; $f_1(x)$, $f_2(x)$* are prescribed functions which can be thought of as the initial position and velocity of an infinite vibrating string. We follow the second method of Sec. 5.2.

To solve (5.3.1), we introduce the Fourier transforms

$$F_i(\mu) = \frac{1}{2\pi} \int_{-\infty}^{\infty} e^{-i\mu x} f_i(x) \, dx \qquad i = 1, 2 \tag{5.3.2}$$

with the inversion formulas

$$f_i(x) = \int_{-\infty}^{\infty} e^{i\mu x} F_i(\mu) \, d\mu \qquad i = 1, 2 \tag{5.3.3}$$

Finally we need the Fourier representation of the solution y

$$y(x; t) = \int_{-\infty}^{\infty} Y(\mu; t) e^{i\mu x} \, d\mu \tag{5.3.4}$$

where $Y(\mu; t)$ is an unknown function, which we will now determine. To do this, we substitute (5.3.4) into the wave equation (5.3.1). Thus

$$0 = \int_{-\infty}^{\infty} [Y_{tt}(\mu; t) + c^2 \mu^2 Y(\mu; t)] e^{i\mu x} \, d\mu$$

This requires that Y be a solution of the ordinary differential equation

$$Y_{tt} + c^2 \mu^2 Y = 0$$

i.e.,

$$Y(\mu; t) = A(\mu) \cos \mu ct + B(\mu) \sin \mu ct \tag{5.3.5}$$

To find $A(\mu)$, $B(\mu)$, we set $t = 0$ in (5.3.4):

$$y(x; 0) = \int_{-\infty}^{\infty} A(\mu) e^{i\mu x} \, d\mu \qquad y_t(x; 0) = \int_{-\infty}^{\infty} \mu c B(\mu) e^{i\mu x} \, d\mu$$

Comparing this with (5.3.3) and the initial condition (5.3.1), we see that we must have

$$F_1(\mu) = A(\mu) \qquad F_2(\mu) = \mu c B(\mu)$$

Substituting these into (5.3.5) and returning to (5.3.4), we have the *Fourier representation*

$$y(x; t) = \int_{-\infty}^{\infty} \left[F_1(\mu) \cos \mu ct + F_2(\mu) \frac{\sin \mu ct}{\mu c} \right] e^{i\mu x} d\mu \qquad (5.3.6)$$

This is the desired representation by Fourier transforms. Using this, we can also obtain an explicit representation in terms of the given functions $f_1(x)$, $f_2(x)$. To do this, recall that

$$\cos \theta = \tfrac{1}{2}(e^{i\theta} + e^{-i\theta}) \qquad \sin \theta = \frac{1}{2i}(e^{i\theta} - e^{-i\theta})$$

Thus,

$$\int_{-\infty}^{\infty} F_1(\mu)(\cos \mu ct)e^{i\mu x} d\mu = \frac{1}{2} \int_{-\infty}^{\infty} F_1(\mu)(e^{i\mu ct} + e^{-i\mu ct})e^{i\mu x} d\mu$$

$$= \frac{1}{2} \int_{-\infty}^{\infty} F_1(\mu)(e^{i\mu(x+ct)} + e^{i\mu(x-ct)} d\mu$$

$$= \tfrac{1}{2}[f_1(x + ct) + f_1(x - ct)]$$

Similarly,

$$\int_{-\infty}^{\infty} F_2(\mu) \frac{\sin \mu ct}{\mu c} e^{i\mu x} d\mu = \frac{1}{2} \int_{-\infty}^{\infty} F_2(\mu) \frac{e^{i\mu ct} - e^{-i\mu ct}}{i\mu c} e^{i\mu x} d\mu$$

$$= \frac{1}{2} \int_{-\infty}^{\infty} F_2(\mu) \frac{e^{i\mu(x+ct)} - e^{i\mu(x-ct)}}{i\mu c} d\mu$$

$$= \frac{1}{2c} \int_{-\infty}^{\infty} F_2(\mu) \left(\int_{x-ct}^{x+ct} e^{i\mu\xi} d\xi \right) d\mu$$

$$= \frac{1}{2c} \int_{x-ct}^{x+ct} \left[\int_{-\infty}^{\infty} e^{i\mu\xi} F_2(\mu) d\mu \right] d\xi$$

$$= \frac{1}{2c} \int_{x-ct}^{x+ct} f_2(\xi) d\xi$$

Putting these together yields *d'Alembert's formula*

$$y(x; t) = \tfrac{1}{2}[f_1(x + ct) + f_1(x - ct] + \frac{1}{2c} \int_{x-ct}^{x+ct} f_2(\xi) d\xi \qquad (5.3.7)$$

This was derived in Chap. 2 in the context of the vibrating string, where we had the boundary conditions $y(0; t) = 0 = y(L; t)$. We now see that the formula is, in fact, more general. The first term corresponds to a superposition of two traveling waves, with velocities $\pm c$. The second term has a similar interpretation.

We can show that (5.3.7) solves the initial-value problem (5.3.1) without any recourse to the Fourier transform. For this purpose we assume that f_1 has two continuous derivatives and f_2 has one continuous derivative. (This is in marked contrast with the heat equation, where piecewise smooth initial data are sufficient to ensure that the solution is differentiable to all orders whenever $t > 0$.) The details are left for Exercise 3.

In many applications of the wave equation we would like to apply d'Alembert's formula in the case where f_1, f_2 are only piecewise smooth. For this purpose we enlarge the concept of solution in the following way. A function $y(x; t)$ is said to be a *weak solution* of the wave equation if there exists a sequence of ordinary solutions $y_n(x; t)$ such that $y(x; t) = \lim_{n \to \infty} y_n(x; t)$. With this extended concept of solution, we can say that (5.3.7) provides a weak solution of the wave equation for *any piecewise smooth* f_1, f_2. The solutions $y_n(x; t)$ may be taken to be the integrals (5.3.6) where the range of integration is restricted to $-n \leq \mu \leq n$.

The d'Alembert formula can be combined with the method of images to solve initial-value problems for the wave equation in the half-line $x > 0$. This corresponds to an infinitely long vibrating string with boundary conditions imposed at one end. For example, we consider the initial-value problem

$$y_{tt} = c^2 y_{xx} \qquad t > 0, x > 0$$

$$y(0; t) = 0 \qquad t > 0$$

$$y(x; 0) = f(x) \qquad x > 0$$

$$y_t(x; 0) = 0 \qquad x > 0$$

corresponding to an infinite string which is tied down at the end $x = 0$. To solve this problem, we extend f as an odd function to $x < 0$ by defining $\overline{f}(-x) = -\overline{f}(x)$ for all x and $\overline{f}(x) = f(x)$ for $x > 0$. Substituting in d'Alembert's formula, we have

$$y(x; t) = \tfrac{1}{2}[\overline{f}(x + ct) + \overline{f}(x - ct)]$$

This function satisfies the boundary conditions, since $y(0; t) = \tfrac{1}{2}[\overline{f}(ct) + \overline{f}(-ct)] = 0$ and $\overline{f}$ is an odd function.

If, instead of the boundary condition $y(0; t) = 0$, we have the boundary condition $y_x(0; t) = 0$ (meaning that the end of the string at $x = 0$ is free to move), then the solution of the wave equation with the same initial conditions is given by the same formula $y(x; t) = \tfrac{1}{2}\overline{f}(x + ct) + \tfrac{1}{2}\overline{f}(x - ct)$, where now $\overline{f}$ is the *even* extension of f, that is, the function $\overline{f}$ defined for all x and satisfying $\overline{f}(-x) = \overline{f}(x)$ for all x while $\overline{f}(x) = f(x)$ for $x > 0$.

General Solution of the Wave Equation

The Fourier method can be used to find the *general solution* of the wave equation $y_{tt} = c^2 y_{xx}$. Writing the solution in the Fourier representation $y(x; t) =$

$\int_{-\infty}^{\infty} Y(\mu; t)e^{i\mu x} \, d\mu$, we substitute in the wave equation to obtain

$$0 = y_{tt} - c^2 y_{xx} = \int_{-\infty}^{\infty} (Y_{tt} + c^2\mu^2 Y)e^{i\mu x} \, d\mu$$

The equation $Y_{tt} + c^2\mu^2 Y = 0$ has the general solution $Y(\mu; t) = C(\mu)e^{i\mu ct} + D(\mu)e^{-i\mu ct}$. This will be the Fourier transform of a real function if $C(-\mu) = \overline{C}(\mu)$, $D(-\mu) = \overline{D}(\mu)$. The solution $y(x; t)$ can be computed as

$$y(x; t) = \int_{-\infty}^{\infty} Y(\mu; t)e^{i\mu x} \, d\mu$$

$$= \int_{-\infty}^{\infty} [C(\mu)e^{i\mu ct} + D(\mu)e^{-i\mu ct}]e^{i\mu x} \, d\mu$$

$$= \int_{-\infty}^{\infty} [C(\mu)e^{i\mu(x+ct)} + D(\mu)e^{i\mu(x-ct)}] \, d\mu$$

$$= f(x + ct) + g(x - ct)$$

where $f(x) = \int_{-\infty}^{\infty} C(\mu)e^{i\mu x} \, d\mu$, $g(x) = \int_{-\infty}^{\infty} D(\mu)e^{i\mu x} \, d\mu$ are the inverse Fourier transforms of $C(\mu)$, $D(\mu)$. This formula, which has been derived by the Fourier transform, is in fact more general, as the following proposition states.

Proposition 5.3.1. Let f, g be continuous functions for which f', g', f'', g'' exist. Then $y(x; t) = f(x + ct) + g(x - ct)$ is a solution of the wave equation. Furthermore, any solution is of this form.

Proof. The first part is an immediate verification, since $y_x(x; t) = f'(x + ct) + g'(x - ct)$, $y_{xx}(x; t) = f''(x + ct) + g''(x - ct)$, $y_t(x; t) = cf'(x + ct) - cg'(x - ct)$, $y_{tt}(x; t) = c^2 f''(x + ct) + c^2 g''(x - ct)$. The second part is proved by introducing characteristic coordinates $\xi = x - ct$, $\eta = x + ct$. If $y(x; t)$ is a solution of the wave equation with continuous second-order partial derivatives, let $z(\xi; \eta) = y(x; t)$. By the chain rule for partial derivatives,

$$y_t = \frac{\partial y}{\partial t} = \frac{\partial z}{\partial \xi}\frac{\partial \xi}{\partial t} + \frac{\partial z}{\partial \eta}\frac{\partial \eta}{\partial t} = -c\frac{\partial z}{\partial \xi} + c\frac{\partial z}{\partial \eta}$$

$$y_{tt} = \frac{\partial^2 y}{\partial t^2}$$

$$= -c\left(\frac{\partial^2 z}{\partial \xi^2}\frac{\partial \xi}{\partial t} + \frac{\partial^2 z}{\partial \xi \partial \eta}\frac{\partial \eta}{\partial t} + c\frac{\partial^2 z}{\partial \eta \partial \xi}\frac{\partial \xi}{\partial t} + \frac{\partial^2 z}{\partial \eta^2}\frac{\partial \eta}{\partial t}\right)$$

$$= c^2\frac{\partial^2 z}{\partial \xi^2} - c^2\frac{\partial^2 z}{\partial \xi \partial \eta} - c^2\frac{\partial^2 z}{\partial \eta \partial \xi} + c^2\frac{\partial^2 z}{\partial \eta^2}$$

$$y_x = \frac{\partial y}{\partial x} = \frac{\partial z}{\partial \xi}\frac{\partial \xi}{\partial x} + \frac{\partial z}{\partial \eta}\frac{\partial \eta}{\partial x} = \frac{\partial z}{\partial \xi} + \frac{\partial z}{\partial \eta}$$

$$y_{xx} = \frac{\partial^2 y}{\partial x^2} = \frac{\partial^2 z}{\partial \xi^2} + \frac{\partial^2 z}{\partial \xi \partial \eta} + \frac{\partial^2 z}{\partial \eta \partial \xi} + \frac{\partial^2 z}{\partial \eta^2}$$

The equation $y_{tt} - c^2 y_{xx} = 0$ translates to $\partial^2 z/(\partial \xi\, \partial \eta) = 0$. This may be solved by two integrations. We have $(\partial/\partial\eta)(\partial z/\partial\xi) = 0$, which requires $\partial z/\partial\xi = \varphi(\xi)$, independent of η. A further integration gives $z(\xi; \eta) = \varphi(\eta) + \int_{\xi_0}^{\xi} \varphi(\xi') \, d\xi'$. Recalling that $\xi = x - ct$, $\eta = x + ct$, we see that $z(\xi; \eta) = f(x + ct) + g(x - ct)$, where $f(\eta) = \psi(\eta)$, $g'(\xi) = \varphi(\xi)$. The proof is complete. ∎

The general solution of the one-dimensional wave equation may be used to solve problems with a time-dependent boundary condition.

Example 5.3.1. Find the solution of the wave equation $y_{tt} - c^2 y_{xx} = 0$ for $x > 0$, $t > 0$ satisfying the initial conditions $u(x; 0) = 0$, $u_t(x; 0) = 0$ for $x > 0$ and the boundary condition $u(0; t) = s(t)$.

Solution. We look for y in the form $f(x + ct) + g(x - ct)$. Substituting the initial conditions and boundary condition, we must have

$$f(x) + g(x) = 0 \qquad x > 0$$

$$cf'(x) - cg'(x) = 0 \qquad x > 0$$

$$f(ct) + g(-ct) = s(t) \qquad t > 0$$

Differentiating the first equation and using the second, we see that $f'(x) = 0$ and $g'(x) = 0$ for $x > 0$; hence both $f(x)$ and $g(x)$ are constant for $x > 0$. From the first equation we have $f(x) = C$, $g(x) = -C$ for $x > 0$. The third equation gives $g(-ct) = s(t) - f(ct) = s(t) - C$ for $t > 0$. Thus $g(x) = -C$ for $x > 0$ and $g(x) = s(-x/c) - C$ for $x < 0$. Substituting these in the general solution form for $y(x; t)$, we have $f(x + ct) + g(x - ct) = C - C = 0$ for $x - ct > 0$ and $f(x + ct) + g(x - ct) = C + s[(ct - x)/c] - C = s(t - x/c)$ for $x - ct < 0$. Thus,

$$y(x; t) = \begin{cases} 0 & x - ct > 0 \\ s\left(t - \dfrac{x}{c}\right) & x - ct < 0 \end{cases}$$

If we think of $s(t)$ as a signal emitted from the source $x = 0$, the result states that an observer positioned at $x > 0$ does not sense the signal until $t = x/c$ time units have elapsed and that, after that, the signal is received verbatim, without distortion or damping. The graphs in Fig. 5.3.1 give the solution for several values of t in the case $s(t) = 3 \sin 2t$. ●

Example 5.3.1 can be combined with d'Alembert's formula to solve a general initial-value problem for the wave equation in the half-space $x > 0$. Consider, e.g., the problem

$$y_{tt} = c^2 y_{xx} \qquad t > 0, \, x > 0$$

$$y(0; t) = s(t) \qquad t > 0$$

$$y(x; 0) = f(x) \qquad x > 0$$

$$y_t(x; 0) = 0 \qquad x > 0$$

The solution is $y(x; t) = \frac{1}{2}\overline{f}(x + ct) + \frac{1}{2}\overline{f}(x - ct)$ for $x > ct$ and $y(x; t) = \frac{1}{2}\overline{f}(x + ct) + \frac{1}{2}\overline{f}(x - ct) + s(t - x/c)$ for $x < ct$.

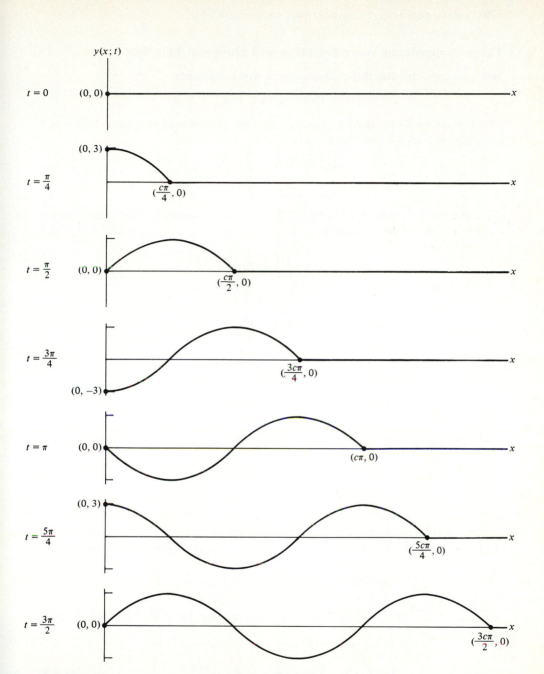

FIGURE 5.3.1
Solution of the wave equation.

Three-Dimensional Wave Equation and Huygens' Principle

We now consider the three-dimensional wave equation

$$u_{tt} = c^2 \nabla^2 u = c^2(u_{xx} + u_{yy} + u_{zz})$$

which is satisfied by the components of the electric and magnetic fields in a vacuum. The initial conditions are

$$u(x, y, z; 0) = f_1(x, y, z)$$

$$u_t(x, y, z; 0) = f_2(x, y, z)$$

To solve this problem, we introduce the *three-dimensional Fourier transforms* F_1, F_2, U through the formulas

$$f_1(x, y, z) = \int\!\!\!\int\!\!\!\int_{-\infty}^{\infty} F_1(\mu)e^{i\langle\mu,x\rangle}\, d\mu$$

$$f_2(x, y, z) = \int\!\!\!\int\!\!\!\int_{-\infty}^{\infty} F_2(\mu)e^{i\langle\mu,x\rangle}\, d\mu$$

$$u(x, y, z; t) = \int\!\!\!\int\!\!\!\int_{-\infty}^{\infty} U(\mu; t)e^{i\langle\mu,x\rangle}\, d\mu$$

where we have used the inner-product notation $\langle\mu, x\rangle = \mu_1 x + \mu_2 y + \mu_3 z$ and $d\mu = d\mu_1\, d\mu_2\, d\mu_3$. Substituting this form for u in the wave equation, we must have

$$0 = \int\!\!\!\int\!\!\!\int_{-\infty}^{\infty} (U_{tt} + c^2|\mu|^2 U)e^{i\langle\mu,x\rangle}\, d\mu = 0 \qquad |\mu|^2 = \mu_1^2 + \mu_2^2 + \mu_3^2$$

Thus U is the solution of the second-order equation $U_{tt} + c^2|\mu|^2 U = 0$ with the initial conditions $U(\mu; 0) = F_1(\mu)$, $U_t(\mu; 0) = F_2(\mu)$. Solving, we have

$$U(\mu; t) = F_1(\mu) \cos ct|\mu| + F_2(\mu) \frac{\sin ct|\mu|}{c|\mu|}$$

Substituting in the formula for $u(x, y, z; t)$, we have the desired Fourier representation

$$u(x, y, z; t) = \int\!\!\!\int\!\!\!\int_{-\infty}^{\infty} \left[F_1(\mu) \cos ct|\mu| + F_2(\mu) \frac{\sin ct|\mu|}{c|\mu|} \right] e^{i\langle\mu,x\rangle}\, d\mu \qquad (5.3.8)$$

Example 5.3.2. Use the Fourier representation (5.3.8) to solve the three-dimensional wave equation $u_{tt} = c^2 \nabla^2 u$ if $u(x, y, z; 0) = 0$, $u_t(x, y, z; 0) = 1$ for $x^2 + y^2 + z^2 < R^2$ and zero otherwise.

Solution. We must compute the three-dimensional Fourier transform F_2. This is given by

$$F_2(\mu) = \left(\frac{1}{2\pi}\right)^3 \iiint\limits_{|x| \le R} \exp\left(-i\langle\mu, x\rangle\right) dx$$

$$= \left(\frac{1}{2\pi}\right)^3 \int_{-\pi}^{\pi} \int_0^{\pi} \int_0^R \exp\left(-i|\mu|r \cos\theta\right) r^2 \sin\theta \, dr \, d\theta \, d\varphi$$

$$= \left(\frac{1}{2\pi}\right)^2 \int_0^R \frac{\exp\left(-i|\mu|r \cos\theta\right)}{i|\mu|r}\bigg|_{\theta=0}^{\theta=\pi} r^2 \, dr$$

$$= \frac{1}{2\pi^2|\mu|} \int_0^R (\sin |\mu|r) r \, dr$$

$$= \frac{1}{2\pi^2|\mu|} \left(\frac{\sin |\mu|R}{|\mu|^2} - \frac{R \cos |\mu|R}{|\mu|}\right)$$

Substituting this in (5.3.8) gives the desired Fourier representation. ●

We now derive an explicit representation, the three-dimensional generalization of d'Alembert's formula. To do this, we begin with the following surface integral formula on a sphere of radius R:

$$\iint\limits_{|\xi|=R} e^{i\langle\mu,\xi\rangle} d\xi = 4\pi R^2 \frac{\sin |\mu|R}{|\mu|R} \qquad |\mu| \ne 0$$

If $|\mu| = 0$, the surface integral is $4\pi R^2$, the area of the surface of the sphere. We use this with $R = ct$ to write

$$e^{i\langle\mu,x\rangle} \frac{\sin ct|\mu|}{c|\mu|} = \frac{1}{4\pi tc^2} \iint\limits_{|\xi|=ct} e^{i\langle\mu,x+\xi\rangle} d\xi$$

We multiply this equation by $F_i(\mu)$, integrate $d\mu$, and formally interchange the order of integration to obtain†

$$\int\!\!\!\int\!\!\!\int_{-\infty}^{\infty} F_i(\mu) \frac{\sin ct|\mu|}{c|\mu|} e^{i\langle\mu,x\rangle} d\mu = \frac{1}{4\pi tc^2} \iint\limits_{|\xi|=ct} \left[\int\!\!\!\int\!\!\!\int_{-\infty}^{\infty} e^{i\langle\mu,x+\xi\rangle} F_i(\mu) \, d\mu\right] d\xi$$

$$= \frac{1}{4\pi tc^2} \iint\limits_{|\xi|=ct} f_i(x + \xi) \, d\xi$$

$$= t M_{ct} f_i \qquad i = 1, 2 \tag{5.3.9}$$

---

† This can be made rigorous by inserting the factor $e^{-\epsilon|\mu|2}$ inside the integral and taking the limit $\epsilon \downarrow 0$ after the interchange.

where we have defined the *mean-value operator* by

$$M_R f(x) = \frac{1}{4\pi R^2} \iint\limits_{|\xi|=R} f(x + \xi)\, d\xi$$

Thus we have obtained an explicit representation of the f_2 term of the solution (5.3.8); to handle the f_1 term, we differentiate the above identity (5.3.9) with respect to t:

$$\int\limits_{-\infty}^{\infty}\!\!\!\int\!\!\int F_i(\mu)\, (\cos |\mu| ct) e^{i\langle \mu, x\rangle}\, d\mu = \frac{d}{dt}\, (tM_{ct}f_i) \qquad i = 1, 2 \qquad (5.3.10)$$

Combining this with the first formula, we have the desired explicit representation:

$$\boxed{u(x, y, z; t) = \frac{d}{dt}\, (tM_{ct}f_1) + tM_{ct}f_2} \qquad (5.3.11)$$

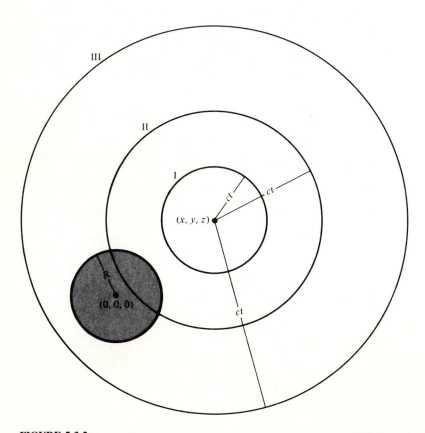

FIGURE 5.3.2
Illustrating Huygens' principle; f_1 and f_2 are zero outside the sphere of radius R about $(0, 0, 0)$. If $ct > R + \sqrt{x^2 + y^2 + z^2}$, then f_1 and f_2 are zero on the surface of integration (labeled III).

This representation displays an important property of the three-dimensional wave equation, the *Huygens' principle*, stated as follows: If f_1 and f_2 are zero for $(x^2 + y^2 + z^2)^{1/2} \geq R$, then $u(x, y, z; t) = 0$ whenever $ct > R + (x^2 + y^2 + z^2)^{1/2}$. This results from the fact that (5.3.11) contains only *surface integrals* over a sphere of a radius ct and that for t sufficiently large the surface of integration does not intersect the set where $f_1 \neq 0$, $f_2 \neq 0$. This is illustrated in Figure 5.3.2.

Application to One- and Two-Dimensional Wave Equations

Formula (5.3.11) can be used to solve the two-dimensional wave equation $u_{tt} = c^2(u_{xx} + u_{yy})$ with initial conditions $u(x, y; 0) = f_1(x, y)$, $u_t(x, y; 0) = f_2(x, y)$. Considering these as functions of (x, y, z), we substitute in (5.3.11) and perform the required integrations. If these are written as integrals in the xy plane, we obtain integrals over the *interior* of the circle of radius ct. This has the consequence that Huygens' principle is *not valid* for wave motion in two dimensions. For example, a pebble dropped in a pond of shallow water will create a wave motion on the surface of the water; an observer positioned r units away from the initial disturbance will sense the disturbance at $t = r/c$ time units later. After this time the solution will continue to be nonzero for all later times, according to the wave equation. This is the phenomenon of a *wake* behind the initial disturbance, present in two-dimensional wave motion. Huygens' principle can be restated to say that, in three-dimensional wave motion, no wake is present.

We now give the details of the passage from the three-dimensional wave equation to the two-dimensional wave equation, known as the *method of descent*. We are given two functions $f_1(x, y)$ and $f_2(x, y)$, and we are looking for the solution $u(x, y; t)$ of the wave equation with $u(x, y; 0) = f_1(x, y)$ and $u_t(x, y; 0) = f_2(x, y)$. To do this, we substitute in the explicit representation formula (5.3.11). Each of the terms involves the mean-value operator M_{ct}. To study its effect on a function of two variables, we note that both upper and lower hemispheres of the sphere $|\xi| = ct$ can be parameterized by the formulas $\xi_3 = \pm\sqrt{(ct)^2 - \xi_1^2 - \xi_2^2}$, and the element of area is transformed as

$$d\xi = \sqrt{\left(1 + \left(\frac{\partial\xi_3}{\partial\xi_1}\right)^2 + \left(\frac{\partial\xi_3}{\partial\xi_2}\right)^2\right)} \, d\xi_1 \, d\xi_2$$

$$= \frac{ct \, d\xi_1 \, d\xi_2}{\sqrt{(ct)^2 - \xi_1^2 - \xi_2^2}}$$

The mean-value operator is a sum of the integral over the upper hemisphere and the integral over the lower hemisphere, which produces a factor of 2 and the formula

$$(M_{ct}f)(P) = \frac{1}{2\pi ct} \iint\limits_{|\xi| < ct} \frac{f(P + \xi) \, d\xi}{\sqrt{(ct)^2 - |\xi|^2}} \tag{5.3.12}$$

This is then substituted into (5.3.11) to obtain the solution. In detail, we have

$$u(P; t) = \frac{1}{2\pi c} \left[\frac{d}{dt} \iint\limits_{|\xi| < ct} \frac{f_1(P + \xi)}{(ct)^2 - |\xi|^2} \, d\xi + \iint\limits_{|\xi| < ct} \frac{f_2(P + \xi)}{\sqrt{(ct)^2 - |\xi|^2}} \, d\xi \right]$$

Example 5.3.3. Show that formula (5.3.11) reduces to d'Alembert's formula in the case where $f_1 = 0$, $f_2(x, y, z) = f(x)$.

Solution. If $f(x, y, z) = f(x)$, then the surface integral can be written as a single integral as follows: For any R,

$$M_R f(x, y, z) = \frac{1}{4\pi R^2} \iint\limits_{|\xi| = R} f(x + \xi_1) \, d\xi$$

$$= \frac{1}{4\pi} \int_{-\pi}^{\pi} \int_0^\pi f(x + R \cos \theta) \sin \theta \, d\theta \, d\varphi$$

$$= \frac{1}{2R} \int_{-R}^{R} f(x + \xi_1) \, d\xi_1$$

Thus, $$t M_{ct} f_2(x, y, z) = \frac{1}{2c} \int_{-ct}^{ct} f(x + \xi_1) \, d\xi_1 = \frac{1}{2c} \int_{x-ct}^{x+ct} f(x') \, dx'$$

which is the appropriate form of d'Alembert's formula in this case.   ●

Laplace's Equation in a Half-Plane; Poisson's Formula

We now illustrate the solution of Laplace's equation by Fourier transforms. We will solve the following boundary-value problem for Laplace's equation in the half-plane $y > 0$

$$u_{xx} + u_{yy} = 0 \qquad -\infty < x < \infty, \, y > 0$$

$$u(x, 0) = f(x) \qquad -\infty < x < \infty$$

$$|u(x, y)| \leq M$$

where $f(x)$ is a piecewise smooth function with $\int_{-\infty}^{\infty} |f(x)| \, dx < \infty$.

To do this, we use the second method of Sec. 5.2. Introducing the Fourier transform formulas $f(x) = \int_{-\infty}^{\infty} F(\mu) e^{i\mu x} \, d\mu$, $F(\mu) = [1/(2\pi)] \int_{-\infty}^{\infty} f(x) e^{-i\mu x} \, dx$, we look for the solution in the form $u(x, y) = \int_{-\infty}^{\infty} F(\mu, y) e^{i\mu x} \, d\mu$. Laplace's equation is

$$0 = u_{xx} + u_{yy} = \int_{-\infty}^{\infty} [-\mu^2 F(\mu, y) + F_{yy}(\mu, y)] e^{i\mu x} \, d\mu$$

Thus F must satisfy the ordinary differential equation $F_{yy}(\mu, y) = \mu^2 F(\mu, y)$ and the initial condition $F(\mu, 0) = F(\mu)$ for each μ. The general solution of the ordinary differential equation is of the form $Ae^{\mu y} + Be^{-\mu y}$. If we impose the initial condition and the boundedness condition, the solution must be

$$F(\mu, y) = \begin{cases} F(\mu)e^{-\mu y} & \text{if } \mu \geq 0 \\ F(\mu)e^{\mu y} & \text{if } \mu < 0 \end{cases}$$

This can be succinctly expressed by using the absolute value as $F(\mu)e^{-|\mu|y}$; thus,

$$u(x, y) = \int_{-\infty}^{\infty} F(\mu)e^{-|\mu|y}e^{i\mu x} \, d\mu$$

This is the desired Fourier representation of the solution.

To obtain an explicit representation, we insert the formula $F(\mu) = [1/(2\pi)] \int_{-\infty}^{\infty} f(\xi)e^{-i\mu\xi} \, d\xi$ into the Fourier representation and formally interchange the order of integration. Thus

$$u(x, y) = \frac{1}{2\pi} \int_{-\infty}^{\infty} \left[\int_{-\infty}^{\infty} f(\xi)e^{-i\mu\xi} \, d\xi \right] e^{-|\mu|y}e^{i\mu x} \, d\mu$$

$$= \frac{1}{2\pi} \int_{-\infty}^{\infty} \left(\int_{-\infty}^{\infty} e^{i\mu(x-\xi)}e^{-|\mu|y} \, d\mu \right) f(\xi) \, d\xi$$

The inner integral is

$$\int_{-\infty}^{\infty} e^{i\mu(x-\xi)}e^{-|\mu|y} \, d\mu = 2 \operatorname{Re} \int_{0}^{\infty} e^{i\mu(x-\xi)}e^{-\mu y} \, d\mu$$

$$= 2 \operatorname{Re} \frac{1}{y - i(x - \xi)}$$

$$= \frac{2y}{y^2 + (x - \xi)^2}$$

Therefore the explicit representation is

$$\boxed{u(x, y) = \frac{1}{\pi} \int_{-\infty}^{\infty} \frac{y}{y^2 + (x - \xi)^2} f(\xi) \, d\xi} \qquad (5.3.13)$$

This is closely related to the Poisson integral formula, obtained in Chap. 2. The final result (5.3.13) makes sense if the function $f(x)$, $-\infty < x < \infty$, is merely bounded and piecewise continuous. The solution is valid in this greater generality, which we summarize as follows.

Theorem 5.3.1. Let $f(x)$, $-\infty < x < \infty$, be bounded and piecewise continuous. Then the integral (5.3.13) defines a solution of Laplace's equation $u_{xx} + u_{yy} = 0$ for $y > 0$, $-\infty < x < \infty$, with $\lim_{y \downarrow 0} u(x, y) = \frac{1}{2}f(x + 0) + \frac{1}{2}f(x - 0)$. This solution is unique among all bounded solutions: $|u(x, y)| \leq M$ for $y > 0$, $-\infty < x < \infty$.

If we allow unbounded solutions, then we lose uniqueness. For example,

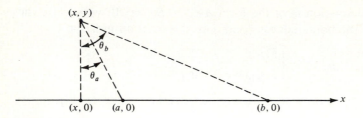

FIGURE 5.3.3
Solution of Laplace's equation.

the function $u(x, y) = y$ satisfies Laplace's equation in the half-plane and $\lim_{y \downarrow 0} u(x, y) = 0$. This can be added to any solution to violate uniqueness.

The proof of this theorem involves differentiation under the integral sign, as in the case of the Gauss-Weierstrass integral (5.2.5). The details are left for Exercises 19 and 20.

Example 5.3.4. Find the bounded solution of Laplace's equation $u_{xx} + u_{yy} = 0$ in the half-plane $y > 0$ and satisfying the boundary conditions $u(x, 0) = 1$ if $a < x < b$ and $u(x, 0) = 0$ otherwise.

Solution. Referring to formula (5.3.13), we have

$$u(x, y) = \frac{1}{\pi} \int_a^b \frac{y}{(x - \xi)^2 + y^2} \, d\xi$$

Making the substitution $v = (\xi - x)/y$, we have $d\xi = y \, dv$, and the denominator of the integrand is $y^2(1 + v^2)$. Changing the limits of integration accordingly, we have

$$u(x, y) = \frac{1}{\pi} \int_{(a-x)/y}^{(b-x)/y} \frac{1}{1 + v^2} \, dv$$

$$= \frac{1}{\pi} \left(\arctan \frac{b - x}{y} - \arctan \frac{a - x}{y} \right)$$

$$= \frac{1}{\pi} (\theta_b - \theta_a)$$

where the angles θ_a, θ_b are depicted in Figure 5.3.3. ●

EXERCISES 5.3

1. Use d'Alembert's formula to solve the wave equation $y_{tt} = c^2 y_{xx}$ with the initial conditions $y(x; 0) = 3 \sin 2x$, $y_t(x; 0) = 0$.

2. Use d'Alembert's formula to solve the wave equation $y_{tt} = c^2 y_{xx}$ with the initial conditions $y(x; 0) = 0$, $y_t(x; 0) = 4 \cos 5x$.

3. Suppose that f_1 has two continuous derivatives and f_2 has one continuous derivative. Show that (5.3.7) is a solution of the initial-value problem (5.3.1).

4. Find the solution of the wave equation $y_{tt} = c^2 y_{xx}$ for $t > 0$, $x > 0$ satisfying the boundary conditions $y(0; t) = 0$ and the initial conditions $y(x; 0) = 0$, $y_t(x; 0) = g(x)$.

5. Find the solution of the wave equation $y_{tt} = c^2 y_{xx}$ for $t > 0$, $x > 0$ satisfying the boundary conditions $y(0; t) = s(t)$ and the initial conditions $y(x; 0) = 0$, $y_t(x; 0) = g(x)$.

6. Show that formula (5.3.11) reduces to d'Alembert's formula in the case where $f_1(x, y, z) = f_1(x)$, $f_2 = 0$.

7. Use formula (5.3.11) to solve the initial-value problem for the three-dimensional wave equation when the initial data f_1, f_2 depend only on $r = \sqrt{x^2 + y^2 + z^2}$: $f_1(x, y, z) = f_1(r)$, $f_2(x, y, z) = f_2(r)$.

8. The oscillations of a gas satisfy the three-dimensional wave equation $u_{tt} = c^2 \nabla^2 u$ with $u(x, y, z; 0) = 0$, $u_t(x, y, z; 0) = T$ if $a^2 \geq x^2 + y^2 + z^2$ and zero otherwise. Find the solution of this initial-value problem.

9. Find the solution of the wave equation $y_{tt} = c^2 y_{xx}$ for $t > 0$, $x > 0$ with the boundary condition $y(0; t) = 4 \cos t$ and the initial conditions $y(x; 0) = 0$, $y_t(x; 0) = 0$. Sketch the solution for several different values of t.

10. Find the solution of the wave equation $y_{tt} = c^2 y_{xx}$ for $t > 0$, $x > 0$ with the boundary conditions $y(0; t) = 3$ for $0 < t < 5$ and $y(0; t) = 0$ for $t > 5$ and the initial conditions $y(x; 0) = 0$, $y_t(x; 0) = 0$. Sketch the solution for several different values of t.

11. Find the solution of the wave equation $y_{tt} = c^2 y_{xx}$ for $t > 0$, $x > 0$ with the boundary condition $y_x(0; t) = 0$ and the initial conditions $y(x; 0) = 1$ for $0 < x < 3$, $y(x; 0) = 0$ for $x > 3$, and $y_t(x; 0) = 0$ for all $x > 0$. Sketch the solution for several different values of t.

12. Find the bounded solution of Laplace's equation $u_{xx} + u_{yy} = 0$ in the half-plane $y > 0$ and satisfying the boundary conditions $u(x, 0) = 2$ if $-4 < x < 4$ and $u(x, 0) = 0$ otherwise.

13. Consider the problem of solving Laplace's equation $u_{xx} + u_{yy} = 0$ in the quadrant $x > 0$, $y > 0$ with the boundary conditions $u(x, 0) = f(x)$, $u(0, y) = 0$. By combining the method of images with (5.3.12), find an explicit representation of the solution.

14. Consider the problem of solving Laplace's equation $u_{xx} + u_{yy} = 0$ in the quadrant $x > 0$, $y > 0$ with the boundary conditions $u(x, 0) = f(x)$, $u_x(0, y) = 0$. By combining the method of images with (5.3.13), find an explicit representation of the solution.

15. Consider the problem of solving Laplace's equation $u_{xx} + u_{yy} = 0$ in the quadrant $x > 0$, $y > 0$ with the boundary conditions $u(x, 0) = f(x)$, $u(0, y) = f(y)$. Apply Exercise 13 twice to find an explicit representation of the solution.

16. Find the separated solutions of Laplace's equation $u_{xx} + u_{yy} = 0$ in the strip $0 < x < L$, $0 < y < \infty$ and satisfying the boundary conditions $u(0, y) = 0$, $u(L, y) = 0$.

17. Consider the problem of solving Laplace's equation $u_{xx} + u_{yy} = 0$ in the strip $0 < x < L$, $0 < y < \infty$ with the boundary conditions $u(x, 0) = f(x)$, $u(0, y) = 0$, $u(L, y) = 0$. Find the Fourier (series) representation of the bounded solution of this problem.

18. Consider the problem of solving Laplace's equation $u_{xx} + u_{yy} = 0$ in the strip $0 < x < L$, $-\infty < y < \infty$, with the boundary conditions $u(0, y) = 0$, $u(L, y) = g(y)$. Find a Fourier representation of the bounded solution of this problem.

19. Let $P(x, y; \xi) = y/[y^2 + (x - \xi)^2]$. Show that P satisfies Laplace's equation $P_{xx} + P_{yy} = 0$ for each ξ, $-\infty < \xi < \infty$, when $y > 0$, $-\infty < x < \infty$.

20. Let $f(x)$, $-\infty < x < \infty$, be a bounded and piecewise continuous function. Define $u(x, y)$ by the integral (5.3.13).

    (a) Prove that

$$u_x(x, y) = \frac{1}{\pi} \int_{-\infty}^{\infty} f(\xi) P_x(x, y; \xi) \, d\xi$$

$$u_{xx}(x, y) = \frac{1}{\pi} \int_{-\infty}^{\infty} f(\xi) P_{xx}(x, y; \xi) \, d\xi$$

$$u_y(x, y) = \frac{1}{\pi} \int_{-\infty}^{\infty} f(\xi) P_y(x, y; \xi) \, d\xi$$

$$u_{yy}(x, y) = \frac{1}{\pi} \int_{-\infty}^{\infty} f(\xi) P_{yy}(x, y; \xi) \, d\xi$$

(b) Conclude that u satisfies Laplace's equation $u_{xx} + u_{yy} = 0$.

(c) Make a change of variable to show that

$$u(x, y) = \frac{1}{\pi} \int_{0}^{\infty} \frac{1}{1 + z^2} [f(x + yz) + f(x - yz)] \, dz$$

(d) Conclude that $\lim_{y \downarrow 0} u(x, y) = \frac{1}{2} f(x + 0) + \frac{1}{2} f(x - 0)$ for each x.

21. Let $f(x) = 1$ for $x > 0$ and $f(x) = 0$ for $x < 0$. Show that $\lim_{x \downarrow 0} u(x, cx)$ depends on c. Does this contradict Exercise 20d?

22. (a) Find the Fourier representation of the solution of the two-dimensional wave equation $u_{tt} - c^2(u_{xx} + u_{yy}) = 0$ with $u(x, y; 0) = 0$, $u_t(x, y; 0) = f(x, y)$.

    (b) By comparing the Fourier representation with the explicit representation in terms of the mean-value operator (5.3.12), deduce the following two-dimensional Fourier transform:

$$\frac{1}{2\pi ct} \iint_{|\xi| < ct} \frac{e^{i(\mu_1 \xi_1 + \mu_2 \xi_2)}}{\sqrt{(ct)^2 - \xi_1^2 - \xi_2^2}} \, d\xi_1 \, d\xi_2 = \frac{\sin t \sqrt{\mu_1^2 + \mu_2^2}}{t \sqrt{\mu_1^2 + \mu_2^2}}$$

5.4 SOLUTION OF THE TELEGRAPH EQUATION FOR AN INFINITE CABLE

The flow of electricity in a cable is described by the partial differential equation

$$LCu_{tt} + (GL + RC)u_t + RGu = u_{xx}$$

where R is the resistance, L is the inductance, C is the capacitance, and G is the leakage, all measured per unit length of cable. The unknown function $u(x; t)$ may be the voltage or the current at time t, measured at the x coordinate of the cable. The derivation of the telegraph equation is carried out in Appendix 5A.

To put this in a more convenient form, let

$$2\beta = \frac{R}{L} - \frac{G}{C} \qquad c^2 = \frac{1}{CL} \qquad \alpha = \frac{RG}{CL}$$

resulting in the equation

$$u_{tt} + 2\beta u_t + \alpha u = c^2 u_{xx} \qquad (5.4.1a)$$

From the defining equations, we have $\beta^2 - \alpha = \frac{1}{4}(R/L - S/G)^2$. Therefore we need only solve equation (5.4.1a) for values of α and β for which $\beta^2 - \alpha \geq 0$. To illustrate the mathematical methods, we will solve (5.4.1a) for arbitrary values of α, β. This will include, for example, the Klein-Gordon equation, which occurs in quantum mechanics, as well as the wave equation, which has been treated in Sec. 5.3.

Since the telegraph equation is second-order in time, it is natural to specify two initial conditions:

$$u(x; 0) = f_1(x) \qquad u_t(x; 0) = f_2(x) \qquad (5.4.1b)$$

Since the equation is linear and homogeneous, we can solve first with $f_1 = 0$ then with $f_2 = 0$ and add the results.

Solution by Fourier Transforms†

To solve the equation, we apply the second method from Sec. 5.2. We write $u(x; t)$ in terms of its Fourier transform

$$u(x; t) = \int_{-\infty}^{\infty} U(\mu; t)e^{i\mu x} d\mu$$

and formally apply the operations (5.4.1a):

$$u_{tt} + 2\beta u_t + \alpha u - c^2 u_{xx} = \int_{-\infty}^{\infty} (U_{tt} + 2\beta U_t + \alpha U + c^2\mu^2 U)e^{i\mu x} d\mu$$

Therefore we solve the ordinary differential equation

$$U_{tt} + 2\beta U_t + (\alpha + c^2\mu^2)U = 0$$
$$U(\mu; 0) = 0 \qquad U_t(\mu; 0) = F_2 \qquad (5.4.2)$$

To solve this equation with constant coefficients, we look for solutions of the form $e^{\gamma t}$ and obtain the quadratic equation

$$\gamma^2 + 2\beta\gamma + (\alpha + c^2\mu^2) = 0$$

with the solutions

$$\gamma = -\beta \pm (\beta^2 - c^2\mu^2 - \alpha)^{1/2}$$

---

† The explicit representation of the solution is developed in Sec. 8.5.

We consider separately three cases:

> **Case 1:** $\beta^2 - \alpha < 0$
>
> **Case 2:** $\beta^2 - \alpha = 0$
>
> **Case 3:** $\beta^2 - \alpha > 0$

In case 1 the solution of (5.4.2) is

$$U(\mu;\, t) = F_2(\mu)e^{-\beta t}\, \frac{\sin t[\alpha - \beta^2 + (c\mu)^2]^{1/2}}{[\alpha - \beta^2 + (c\mu)^2]^{1/2}}$$

Thinking of the separated solutions $U(\mu;\, t)e^{i\mu x}$ as a difference of two plane waves

$$F_2(\mu)e^{-\beta t} \exp i\{\mu x \pm t[\alpha - \beta^2 + (c\mu)^2]^{1/2}\}$$

we see that each of these propagates with phase velocity

$$v_\mu = \frac{1}{\mu}\, [\alpha - \beta^2 + (c\mu)^2]^{1/2}$$

while damped by the factor $e^{-\beta t}$.

In case 2 the Fourier transform is

$$U(\mu;\, t) = F_2(\mu)e^{-\beta t}\, \frac{\sin c\mu t}{c\mu}$$

In this case we have the plane waves

$$e^{-\beta t}e^{i\mu(x \pm ct)}$$

each of which propagates with phase velocity c, damped by the factor $e^{-\beta t}$.

In case 3 the Fourier transform is

$$U(\mu;\, t) = \begin{cases} F_2(\mu)e^{-\beta t}\, \dfrac{\sin t[(c\mu)^2 - (\beta^2 - \alpha)]^{1/2}}{[(c\mu)^2 - (\beta^2 - \alpha)]^{1/2}} & |c\mu| \geq \sqrt{\beta^2 - \alpha} \\[4mm] F_2(\mu)e^{-\beta t}\, \dfrac{\sinh t[\beta^2 - \alpha - (c\mu)^2]^{1/2}}{[\beta^2 - \alpha - (c\mu)^2]^{1/2}} & |c\mu| < \sqrt{\beta^2 - \alpha} \end{cases}$$

In this case we have two kinds of solutions. If $|c\mu| \geq \sqrt{\beta^2 - \alpha}$, the solutions are plane waves with phase velocity $(1/\mu)[(c\mu)^2 - (\beta^2 - \alpha)]^{1/2}$, damped by the factor $e^{-\beta t}$; if $|c\mu| < \sqrt{\beta^2 - \alpha}$, the solutions are simply damped exponentials

$$\exp t\{-\beta \pm [\beta^2 - \alpha - (c\mu)^2]^{1/2}\}$$

Using this information, we can now write the solution in each of the three cases.

Case 1. $\beta^2 < \alpha$

$$u(x;\, t) = e^{-\beta t} \int_{-\infty}^{\infty} F_2(\mu)\, \frac{\sin t[(\alpha - \beta^2) + (c\mu)^2]^{1/2}}{[(\alpha - \beta^2) + (c\mu)^2]^{1/2}}\, e^{i\mu x}\, d\mu$$

Case 2. $\beta^2 = \alpha$

$$u(x; t) = e^{-\beta t} \int_{-\infty}^{\infty} F_2(\mu) \frac{\sin \mu c t}{\mu c} e^{i\mu x} \, d\mu$$

Case 3. $\beta^2 > \alpha$

$$u(x; t) = e^{-\beta t} \int_{|c\mu| \geq \sqrt{\beta^2 - \alpha}} F_2(\mu) \frac{\sin t[(c\mu)^2 - (\beta^2 - \alpha)]^{1/2}}{[(c\mu)^2 - (\beta^2 - \alpha)]^{1/2}} e^{i\mu x} \, d\mu$$

$$+ e^{-\beta t} \int_{|c\mu| < \sqrt{\beta^2 - \alpha}} F_2(\mu) \frac{\sinh t[\beta^2 - \alpha - (c\mu)^2]^{1/2}}{[\beta^2 - \alpha - (c\mu)^2]^{1/2}} e^{i\mu x} \, d\mu$$

We have thus obtained the desired Fourier representation of the solution of the telegraph equation.

We now verify that these formal solutions are indeed rigorous solutions of the problem (5.4.1a,b) with $f_1 = 0$. For this purpose we assume that the Fourier transform of the initial data satisfies

$$\int_{-\infty}^{\infty} |\mu| |F_2(\mu)| \, d\mu < \infty$$

(This will happen if, for example, f_2 has three continuous derivatives which are absolutely integrable.) With this hypothesis we can follow the arguments of Sec. 5.2 and take the derivatives under the integral sign. For example, in case 1,

$$(e^{\beta t} u)_x = \int_{-\infty}^{\infty} F_2(\mu) \frac{\sin t[\alpha - \beta^2 + (c\mu)^2]^{1/2}}{[\alpha - \beta^2 + (c\mu)^2]^{1/2}} i\mu e^{i\mu x} \, d\mu$$

$$(e^{\beta t} u)_{xx} = \int_{-\infty}^{\infty} F_2(\mu) \frac{\sin t[\alpha - \beta^2 + (c\mu)^2]^{1/2}}{[\alpha - \beta^2 + (c\mu)^2]^{1/2}} (i\mu)^2 e^{i\mu x} \, d\mu$$

$$(e^{\beta t} u)_{tt} = \int_{-\infty}^{\infty} F_2(\mu) \frac{\sin t[\alpha - \beta^2 + (c\mu)^2]^{1/2}}{[\alpha - \beta^2 + (c\mu)^2]^{1/2}} [\alpha - \beta^2 + (c\mu)^2] e^{i\mu x} \, d\mu$$

Clearly $(e^{\beta t} u)_{tt} + c^2 (e^{\beta t} u)_{xx} + (\alpha - \beta^2)(e^{\beta t} u) = 0$, which is equivalent to the telegraph equation.

In case 3 the analysis is the same. In case 2 we can do better, since, apart from the factor $e^{-\beta t}$, this is just the Fourier representation (5.3.6) of the solution of the wave equation, which can be rewritten as

$$u(x; t) = \frac{e^{-\beta t}}{2c} \int_{x-ct}^{x+ct} f_2(\xi) \, d\xi$$

If f_2' exists, this is a solution of the telegraph equation.

Uniqueness of the Solution

We now discuss the *uniqueness* of our solution. For this purpose, let u_1, u_2 be two solutions which have the same initial data and are zero for large x, depending on t. (This is a natural assumption since telegraph signals are expected to move with a finite velocity.) Letting $v = (u_1 - u_2)e^{\beta t}$, we see that v satisfies the equation $v_{tt} + (\alpha - \beta^2)v = c^2 v_{xx}$ with zero initial conditions. We now introduce the *energy functional*

$$E(t) = \frac{1}{2} \int_{-\infty}^{\infty} (v_t^2 + c^2 v_x^2 + v^2) \, dx$$

Differentiating with respect to t, using the equation for v, and integrating by parts, we have

$$E'(t) = \int_{-\infty}^{\infty} (v_t v_{tt} + c^2 v_x v_{xt} + v v_t) \, dx$$

$$= \int_{-\infty}^{\infty} \{v_t[c^2 v_{xx} - (\alpha - \beta^2)v] + c^2 v_x v_{xt} + v v_t\} \, dx$$

$$= -c^2 \int_{-\infty}^{\infty} v_x v_{tx} \, dx + (1 + \beta^2 - \alpha) \int_{-\infty}^{\infty} v v_t \, dx + c^2 \int_{-\infty}^{\infty} v_x v_{xt} \, dx$$

$$= (1 + \beta^2 - \alpha) \int_{-\infty}^{\infty} v v_t \, dx$$

But

$$v v_t \leq \tfrac{1}{2}(v^2 + v_t^2)$$

Therefore,

$$E'(t) \leq bE(t) \qquad b = \frac{1 + \beta^2 - \alpha}{2}$$

Thus,

$$\frac{d}{dt}[e^{-bt}E(t)] = [E'(t) - bE(t)]e^{-bt} \leq 0$$

But v and v_t are both zero for $t = 0$; hence

$$E(0) = 0$$

Applying the fundamental theorem of calculus gives

$$e^{-bt}E(t) \leq 0$$

But $E(t)$ is the integral of a nonnegative function, hence nonnegative. Therefore we conclude that

$$E(t) = 0$$

This means that all the terms in the integrand of $E(t)$ are also zero, in particular v_t. Integrating once more and using the fundamental theorem, we see that v itself is identically zero, and we have proved uniqueness.   ∎

EXERCISES 5.4

1. Let $u(x; t)$ be a solution of the telegraph equation $u_{tt} + 2\beta u_t + \alpha u = c^2 u_{xx}$. Show that $v(x; t) = e^{\beta t} u(x; t)$ is a solution of the equation $v_{tt} - c^2 v_{xx} + (\alpha - \beta^2)v = 0$.
2. Let $v(x; t)$ be a solution of the equation $v_{tt} - c^2 v_{xx} + (\alpha - \beta^2)v = 0$. Show that $u(x; t) = e^{-\beta t} v(x; t)$ is a solution of the telegraph equation $u_{tt} + 2\beta u_t + \alpha u = c^2 u_{xx}$.
3. Solve the initial-value problem for the telegraph equation

$$u_{tt} + 2\beta u_t + \alpha u = c^2 u_{xx} \qquad t > 0, \ -\infty < x < \infty$$

$$u(x; 0) = f_1(x) \qquad -\infty < x < \infty$$

$$u_t(x; 0) = 0 \qquad -\infty < x < \infty$$

Consider separately the cases $\alpha > \beta^2$, $\alpha = \beta^2$, $\alpha < \beta^2$. [*Hint:* Solve (5.4.2) with $U(\mu; 0) = F_1(\mu)$, $U_t(\mu; 0) = 0$.]
4. Let $f(x)$ be a smooth function whose derivatives $f^{(i)}$ satisfy

$$\int_{-\infty}^{\infty} |f^{(i)}(x)| \, dx < \infty \qquad i = 0, 1, 2, 3$$

where $f^{(0)} = f$. Let F be the Fourier transform of f. Show that

$$\int_{-\infty}^{\infty} |\mu||F(\mu)| \, d\mu < \infty$$

Hint: Integrate by parts the formula (5.1.6) three times to obtain $F(\mu) = O(|\mu|^{-3})$, $|\mu| \to \infty$.]
*5. Consider the three-dimensional telegraph equation

$$u_{tt} + 2\beta u_t + \alpha u = c^2 \nabla^2 u$$

with initial conditions

$$u(x, y, z; 0) = 0 \qquad u_t(x, y, z; 0) = f_2(x, y, z)$$

Find a Fourier representation of the solution.
6. Find all complex separated solutions $u(x; t) = e^{i\omega t + \gamma x}$ of the telegraph equation $u_{tt} + 2\beta u_t + \alpha u = c^2 u_{xx}$ where ω is real and positive.
7. Find the bounded solution of the telegraph equation $u_{tt} + 2\beta u_t + \alpha u = c^2 u_{xx}$ for $x > 0$, $-\infty < t < \infty$, with $u(0; t) = A \cos \omega t$, where A, ω are real and positive.

APPENDIX 5A
DERIVATION OF THE TELEGRAPH EQUATION

In this appendix we derive a system of two equations satisfied by the current and voltage along a one-dimensional conducting medium, which may be thought of as a cable along the x axis.

We are given the resistance, inductance, capacitance, and leakage parameters, denoted, respectively, R, L, C, G. These are supposed to be constants, independent of position and time. The unknown functions are the voltage $v(x; t)$ and the current $i(x; t)$, measured at a point x of the cable at time t. We suppose that these are continuous functions of both variables with continuous first partial derivatives v_x, v_t, i_x, i_t. To derive the necessary equations, we consider the voltage loss and the current loss along a section of the cable $x_1 \leq x \leq x_2$.

The voltage loss comes in two parts. First, there is a voltage loss that is proportional to the current in the cable, where the proportionality factor is (defined to be) the resistance. Second, there is a voltage loss which is proportional to the rate of change of the current, where the proportionality factor is (defined to be) the inductance. Putting these together, we have

$$v(x_1; t) - v(x_2; t) = R \int_{x_1}^{x_2} i(x; t) \, dx + L \int_{x_1}^{x_2} i_t(x; t) \, dx$$

But the fundamental theorem of calculus gives the alternative expression

$$v(x_1; t) - v(x_2; t) = - \int_{x_1}^{x_2} v_x(x; t) \, dx$$

Subtracting these two equations, we have for any segment of the cable

$$\int_{x_1}^{x_2} [v_x(x; t) + Ri(x; t) + Li_t(x; t)] \, dx = 0$$

Since the functions are assumed to be continuous, we conclude that the integrand is identically zero, leading to the first equation of the system

$$v_x(x; t) + Ri(x; t) + Li_t(x; t) = 0$$

To deduce the second equation of the system, we consider the change in current in a section of the cable. The current changes on two accounts. First, there is a current loss that is proportional to the voltage level, where the proportionality factor is (defined to be) the leakage. Second, there is a current loss that is proportional to the rate of change of the voltage, where the constant of proportionality is (defined to be) the capacitance. Putting these together, we have for the current loss along this section $G \int_{x_1}^{x_2} v(x; t) \, dx + C \int_{x_1}^{x_2} v_t(x; t) \, dx$. But the basic relation between charge and current indicates that the above quantity is the difference in the current. Thus

$$i(x_1; t) - i(x_2; t) = G \int_{x_1}^{x_2} v(x; t) \, dx + C \int_{x_1}^{x_2} v_t(x; t) \, dx$$

The fundamental theorem of calculus gives us the alternative formula

$$i(x_1; t) - i(x_2; t) = - \int_{x_1}^{x_2} i_x(x; t) \, dx$$

Subtracting these two equations, we have for any segment of the cable

$$\int_{x_1}^{x_2} [i_x(x; t) + Gv(x; t) + Cv_t(x; t)] \, dx = 0$$

Since the functions are assumed to be continuous, we conclude that the integrand is identically zero, leading to the second equation of the system $i_x + Gv + Cv_t = 0$.

The system of equations may be written as

$$i_x + Cv_t + Gv = 0 \qquad (5A.1)$$

$$v_x + Li_t + Ri = 0 \qquad (5A.2)$$

To obtain the desired second-order equation, we assume that both $v(x; t)$ and $i(x; t)$ have continuous second partial derivatives v_{xx}, v_{tt}, v_{xt} ($= v_{tx}$), i_{xx}, i_{tt}, i_{xt} ($= i_{tx}$). To obtain the equation satisfied by the voltage $v(x; t)$, we take the t derivative of (5A.1), multiply by L, and subtract the x derivative of (5A.2). The term Li_{xt} cancels, and we are left with the equation $LCv_{tt} + GLv_t = v_{xx} + Ri_x$. But Ri_x can be obtained from (5A.1) in terms of v alone as $Ri_x = -RCv_t - RGv$, leading to the telegraph equation

$$LCv_{tt} + (GL + RC)v_t + RGv = v_{xx}$$

To obtain the equation satisfied by the current $i(x; t)$, we proceed similarly, taking the x derivative of (5A.1) and subtracting C times the t derivative of (5A.2). We obtain the equation $i_{xx} + Gv_x = Li_{tt} + Ri_t$. Using (5A.2) again to solve for v_x and substituting, we get the same telegraph equation:

$$LCi_{tt} + (GL + RC)i_t + RGi = i_{xx}$$

CHAPTER
6

ASYMPTOTIC
SOLUTIONS

Introduction

Several times in the previous chapters we have encountered notions from asymptotic analysis. This refers to the simplification of a complicated formula which contains a parameter. When the parameter becomes very large, it is often possible to find a simpler formula which gives an extremely accurate approximation to the original formula.

In all our previous examples we were able to analyze infinite *series* by examining the largest term of the series when the time parameter became large.

In the case of functions represented by *integrals* (such as in the Fourier integral representation of the solution of the heat equation), there is no single "largest term" in the integrand which gives the asymptotic behavior. Hence we must develop more systematic methods of analysis.

In Sec. 6.2 we show that the elementary technique of integration by parts can provide asymptotic statements in many cases of interest. In more refined cases one must use *Laplace's method,* developed in Sec. 6.3. There is also a counterpart of Laplace's method for oscillatory integrands represented by imaginary exponentials, the *method of stationary phase,* developed in Sec. 6.5. These three methods are systematically applied to various integral representations of solutions of the heat equation and Laplace's equation to deduce one-term asymptotic formulas. If one desires a formula with additional terms, there is the *method of asymptotic expansions,* developed in Sec. 6.4, which extends each of the three previous methods to obtain additional terms in an asymptotic expansion. In Sec. 6.6 we combine the previous methods to obtain the asymptotic analysis of a "wave packet," obtained from the Fourier representation of the solution of the telegraph equation. Section 6.1 gives an illustration of asymptotic analysis in the familiar case of the factorial function, leading eventually to Stirling's formula, which is finally obtained rigorously in Sec. 6.4.

6.1 ASYMPTOTIC ANALYSIS OF THE FACTORIAL FUNCTION

In this section we illustrate some elementary asymptotic methods on a familiar mathematical example, the factorial function

$$n! = 1 \cdot 2 \cdot 3 \cdots n$$

This formula, although elementary in appearance, contains $n - 1$ multiplications, and its computation can be time-consuming. If we take logarithms to change the multiplications to additions, then we must have a table (or program) to compute the necessary logarithms and finally to compute the exponential. So it is clearly desirable to have a simpler formula which gives accurate approximations for large n.

To do the analysis, we write $n! = n^n e^{A(n)}$ and try to find a difference equation satisfied by the function $A(n)$ defined thus.† We have

$$\frac{e^{A(n)}}{e^{A(n-1)}} = \frac{n!/n^n}{(n-1)!/(n-1)^{n-1}}$$

$$= \left(1 - \frac{1}{n}\right)^{n-1}$$

Therefore the function $A(n)$ satisfies the difference equation

$$A(n) - A(n-1) = (n-1) \log\left(1 - \frac{1}{n}\right)$$

The Taylor expansion of the logarithm at $x = 0$ is $\log(1 + x) = x - \frac{1}{2}x^2 + O(x^3)$ which yields

$$A(n) - A(n-1) = (n-1)\left[\frac{-1}{n} - \frac{1}{2n^2} + O\left(\frac{1}{n^3}\right)\right]$$

$$= -1 + \frac{1}{2n} + O\left(\frac{1}{n^2}\right)$$

Therefore we can retrieve the function $A(n)$ as a telescoping sum:

$$A(n) = A(1) + \sum_{k=1}^{n} [A(k) - A(k-1)]$$

$$= 0 + \sum_{k=1}^{n}\left[-1 + \frac{1}{2k} + O\left(\frac{1}{k^2}\right)\right]$$

---

† A. C. King and M. E. Mortimer, "Approximating Factorials: A Difference Equation Approach," *SIAM Review*, vol. 27, 1985, pp. 565–567.

But the series $\Sigma_1^\infty (1/k^2)$ converges, and the sum $\Sigma_{k=1}^n (1/k)$ is log n plus a constant plus a term which tends to zero. Therefore

$$A(n) = -n + \tfrac{1}{2} \log n + C + O\left(\frac{1}{n}\right)$$

where the constant C is to be determined. By exponentiation we have

$$n! = n^n \exp\left[-n + \tfrac{1}{2} \log n + C + O\left(\frac{1}{n}\right)\right]$$

$$= n^{n+1/2} e^{-n} e^C \left[1 + O\left(\frac{1}{n}\right)\right]$$

The well-known form of Stirling's formula asserts that $e^C = \sqrt{2\pi}$. Our elementary method fails to reveal this fact. To make this precise identification of the constant here and in other asymptotic problems involving partial differential equations, it is extremely useful to have the desired quantity as an *integral representation*. In the case of $n!$, one suitable integral representation is the elementary integral formula

$$n! = \int_0^\infty x^n e^{-x} \, dx$$

In Sec. 6.4 we use this to obtain the precise version of Stirling's formula with $e^C = \sqrt{2\pi}$.

In the case of solutions of partial differential equations, the desired integral representation may be either the Fourier representation or the explicit representation, if it is available.

EXERCISES 6.1

1. Let $B(n)$, $n \geq 1$, be a real-valued sequence which satisfies a different equation of the form $B(n) - B(n - 1) = a + b/n + O(1/n^2)$. Prove that we have the asymptotic formula

$$B(n) = an + b \log n + C + O\left(\frac{1}{n}\right)$$

for a suitable constant C.

2. Let $B(n)$, $n \geq 1$, be a real-valued sequence which satisfies a difference equation of the form $B(n) - B(n - 1) = an + b + c/n + O(1/n^2)$. Prove that we have the asymptotic formula $B(n) = \bar{a}n^2 + \bar{b}n + \bar{c} \log n + C + O(1/n^2)$ for suitable constants $\bar{a}, \bar{b}, \bar{c}, C$, and identify $\bar{a}, \bar{b}$ in terms of (a, b). [*Hint:* Recall the formula for the sum of an arithmetic progression: $\Sigma_{k=1}^n k = \tfrac{1}{2}n(n + 1)$.]

3. This exercise is devoted to giving an independent proof of the "nth-root version" of Stirling's formula: $\sqrt[n]{n!}/n \rightarrow e^{-1}$, when $n \rightarrow \infty$.
   (a) Use the graph of $y = \log x$ to show that

$$\int_1^{n-1} \log x \, dx \leq \log 1 + \cdots + \log n \int_1^n \log x \, dx$$

TABLE 6.1.1

n	$n!$	$(n/e)^n$	$\sqrt[n]{n!}$	n/e
1	1	0.37	1	0.37
10	3.63×10^6	4.54×10^5	4.53	3.68
100	9.33×10^{157}	3.72×10^{156}	37.99	36.79
1000	4.08×10^{2567}	5.08×10^{2565}	369.49	367.88

(b) Use elementary calculus to compute $\int_1^n \log x \, dx$.

(c) Use parts (a) and (b) to show that when $n \to \infty$,

$$\frac{\log 1 + \cdots + \log n}{n} - \log n \to -1$$

(d) Conclude that $\sqrt[n]{n!}/n \to e^{-1}$ when $n \to \infty$.

4. By the use of suitable tables or other numerical aids, verify the entries in Table 6.1.1.

5. Complete Table 6.1.1 by computing the value of $(n/e)^n (2\pi n)^{1/2}$ for $n = 1, 10, 100, 1000$.

6.2   INTEGRATION BY PARTS

Statement and Proof of Result

We will study expressions of the form

$$f(t) = \int_a^b g(x) e^{th(x)} \, dx \tag{6.2.1}$$

when $t \to \infty$. Here a and b are fixed limits of integration (independent of t), one of which may be infinite, but not both; $g(x)$ and $h(x)$ are smooth functions with $h(x)$ real-valued, while $g(x)$ may be either real- or complex-valued. The main hypothesis of this section is

$$h'(x) \neq 0 \qquad a \leq x \leq b$$

In particular, $h(x)$ has no interior maximum for $a < x < b$. If $a = -\infty$ or $b = +\infty$, we also require that $h(x) \to -\infty$ fast enough that the resulting improper integral converges. The main result is

$$\boxed{f(t) = \frac{e^{tH}}{t} \left[C + O\left(\frac{1}{t}\right) \right] \qquad t \to \infty} \tag{6.2.2}$$

where

$$H = \max_{a \leq x \leq b} h(x) = \max \, [h(a), h(b)]$$

$$C = \begin{cases} \dfrac{g(b)}{h'(b)} & \text{if } h'(x) > 0 \\[2mm] \dfrac{-g(a)}{h'(a)} & \text{if } h'(x) < 0 \end{cases}$$

In other words, the integral behaves as the largest exponential in the integrand does, softened by a factor of $1/t$. We are already familiar with such phenomena for a discrete sum containing a large parameter, e.g.,

$$2^t + 3^t + 5^t = 5^t[1 + O(\tfrac{3}{5})^t] \qquad t \to \infty$$

In the continuous case we have infinitely many summands, hence the additional complexity of the factor $1/t$.

To prove (6.2.2), we multiply and divide the integrand by $th'(x)$ and obtain

$$f(t) = \int_a^b \frac{g(x)}{th'(x)} \, d(e^{th(x)}) = \frac{g(x)e^{th(x)}}{th'(x)} \bigg|_a^b - \frac{1}{t} \int_a^b e^{th(x)} \left(\frac{g}{h'} \right)' dx$$

The second integral is of the same form as the original integral (6.2.1) for $f(t)$, but with g replaced by $(g/h')'$. If we now apply the identical procedure to this integral, we obtain

$$\int_a^b e^{th(x)} \left(\frac{g}{h'} \right)' dx = \frac{1}{th'} \left(\frac{g}{h'} \right)' e^{th(x)} \bigg|_a^b - \frac{1}{t} \int_a^b e^{th(x)} \left[\frac{1}{h'} \left(\frac{g}{h'} \right)' \right]' dx$$

But $e^{th(x)} \le e^{tH}$ for all $a \le x \le b$, and therefore the above expression is $O(e^{tH}/t)$ when $t \to \infty$. Therefore we have shown that

$$f(t) = \frac{g(x)e^{th(x)}}{th'(x)} \bigg|_a^b + O\left(\frac{e^{tH}}{t^2} \right) \qquad t \to \infty$$

where $h(x)$ is a continuous function which assumes its maximum at one of the endpoints $x = a$ or $x = b$ but not at both. If $h'(x) < 0$, the maximum is assumed at $x = a$; otherwise, at $x = b$. In either case we have the stated result.

Two Applications

Example 6.2.1. Find an asymptotic formula for the *complementary error function*, defined by the integral

$$1 - \Phi(y) = \frac{1}{\sqrt{2\pi}} \int_y^\infty e^{-x^2/2} \, dx$$

Solution. The integral is not presented in the form (6.2.1). To transform it to this form, make the change of variable $u = x/y$, with the result

$$1 - \Phi(y) = \frac{y}{\sqrt{2\pi}} \int_1^\infty e^{-u^2 y^2/2} \, du$$

In this integral we make the identifications $a = 1$, $b = \infty$, $t = y^2$, $h(u) = -\tfrac{1}{2}u^2$, $g(u) = 1$. The maximum of $h(u)$ is $H = h(1) = -\tfrac{1}{2}$, while $C = -g(1)/h'(1) = 1$.

Applying (6.2.2), we have

$$1 - \Phi(y) = \frac{y}{\sqrt{2\pi}} \frac{e^{-t/2}}{t} \left[1 + O\left(\frac{1}{t}\right) \right]$$

$$= \frac{1}{y\sqrt{2\pi}} e^{-y^2/2} \left[1 + O\left(\frac{1}{y^2}\right) \right] \qquad y \to \infty$$

The accuracy of this approximation can be inferred from the following table of values:†

y	$1 - \Phi(y)$	$e^{-y^2/2}/(y\sqrt{2\pi})$
2.25	1.22×10^{-2}	1.41×10^{-2}
3.00	1.30×10^{-3}	1.48×10^{-3}
3.75	8.82×10^{-5}	9.41×10^{-5}
4.50	3.45×10^{-6}	3.56×10^{-6}

●

We now turn to an example involving the solution of the heat equation when $t \to \infty$.

Example 6.2.2. Two materials of the same conductivity K are initially at temperatures T_1, T_2. Find asymptotic formulas for the temperature $u(x; t)$ when $t \to 0$, $t \to \infty$.

Solution. In Example 5.2.3 we found the exact solution

$$u(x; t) = T_2\Phi\left(\frac{x}{\sqrt{2Kt}}\right) + T_1\left[1 - \Phi\left(\frac{x}{\sqrt{2Kt}}\right)\right]$$

To analyze the solution for $t \to \infty$, we use the Taylor expansion of $\Phi(z)$ about $z = 0$:

$$\Phi(z) = \Phi(0) + z\Phi'(0) + O(z^2) \qquad z \to 0$$

Taking $z = x/\sqrt{2Kt}$, we have

$$\Phi\left(\frac{x}{\sqrt{2Kt}}\right) = \frac{1}{2} + \frac{x}{\sqrt{2Kt}} \frac{1}{\sqrt{2\pi}} + O\left(\frac{1}{t}\right) \qquad t \to \infty$$

Substituting this in the exact solution, we have

$$u(x; t) = T_2\left[\frac{1}{2} + \frac{x}{\sqrt{4\pi Kt}} + O\left(\frac{1}{t}\right)\right] + T_1\left[\frac{1}{2} - \frac{x}{\sqrt{4\pi Kt}} + O\left(\frac{1}{t}\right)\right]$$

$$= \tfrac{1}{2}(T_1 + T_2) + O\left(\frac{1}{\sqrt{t}}\right) \qquad t \to \infty$$

---

† The values for $1 - \Phi(y)$ are from *Handbook of Mathematical Functions*, by Abramowitz and Stegun, p. 316, where $1 - \Phi(y) = \tfrac{1}{2} \text{erf}_c (y/\sqrt{2})$.

To analyze the solution for $t \to 0$, we use Example 6.2.1:

$$\Phi(-z) = 1 - \Phi(z) = \frac{1}{z\sqrt{2\pi}} e^{-z^2/2} \left[1 + O\left(\frac{1}{z^2}\right)\right] \qquad z \to \infty$$

Taking $z = \pm x/\sqrt{2Kt}$, we have

$$1 - \Phi\left(\frac{x}{\sqrt{2Kt}}\right) = \left(\frac{Kt}{\pi x^2}\right)^{1/2} e^{-x^2/(4Kt)}[1 + O(t)] \qquad x > 0, t \to 0$$

$$\Phi\left(\frac{x}{\sqrt{2Kt}}\right) = \left(\frac{Kt}{\pi x^2}\right)^{1/2} e^{-x^2/(4Kt)}[1 + O(t)] \qquad x < 0, t \to 0$$

Substituting this, we have

$$u(x; t) = T_2 \left\{1 - \left(\frac{Kt}{\pi x^2}\right)^{1/2} e^{-x^2/(4Kt)}[1 + O(t)]\right\}$$

$$+ T_1 \left(\frac{Kt}{\pi x^2}\right)^{1/2} e^{-x^2/(4Kt)}[1 + O(t)]$$

$$= T_2 + O(e^{-x^2/(4Kt)}) \qquad x > 0, t \to 0$$

$$u(x; t) = T_2 \left(\frac{Kt}{\pi x^2}\right)^{1/2} e^{-x^2/(4Kt)}[1 + O(t)]$$

$$+ T_1 \left\{1 - \left(\frac{Kt}{\pi x^2}\right)^{1/2} e^{-x^2/(4Kt)}[1 + O(t)]\right\}$$

$$= T_1 + O(e^{-x^2/(4Kt)}) \qquad x < 0, t \to 0$$

At $x = 0$, $u(x; t) = \frac{1}{2}T_1 + \frac{1}{2}T_2$ for all $t > 0$.   ●

The above analysis shows that when $t \to 0$, $u(x; t)$ tends to $u(x; 0)$ faster than any power of t, that is, $u(x; t) - u(x; 0) = O(t^n)$, $t \to 0$, for $n = 1, 2, \ldots$. This shows that $\partial^n u/\partial t^n|_{t=0} = 0$, but $u(x; t)$ is not identically zero. We have a "real" example of a function which is not represented by its Taylor series about $t = 0$.

EXERCISES 6.2

Obtain asymptotic expressions for the following integrals when $t \to \infty$.

1. $f(t) = \displaystyle\int_0^1 e^{tx} \sin x \, dx$

2. $f(t) = \displaystyle\int_1^\infty e^{-tx^2} x^4 \, dx$

3. $f(t) = \displaystyle\int_t^\infty \frac{e^{-u}}{u} \, du$

4. $f(t) = \displaystyle\int_0^\infty \frac{e^{-tx}}{1 + x^2} \, dx$

5. By an appropriate use of integration by parts, find an asymptotic expression for

$$f(t) = \int_0^\infty \frac{\sin x}{x + t} \, dx$$

6. Two materials of the same conductivity K are initially at the temperatures $T_1 = 0$, $T_2 = 100$. At time $t = 0$ they are brought together. Find asymptotic formulas for the temperature $u(x; t)$ when $t \to 0$, $t \to \infty$.

7. Two materials of the same conductivity K are initially at the temperatures $T_1 = 0$, $T_2 = 100$. At time $t = 0$ they are brought together. Find an asymptotic formula for $u(at; t)$ when $t \to \infty$, where a is a positive constant.

8. Let $u(x, y)$ be the solution of Laplace's equation $u_{xx} + u_{yy} = 0$ for $y > 0$, $-\infty < x < \infty$, with the boundary conditions $u(x, 0) = f(x) = \int_{-\infty}^{\infty} F(\mu)e^{i\mu x} \, d\mu$. Show that for each x, $u(x, y) = (2/y)F(0) + O(1/y^2)$, $y \to \infty$. Equivalently $u(x, y) = [1/(\pi y)] \int_{-\infty}^{\infty} f(z) \, dz + O(1/y^2)$, $y \to \infty$. [*Hint:* Use the Fourier representation (5.2.13).]

6.3 LAPLACE'S METHOD

Statement and Proof of Result

We continue the asymptotic analysis of integrals of the form

$$f(t) = \int_a^b g(x)e^{th(x)} \, dx \tag{6.3.1}$$

now with the possibility that $h'(x) = 0$ at one or more points. In this case it is still true that $f(t) \sim e^{tH}$ when $t \to \infty$, where H is the maximum of $h(x)$, $a \le x \le b$. The new feature results from the possibility of points x_i, where $h(x_i) = H$ and $h'(x_i) = 0$. We assume that $h''(x_i) \neq 0$ at each of these points. [Of course, it follows that $h''(x_i) < 0$ since we are at a maximum of h.] These points fall into two groups: interior global maxima of h and boundary maxima where $h'(x_i) = 0$. The exact contribution of the second type of point is one-half that of the first type. We now state the result of Laplace's method:

$$f(t) = \frac{e^{tH}}{\sqrt{t}}\left[C + O\left(\frac{1}{\sqrt{t}}\right)\right] \qquad t \to \infty \tag{6.3.2}$$

where

$$C = \sqrt{2\pi}\left\{\sum_{\substack{a < x_i < b \\ h(x_i) = H}} \frac{g(x_i)}{[-h''(x_i)]^{1/2}} + \frac{1}{2}\sum_{\substack{x_i = a \text{ or } b \\ h(x_i) = H \\ h'(x_i) = 0}} \frac{g(x_i)}{[-h''(x_i)]^{1/2}}\right\} \tag{6.3.3}$$

The proof, which is divided into three steps, may be omitted at a first reading.

For simplicity in the proof, let us assume that the maximum is at *exactly one interior point* x_1. (The other cases, which are not essentially different, are dealt with in the exercises.) Now we can write the integral (6.3.1) in the form

$$f(t) = \text{I} + \text{II} + \text{III}$$

where

$$\text{I} = \int_a^{x_1 - \delta} g(x)e^{th(x)} \, dx$$

$$\text{II} = \int_{x_1 - \delta}^{x_1 + \delta} g(x)e^{th(x)} \, dx$$

$$\text{III} = \int_{x_1 + \delta}^b g(x)e^{th(x)} \, dx$$

and δ is a positive number which will be specified. The first and third integrals are straightforward. Indeed, let

$$h_1 = \max_{a \leq x \leq x_1 - \delta} h(x) \qquad h_2 = \max_{x_1 + \delta \leq x \leq b} h(x)$$

Then   I $= O(e^{th_1})$ III $= O(e^{th_2})$ $t \to \infty$

In addition, we must have $h_1 < H$, $h_2 < H$. Indeed, otherwise the maximum of the continuous function h would be assumed on the interval $[a, x_1 - \delta]$ or the interval $[x_1 + \delta, b]$, contrary to the hypothesis. Therefore it remains to analyze the middle integral when $t \to \infty$.

To do this, we begin with the Taylor expansions of $g(x)$, $h(x)$ about $x = x_1$, written in the form

$$|g(x) - g(x_1)| \leq M_1|x - x_1| \tag{6.3.4}$$

$$|h(x) - H - \tfrac{1}{2}h''(x_1)(x - x_1)^2| \leq M_2|x - x_1|^3 \tag{6.3.5}$$

The Taylor expansion (6.3.5) implies that

$$h(x) \leq H - M_3(x - x_1)^2 \qquad |x - x_1| < \delta \tag{6.3.5a}$$

Here M_1, M_2, M_3 are suitable positive constants. This is proved in detail at the end of this section.

In the following steps we show that $g(x)$ may be replaced by $g(x_1)$ and that $h(x)$ may be replaced by its second-order Taylor expansion.

Step 1

$$\int_{x_1-\delta}^{x_1+\delta} g(x)e^{th(x)} \, dx = g(x_1) \int_{x_1-\delta}^{x_1+\delta} e^{th(x)} \, dx + O\left(\frac{e^{tH}}{t}\right) \qquad t \to \infty \tag{6.3.6}$$

To prove this, we use (6.3.4) to write

$$\left| \int_{x_1-\delta}^{x_1+\delta} g(x)e^{th(x)} \, dx - g(x_1) \int_{x_1-\delta}^{x_1+\delta} e^{th(x)} \, dx \right| \leq M_1 \int_{x_1-\delta}^{x_1+\delta} |x - x_1|e^{th(x)} \, dx$$

From (6.3.5a) the last integral is less than or equal to

$$M_1 e^{tH} \int_{x_1-\delta}^{x_1+\delta} |x - x_1| \exp\left[-tM_3(x - x_1)^2\right] dx$$

By making the change of variables $y = \sqrt{t}(x - x_1)$, this is

$$M_1 \frac{e^{tH}}{t} \int_{-\delta\sqrt{t}}^{\delta\sqrt{t}} |y|e^{-M_3y^2} \, dy = O\left(\frac{e^{tH}}{t}\right)$$

This completes the proof of (6.3.6).

Step 2

$$\int_{x_1-\delta}^{x_1+\delta} e^{th(x)} \, dx = e^{tH} \int_{x_1-\delta}^{x_1+\delta} \exp\left[\frac{th''(x_1)(x - x_1)^2}{2}\right] dx + O\left(\frac{e^{tH}}{t}\right) \quad (6.3.7)$$

To prove this, we use the following inequality, valid for real numbers A, B:

$$|e^A - e^B| \le |A - B|e^C$$

$$C = \max(A, B)$$

We apply this with $A = h(x) - H$, $B = \frac{1}{2}h''(x_1)(x - x_1)^2$. Clearly we have $\max(A, B) \le -M_3(x - x_1)^2$ for $|x - x_1| < \delta$, $M_3 = \frac{1}{3}|h''(x_1)|$. Thus

$$\left|\int_{x_1-\delta}^{x_1+\delta} e^{th(x)} \, dx - e^{tH} \int_{x_1-\delta}^{x_1+\delta} \exp\left[\frac{th''(x_1)(x - x_1)^2}{2}\right] dx\right|$$

$$= e^{tH} \left|\int_{x_1-\delta}^{x_1+\delta} (e^{tA} - e^{tB}) \, dx\right|$$

$$\le e^{tH} \int_{x_1-\delta}^{x_1+\delta} tM_2|x - x_1|^3 \exp[-tM_3(x - x_1)^2] \, dx$$

We again make the change of variable $y = \sqrt{t}(x - x_1)$ in this integral and see that the integral is

$$\frac{e^{tH}}{t^2} \int_{-\delta\sqrt{t}}^{\delta\sqrt{t}} M_2|y|^3 e^{-M_3 y^2} \, dy = O\left(\frac{e^{tH}}{t}\right)$$

This completes the proof of (6.3.7).

Step 3

$$e^{tH} \int_{x_1-\delta}^{x_1+\delta} \exp\left[\frac{th''(x_1)(x - x_1)^2}{2}\right] dx$$

$$= e^{tH} \left[\frac{2\pi}{-th''(x_1)}\right]^{1/2} \left[1 + O\left(\frac{1}{t}\right)\right] \quad (6.3.8)$$

To prove this, we again apply the change of variable $y = \sqrt{t}(x - x_1)$ and transform the original integral to

$$\frac{e^{tH}}{\sqrt{t}} \int_{-\delta\sqrt{t}}^{\delta\sqrt{t}} \exp\left[\frac{y^2 h''(x_1)}{2}\right] dy$$

When $t \to \infty$, the integral $= \{2\pi/[-h''(x_1)]\}^{1/2} + O(1/t)$, which is in the required form. Combining steps 1, 2, and 3 completes the proof of Laplace's method.

If we also have a maximum at a boundary point, for example, $x = a$, where $h'(a) = 0$, $h''(a) < 0$, then we must modify the analysis by taking the integral

$$II = \int_a^{a+\delta} g(x)e^{th(x)} \, dx$$

By repeating the same steps, we will find

$$II \sim \frac{g(a)e^{th(a)}}{\sqrt{-th''(a)}} \int_0^{\delta\sqrt{t}} e^{-u^2/2} \, du$$

When $t \to \infty$, the integral has the limiting value $\frac{1}{2}\sqrt{2\pi}$, which explains the factor of $\frac{1}{2}$ for the boundary terms in (6.3.3).

Three Applications to Integrals

Example 6.3.1. Find an asymptotic formula for $f(t) = \int_0^1 e^{tx(1-x)} \, dx$.

Solution. In this case $h(x) = x(1 - x)$, and the maximum is attained at $x = \frac{1}{2}$, where $h'(\frac{1}{2}) = 0$, $h''(\frac{1}{2}) = -2$, which is nonzero. Thus we may apply Laplace's method to obtain

$$\int_0^1 e^{tx(1-x)} \, dx = \sqrt{\frac{\pi}{t}} \, e^{t/4} \left[1 + O\left(\frac{1}{\sqrt{t}} \right) \right] \qquad t \to \infty$$

●

Example 6.3.2. Find an asymptotic formula for $f(t) = \int_0^1 e^{t(2x-x^2)} \, dx$.

Solution. In this case $h(x) = 2x - x^2$, and the maximum is assumed at $x = 1$, where $h'(1) = 0$, $h''(1) = -2 < 0$. Thus we may apply Laplace's method at the endpoint $x = 1$, with the result

$$\int_0^1 e^{t(2x-x^2)} \, dx = e^t \sqrt{\frac{\pi}{4t}} \left[1 + O\left(\frac{1}{\sqrt{t}} \right) \right] \qquad t \to \infty$$

●

Example 6.3.3. Find an asymptotic formula for $f(t) = \int_0^\infty e^{t \ln x} e^{-x} \, dx$.

Solution. This is the integral which defines the factorial function $f(t) = t!$ for $t = 1, 2, 3, \ldots$. We will apply Laplace's method to obtain Stirling's formula. To do this, we might try $h(x) = \ln x$, $g(x) = e^{-x}$; but $\ln x$ has no maximum for $0 < x < \infty$; therefore Laplace's method is not applicable in this form. Nevertheless the function $x \to e^{t \ln x - x}$ has a maximum at $x = t$, so we make the change of variable $y = x/t$. This gives

$$f(t) = t \int_0^\infty \exp\left(t \ln y + t \ln t - yt \right) dy$$

$$= t^{t+1} \int_0^\infty e^{th(y)} \, dy$$

where
$$h(y) = \ln y - y$$

$$h'(y) = \frac{1}{y} - 1$$

$$h''(y) = -\frac{1}{y^2}$$

We can now apply Laplace's method: h has a global maximum at $y = 1$, where $h(1) = -1$, $h''(1) = -1$. Therefore

$$f(t) = t^{t+1} \sqrt{\frac{2\pi}{t}} e^{-t} \left[1 + O\left(\frac{1}{\sqrt{t}}\right) \right] \qquad t \to \infty$$

This is the required form of Stirling's formula. ●

Applications to the Heat Equation

Our next example gives an application of Laplace's method to the Fourier representation of the solution of the heat equation, to obtain an asymptotic formula when $t \to \infty$.

Example 6.3.4. Let $u(x; t)$ be the solution of the heat equation $u_t = K u_{xx}$ with initial condition $u(x; 0) = f(x)$ for $t > 0$, $-\infty < x < \infty$. Use Laplace's method to obtain asymptotic formulas for the temperature and heat flux when $t \to \infty$.

Solution. We use the Fourier representation (5.2.2)

$$u(x; t) = \int_{-\infty}^{\infty} F(\mu) e^{i\mu x} e^{-\mu^2 K t} \, d\mu$$

where $F(\mu)$ is the Fourier transform of $f(x)$. We apply Laplace's method with $h(\mu) = -K\mu^2$, $a = -\infty$, $b = \infty$, $g(\mu) = F(\mu) e^{i\mu x}$. The function $h(\mu)$ has a single maximum at $\mu = 0$, where $H = h(0) = 0$, $h'(0) = 0$, $h''(0) = -2K$. Applying (6.3.2), we have

$$u(x; t) = \frac{1}{\sqrt{t}} \left[F(0) \sqrt{\frac{2\pi}{2K}} + O\left(\frac{1}{\sqrt{t}}\right) \right] \qquad t \to \infty$$

Recalling that $F(\mu) = [1/(2\pi)] \int_{-\infty}^{\infty} f(x) e^{-i\mu x} \, dx$, we have

$$u(x; t) = \frac{1}{\sqrt{4\pi K t}} \int_{-\infty}^{\infty} f(x) \, dx + O\left(\frac{1}{t}\right) \qquad t \to \infty$$

To study the heat flux, we differentiate and get

$$u_x(x; t) = \int_{-\infty}^{\infty} i\mu e^{i\mu x} F(\mu) e^{-\mu^2 K t} \, d\mu$$

Applying Laplace's method again with $h(\mu) = -K\mu^2$, $g(\mu) = i\mu e^{i\mu x} F(\mu)$, we have a single maximum at $\mu = 0$, where $h(0) = 0$, $g(0) = 0$. Applying (6.3.2), we have

$C = 0$, and thus

$$u_x(x; t) = O\left(\frac{1}{t}\right) \qquad t \to \infty$$

The heat flux tends to zero faster than the temperature does. ●

In the next example we apply the asymptotic methods to the explicit (Gauss-Weierstrass) representation of $u(x; t)$ to deduce information for small times.

Example 6.3.5. Let $u(x; t)$ be the solution of the heat equation $u_t = Ku_{xx}$ for $t > 0$, $-\infty < x < \infty$, with the initial conditions $u(x; 0) = f(x)$ for $a \le x \le b$ and $u(x; 0) = 0$ otherwise. Find an asymptotic formula for $u(x; t)$ when $t \downarrow 0$, considering separately the cases $x < a$, $x = a$, $a < x < b$, $x = b$, and $x > b$.

Solution. The Gauss-Weierstrass representation (5.2.5) takes the form

$$u(x; t) = \frac{1}{\sqrt{4\pi Kt}} \int_a^b e^{-(x-z)^2/(4Kt)} f(z) \, dz$$

We introduce a new variable $s = 1/(4Kt)$ and analyze the integral

$$I = \int_a^b f(z) e^{-s(x-z)^2} \, dz$$

For $x < a$ the maximum of the exponential is assumed at $z = a$, where the derivative is nonzero. Hence we may apply integration by parts to obtain the result $I = \{f(a)/[2s(x - a)]\}e^{-s(x-a)^2} [1 + O(1/s)]$. Similarly for $x > b$ we obtain $I = \{f(b)/[2(b - x)]\}e^{-s(x-b)^2} [1 + O(1/s)]$. If $a < x < b$, the maximum of the exponential is assumed at $z = x$ where the derivative is zero. Hence we apply Laplace's method to obtain $I = \sqrt{2\pi} f(x)/\{\sqrt{2s}[1 + O(1/\sqrt{s})]\}$. The endpoints $x = a$ and $x = b$ give the same contribution with a factor of $\frac{1}{2}$. Combining these, we obtain

$$u(x; t) = \begin{cases} \sqrt{\dfrac{Kt}{\pi}}\, e^{-(x-a)^2/(4Kt)}[f(a) + O(t)] & x < a \\[2ex] \frac{1}{2}f(x) + O(\sqrt{t}) & x = a, b \\[2ex] f(x) + O(\sqrt{t}) & a < x < b \\[2ex] \sqrt{\dfrac{Kt}{\pi}}\, e^{-(x-b)^2/(4Kt)} [f(b) + O(t)] & x > b \end{cases}$$

●

Proof of (6.3.5a)

By the extended mean-value theorem, we may write

$$h(x) = H + \tfrac{1}{2}h''(x_1)(x - x_1)^2 + \tfrac{1}{6}h'''(\xi)(x - x_1)^3$$

for some ξ with $|\xi - x_1| \le |x - x_1|$. If h''' is identically zero, there is nothing to prove. Otherwise let $M = \max |h'''|$, where the maximum is taken over any interval $(x_1 - \delta, x_1 + \delta)$. Write

$$H - h(x) = \tfrac{1}{2}|h''(x_1)|(x - x_1)^2 \left[1 + \frac{h'''(\xi)(x - x_1)}{3|h''(x_1)|} \right]$$

Let $|x - x_1| < |h''(x_1)|/M$. Then $|h'''(\xi)(x - x_1)/h''(x_1)| < 1$, and we have

$$H - h(x) \geq \tfrac{1}{2}|h''(x_1)|(x - x_1)^2(1 - \tfrac{1}{3})$$
$$= \tfrac{1}{3}|h''(x_1)|(x - x_1)^2$$

which was to be proved.

EXERCISES 6.3

In Exercises 1 to 3, apply Laplace's method to obtain an asymptotic formula for $f(t)$, $t \to \infty$.

1. $f(t) = \int_{-\pi/2}^{\pi/2} (3x + 2)e^{-t \sin^2 x}\, dx$
2. $f(t) = \int_{-2}^{2} (3 + 2 \cos x)e^{-tx^2}\, dx$
3. $f(t) = \int_{-1}^{1} e^{-tP_4(x)}\, dx$, where P_4 is the fourth Legendre polynomial.

4. Let $I_0(t)$ be the modified Bessel function

$$I_0(t) = \frac{1}{2\pi} \int_0^{2\pi} e^{t \sin \theta}\, d\theta$$

   Apply Laplace's method to find an asymptotic formula for $I_0(t)$, $t \to \infty$.

5. Let $h(x)$, $a < x < b$, be a differentiable function with a single maximum at x_1, $a < x_1 < b$. Assume that $h'(x_1) = 0$, $h''(x_1) = 0$, $h'''(x_1) = 0$, $h^{(4)}(x_1) < 0$. Show that Laplace's method can be modified to obtain an asymptotic formula of the form

$$\int_a^b e^{th(x)}\, dx \approx C \frac{e^{th(x_1)}}{[-th^{(4)}(x_1)]^{1/4}}$$

   with $C = (24)^{1/4} \int_{-\infty}^{\infty} e^{-u^4}\, du$.

6. Apply the method of Exercise 5 to obtain an asymptotic formula for the integral

$$\int_{-\pi/2}^{\pi/2} \exp t \left(1 - \cos x - \frac{x^2}{2} \right) dx \qquad t \to \infty$$

7. Let $u(x; t)$ be the solution of the heat equation $u_t = Ku_{xx}$ for $x > 0$, $t > 0$ with the initial condition $u(x; 0) = f(x)$ and the boundary condition $u(0; t) = 0$. Use Laplace's method to obtain asymptotic formulas for the temperature and heat flux when $t \to \infty$.

8. Let $u(x; t)$ be the solution of the heat equation $u_t = Ku_{xx}$ for $x > 0$, $t > 0$ with the initial condition $u(x; 0) = f(x)$ and the boundary condition $u_x(0; t) = 0$. Use Laplace's method to obtain asymptotic formulas for the temperature and heat flux when $t \to \infty$.

9. Let $u(x; t)$ be the solution of the heat equation $u_t = Ku_{xx}$ for $t > 0$, $-\infty < x < \infty$, with initial condition $u(x; 0) = 100/(1 + x^2)$. Use Laplace's method to obtain asymptotic formulas for the temperature and heat flux when $t \to \infty$.

10. Let $u(x; t)$ be the solution of the heat equation $u_t = Ku_{xx}$ for $t > 0$, $x > 0$ with initial condition $u(x; 0) = xe^{-x^2}$ for $x > 0$ and the boundary condition $u_x(0; t) = 0$ for $t > 0$. Use Laplace's method to find asymptotic formulas for the temperature and heat flux when $t \to \infty$.

11. Let $u(x; t)$ be the solution of the heat equation $u_t = Ku_{xx}$ for $t > 0$, $-\infty < x < \infty$, with the initial conditions $u(x; 0) = f(x)$ for $a < x < b$ and $u(x; 0) = 0$ otherwise. Find an asymptotic formula for the heat flux $u_x(x; t)$ when $t \downarrow 0$, considering separately the cases $x < a$, $x = a$, $a < x < b$, $x = b$, and $x > b$.

12. Let $u(x; t)$ be the solution of the heat equation $u_t = Ku_{xx}$ for $t > 0$, $x > 0$ with the initial conditions $u(x; 0) = f(x)$ for $a < x < b$ and $u(x; 0) = 0$ otherwise and the boundary condition $u(0; t) = 0$. Find an asymptotic formula for $u(x; t)$ when $t \downarrow 0$, considering separately the cases $x < a$, $x = a$, $a < x < b$, $x = b$, and $x > b$.

13. Under the same conditions as in Exercise 12, find an asymptotic formula for the heat flux $u_x(x; t)$ when $t \downarrow 0$, considering the five separate cases.

*6.4   ASYMPTOTIC EXPANSIONS†

In preceding sections we have used integration by parts and Laplace's method to obtain asymptotic formulas of the form

$$\int_a^b e^{th(x)}g(x)\, dx = \frac{e^{tH}}{t^\alpha}\left[C + O\left(\frac{1}{t^\alpha}\right)\right] \qquad t \to \infty \qquad (6.4.1)$$

where $\alpha = \frac{1}{2}$ or 1. In many cases we may wish to obtain an *asymptotic expansion*, i.e., a formula of the form

$$\int_a^b e^{th(x)}g(x)\, dx = \frac{e^{tH}}{t^\alpha}\left[C_0 + \frac{C_1}{t} + \frac{C_2}{t^2} + \cdots + \frac{C_N}{t^N} + O\left(\frac{1}{t^{N+1}}\right)\right] \qquad t \to \infty$$

This can be obtained by extending the methods of previous sections.

Extension of Integration-by-Parts Method

First we extend the method of integration by parts. Indeed, if $h'(x) \neq 0$, we can write

$$\int_a^b e^{th(x)}g(x)\, dx = \frac{g(x)e^{th(x)}}{th'(x)}\bigg|_a^b - \frac{1}{t}\int_a^b e^{th(x)}\left(\frac{g}{h'}\right)'\, dx \qquad (6.4.2)$$

The second integral is of the same form as the first with g replaced by $(g/h')'$. By repeating this procedure N times, we can obtain N terms in the asymptotic expansion of the original integral.

---

† This section is optional. The results are not used in the following sections.

Example 6.4.1. Obtain an asymptotic expansion of the integral

$$f(t) = \int_t^\infty e^{-u^2/2} \, du$$

Solution. Having introduced the method, we bypass the transformation to the standard form (6.4.2) and work directly. Thus

$$\int_t^\infty e^{-u^2/2} \, du = -\int_t^\infty \frac{1}{u} \, d(e^{-u^2/2})$$

$$= \frac{1}{t} e^{-t^2/2} - \int_t^\infty \frac{e^{-u^2/2}}{u^2} \, du$$

$$\int_t^\infty \frac{e^{-u^2/2}}{u^2} \, du = -\int_t^\infty \frac{1}{u^3} \, d(e^{-u^2/2})$$

$$= \frac{1}{t^3} e^{-t^2/2} - 3 \int_t^\infty \frac{e^{-u^2/2}}{u^4} \, du$$

$$\int_t^\infty \frac{1}{u^4} e^{-u^2/2} \, du = -\int_t^\infty \frac{1}{u^5} \, d(e^{-u^2/2})$$

$$= \frac{1}{t^5} e^{-t^2/2} - 5 \int_t^\infty \frac{e^{-u^2/2}}{u^6} \, du$$

Repeating this, we see that

$$f(t) = \frac{e^{-t^2/2}}{t}$$

$$\times \left[1 - \frac{1}{t^2} + \frac{3}{t^4} - \frac{15}{t^6} + \cdots + (-1)^N \frac{1 \cdot 3 \cdot 5 \cdots (2N-1)}{t^{2N}} + O\left(\frac{1}{t^{2N+2}}\right) \right]$$

This is the desired expansion. ●

Extension of Laplace's Method

We now turn to Laplace's method, with the goal of an asymptotic expansion. For this purpose, we assume a single maximum at $x = x_1$, where $a < x_1 < b$, $h''(x_1) < 0$. We may restrict our attention to the integral for $|x - x_1| < \delta$ since

$$\int_a^b e^{th(x)} g(x) \, dx = \int_{x_1-\delta}^{x_1+\delta} e^{th(x)} g(x) \, dx + O(e^{t(H-\epsilon)}) \qquad t \to \infty \qquad (6.4.3)$$

To obtain an asymptotic expansion, we replace $g(x)$, $h(x)$ by their Taylor expansions about $x = x_1$:

$$g(x) \approx g_1 + (x - x_1)g_1' + \tfrac{1}{2}(x - x_1)^2 g_1'' + \cdots + \frac{1}{N!}(x - x_1)^N g_1^{(N)}$$

$$h(x) \approx H + \tfrac{1}{2}(x - x_1)^2 h_1'' + \cdots + \frac{1}{N!}(x - x_1)^N h_1^{(N)}$$

We substitute these into (6.4.3) and make the change of variable $y = \sqrt{t}(x - x_1)$, with the result

$$\int_{x_1-\delta}^{x_1+\delta} e^{th(x)}g(x)\,dx \approx e^{tH} \int_{-\delta\sqrt{t}}^{\delta\sqrt{t}} e^{h_1''y^2/2} \exp\left(\frac{h_1'''y^3}{6t^{1/2}} + \cdots + \frac{h_1^{(N)}y^N}{N!t^{N/2-1}}\right)$$

$$\times \left(g_1 + \frac{yg_1'}{t^{1/2}} + \cdots + \frac{y^N g_1^{(N)}}{N!t^{N/2}}\right)\frac{dy}{\sqrt{t}}$$

Now we expand the second exponential in a power series and collect like powers of $1/t$. The resulting expression has the form

$$e^{tH} \int_{-\delta\sqrt{t}}^{\delta\sqrt{t}} e^{h_1''y^2/2} \left\{ g_1 + \frac{1}{\sqrt{t}}\left(yg_1' + \frac{y^3 h_1'''}{6}\right)\right.$$

$$\left. + \frac{1}{t}\left[\frac{y^2 g_1''}{2} + \frac{y^4 g_1' h_1'''}{6} + \frac{y^4 g_1 h_1^{(4)}}{24} + \frac{y^6 g_1 (h_1''')^2}{72}\right] + O\left(\frac{1}{t^{3/2}}\right)\right\}\frac{dy}{\sqrt{t}}$$

To evaluate these integrals when $t \to \infty$, we recall that

$$\frac{1}{\sqrt{2\pi}}\int_{-\infty}^{\infty} y^n e^{-y^2/2}\,dy = \begin{cases} 0 & n \text{ odd} \\ 1 \cdot 3 \cdot 5 \cdots (2m-1) & n = 2m \end{cases}$$

Using these, we may compute as many terms as we like in an asymptotic expansion. The fractional powers of $1/t$ are multiplied by odd powers of y, hence do not appear in the final result. We illustrate the method with Stirling's formula.

Application to Stirling's Formula

Example 6.4.2. Find the asymptotic expansion of the integral

$$f(t) = \int_0^\infty e^{t(\ln x - x)}\,dx$$

Solution. For this function, $g(x) = 1$, $h(x) = \ln x - x$ with a single global maximum at $x = 1$, where we have the Taylor expansion

$$\ln x - x \approx -1 - \frac{1}{2}(x-1)^2 + \frac{1}{3}(x-1)^3 - \cdots + (-1)^{N-1}\frac{1}{N}(x-1)^N$$

$$f(t) \approx \int_{1-\delta}^{1+\delta} \exp\left\{t\left[-1 - \frac{(x-1)^2}{2} + \frac{(x-1)^3}{3} + \cdots\right.\right.$$

$$\left.\left. + (-1)^{N-1}\frac{(x-1)^N}{N}\right]\right\}\,dx$$

$$= \frac{e^{-t}}{\sqrt{t}}\int_{-\delta\sqrt{t}}^{\delta\sqrt{t}} e^{-y^2/2}\exp\left[\frac{y^3}{3t^{1/2}} - \frac{y^4}{4t} + \cdots + (-1)^{N-1}\frac{y^N}{Nt^{-1+N/2}}\right]\,dy$$

$$= \frac{e^{-t}}{\sqrt{t}} \int_{-\delta\sqrt{t}}^{\delta\sqrt{t}} e^{-y^2/2} \left[1 + \frac{y^3}{3t^{1/2}} + \frac{1}{t} \left(\frac{y^6}{18} - \frac{y^4}{4} \right) + O\left(\frac{1}{t^2}\right) \right] dy$$

$$= e^{-t} \sqrt{\frac{2\pi}{t}} \left[1 + 0 + \frac{1}{t} \left(\frac{15}{18} - \frac{3}{4} \right) + O\left(\frac{1}{t^2}\right) \right]$$

Putting these together, we have established a more precise form of Stirling's formula:

$$n! = n^{n+1/2} e^{-n} \sqrt{2\pi} \left[1 + \frac{1}{12n} + O\left(\frac{1}{n^2}\right) \right] \qquad n \to \infty \qquad \bullet$$

EXERCISES 6.4

1. Show that the asymptotic expansion of Stirling's integral for $n!$ contains only *integral powers* of n^{-1}.
2. Obtain the coefficient of $1/n^2$ in Stirling's formula.
3. Using integration by parts, show that

$$\left(\frac{1}{t} - \frac{1}{t^3} \right) e^{-t^2/2} \le \int_t^\infty e^{-u^2/2} \, du \le \frac{1}{t} e^{-t^2/2} \qquad 0 < t < \infty$$

(This is not an asymptotic statement!)

4. Obtain an asymptotic expansion of the exponential integral

$$f(t) = \int_t^\infty \frac{e^{-u}}{u} \, du \qquad \text{when } t \to \infty$$

5. Obtain three terms in the asymptotic expansion of

$$f(t) = \int_0^\infty \frac{e^{-tx}}{1 + x^2} \, dx \qquad \text{when } t \to \infty$$

6. Let $u(x; t)$ be the solution of the heat equation $u_t = K u_{xx}$ for $t > 0$, $-\infty < x < \infty$, with initial condition $u(x; 0) = f(x) = \int_{-\infty}^\infty F(\mu) e^{-i\mu x} \, d\mu$. Suppose, in addition, that the Fourier transform $F(\mu)$ has the form $F(\mu) = F(0) e^{-\mu^2} + O(\mu^4)$ when $\mu \to 0$. Find an asymptotic expansion of $u(x; t)$ when $t \to \infty$.

7. Let $u(x, y)$ be the solution of Laplace's equation $u_{xx} + u_{yy} = 0$ for $y > 0$, $-\infty < x < \infty$, with the boundary condition $u(x, 0) = f(x) = \int_{-\infty}^\infty F(\mu) e^{i\mu x} \, d\mu$. Show that when $y \to \infty$, we have an asymptotic expansion of the form

$$u(x, y) = \sum_{k=1}^N \frac{1}{y^k} R_k(x) + O\left(\frac{1}{y^{N+1}}\right)$$

and identify the functions $R_k(x)$ in terms of the Fourier transform $F(\mu)$ and its derivatives.

6.5 THE METHOD OF STATIONARY PHASE

Statement of Result

In many problems we have to deal with integrals of the form

$$\int_a^b e^{it\varphi(\mu)}g(\mu)\, d\mu \tag{6.5.1}$$

where φ is a real-valued function, called the *phase function,* while g may be either real- or complex-valued. In contrast to Laplace's method, the exponent is now purely imaginary; hence the integrand is an oscillatory function of t. As long as $\varphi'(\mu) \neq 0$, we may integrate by parts and conclude that the integral is $O(1/t)$ when $t \to \infty$. The main contribution comes from the points μ_j, where $\varphi'(\mu_j) = 0$. These are called *stationary points.* We assume a finite number of stationary points (μ_j) with $a < \mu_j < b$, $\varphi''(\mu_j) \neq 0$ and $\int_a^b |g(\mu)|\, d\mu < \infty$. Then when $t \to \infty$,

$$\int_a^b e^{it\varphi(\mu)}g(\mu)\, d\mu = \sum_{j:\varphi''(\mu_j)>0} \left[\frac{2\pi}{t\varphi''(\mu_j)}\right]^{1/2} e^{it\varphi(\mu_j)}e^{i\pi/4}g(\mu_j)$$

$$+ \sum_{j:\varphi''(\mu_j)<0} \left[\frac{2}{t|\varphi''(\mu_j)|}\right]^{1/2} e^{it\varphi(\mu_j)}e^{-i\pi/4}g(\mu_j) + O\left(\frac{1}{t}\right) \tag{6.5.2}$$

In contrast to Laplace's method, *we must sum over all stationary points of* φ, not simply those where φ is maximum.

If the endpoints $\mu = a$ and $\mu = b$ are stationary points, they contribute to (6.5.2) with a factor of $\frac{1}{2}$, just as in Laplace's method.

This complicated-looking formula becomes easier to remember if we restate it in the following fashion: Replace $\varphi(\mu)$ by its second-order Taylor expansion, and replace $g(\mu)$ by its value at the stationary point. Do the resulting integrals, one for each stationary point, and sum over all stationary points.

Example 6.5.1. Obtain an asymptotic formula for the integral

$$\int_{-\pi/2}^{\pi/2} e^{it\cos\mu}\, d\mu$$

Solution. In this case $\varphi(\mu) = \cos\mu$, $\varphi'(\mu) = -\sin\mu$, $\varphi''(\mu) = -\cos\mu$. The only stationary point is at $\mu = 0$, where $\varphi''(0) = -1$. Thus we have

$$\int_{-\pi/2}^{\pi/2} e^{it\cos\mu}\, d\mu = \sqrt{\frac{2\pi}{t}}\, e^{it}e^{-i\pi/4} + O\left(\frac{1}{t}\right) \qquad t \to \infty \qquad \bullet$$

We now sketch the proof of (6.5.2), breaking the analysis into several steps. The proof may be omitted at a first reading.

Proof

Step 1 It suffices to consider integrals of the form

$$\int_{\mu_j-\delta}^{\mu_j+\delta} e^{it\varphi(\mu)} g(\mu) \, d\mu$$

Indeed, to prove this, notice that the remaining contributions to the integral (6.5.1) are over intervals where $\varphi'(\mu) \neq 0$. Each of these can be written in the form

$$\int_\alpha^\beta e^{it\varphi(\mu)} g(\mu) \, d\mu = \int_\alpha^\beta \frac{g(\mu)}{it\varphi'(\mu)} \, d(e^{it\varphi(\mu)})$$

$$= e^{it\varphi(\mu)} \left.\frac{g(\mu)}{it\varphi'(\mu)}\right|_\alpha^\beta - \frac{1}{it} \int_\alpha^\beta e^{it\varphi(\mu)} \frac{d}{d\mu}\left[\frac{g(\mu)}{\varphi'(\mu)}\right] d\mu$$

Both terms are $O(1/t)$, $t \to \infty$, and can therefore be included in the remainder term in (6.5.2).

Step 2 It suffices to consider integrals of the form

$$\int_{-\bar\delta_1}^{\bar\delta_2} e^{\pm itv^2} \bar g(v) \, dv$$

Indeed, we can introduce the new variable of integration

$$v = \sqrt{\pm \frac{\varphi(\mu) - \varphi(\mu_j)}{(\mu - \mu_j)^2}} \, (\mu - \mu_j)$$

The plus sign is chosen if $\varphi''(\mu_j) > 0$, the minus sign if $\varphi''(\mu_j) < 0$. With this change of variable,

$$\varphi(\mu) = \varphi(\mu_j) \pm v^2$$

$$\left.\frac{dv}{d\mu}\right|_{\mu=\mu_i} = \sqrt{\frac{|\varphi''(\mu_j)|}{2}}$$

$$\bar\delta_2 = v(\mu_j + \delta) \qquad -\bar\delta_1 = v(\mu_j - \delta)$$

Therefore,

$$\int_{\mu_j-\delta}^{\mu_j+\delta} e^{it\varphi(\mu)} g(\mu) \, d\mu = e^{it\varphi(\mu_j)} \int_{-\bar\delta_1}^{\bar\delta_2} e^{\pm itv^2} g(\mu(v)) \frac{d\mu}{dv} \, dv$$

Thus,

$$\bar g(v) = g(\mu(v)) \frac{d\mu}{dv}$$

$$\bar g(0) = g(\mu_j) \sqrt{\frac{2}{|\varphi''(\mu_j)|}}$$

Step 3 It suffices to consider

$$\int_{-\tilde{\delta}_1}^{\tilde{\delta}_2} e^{\pm itv^2} \, dv$$

Indeed, the integral obtained in step 2 can be written in the form

$$\overline{g}(0) \int_{-\tilde{\delta}_1}^{\tilde{\delta}_2} e^{\pm itv^2} \, dv + \int_{-\tilde{\delta}_1}^{\tilde{\delta}_2} e^{\pm itv^2} v h(v) \, dv$$

where
$$h(v) = \frac{\overline{g}(v) - \overline{g}(0)}{v}$$

The second integral can be integrated by parts, as we now show for the positive case:

$$\int_{-\tilde{\delta}_1}^{\tilde{\delta}_2} h(v) v e^{itv^2} \, dv = \frac{1}{2it} \int_{-\tilde{\delta}_1}^{\tilde{\delta}_2} h(v) \, d(e^{itv^2})$$

$$= \frac{1}{2it} \left[h(v) e^{itv^2} \Big|_{-\tilde{\delta}_1}^{\tilde{\delta}_2} - \int_{-\tilde{\delta}_1}^{\tilde{\delta}_2} h'(v) e^{itv^2} \, dv \right]$$

$$= O\left(\frac{1}{t}\right)$$

The integral with the negative exponent is handled in exactly the same way.

Step 4 We evaluate $\int_{-\tilde{\delta}_1}^{\tilde{\delta}_2} e^{\pm itv^2} \, dv$. This is called a *Fresnel integral*. In the appendix to this section we show that $\int_0^\infty e^{\pm ix^2} \, dx = \sqrt{\pi/4} e^{\pm i\pi/4}$. To obtain a result with remainder, we integrate by parts:

$$\int_M^\infty e^{ix^2} \, dx = -\frac{e^{iM^2}}{2iM} + \left(\frac{1}{2i}\right) \int_M^\infty \frac{e^{ix^2}}{x^2} \, dx = O\left(\frac{1}{M}\right)$$

Applying this with the change of variable $x^2 = tv^2$, we have

$$\int_{-\tilde{\delta}_1}^{\tilde{\delta}_2} e^{\pm itv^2} \, dv = \sqrt{\frac{\pi}{t}} e^{\pm i\pi/4} + O\left(\frac{1}{t}\right)$$

Combining the results of steps 1 to 4 completes the proof.   ∎

Application to Bessel Functions*

As an application of the method of stationary phase, we deduce the following asymptotic behavior of the Bessel function:

$$\boxed{J_0(t) = \sqrt{\frac{2}{\pi t}} \cos\left(t - \frac{\pi}{4}\right) + O\left(\frac{1}{t}\right) \qquad t \to \infty}$$

---

* See also the Appendix on "Using Mathematica."

To do this, we begin with the integral representation

$$J_0(t) = \frac{1}{2\pi} \int_{-\pi}^{\pi} e^{it \cos \theta} \, d\theta = \frac{1}{\pi} \int_0^{\pi} e^{it \cos \theta} \, d\theta$$

since the real and imaginary parts of $e^{it \cos \theta}$ are even functions. We take $\varphi(\theta) = \cos \theta$ in the method of stationary phase. We have

$$\varphi'(\theta) = -\sin \theta \qquad \varphi''(\theta) = -\cos \theta$$

Therefore we have stationary points at $\theta = 0$, $\theta = \pi$, the endpoints of the interval. At these points we have

$$\varphi(0) = 1 \qquad \varphi''(0) = -1$$

$$\varphi(\pi) = -1 \qquad \varphi''(\pi) = 1$$

Since we are dealing with endpoints, the contribution is one-half the normal value, and we obtain

$$\int_0^{\pi} e^{it \cos \theta} \, d\theta = \frac{1}{2} \sqrt{\frac{2\pi}{t}} \, e^{it} e^{-i\pi/4} + \frac{1}{2} \sqrt{\frac{2\pi}{t}} \, e^{-it} e^{i\pi/4} + O\left(\frac{1}{t}\right)$$

$$= \sqrt{\frac{2\pi}{t}} \cos \left(t - \frac{\pi}{4}\right) + O\left(\frac{1}{t}\right)$$

This is the stated result.

Using this result, we have a new proof of the fact that $J_0(t)$ has infinitely many zeros. Indeed, if $t = n\pi + \pi/4$, then $\cos(t - \pi/4) = \cos n\pi = (-1)^n$ and

$$J_0\left(n\pi + \frac{\pi}{4}\right) = \frac{(-1)^n}{\pi\sqrt{n/2}} + O\left(\frac{1}{n}\right)$$

If n is odd, this is negative for all sufficiently large n, whereas if n is even, this is positive for all sufficiently large n. Hence $J_0(t)$ must change sign infinitely often when $t \to \infty$.

Appendix: Evaluation of $\int_0^{\infty} e^{ix^2} \, dx$

We now give the proof that

$$I = \int_0^{\infty} \sin x^2 \, dx = \int_0^{\infty} \cos x^2 \, dx = \sqrt{\frac{\pi}{8}}$$

or equivalently that $\int_0^{\infty} e^{ix^2} \, dx = \frac{1}{2} \sqrt{\pi} e^{i\pi/4} = (1 + i)\sqrt{(\pi/8)}$.

Step 1 *The improper integral converges to a complex number with positive imaginary part.*

To prove this, we first make the change of variable $x = u^2$ to reduce the problem to the improper integral $\int_0^{\infty} (e^{i\pi}/u^{1/2}) \, du$. For the imaginary part we write

$$\int_0^{2N\pi} \frac{\sin u}{u^{1/2}} \, du = a_1 - a_2 + a_3 - a_1 + \cdots + a_{2N-1} - a_{2N}$$

where $a_k = \int_{(k-1)\pi}^{k\pi} (|\sin u|/u^{1/2}) \, du$. Making the further change of variable $u = \pi + v$ in the integral for a_k, we have

$$a_{k-1} - a_k = \int_{(k-2)\pi}^{(k-1)\pi} |\sin u| \left\{ \frac{1}{u^{1/2}} - \frac{1}{(u + \pi)^{1/2}} \right\} du > 0 \qquad (k = 2, 3, \ldots)$$

This shows that $a_1 > a_2 > a_3 > \cdots > 0$. Therefore the improper integral converges to a positive value along the sequence $\{2N\pi\}$. To show unrestricted convergence, note that

$$\left| \int_0^x \frac{\sin u}{u^{1/2}} \, du - \int_0^{2\pi[x/2\pi]} \frac{\sin u}{u^{1/2}} \, du \right| \leq \frac{1}{[x/2\pi]}$$

which tends to zero when $x \uparrow \infty$.

To show convergence of the cosine integral write

$$\int_{1/2\pi}^{(2N+1/2)\pi} \frac{\cos u}{u^{1/2}} \, du = b_1 - b_2 + \cdots + b_{2N-1} - b_{2N}$$

with $b_1 < b_2 < b_3 < 0$. The details are left to the reader.

Step 2

$$I = \lim_{p \to 0} \int_0^\infty e^{-px^2} \sin x^2 \, dx = \lim_{p \to 0} \int_0^\infty e^{-px^2} \cos x^2 \, dx$$

To prove this, let

$$I_p = \int_0^\infty e^{-px^2} \sin x^2 \, dx \qquad p > 0$$

$$f(x) = \int_0^x \sin t^2 \, dt \qquad x > 0$$

$$g(x) = e^{-px^2} \qquad p > 0, x > 0$$

Since the improper integral $\int_0^\infty \sin t^2 \, dt$ has been shown convergent, given $\epsilon > 0$, there exists $N = N(\epsilon)$ such that $|f(x) - f(N)| < \epsilon$ for $x \geq N$. Now, for any $M > N$, we integrate the product $(fg)'$ for $N < x < M$, with the result

$$f(M)g(M) - f(N)g(N) = \int_N^M (fg' + f'g) \, dx$$

$$= -2p \int_N^M f(x)xe^{-px^2} \, dx + \int_N^M \sin x^2 \, e^{-px^2} \, dx$$

The first integral can be handled by writing

$$[f(N) - \epsilon](e^{-pN^2} - e^{-pM^2})$$

$$\leq \int_N^M f(x)2pxe^{-px^2}\, dx \leq [f(N) + \epsilon](e^{-pN^2} - e^{-pM^2})$$

Letting $M \to \infty$, we have

$$-\epsilon e^{-pN^2} \leq \int_N^\infty \sin x^2\, e^{-px^2}\, dx \leq \epsilon e^{-pN^2}$$

Using this in the definition of I_p, we have

$$\int_0^N e^{-px^2} \sin x^2\, dx - \epsilon e^{-pN^2} \leq I_p \leq \int_0^N e^{-px^2} \sin x^2\, dx + \epsilon e^{-pN^2}$$

Letting $p \to 0$, we have

$$f(N) - \epsilon \leq \varliminf_{p \to 0} I_p \leq \varlimsup_{p \to 0} I_p \leq f(N) + \epsilon$$

But the middle terms are independent of N. Letting $N \to \infty$, we have

$$\int_0^\infty \sin x^2\, dx - \epsilon \leq \varliminf_{p \to 0} I_p \leq \varlimsup_{p \to 0} I_p \leq \int_0^\infty \sin x^2\, dx + \epsilon \qquad \forall \epsilon > 0$$

Therefore we have proved that $\lim_{p \to 0} I_p = \int_0^\infty \sin x^2\, dx$. The cosine integral is handled in the same way.

Step 3

$$J_p = \left(\int_0^\infty e^{-px^2} e^{ix^2}\, dx \right)^2 = \frac{\pi}{4(p - i)} \qquad p > 0$$

This is done by the familiar method of writing the result as a double integral and taking polar coordinates. Thus

$$J_p = \int_0^\infty \int_0^\infty e^{-p(x^2+y^2)} e^{i(x^2+y^2)}\, dx\, dy$$

$$= \int_0^{\pi/2} \int_0^\infty e^{-pr^2} e^{ir^2} r\, dr\, d\theta$$

$$= \frac{\pi}{2} \int_0^\infty e^{-r^2(p-i)} r\, dr$$

$$= \frac{\pi}{2} \frac{1}{2(p - i)}$$

which was to be proved.

To complete the evaluation of the integrals, from steps 2 and 3 we have

$$I^2 = \lim_{p \to 0} J_p = \frac{\pi i}{4}$$

In step 1 we showed that $I > 0$. Therefore we must take the square root with positive imaginary part $i^{1/2} = (1 + i)/\sqrt{2}$, which gives the result as stated.

The author is indebted to Professor R. P. Boas for providing the argument just presented, which avoids any use of complex contour integration.†

EXERCISES 6.5

Apply the method of stationary phase to find asymptotic formulas for the following functions when $t \to \infty$.

1. $f(t) = \int_{-2}^{2} e^{it\mu^2} \, d\mu$
2. $f(t) = \int_{-\pi/2}^{\pi/2} e^{it \cos \theta} \cos^2 \theta \, d\theta$
3. $f(t) = \int_{0}^{\pi/2} e^{it \cos \theta} \cos \theta \, d\theta$
4. $f(t) = \int_{-\pi}^{\pi} e^{it \cos \theta} e^{-im\theta} \, d\theta$, $m = 1, 2, 3, \ldots$

5. Use Exercise 4, together with the integral representation (3.2.9), to obtain an asymptotic formula for the Bessel function $J_m(t)$ when $t \to \infty$.

6.6 ASYMPTOTIC ANALYSIS OF THE TELEGRAPH EQUATION

We now apply the asymptotic methods to an initial-value problem for the telegraph equation. This example illustrates all the methods discussed in this chapter.

We consider the initial-value problem

$$u_{tt} + 2\beta u_t + \alpha u = c^2 u_{xx}$$

$$u(x; 0) = 0$$

$$u_t(x; 0) = f_2(x)$$

where f_2 is supposed to be a smooth, even, real-valued, absolutely integrable function with an absolutely integrable Fourier transform, for example, $f_2(x) = e^{-x^2/2}$. Since the telegraph equation describes signals in a transmission cable, it is natural to study the asymptotic behavior of

$$u(x + vt; t) \qquad t \to \infty$$

In other words, we study the behavior of a signal when we move backward along the cable at constant speed v.

---

† See also the article by Harley Flanders, "On the Fresnel Integrals," *American Mathematical Monthly*, vol. 89, 1982, pp. 264–266.

In Chap. 5 we found that the solution of the telegraph equation takes different forms depending on whether $\beta^2 < \alpha$, $\beta^2 = \alpha$, or $\beta^2 > \alpha$.

Solution in the Critical Case

In the second case ($\beta^2 = \alpha$), the solution of the initial-value problem has the form

$$u(x; t) = \frac{e^{-\beta t}}{2c} \int_{x-ct}^{x+ct} f_2(\xi)\, d\xi$$

which is a superposition of two damped but undistorted plane waves. This solution can be analyzed without asymptotic methods. We now proceed to analyze case 1 and case 3 separately.

Solution in Case 1: $\beta^2 < \alpha$

In case 1, $\beta^2 < \alpha$ and the solution has the Fourier representation

$$u(x; t) = e^{-\beta t} \int_{-\infty}^{\infty} F(\mu) \sin \left(t\sqrt{\alpha - \beta^2 + (c\mu)^2}\right) e^{i\mu x}\, d\mu \qquad (6.6.1)$$

where

$$F(\mu) = \frac{F_2(\mu)}{\sqrt{\alpha - \beta^2 + (c\mu)^2}}$$

We can apply the method of stationary phase to the integral defining $u(x + vt; t)$. This is written in the form

$$e^{\beta t} u(x + vt; t) = \frac{1}{2i} \int_{-\infty}^{\infty} e^{it\varphi_+(\mu)} F(\mu) e^{i\mu x}\, d\mu - \frac{1}{2i} \int_{-\infty}^{\infty} e^{it\varphi_-(\mu)} F(\mu) e^{i\mu x}\, d\mu$$

where

$$\varphi_\pm(\mu) = v\mu \pm \sqrt{\alpha - \beta^2 + (c\mu)^2}$$

To examine the first integral, we must find the stationary point(s) of the phase function φ_+. Computing the first and second derivatives, we have

$$\varphi'_+(\mu) = v + \frac{c^2\mu}{\sqrt{\alpha - \beta^2 + (c\mu)^2}}$$

$$\varphi''_+(\mu) = \frac{c^2(\alpha - \beta^2)}{[\alpha - \beta^2 + (c\mu)^2]^{3/2}}$$

Solving the equation $\varphi'_+(\mu) = 0$, we see that the stationary point is given by

$$\mu_v^+ = -\frac{v}{c} \sqrt{\frac{\alpha - \beta^2}{c^2 - v^2}} \qquad |v| < c$$

If $|v| > c$, the phase function φ_+ has no stationary points. Substituting into the above formulas, we have

$$\varphi_+(\mu_v^+) = \frac{1}{c} \sqrt{(\alpha - \beta^2)(c^2 - v^2)}$$

$$\varphi_+''(\mu_v^+) = \frac{1}{c} (c^2 - v^2)^{3/2}(\alpha - \beta^2)^{-1/2}$$

With this information we may use the method of stationary phase:

$$\int_{-\infty}^{\infty} e^{it\varphi_+(\mu)} F(\mu) e^{i\mu x} \, d\mu = \sqrt{\frac{2\pi}{t\varphi_+''(\mu_v^+)}} \, e^{it\varphi_+(\mu_v^+)} e^{i\pi/4} F(\mu_v^+) e^{i\mu_v^+ x} + O\left(\frac{1}{t}\right)$$

We must also consider the phase function $\varphi_-(\mu) = v\mu - \sqrt{\alpha - \beta^2 + (c\mu)^2}$ for which the stationary point is

$$\mu_v^- = \frac{v}{c} \sqrt{\frac{\alpha - \beta^2}{c^2 - v^2}} \qquad |v| < c$$

$$\varphi_-(\mu_v^-) = -\frac{1}{c} \sqrt{(\alpha - \beta^2)(c^2 - v^2)} = -\varphi_+(\mu_v^+)$$

$$\varphi_-''(\mu_v^-) = -\frac{1}{c} (c^2 - v^2)^{3/2}(\alpha - \beta^2)^{-1/2} = -\varphi_+''(\mu_v^+)$$

The method of stationary phase now gives

$$\int_{-\infty}^{\infty} e^{it\varphi_-(\mu)} F(\mu) e^{i\mu x} \, d\mu = \sqrt{\frac{2\pi}{t|\varphi_-''(\mu_v^-)|}} \, e^{it\varphi_-(\mu_v^-)} e^{-i\pi/4} F(\mu_v^-) e^{i\mu_v^- x} + O\left(\frac{1}{t}\right)$$

Subtracting the two results gives

$$\boxed{u(x + vt; t) = C_1 \frac{e^{-\beta t}}{\sqrt{t}} \sin\left(\omega t + \mu_v^+ x + \frac{\pi}{4}\right) + O\left(\frac{e^{-\beta t}}{t}\right) \qquad |v| < c} \qquad (6.6.2)$$

where we have set

$$C_1 = (2\pi c)^{1/2}(\alpha - \beta^2)^{1/4}(c^2 - v^2)^{-3/4} F(\mu_v^+)$$

$$\omega = \frac{1}{c}(\alpha - \beta^2)^{1/2}(c^2 - v^2)^{1/2} = \varphi_+(\mu_v^+)$$

and used the fact that F is an even function in the form $F(\mu_v^+) = F(\mu_v^-)$. If $|v| > c$, the phase function has no stationary points and we may apply the method of integration by parts, with the result

$$\boxed{u(x + vt; t) = O\left(\frac{e^{-\beta t}}{t}\right) \qquad |v| > c} \qquad (6.6.3)$$

Thus we see that when $t \to \infty$, the larger values of the solution are contained within $x \pm ct$. Thus "most" of the disturbance moves at speeds less than or equal to c. This completes the analysis of case 1, where $\beta^2 < \alpha$.

Solution in Case 3: $\beta^2 > \alpha$

We now analyze case 3, where $\beta^2 > \alpha$. We have shown in Chap. 5 that in this case the solution has the Fourier representation

$$e^{\beta t} u(x; t) = I_1(x; t) + I_2(x; t)$$

where

$$I_1(x; t) = \int_{|c\mu| > \sqrt{\beta^2 - \alpha}} F(\mu) \sin t\sqrt{(c\mu)^2 - (\beta^2 - \alpha)}\, e^{i\mu x}\, d\mu$$

$$I_2(x; t) = \int_{|c\mu| < \sqrt{\beta^2 - \alpha}} F(\mu) \sinh t\sqrt{\beta^2 - \alpha - (c\mu)^2}\, e^{i\mu x}\, d\mu$$

$$F(\mu) = \begin{cases} \dfrac{F_2(\mu)}{\sqrt{(c\mu)^2 - (\beta^2 - \alpha)}} & |c\mu| > \sqrt{\beta^2 - \alpha} \\[2ex] \dfrac{F_2(\mu)}{\sqrt{\beta^2 - \alpha - (c\mu)^2}} & |c\mu| < \sqrt{\beta^2 - \alpha} \end{cases}$$

To obtain a preliminary idea of the result, notice that at $\mu = 0$ the second integrand $= F(0) \sinh t\sqrt{\beta^2 - \alpha}$, which becomes exponentially large when $t \to \infty$. The first integral is much smaller, since

$$|I_1(x; t)| \leq \int_{|c\mu| > \sqrt{\beta^2 - \alpha}} |F(\mu)|\, d\mu$$

$$= O(1) \qquad t \to \infty$$

To analyze the second integral, write $\sinh \theta = \frac{1}{2}(e^\theta - e^{-\theta})$. The negative exponential contributes at most $O(1)$ when $t \to \infty$. To analyze the remaining term, we apply Laplace's method to the integral

$$\int_{|c\mu| < \sqrt{\beta^2 - \alpha}} e^{th(\mu)} e^{i\mu x} F(\mu)\, d\mu$$

where

$$h(\mu) = \sqrt{\beta^2 - \alpha - (c\mu)^2}$$

$$h'(\mu) = -\frac{c^2\mu}{\sqrt{\beta^2 - \alpha - (c\mu)^2}}$$

$$h''(\mu) = -\frac{c^2(\beta^2 - \alpha)}{[\beta^2 - \alpha - (c\mu)^2]^{3/2}}$$

The maximum of h occurs when $\mu = 0$, where

$$h(0) = \sqrt{\beta^2 - \alpha}$$

$$h''(0) = -\frac{c^2}{\sqrt{\beta^2 - \alpha}}$$

Therefore Laplace's method gives

$$\int_{|c\mu| < \sqrt{\beta^2 - \alpha}} e^{th(\mu)} F(\mu) e^{i\mu x}\, d\mu = \sqrt{2\pi}\, \frac{e^{t\sqrt{\beta^2 - \alpha}}}{\sqrt{-th''(0)}} \left[F(0) + O\left(\frac{1}{\sqrt{t}}\right) \right]$$

Therefore we have shown that

$$\boxed{\; u(x; t) = \frac{e^{-qt}}{\sqrt{t}} \left[C_2 + O\left(\frac{1}{\sqrt{t}}\right) \right] \qquad t \to \infty \;} \qquad (6.6.4)$$

where

$$q = \beta - \sqrt{\beta^2 - \alpha}$$

$$C_2 = \frac{\sqrt{2\pi}}{c} (\beta^2 - \alpha)^{1/4} F(0)$$

If $v \neq 0$, we have to deal with the integral

$$e^{\beta t} u(x + vt; t) = \int_{|c\mu| < \sqrt{\beta^2 - \alpha}} \exp\{t[\sqrt{\beta^2 - \alpha - (c\mu)^2} + iv\mu]\}\, F(\mu) e^{i\mu x}\, d\mu$$

We cannot apply Laplace's method because of the imaginary term $iv\mu$ in the exponent. Nevertheless, we can analyze this integral by a direct analysis and integration by parts.

To do this, we note that for any $\delta > 0$, the contribution to the integral from the values of μ with $\delta < |c\mu| < \sqrt{\beta^2 - \alpha}$ can be ignored; specifically

$$\int_{\delta < |c\mu| < \sqrt{\beta^2 - \alpha}} \exp\{t[\sqrt{\beta^2 - \alpha - (c\mu)^2} + iv\mu]\}\, F(\mu) e^{i\mu x}\, d\mu$$

$$= O(e^{t\sqrt{\beta^2 - \alpha - \delta^2}}) \qquad t \to \infty$$

However, at $\mu = 0$ the integrand is $e^{t\sqrt{\beta^2 - \alpha}}$, which is larger. Therefore we may ignore this contribution to the integral.

To analyze the remaining contribution to the integral, we let $h(\mu) = iv\mu + \sqrt{\beta^2 - \alpha - (c\mu)^2}$ and integrate by parts:

$$\int_{-\delta}^{\delta} e^{th(\mu)} F(\mu) e^{i\mu x}\, d\mu = \int_{-\delta}^{\delta} \frac{F(\mu) e^{i\mu x}}{th'(\mu)}\, d(e^{th(\mu)})$$

$$= \frac{F(\mu) e^{i\mu x}}{th'(\mu)} e^{th(\mu)} \Bigg|_{-\delta}^{\delta} - \frac{1}{t} \int_{-\delta}^{\delta} e^{th(\mu)} \frac{d}{d\mu} \left[\frac{F(\mu) e^{i\mu x}}{h'(\mu)} \right] d\mu$$

But $e^{th(\mu)} = \exp\left[t\sqrt{\beta^2 - \alpha - (c\mu)^2}\right] \leq e^{t\sqrt{\beta^2 - \alpha}}$ for $-\delta \leq \mu \leq \delta$. Therefore the above integral is $O(e^{t\sqrt{\beta^2 - \alpha}}/t)$, $t \to \infty$, and we have proved that

$$\boxed{u(x + vt;\, t) = O\left(\frac{e^{-qt}}{t}\right) \qquad v \neq 0,\, t \to \infty} \qquad (6.6.5)$$

where $q = \beta - \sqrt{\beta^2 - \alpha}$. This completes the asymptotic analysis of the telegraph equation.

EXERCISES 6.6

1. Let $u(x;\, t)$ be the solution of the equation $u_{tt} + m^2 u = c^2 u_{xx}$ with $u(x;\, 0) = 0$, $u_t(x;\, 0) = f_2(x)$, and $m,\, c > 0$. Show that we have the asymptotic formula

$$u(x;\, t) = \frac{C_1}{\sqrt{t}} \sin\left(mt + \frac{\pi}{4}\right) + O\left(\frac{1}{t}\right) \qquad t \to \infty$$

where $\qquad C_1 = \left(\frac{1}{2\pi m c^2}\right)^{1/2} \int_{-\infty}^{\infty} f_2(x)\, dx$

[*Hint:* Take $\beta = 0$, $v = 0$ in (6.6.2).]

2. Consider the telegraph equation in the limiting case $\alpha = 0$:

$$u_{tt} + 2\beta u_t = c^2 u_{xx}$$

$$u(x;\, 0) = 0 \qquad u_t(x;\, 0) = f_2(x)$$

Show that we have the asymptotic formula

$$u(x;\, t) = \frac{C_2}{\sqrt{t}} + O\left(\frac{1}{t}\right) \qquad t \to \infty$$

$$u(x + vt;\, t) = O\left(\frac{1}{t}\right) \qquad t \to \infty,\, v \neq 0$$

for an appropriate constant C_2.

*3. Consider the initial-value problem for the telegraph equation

$$u_{tt} + 2\beta u_t + \alpha u = c^2 u_{xx}$$

$$u(x;\, 0) = f_1(x) \qquad u_t(x;\, 0) = 0$$

Repeat the analysis of this section to determine an asymptotic formula for $u(x;\, t)$ when $t \to \infty$.

CHAPTER
7

NUMERICAL
SOLUTIONS

Introduction

In this chapter we present methods for obtaining approximate solutions of certain boundary-value problems for partial differential equations. In practice, this may be necessary for a variety of reasons, e.g., the geometry of the boundary or the explicit form of the coefficients in the equation. Even if the problem admits an explicit solution by a series or an integral, it may be difficult to compute the numerical value of the solution by using the series or integral.

The method of finite differences is completely general. We replace the continuous variables x, y, z, t by discrete variables x_i, y_j, z_k, t_m, which assume a finite number of values. Then we replace each of the derivatives in the equation by a suitable difference quotient. This converts the differential equation to a system of algebraic equations, which we may try to solve. Having obtained the solution, we may inquire about the accuracy of the approximation, i.e., the magnitude of the difference between the true solution and the approximation obtained.

7.1 CALCULUS OF FINITE DIFFERENCES
FOR ORDINARY DIFFERENTIAL
EQUATIONS

As a simple illustration of the method, we consider the following oversimplified problem:

$$u'(x) = e^{x^2} \qquad \text{for } 0 \le x \le 1, \, u(0) = 0$$

324

[Of course, the exact solution can be written as an integral: $u(x) = \int_0^x e^{t^2} \, dt$.] To employ the method of finite differences, we select a mesh of points $0 = x_0 < x_1 < x_2 < \cdots < x_n = 1$ and replace each of the derivatives by the corresponding difference quotient; thus

$u(x_i)$ is replaced by u_i $0 \le i \le n$

$u'(x_i)$ is replaced by $\dfrac{u_i - u_{i-1}}{x_i - x_{i-1}}$ $1 \le i \le n$

Thus we have the equations

$$u_0 = 0$$

$$\frac{u_i - u_{i-1}}{x_i - x_{i-1}} = e^{x_i^2} \qquad 1 \le i \le n$$

To solve these equations, we write the telescoping sum

$$u_j = u_j - u_0 = \sum_{i=1}^{j} (u_i - u_{i-1}) = \sum_{i=1}^{j} e^{x_i^2}(x_i - x_{i-1})$$

This is an approximating sum for the riemannian integral $\int_0^{x_j} e^{x^2} \, dx$, which is the exact solution. The difference between the approximate solution and the exact solution is

$$u(x_j) - \sum_{i=1}^{j} e^{x_i^2}(x_i - x_{i-1}) = \sum_{i=1}^{j} \int_{x_{i-1}}^{x_i} (e^{x^2} - e^{x_i^2}) \, dx$$

which is less than x_j times the largest of the differences $e^{x_i^2} - e^{x_{i-1}^2}$, for $i = 1, \ldots, n$.

Example 7.1.1. Find the approximate solution for $u(\tfrac{1}{2})$ of the equation $u'(x) = e^{x^2}$, $u(0) = 0$, corresponding to the mesh $x_0 = 0$, $x_1 = 0.1$, $x_2 = 0.2$, $x_3 = 0.3$, $x_4 = 0.4$, $x_5 = 0.5$. Estimate the error.

Solution. The approximate solution for $u(\tfrac{1}{2})$ is $\sum_{i=1}^{5} e^{x_i^2}(x_i - x_{i-1})$. We have the following values, correct to five decimals:

$$e^{(0.1)^2} = e^{0.01} = 1.01005$$

$$e^{(0.2)^2} = e^{0.04} = 1.04081$$

$$e^{(0.3)^2} = e^{0.09} = 1.09417$$

$$e^{(0.4)^2} = e^{0.16} = 1.17351$$

$$e^{(0.5)^2} = e^{0.25} = 1.28403$$

The approximation for $u(\tfrac{1}{2})$ is thus $(0.1)(5.60257) = 0.56026$, correct to four decimals. To estimate the error, we note that the largest of the differences $e^{x_i^2} - e^{x_{i-1}^2}$ occurs for $i = 5$; thus $e^{0.25} - e^{0.16} = 0.11052$, and the error is less than

$(0.5)(0.11052) = 0.05526$. Thus the approximation $u(\tfrac{1}{2}) \cong 0.56026$ is correct to within 0.06, slightly more than a 10 percent error. ●

In our next example we choose a more general first-order linear differential equation.

Example 7.1.2. Use the method of finite differences to obtain an approximate solution of the equation $u'(x) + xu(x) = 1$ for $0 \le x \le 1$ with $u(0) = 0$. Use the mesh $x_i = i/10$ for $i = 1, 2, \ldots, 10$.

Solution. We let $u_0 = 0$ and $u_i = u(i/10)$ for $i = 1, 2, \ldots, 10$. The differential equation is replaced by the finite difference equations

$$10(u_i - u_{i-1}) + \frac{i}{10} u_i = 1 \qquad i = 1, \ldots, 10$$

This is written in the form $(10 + i/10)u_i = 1 + 10u_{i-1}$, and thus the solution is

$$u_i = \frac{10(1 + 10u_{i-1})}{100 + i} \qquad i = 1, \ldots, 10$$

Recalling that $u_0 = 0$, we have, correct to four decimals,

$$u_1 = \frac{10}{101} = 0.09901 \qquad u_6 = \frac{10(5.44918)}{106} = 0.51407$$

$$u_2 = \frac{10(1.9901)}{102} = 0.19511 \qquad u_7 = \frac{10(6.1407)}{107} = 0.57389$$

$$u_3 = \frac{10(2.9511)}{103} = 0.28651 \qquad u_8 = \frac{10(6.7389)}{108} = 0.62397$$

$$u_4 = \frac{10(3.8651)}{104} = 0.37164 \qquad u_9 = \frac{10(7.2397)}{109} = 0.66419$$

$$u_5 = \frac{10(4.7164)}{105} = 0.44918 \qquad u_{10} = \frac{10(7.6419)}{110} = 0.69472$$ ●

The method of finite differences may be applied to differential equations of higher order. To handle problems of this type, we replace the higher derivatives by appropriate difference quotients. For example, if the mesh consists of equally spaced points $x_1, \ldots, x_n$ with $x_i - x_{i-1} = \Delta x$, then we replace the second derivative $u''(x_i)$ by the *symmetric difference quotient*

$$\frac{u(x_i + \Delta x) - 2u(x_i) + u(x_i - \Delta x)}{(\Delta x)^2}$$

Example 7.1.3. Use the method of finite differences to obtain an approximation to the solution of $u''(x) = e^{x^2}$ with the boundary conditions $u(0) = 0$, $u(1) = 0$. Use the mesh $x_i = i/4$ for $i = 1, 2, 3, 4$.

Solution. We have the following system of equations for u_i:

$$u_0 = 0$$

$$16(u_2 - 2u_1 + u_0) = e^{0.0625}$$

$$16(u_3 - 2u_2 + u_1) = e^{0.25}$$

$$16(u_4 - 2u_3 + u_2) = e^{0.5625}$$

$$u_4 = 0$$

These may be rewritten in the form

$$-2u_1 + u_2 = (0.0625)e^{0.0625} = 0.06653$$

$$u_1 - 2u_2 + u_3 = (0.0625)e^{0.25} = 0.08025$$

$$u_2 - 2u_3 = (0.0625)e^{0.5625} = 0.10969$$

This system of three simultaneous equations may be solved by row reduction, with the result $u_1 = -0.11745$, $u_2 = -0.16837$, $u_3 = -0.13903$. ●

EXERCISES 7.1

1. For an arbitrary smooth function $u(x)$, define the *forward replacement error* by

$$e^+(x; h) = u'(x) - \frac{u(x + h) - u(x)}{h}$$

Compute $e^+(x; h)$ for the following cases.
(a) $u(x) = \cos x$, $x = 0$, $h = 0.1$
(b) $u(x) = x^2$, $x = 1$, $h = 0.1$
(c) $u(x) = e^{3x}$, $x = 0$, $h = 0.1$
(d) $u(x) = x \sin 15x$, $x = 0$, $h = 0.1$

2. Define the *backward replacement error* by

$$e^-(x; h) = u'(x) - \frac{u(x) - u(x - h)}{h}$$

Compute $e^-(x; h)$ for the cases of Exercise 1.

3. Define the *symmetric replacement error* by

$$e^0(x; h) = u'(x) - \frac{u(x + h/2) - u(x - h/2)}{h}$$

Compute $e^0(x; h)$ for all cases in Exercise 1.

4. With reference to Exercises 1 to 3, show that the following bounds hold for an arbitrary smooth function $u(x)$:
(a) $\left| e^+(x; h) \right| \le \frac{1}{2} h \max_{x \le t \le x+h} \left| u''(t) \right|$
(b) $\left| e^-(x; h) \right| \le \frac{1}{2} h \max_{x-h \le t \le x} \left| u''(t) \right|$
(c) $\left| e^0(x; h) \right| \le \frac{1}{4} h \max_{x-h/2 \le t \le x+h/2} \left| u''(t) \right|$

Hint: Begin with the Taylor formula with remainder

$$u(x + h) - u(x) = hu'(x) + \int_x^{x+h} (t - x)u''(t)\, dt$$

5. Define the *second-order symmetric replacement error* by

$$E^0(x; h) = u''(x) - \frac{u(x + h) + u(x - h) - 2u(x)}{h^2}$$

Compute $E^0(x; h)$ for the following cases:
(a) $u(x) = x^2$, $x = 0$, $h = 0.1$
(b) $u(x) = e^{3x}$, $x = 0$, $h = 0.1$
(c) $u(x) = \cos x$, $x = 0$, $h = 0.1$
(d) $u(x) = x^2 \sin 15x$, $x = 0$, $h = 0.1$

6. Show that the second-order symmetric replacement error satisfies the bound

$$|E^0(x; h)| \le \frac{h^2}{12} \max_{|t-x|\le h} |u^{(iv)}(t)|$$

Hint: Use the Taylor formula of order 4:

$$u(x + h) - u(x) = hu'(x) + \frac{h^2}{2} u''(x) + \frac{h^3}{6} u'''(x) + \int_x^{x+h} \frac{(t - x)^3}{6} u^{(iv)}(t)\, dt$$

7. Define the *second-order forward replacement error* by

$$E^+(x; h) = u''(x) - \frac{u(x + 2h) + u(x) - 2u(x + h)}{h^2}$$

Show that we have the bound $|E^+(x; h)| \le 2h \max_{x \le t \le x+2h} |u'''(t)|$.

8. Find the approximate solution of the equation $u'(x) = e^{-x^2}$, $u(0) = 0$, by the method of finite differences. Use the mesh $x_i = i/10$ for $i = 1, \ldots, 10$.

9. In Exercise 8, estimate the error in computing $u(\frac{1}{2})$.

10. Use the method of finite differences to obtain an approximate numerical solution of the equation $u'(x) + 3xu(x) = 2$ for $0 < x < 1$, with $u(0) = 0$. Use the mesh $x_i = i/10$ for $i = 1, \ldots, 10$.

11. Use the method of finite differences to obtain an approximate solution to the equation $u''(x) = -e^{-x^2}$ with the boundary conditions $u(0) = 0$, $u(1) = 0$. Use the mesh $x_i = i/6$ for $i = 1, \ldots, 6$.

12. Let $u(x)$ be an arbitrary smooth function, and define the *fourth-order symmetric replacement error* by

$$F^0(x; h) = u^{(iv)}(x) - \frac{u(x + 2h) - 4u(x + h) + 6u(x) - 4u(x - h) + u(x - 2h)}{h^4}$$

Show that we have the following bound:

$$|F^0(x; h)| \le \frac{h^2}{45} \max_{|t-x|\le h} |u^{(vi)}(t)|$$

7.2 NUMERICAL SOLUTION OF THE ONE-DIMENSIONAL HEAT EQUATION

In this section we begin the study of numerical solutions of partial differential equations. As our first model, we consider the following problem for the heat equation:

$$u_t = Ku_{xx} \qquad t > 0, 0 < x < L$$

$$u(0; t) = 0 = u(L; t) \qquad t > 0 \qquad\qquad (7.2.1)$$

$$u(x; 0) = f(x) \qquad 0 < x < L$$

Physically this represents heat flow in a slab whose faces are maintained at absolute zero. In Chap. 2 we solved this problem by using Fourier series.

Formulation of a Difference Equation

To solve this problem by the method of finite differences, we choose a mesh $0 = x_0 < x_1 < x_2 \cdots < x_{n+1} = L$ with $x_{i+1} - x_i = \Delta x$, independent of i for $i = 0, \ldots, n$. We replace $u(x_i; t)$ by $u_i(t)$ for $i = 0, 1, \ldots, n$. Similarly the t axis is replaced by a mesh of points (t_i), with $t_{i+1} - t_i = \Delta t$. To employ the method of finite differences, we make the following replacements for the partial derivatives which occur in the heat equation:

$$u_t(x_i; t) \qquad \text{is replaced by} \qquad \frac{u_i(t + \Delta t) - u_i(t)}{\Delta t}$$

$$u_{xx}(x_i; t) \qquad \text{is replaced by} \qquad \frac{u_{i+1}(t) + u_{i-1}(t) - 2u_i(t)}{(\Delta x)^2}$$

Thus the partial differential equation is replaced by the following linear equations with boundary conditions:

$$\frac{u_i(t + \Delta t) - u_i(t)}{\Delta t} = K \frac{u_{i+1}(t) + u_{i-1}(t) - 2u_i(t)}{(\Delta x)^2} \qquad 1 \le i \le n$$

$$u_0(t) = 0 \qquad u_{n+1}(t) = 0$$

We solve these linear equations for $u_i(t + \Delta t)$ to obtain

$$\boxed{u_i(t + \Delta t) = \frac{K \Delta t}{(\Delta x)^2} u_{i+1}(t) + \frac{K \Delta t}{(\Delta x)^2} u_{i-1}(t) + \left[1 - \frac{2K \Delta t}{(\Delta x)^2} \right] u_i(t)} \qquad (7.2.2)$$

Thus $u_i(t + \Delta t)$ is a weighted average of the numbers $u_{i+1}(t)$, $u_{i-1}(t)$, and $u_i(t)$ provided that $2K \Delta t/(\Delta x)^2 \le 1$. When this condition is satisfied, all the coefficients in (7.2.2) are positive and their sum is unity,[†] thus the term *weighted*

[†] For other equations we may require only that the coefficients in the difference scheme be positive and sum to *less* than unity, e.g., the equation $u_t = u_{xx} - u$.

average. For given values of the diffusivity K and the mesh size Δx, this condition can be realized by a suitable choice of the time step Δt. It is to our advantage to choose Δt as large as possible, consistent with this restriction. In particular, if we choose $2K \Delta t/(\Delta x)^2 = 1$, then the coefficient of $u_i(t)$ is zero and the formula for $u_i(t + \Delta t)$ is simplified. It is shown in advanced works on numerical analysis that the condition $2K \Delta t/(\Delta x)^2 \leq 1$ is a *stability condition*, meaning that the numerical approximations will be close to the true solution in a suitable sense.

Computational Molecule

We can illustrate graphically the numerical algorithm implied by equation (7.2.2) in terms of the "computational molecule" depicted in Figure 7.2.1. The coefficient $r = K \Delta t/(\Delta x)^2$ multiplies the current values $u_{i+1}(t)$, $u_{i-1}(t)$, while the coefficient $1 - 2r$ multiplies the current value $u_i(t)$. Together they produce the new value $u_i(t + \Delta t)$. On a larger scale we may graph the boundary values and unknown values as part of a rectangular grid, as depicted in Figure 7.2.2. The initial and boundary values are represented by small cubes while the unknown solution values are represented by black dots. We now illustrate with an example which affords comparison between the numerical and Fourier series solutions.

Examples and Comparison with Fourier Method

Example 7.2.1. Find the numerical solution of the heat equation $u_t = Ku_{xx}$ for $0 < x < 1$, $t > 0$ with the boundary conditions $u(0; t) = 0$, $u(1; t) = 0$ and the initial condition $u(x; 0) = 100x(1 - x)$. Use the mesh with $\Delta x = 0.1$, $K \Delta t/(\Delta x)^2 = \frac{1}{3}$, and compute the solution for $0 \leq t \leq 20 \Delta t$.

Solution. In this case the basic equations (7.2.2) take the form

$$u_i(t + \Delta t) = \frac{K \Delta t}{(\Delta x)^2} u_{i+1}(t) + \frac{K \Delta t}{(\Delta x)^2} u_{i-1}(t) + \left[1 - \frac{2K \Delta t}{(\Delta x)^2} \right] u_i(t)$$

$$= \tfrac{1}{3}u_{i-1}(t) + \tfrac{1}{3}u_{i+1}(t) + \tfrac{1}{3}u_i(t)$$

Since the initial conditions are symmetric about $x = \frac{1}{2}$, we may restrict our attention to the values $x = 0$, 0.1, 0.2, 0.3, 0.4, 0.5. The computational grid is depicted in Figure 7.2.3.   ●

It is instructive to compare the numerical solution with the exact solution obtained by Fourier series. To do this, we recall the Fourier sine series for $0 < x < 1$

$$100x(1 - x) = \frac{800}{\pi^3} \left(\sin \pi x + \frac{1}{27} \sin 3\pi x + \cdots \right)$$

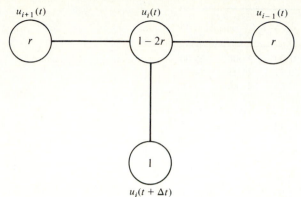

FIGURE 7.2.1
Computational molecule for the heat equation.

The solution of the heat equation with this initial condition is

$$u(x; t) = \frac{800}{\pi^3}\left(\sin \pi x \, e^{-\pi^2 Kt} + \frac{1}{27} \sin 3\pi x \, e^{-(3\pi)^2 Kt} + \cdots\right)$$

We compute the values of the various terms for $x = 0, 0.1, 0.2, 0.3, 0.4, 0.5$ and $Kt = K(20 \, \Delta t) = \frac{20}{3}(\Delta x)^2 = \frac{1}{15}$.

x	0	0.1	0.2	0.3	0.4	0.5
$\sin \pi x$	0	0.309	0.588	0.809	0.951	1.000
$\frac{1}{27} \sin 3\pi x$	0	0.011	0.035	0.030	-0.022	-0.037
$\frac{800}{\pi^3} \sin \pi x \, e^{-\pi^2/15}$	0	4.128	7.856	10.808	12.705	13.363
$\frac{800}{\pi^3} \frac{1}{27} \sin 3\pi x \, e^{-9\pi^2/15}$	0	0.000	0.002	0.002	-0.002	-0.002
$u(x; t)$	0	4.128	7.858	10.810	12.703	13.361

Thus we see that the numerical solution and the Fourier solution agree to within 2 percent for the tabulated values!

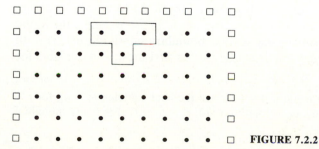

FIGURE 7.2.2

t \ x	0.0	0.1	0.2	0.3	0.4	0.5
0	0	9.00	16.00	21.00	24.00	25.00
Δt	0	8.33	15.33	20.33	23.33	24.33
2 Δt	0	7.89	14.66	19.66	22.66	23.66
3 Δt	0	7.52	14.07	18.99	21.99	22.99
4 Δt	0	7.20	13.53	18.35	21.32	22.32
5 Δt	0	6.91	13.03	17.73	20.66	21.65
6 Δt	0	6.65	12.56	17.14	20.01	20.99
7 Δt	0	6.40	12.12	16.57	19.38	20.34
8 Δt	0	6.17	11.70	16.02	18.76	19.70
9 Δt	0	5.96	11.30	15.49	18.16	19.07
10 Δt	0	5.75	10.92	14.98	17.57	18.46
11 Δt	0	5.56	10.55	14.49	17.00	17.87
12 Δt	0	5.37	10.20	14.00	16.45	17.29
13 Δt	0	5.19	9.86	13.55	15.91	16.73
14 Δt	0	5.02	9.53	13.11	15.40	16.18
15 Δt	0	4.85	9.22	12.68	14.90	15.66
16 Δt	0	4.69	8.92	12.27	14.41	15.15
17 Δt	0	4.54	8.63	11.87	13.61	14.66
18 Δt	0	4.39	8.35	11.37	13.38	13.96
19 Δt	0	4.25	8.02	11.03	12.90	13.57
20 Δt	0	4.09	7.77	10.65	12.50	13.12

FIGURE 7.2.3

This example suggests that we can expect close agreement between the numerical solution and the exact solution of the heat equation. The following general theorem affirms this result in general.

> **Theorem 7.2.1.**[†] Let $u(x; t)$ be the solution of the heat equation $u_t = K u_{xx}$ for $t > 0$, $0 < x < L$ with the boundary conditions $u(0; t) = \alpha(t)$, $u(L; t) = \beta(t)$ and the initial condition $u(x; 0) = f(x)$. Let $u_i(t)$ be the numerical solution obtained with $2K \, \Delta t/(\Delta x)^2 \le 1$. Then
>
> $$\max_{1 \le i \le n} |u(x_i; t) - u_i(t)| \le \frac{MK \, \Delta t(\Delta x)^4}{135}$$
>
> where M is a bound for $u(x; t)$ and its first six derivatives.

In practice, it may be difficult find the constant M.

To illustrate the role of the stability condition $2K \, \Delta t/(\Delta x)^2 \le 1$, we consider a simple example with two different mesh sizes. In this example the initial data $u(x; 0) = f(x)$ do not satisfy the second boundary condition $u(1; t) = 0$; nevertheless the difference scheme is stable for the first mesh and unstable for the second.

> **Example 7.2.2.** Find the numerical solution of the heat equation $u_t = u_{xx}$ for $t > 0$, $0 < x < 1$ with the boundary condition $u(0; t) = 0$, $u(1; t) = 0$ and the initial

---

† W. E. Milne, *Numerical Solution of Differential Equations*, Dover, New York, 1970, p. 122.

condition $u(x; 0) = 4x$. First use the mesh size $\Delta x = \frac{1}{4}$, $\Delta t = \frac{1}{32}$, then use the mesh size $\Delta x = \frac{1}{4}$, $\Delta t = \frac{1}{8}$.

Solution. For the first choice, we have $K \, \Delta t/(\Delta x)^2 = \frac{1}{32}/\frac{1}{16} = \frac{1}{2}$, and the form of the solution (7.2.2) is $u_i(t + \Delta t) = \frac{1}{2}u_{i+1}(t) + \frac{1}{2}u_{i-1}(t)$. Applying this to the initial data $u(x, 0) = 4x$, we have the following table of values:

t \ x	0	$\frac{1}{4}$	$\frac{1}{2}$	$\frac{3}{4}$	1
0	0	1	2	3	4
$\frac{1}{32}$	0	1	2	3	0
$\frac{1}{16}$	0	1	2	1	0
$\frac{3}{32}$	0	1	1	1	0
$\frac{1}{8}$	0	0.5	1	0.5	0

For the second choice we have $K \, \Delta t/(\Delta x)^2 = \frac{1}{8}/\frac{1}{16} = 2$, the form of the solution (7.2.2) is $u_i(t + \Delta t) = 2u_{i+1}(t) + 2u_{i-1}(t) - 3u_i(t)$, and we have the following table of values:

t \ x	0	$\frac{1}{4}$	$\frac{1}{2}$	$\frac{3}{4}$	1
0	0	1	2	3	4
$\frac{1}{8}$	0	1	2	3	0
$\frac{1}{4}$	0	1	2	-5	0
$\frac{3}{8}$	0	1	-14	19	0
$\frac{1}{2}$	0	-31	82	-85	0

    From this example, we see that the second choice $\Delta x = \frac{1}{4}$, $\Delta t = \frac{1}{8}$ leads to an absurd result, since we expect on physical grounds that the temperature will remain positive and tend to zero as t becomes larger; in fact, this is known from the analysis of Chap. 2, where we solved this problem by a Fourier sine series. The absurd values in the example are not surprising, in view of the fact that the ratio $2K \, \Delta t/(\Delta x)^2$ is larger than unity.

    The method of finite differences can also be applied to solve the heat equation with boundary conditions involving the derivative u_x. Consider, e.g., the problem

$$u_t = Ku_{xx} \qquad\qquad t > 0, 0 < x < L$$

$$u_x(0; t) = 0 \qquad u_x(L; t) = 0 \qquad t > 0$$

$$u(x; 0) = f(x) \qquad\qquad 0 < x < L$$

Physically this represents heat flow in a slab whose faces are insulated. In this problem it is natural to replace the boundary condition by the equations $u_0 = u_1$, $u_n = u_{n+1}$, where $u_i(t) = u(x_i; t)$. The heat equation is treated as before, and the approximations are obtained from equation (7.2.2).

Example 7.2.3. Solve the heat equation $u_t = u_{xx}$ for $t > 0$, $0 < x < 1$ with the boundary conditions $u_x(0; t) = 0$, $u_x(1; t) = 0$, initial condition $u(x; 0) = 4x$, and mesh size $\Delta x = \frac{1}{4}$, $\Delta t = \frac{1}{32}$.

Solution. We have $K \Delta t/(\Delta x)^2 = \frac{1}{32}/\frac{1}{16} = \frac{1}{2}$, and the form of the solution (7.2.2) is $u_i(t + \Delta t) = \frac{1}{2}u_{i+1}(t) + \frac{1}{2}u_{i-1}(t)$ for $i = 1, 2, 3$. When we impose the conditions $u_0 = u_1$, $u_3 = u_4$, we obtain the following table of values:

t \ x	0	$\frac{1}{4}$	$\frac{1}{2}$	$\frac{3}{4}$	1
0	0	1	2	3	4
$\frac{1}{32}$	1.00	1.00	2.00	3.00	3.00
$\frac{1}{16}$	1.50	1.50	2.00	2.50	2.50
$\frac{3}{32}$	1.75	1.75	2.00	2.25	2.25
$\frac{1}{8}$	1.88	1.88	2.00	2.13	2.13

●

We now turn to a problem involving a slab of variable conductivity. To be specific, suppose that the slab is composed of two materials whose diffusivity coefficients are K_1, K_2 and that these are of thickness L_1, L_2, respectively. Thus we have the following problem:

$$u_t = K_1 u_{xx} \qquad\qquad 0 < x < L_1$$

$$u_t = K_2 u_{xx} \qquad\qquad L_1 < x < L_1 + L_2$$

$$u(0; t) = 0 \qquad u(L_1 + L_2; t) = 0 \qquad t > 0$$

$$u(x; 0) = f(x) \qquad\qquad 0 < x < L_1 + L_2$$

This problem is not easily solved by separation of variables. To employ the method of finite differences, we make the additional requirement that the temperature and heat flux be continuous at the interface $x = L_1$. Mathematically, we have the additional boundary conditions

$$u(L_1 - 0; t) = u(L_1 + 0; t)$$

$$K_1 u_x(L_1 - 0; t) = K_2 u_x(L_1 + 0; t)$$

To solve this problem by the method of finite differences, first we select a mesh $0 = x_0 < x_1 < \cdots < x_{n+1} = L_1 < x_{n+2} < \cdots < x_{n+m+1} = L_1 + L_2$.

The various derivatives are replaced by the following difference quotients:

$u_t(x_i; t)$ is replaced by $\dfrac{u_i(t + \Delta t) - u_i(t)}{\Delta t}$

$u_{xx}(x_i; t)$ is replaced by $\dfrac{u_{i+1}(t) + u_{i-1}(t) - 2u_i(t)}{(\Delta x)^2}$

$u_x(L_1 - 0)$ is replaced by $\dfrac{u_{n+1}(t) - u_n(t)}{\Delta x}$

$u_x(L_1 + 0)$ is replaced by $\dfrac{u_{n+2}(t) - u_{n+1}(t)}{\Delta x}$

Substituting these in the previous equations and solving, we obtain the following system of equations for the solution:

$$u_0(t) = 0 \qquad u_{n+m+1}(t) = 0$$

$$u_i(t + \Delta t) = K_1 \frac{\Delta t}{(\Delta x)^2} [u_{i+1}(t) + u_{i-1}(t)]$$

$$+ \left[1 - 2K_1 \frac{\Delta t}{(\Delta x)^2} \right] u_i(t) \qquad i = 1, \ldots, n \quad (7.2.3)$$

$$u_i(t + \Delta t) = K_2 \frac{\Delta t}{(\Delta x)^2} [u_{i+1}(t) + u_{i-1}(t)]$$

$$+ \left[1 - 2K_2 \frac{\Delta t}{(\Delta x)^2} \right] u_i(t) \qquad i = n + 2, \ldots, n + m \quad (7.2.4)$$

$$u_{n+1}(t) = \frac{K_2 u_{n+2}(t) + K_1 u_n(t)}{K_1 + K_2} \qquad (7.2.5)$$

Thus, to obtain the solution, first we obtain $u_1(t), \ldots, u_n(t)$ from (7.2.3). Next we obtain $u_{n+2}(t), \ldots, u_{n+m}(t)$ from (7.2.4). Finally we obtain $u_{n+1}(t)$ from the interface condition (7.2.5).

Example 7.2.4. Solve the following problem of the heat equation $u_t = u_{xx}$ for $0 < x < \frac{1}{2}$, $u_t = \frac{1}{2}u_{xx}$ for $\frac{1}{2} < x < 1$, with the boundary conditions $u(0; t) = 0$, $u(1; t) = 0$ and the initial condition $u(x; 0) = 4x$. Use the mesh size $\Delta x = \frac{1}{4}$, $\Delta t = \frac{1}{32}$.

Solution. We have $K_1 = 1$, $K_2 = \frac{1}{2}$, $K_1 \Delta t/(\Delta x)^2 = \frac{1}{2}$, $K_2 \Delta t/(\Delta x)^2 = \frac{1}{4}$, and the previous equations become

$$u_0(t) = 0 \qquad u_4(t) = 0$$

$$u_1(t + \Delta t) = \tfrac{1}{2}u_0(t) + \tfrac{1}{2}u_2(t)$$

$$u_3(t + \Delta t) = \tfrac{1}{4}u_2(t) + \tfrac{1}{4}u_4(t) + \tfrac{1}{2}u_3(t)$$

$$u_2(t) = \tfrac{2}{3}u_1(t) + \tfrac{1}{3}u_3(t)$$

We obtain the following table of values for $t = 0, \frac{1}{32}, \frac{1}{16}, \frac{3}{32}, \frac{1}{8}$:

t \ x	0	$\frac{1}{4}$	$\frac{1}{2}$	$\frac{3}{4}$	1
0	0	1	2	3	4
$\frac{1}{32}$	0	1.00	1.67	3.00	0
$\frac{1}{16}$	0	0.83	1.19	1.92	0
$\frac{3}{32}$	0	0.60	0.82	1.26	0
$\frac{1}{8}$	0	0.41	0.55	0.83	0

We now briefly discuss the solution of the wave equation $u_{tt} = c^2 u_{xx}$ by the method of finite differences. We choose a mesh, replace the derivatives u_{tt}, u_{xx} by the appropriate difference quotients, and obtain the equation

$$\frac{u_i(t + \Delta t) + u_i(t - \Delta t) - 2u_i(t)}{(\Delta t)^2} = c^2 \frac{u_{i+1} + u_{i-1} - 2u_i}{(\Delta x)^2}$$

If we choose $\Delta t, \Delta x$ so that $\Delta x / \Delta t = c$, these equations become

$$u_i(t + \Delta t) + u_i(t - \Delta t) = u_{i+1}(t) + u_{i-1}(t)$$

This system of equations has a general solution in the form

$$u_i(t) = f(x_i + ct) + g(x_i - ct)$$

which is the same as the general solution of the wave equation obtained in Chap. 5 in connection with the vibrating string. (See Exercise 8.)

EXERCISES 7.2

1. Find the numerical solution of the following initial-value problem for the heat equation $u_t = \frac{1}{2}u_{xx}$ for $t > 0, 0 < x < 1$ with the boundary conditions $u(0; t) = 0, u(1; t) = 0$ and the initial condition $u(x; 0) = \sin \pi x$. Use the mesh size $\Delta x = \frac{1}{4}, \Delta t = \frac{1}{64}$, and compute the solution for $t = 0, \frac{1}{64}, \frac{1}{32}, \frac{3}{64}, \frac{1}{4}$. Compare this with the exact solution.
2. Find the numerical solution of the following initial-value problem of the heat equation $u_t = u_{xx}$ for $t > 0, 0 < x < 1$ with the boundary conditions $u(0; t) = 0, u_x(1; t) = 0$ and the initial condition $u(x; 0) = 4x$. Use the mesh size $\Delta x = \frac{1}{4}, \Delta t = \frac{1}{32}$, and compute the solution for $t = 0, \frac{1}{32}, \frac{1}{16}, \frac{3}{32}, \frac{1}{8}$.
3. Find the numerical solution of the following initial-value problem of the heat equation $u_t = u_{xx}$ for $t > 0, 0 < x < 1$ with time-dependent boundary conditions $u(0; t) = t$, $u(1; t) = t$ and the initial condition $u(x; 0) = 0$. Use the mesh size $\Delta x = 0.1, \Delta t = 0.01$, and compute the solution for $t = 0.1, 0.2, \ldots, 0.9, 1.0$.
4. For each of the following equations, suppose that the mesh size Δx is given. In each case, derive an appropriate stability condition, and find the largest time step Δt which satisfies the stability condition. (You are not required to find the numerical solution.)
   (a) $u = \frac{1}{2}u_{xx}$ (b) $u_t = u_{xx} + 3u_x$
   (c) $u_t = u_{xxxx}$ (d) $u_t = u_{xx} - 4u$

5. Find the numerical solution of the heat equation $u_t = u_{xx}$ for $t > 0$, $0 < x < 1$ with the boundary conditions $u(0; t) = 0$, $u(1; t) = 0$ and the initial condition $u(x; 0) = 4x(1 - x)$. Use the mesh size $\Delta x = 0.1$, $\Delta t = 0.005$, and compute the solution for $0 < t \leq 0.1$. [*Hint:* By symmetry it is only necessary to compute $u_i(t)$ for $i = 1, \ldots, 5$.]

6. Consider the heat equation $u_t = K u_{xx}$ for $t > 0$, $0 < x < L$ with the boundary conditions $u(0; t) = 0$, $u(L; t) = 0$. Recall the relaxation time $T = L^2/(\pi^2 K)$ from Chap. 2. Assuming that a numerical solution has been found with the largest possible time step Δt, how many time steps N are necessary so that $N \Delta t = T$?

7. Solve the following problem of the heat equation $u_t = u_{xx}$ for $0 < x < \frac{1}{2}$, $u_t = \frac{1}{2}u_{xx}$ for $\frac{1}{2} < x < 1$ with the boundary conditions $u(0; t) = 0$, $u_x(1; t) = 0$ and the initial condition $u(x; 0) = 4x$. Use the mesh size $\Delta x = \frac{1}{4}$, $\Delta t = \frac{1}{32}$.

8. Show that any function of the form $u_i(t) = f(x_i + ct) + g(x_i - ct)$ is a solution of the difference equation $u_i(t + \Delta t) + u_i(t - \Delta t) = u_{i+1}(t) + u_{i-1}(t)$ provided that $\Delta x/\Delta t = c$.

7.3   SOLUTIONS IN SEVERAL DIMENSIONS

In this section we formulate and obtain the numerical solution of the heat equation and the Laplace equation in two and three dimensions. The methods are adapted to domains of arbitrary shape, many of which do not lend themselves naturally to a solution by separation of variables. Hence numerical solutions become an indispensable tool in obtaining the solution of the problem.

We consider first the two-dimensional case. To study either the heat equation or the Laplace equation, we must obtain a suitable finite difference replacement for the Laplacian $\nabla^2 u = u_{xx} + u_{yy}$. There are many ways of doing this, and we choose here the simplest, consistent with our treatment in Sec. 7.2. We choose a mesh of points (x_i, y_j), with $x_{i+1} - x_i = \Delta x$, $y_{j+1} - y_j = \Delta y$, and we set $u_{ij} = u(x_i, y_j)$. Making the usual replacements for u_{xx} and u_{yy}, we obtain the rule that we must replace $\nabla^2 u$ by

$$\frac{u_{i+1,j} + u_{i-1,j} - 2u_{i,j}}{(\Delta x)^2} + \frac{u_{i,j+1} + u_{i,j-1} - 2u_{i,j}}{(\Delta y)^2}$$

If these have a common value $\Delta x = \Delta y = h$, then the formula simplifies to

$$\frac{u_{i+1,j} + u_{i-1,j} + u_{i,j+1} + u_{i,j-1} - 4u_{i,j}}{h^2}$$

Thus we take the net difference between the value of u at (i, j) and the four neighboring points $(i + 1, j)$, $(i - 1, j)$, $(i, j + 1)$, $(i, j - 1)$ (see Figure 7.3.1).

Heat Equation in a Triangular Region

We consider now the heat equation $u_t = K \nabla^2 u$. Replacing the time derivative u_t by $u_{i,j}(t + \Delta t) - u_{i,j}(t)$, we obtain the finite difference equation

$$\frac{u_{i,j}(t + \Delta t) - u_{i,j}(t)}{\Delta t} = K \frac{u_{i+1,j}(t) + u_{i-1,j}(t) + u_{i,j+1}(t) + u_{i,j-1}(t) - 4u_{i,j}(t)}{h^2}$$

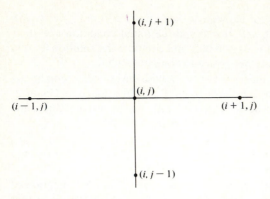

FIGURE 7.3.1
Computational molecule for Laplace's
equation.

which is solved by

$$
\begin{aligned}
u_{i,j}(t + \Delta t) \\
= K \frac{\Delta t}{h^2} \left[u_{i+1,j}(t) + u_{i-1,j}(t) + u_{i,j+1}(t) + u_{i,j-1}(t) \right] + \left(1 - 4K \frac{\Delta t}{h^2} \right) u_{i,j}(t)
\end{aligned}
$$

This is a weighted average of the five values $u_{i+1,j}$, $u_{i-1,j}$, $u_{i,j+1}$, $u_{i,j-1}$, $u_{i,j}$
provided that $4K \Delta t / h^2 \leq 1$. This is the appropriate stability condition for the
two-dimensional heat equation. When this is satisfied, the numerical solutions
obtained will furnish a suitable approximation to the exact solutions when the
mesh size h is small.

Example 7.3.1. Solve the heat equation $u_t = u_{xx} + u_{yy}$ for $t > 0$ in the triangular
region $0 < x < y < 1$ with the boundary conditions that $u(x, y; t) = 0$ on the three
sides $x = 0$, $y = 1$, $x = y$ and the initial condition $u(x, y; 0) = 8x(1 - y)$. Use the
mesh size $\Delta x = \Delta y = \frac{1}{4}$, $\Delta t = \frac{1}{64}$.

Solution. We have $K \Delta t/(\Delta x)^2 = \frac{1}{4}$, and the form of the solution is

$$
u_{ij}(t + \Delta t) = \tfrac{1}{4}(u_{i+1,j} + u_{i-1,j} + u_{i,j+1} + u_{i,j-1})
$$

The numerical values are represented in the following triangular tables corresponding to $t = 0$, $t = \frac{1}{64}$, $t = \frac{1}{32}$, $t = \frac{3}{64}$, $t = \frac{1}{16}$:

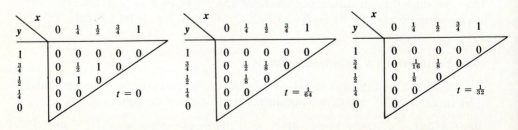

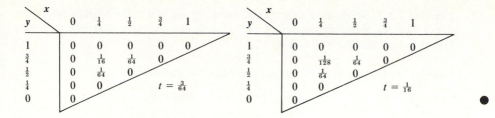

Laplace's Equation in a Triangular Region

The numerical solution of the heat equation is characterized as an *explicit procedure;* i.e., the values of $u_{i,j}(t + \Delta t)$ are obtained explicitly as linear combinations of $u_{i',j'}(t)$, for various values of (i', j'). This feature is no longer present in the numerical solution of Laplace's equation, where we must solve a system of linear equations to obtain the numerical solution. Consider, e.g., the following boundary-value problem for Laplace's equation in the triangular region $0 < x < y < 1$:

$$u_{xx} + u_{yy} = 0 \qquad 0 < x < y < 1$$

$$u(0, y) = f_1(y) \qquad 0 < y < 1$$

$$u(x, 1) = f_2(x) \qquad 0 < x < 1$$

$$u(x, x) = f_3(x) \qquad 0 < x < 1$$

To solve this problem by the method of finite differences, we take a mesh with $\Delta x = \Delta y = 1/N$ and replace $u(i/N, j/N)$ by $u_{i,j}$. The Laplace equation is replaced by the difference equation

$$u_{i+1,j} + u_{i-1,j} + u_{i,j+1} + u_{i,j-1} - 4u_{i,j} = 0 \qquad 0 < i < j < N$$

and the boundary conditions are replaced by the equations

$$u_{0j} = f_1\left(\frac{j}{N}\right) \qquad 0 \le j \le N$$

$$u_{iN} = f_2\left(\frac{i}{N}\right) \qquad 0 < i \le N$$

$$u_{ii} = f_3\left(\frac{i}{N}\right) \qquad 0 < i < N$$

This is a system of $\frac{1}{2}(N + 1)(N + 2)$ equations for the unknowns $u_{i,j}$, $0 \le i \le j \le N$. It can be shown that this system of linear equations has a unique solution for any choice of functions f_1, f_2, f_3. If N is small, the solution can be found by elementary linear algebra. If N is large it may be necessary to do extensive machine computation to obtain the explicit solution.

Example 7.3.2. Solve the Laplace equation $u_{xx} + u_{yy} = 0$ in the triangle $0 < x < y < 1$ with the boundary conditions $u(0, y) = 0$, $u(x, 1) = x(1 - x)$, $u(x, x) = 0$. Use the mesh size $\Delta x = \frac{1}{4} = \Delta y$.

Solution. Replacing $u(i/4, j/4)$ by $u_{i,j}$, we have the equations

$$u_{11} + u_{13} + u_{02} + u_{22} - 4u_{12} = 0$$

$$u_{12} + u_{14} + u_{03} + u_{23} - 4u_{13} = 0$$

$$u_{22} + u_{24} + u_{13} + u_{33} - 4u_{23} = 0$$

$$u_{00} = 0 \qquad u_{01} = 0 \qquad u_{02} = 0 \qquad u_{03} = 0 \qquad u_{04} = 0$$

$$u_{14} = \tfrac{3}{16} \qquad u_{24} = \tfrac{1}{4} \qquad u_{34} = \tfrac{3}{16} \qquad u_{44} = 0$$

$$u_{11} = 0 \qquad u_{22} = 0 \qquad u_{33} = 0$$

Making the appropriate substitutions, we have

$$u_{13} - 4u_{12} = 0 \qquad u_{12} + \tfrac{3}{16} + u_{23} - 4u_{13} = 0 \qquad \tfrac{1}{4} + u_{13} - 4u_{23} = 0$$

The solution of this system of three equations is $u_{12} = \frac{1}{56}$, $u_{13} = \frac{1}{14}$, $u_{23} = \frac{9}{112}$. ●

For larger values of N, this numerical method for solving Laplace's equation leads to large systems of linear equations which may be difficult to solve. To deal with such cases, we regard the solution of Laplace's equation as the limit of the solution of the heat equation when the time t becomes large; in symbols,

$$u(x, y) = \lim_{t \to \infty} u(x, y; t)$$

The function $u(x, y; t)$ is taken to be a solution of the heat equation $u_t = K(u_{xx} + u_{yy})$ satisfying the same boundary conditions as $u(x, y)$ and with an arbitrary initial condition. By an appropriate choice of the initial conditions, we can obtain an effective numerical algorithm for obtaining an approximate solution of Laplace's equation.

We illustrate these ideas by the following numerical example.

Example 7.3.3. Solve the Laplace equation $u_{xx} + u_{yy} = 0$ in the triangle $0 < x < y < 1$ with the boundary conditions $u(0, y) = 0$, $u(x, 1) = 1$, $u(x, x) = 0$. Use the mesh size $\Delta x = \Delta y = \frac{1}{6}$.

Solution. We solve the heat equation $u_t = u_{xx} + u_{yy}$ numerically with the initial condition $u(x, y; 0) = 0$ and with the same boundary conditions $u(0, y; t) = 0$, $u(x, 1; t) = 1$, $u(x, x; t) = 0$. Choosing $\Delta t = \frac{1}{144}$ gives the iterative solution $u_{i,j}(t + \Delta t) = \frac{1}{4}[u_{i+1,j}(t) + u_{i-1,j}(t) + u_{i,j+1}(t) + u_{i,j-1}(t)]$. The following numerical tables give the values of the first five iterations, correct to three decimals, for $t = 0, \frac{1}{72}, \frac{2}{72}, \frac{3}{72}, \frac{4}{72}, \frac{5}{72}, \frac{6}{72}$. The values u_{60}, u_{66}, and u_{00} are not listed, since they are not used in the iterations.

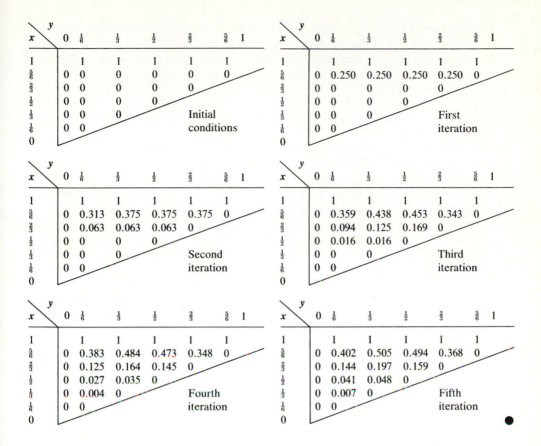

EXERCISES 7.3

1. Obtain the numerical solution of the heat equation $u_t = u_{xx} + u_{yy}$ in the square $0 < x < 1$, $0 < y < 1$ with the boundary condition $u(x, y; t) = 0$ on four sides and the initial condition $u(x, y; 0) = 8xy$. Use the mesh size $\Delta x = \Delta y = \frac{1}{4}$, $\Delta t = \frac{1}{64}$ and solve for $0 \le t \le \frac{1}{16}$.

2. Obtain the numerical solution of the heat equation $u_t = u_{xx} + u_{yy}$ in the L-shaped region formed by removing the square $0 < x < 1$, $0 < y < 1$ from the square $0 < x < 2$, $0 < y < 2$. Use the boundary condition $u(x, y; t) = 0$ on all six sides and the initial condition $u(x, y; 0) = 4x + 2y$. Use the mesh size $\Delta x = \Delta y = \frac{1}{4}$, $\Delta t = \frac{1}{64}$, and solve for $0 \le t \le \frac{1}{16}$.

3. Obtain the numerical solution of the Laplace equation $u_{xx} + u_{yy} = 0$ in the square $0 < x < 1$, $0 < y < 1$ with boundary conditions $u(x, 0) = 0$, $u(x, 1) = 0$, $u(0, y) = 0$, $u(1, y) = 1$. Use the mesh size $\Delta x = \Delta y = \frac{1}{4}$, and solve explicitly the resulting system of linear equations.

4. Solve Exercise 3 by the iterative method of Example 7.3.3 by solving the heat equation $u_t = u_{xx} + u_{yy}$ with the initial condition $u(x, y; 0) = 0$. Compare the result of five iterations with the result obtained in Exercise 3.

5. Solve Exercise 3 with the mesh size $\Delta x = \Delta y = \frac{1}{10}$ by using the iterative method of Example 7.3.3, using five iterations.

6. Without reference to numerical solutions, give a formal proof of the following theorem: Suppose that $u(x, y; t)$ is a solution of the heat equation $u_t = K(u_{xx} + u_{yy})$ in the square $0 < x < 1,\ 0 < y < 1$ with boundary conditions $u(0, y; t) = f_1(y)$, $u(1, y; t) = f_2(y)$, $u(x, 0; t) = f_3(x)$, $u(x, 1; t) = f_4(x)$. Then $\lim_{t \to \infty} u(x, y; t) = u(x, y)$, the solution of Laplace's equation $u_{xx} + u_{yy} = 0$, which satisfies the same boundary conditions.

7. In Exercise 6, replace the square $0 < x < 1,\ 0 < y < 1$ by the triangle $0 < x < y < 1$. Formulate and prove a limit theorem analogous to Exercise 6.

8. Formulate the numerical solution of the heat equation $u_t = K(u_{xx} + u_{yy} + u_{zz})$ in the three-dimensional cube $0 < x < 1, 0 < y < 1, 0 < z < 1$. Formulate an appropriate stability condition for the difference scheme.

9. Obtain the numerical solution of Exercise 8 in the case where the boundary condition is $u(x, y, z; t) = 0$ on all six faces and the initial condition is $u(x, y, z; 0) = 1$. Use the mesh size $\Delta x = \Delta y = \Delta z = \frac{1}{4}$ and the largest possible time step Δt, consistent with the stability condition.

10. Formulate the numerical solution of Laplace's equation $u_{xx} + u_{yy} + u_{zz} = 0$ in the three-dimensional cube $0 < x < 1, 0 < y < 1, 0 < z < 1$ corresponding to the mesh size $\Delta x = \Delta y = \Delta z = \frac{1}{4}$. How many simultaneous linear equations must one solve to obtain the numerical solution by this method?

CHAPTER
8

GREEN'S FUNCTIONS

Introduction

Many boundary-value problems for linear partial differential equations may be solved in terms of integral transforms. These formulas give explicit representations of the solutions as linear transforms of the initial and boundary conditions. For example, the d'Alembert solution of the wave equation (Sec. 2.4), the Poisson integral representation of the solution of Laplace's equation (Sec. 3.2), and the Gauss-Weierstrass representation of the solution of the heat equation (Sec. 5.2) are all of this type. The purpose of this chapter is to pursue this topic more generally.

8.1 GREEN'S FUNCTIONS FOR ORDINARY DIFFERENTIAL EQUATIONS

An Example

We begin with the simplest example, to illustrate the main ideas. This is the boundary-value problem for the ordinary differential equation

$$y'' = -f(x) \qquad 0 < x < L$$
$$y(0) = 0 = y(L)$$

Here $f(x)$, $0 < x < L$, is a given piecewise smooth function. This equation may be used to model the static transverse deflections of a string which is fixed at both ends and subject to a spatially dependent forcing law.

To solve this problem, we use calculus to write $y'(x) = -\int_0^x f(z)\,dz + A$, $y(x) = -\int_0^x \int_0^\xi f(z)\,dz\,d\xi + Ax + B$, where A, B are constants to be determined. The boundary conditions yield

$$0 = 0 + A \cdot 0 + B$$

$$0 = -\int_0^L \int_0^\xi f(z)\,dz\,d\xi + AL + B$$

with the conclusions $B = 0$, $A = (1/L)\int_0^L \int_0^\xi f(z)\,dz\,d\xi$. We can reduce the iterated integrals to single integrals by interchanging the order of integration. Thus $\int_0^x \int_0^\xi f(z)\,dz\,d\xi = \int_0^x \int_z^x f(z)\,d\xi\,dz = \int_0^x (x - z)f(z)\,dz$, and the solution is written as

$$y(x) = -\int_0^x (x - z)f(z)\,dz + \frac{x}{L}\int_0^L (L - z)f(z)\,dz$$

$$= \int_0^x \frac{z}{L}(L - x)f(z)\,dz + \int_x^L \frac{x}{L}(L - z)f(z)\,dz$$

$$= \int_0^L G(x, z)f(z)\,dz$$

where *Green's function* $G(x, z)$ is defined by

$$G(x, z) = \begin{cases} \dfrac{z(L - x)}{L} & 0 \le z \le x \\[2mm] \dfrac{x(L - z)}{L} & x \le z \le L \end{cases}$$

We have obtained an explicit representation of the solution in terms of the right member $f(x)$, $0 < x < L$, and a function $G(x, z)$ which depends on only the differential equation and the boundary conditions. This formula defines a linear transform of the right member as a solution of the given problem.

Green's function has the following characteristic properties:

1. For each z, $x \to G(x, z)$ satisfies $G'' = 0$ except when $x = z$.
2. G satisfies the boundary conditions $G(0, z) = 0 = G(L, z)$.
3. $G(z + 0, z) - G(z - 0, z) = 0$
4. $(\partial G/\partial x)(z + 0, z) - (\partial G/\partial x)(z - 0, z) = -1$
5. $G(x, z) = G(z, x)$

We leave it as an exercise to show that $G(x, z)$ is uniquely determined from conditions 1 through 4. Note that condition 3 signifies that G is a continuous function while 4 signifies that the first derivatives are discontinuous in a precise manner.

Green's function may also be represented by a Fourier sine series. The nth Fourier sine coefficient of G is $B_n(z) = (2/L)\int_0^L G(x, z)\sin(n\pi x/L)\,dx$. By what we have shown, this is the solution of $B''(z) = -(2/L)\sin(n\pi z/L)$ with boundary conditions $B(0) = 0 = B(L)$, which is easily seen to be $B(z) = (2/L)[L/(n\pi)]^2 \sin(n\pi z/L)$, leading to the Fourier representation of Green's function in the form

$$G(x, z) = \frac{2}{L} \sum_{n=1}^{\infty} \left(\frac{L}{n\pi}\right)^2 \left(\sin \frac{n\pi x}{L} \sin \frac{n\pi z}{L}\right)$$

This series converges uniformly for $0 \le x, z \le L$.

The Generic Case

We now turn to the case of a general self-adjoint second-order ordinary differential equation. Consider the boundary-value problem

$$\mathcal{L}\, y := [p(x)y']' + q(x)y \qquad = -f(x) \qquad a < x < b \qquad (8.1.1)$$

$$\cos \alpha\, y(a) + \sin \alpha\, y'(a) \;=\; 0 \qquad\qquad\qquad (8.1.2)$$

$$\cos \beta\, y(b) + \sin \beta\, y'(b) \;=\; 0 \qquad\qquad\qquad (8.1.3)$$

Here f, q, p are given continuous functions with $p(x) > 0$ for $a \le x \le b$, and α, β are real constants. In addition, *we assume that $\lambda = 0$ is not an eigenvalue of the associated Sturm-Liouville system.*

    We solve this problem by the method of variation of parameters, familiar for ordinary differential equations. The solution is sought in the form

$$y(x) = u_1(x)y_1(x) + u_2(x)y_2(x)$$

where y_1, y_2 are solutions of the homogeneous equation and the functions u_1, u_2 are to be chosen. To be specific, we determine y_1, y_2 up to constant multiples by requiring that y_1 satisfy (8.1.2) and that y_2 satisfy (8.1.3). Since $\lambda = 0$ is not an eigenvalue, we conclude that y_1, y_2 must be linearly independent. We have

$$y' = u_1 y_1' + u_2 y_2' + u_1' y_1 + u_2' y_2$$

The method of variation of parameters further requires that $u_1' y_1 + u_2' y_2 = 0$. With this determination, we see that y satisfies (8.1.2) if $u_2(a) = 0$. Similarly, y satisfies (8.1.3) if $u_1(b) = 0$. To satisfy (8.1.1), we write

$$(py')' + qy = [p(u_1 y_1' + u_2 y_2')]' + q(u_1 y_1 + u_2 y_2)$$

$$= u_1[(py_1')' + qy_1] + u_2[(py_2')' + qy_2] + p(u_1' y_1' + u_2' y_2')$$

The first two terms are zero since y_1, y_2 are both solutions to the homogeneous equation. Therefore (8.1.1) can be solved by the method of variation of parameters if we have satisfied the simultaneous system

$$p(x)(u_1' y_1' + u_2' y_2') = -f(x)$$

$$u_1' y_1 + u_2' y_2 = 0$$

These are easily solved to yield

$$u_1'(x) = \frac{f(x)y_2(x)}{p(x)W(x)} \qquad u_2'(x) = -\frac{f(x)y_1(x)}{p(x)W(x)}$$

where $W(x) = y_1(x)y_2'(x) - y_1'(x)y_2(x)$ is the wronskian of the two solutions. We determine u_1, u_2 uniquely by $u_1(b) = 0$, $u_2(a) = 0$ to obtain

$$y(x) = -y_1(x) \int_x^b \frac{f(\xi)y_2(\xi)}{p(\xi)W(\xi)}\, d\xi - y_2(x) \int_a^x \frac{f(\xi)y_1(\xi)}{p(\xi)W(\xi)}\, d\xi$$

This can be written as a single integral if we use the definition

$$G(x, \xi) = \begin{cases} -\dfrac{y_2(x)y_1(\xi)}{p(\xi)W(\xi)} & a \le \xi \le x \\[2mm] -\dfrac{y_1(x)y_2(\xi)}{p(\xi)W(\xi)} & x \le \xi \le b \end{cases} \qquad (8.1.4)$$

to obtain the formula

$$y(x) = \int_a^b G(x, \xi)f(\xi)\, d\xi \qquad (8.1.5)$$

We have obtained Green's function of the general second-order self-adjoint equation (8.1.1) with the separable boundary conditions (8.1.2) and (8.1.3). Green's function depends on only the boundary conditions and the functions $p(x)$, $q(x)$; it makes no reference to the right member $f(x)$. We note the following properties of Green's function in this case:

1. The function $x \to G(x, \xi)$ satisfies the homogeneous equation if $x \ne \xi$.
2. The function $x \to G(x, \xi)$ satisfies the boundary conditions (8.1.2) and (8.1.3).
3. The function $G(x, \xi)$ is continuous for $a \le \xi, x \le b$.
4. The partial derivatives are continuous except for $x = \xi$, where we have

$$\left(\frac{\partial G}{\partial x}\right)(\xi + 0, \xi) - \left(\frac{\partial G}{\partial x}\right)(\xi - 0, \xi) = -\frac{1}{p(\xi)}$$

5. The function $G(x, \xi)$ is symmetric: $G(x, \xi) = G(\xi, x)$ for $a \le \xi, x \le b$.

We can summarize the above discussion in the form of a theorem.

Theorem 8.1.1. Suppose that $\lambda = 0$ is not an eigenvalue of equation (8.1.1) with $f = 0$. Then the unique solution of equation (8.1.1) with boundary conditions (8.1.2) and (8.1.3) is given by the integral (8.1.5) where Green's function is defined by (8.1.4).

Proof. To prove the stated uniqueness, let y, $\bar{y}$ be solutions of $\mathscr{L}y = -f$ satisfying the boundary conditions. Then $y - \bar{y}$ satisfies the boundary conditions, and $\mathscr{L}(y - \bar{y}) = 0$. Since $\lambda = 0$ is not an eigenvalue, we conclude that $y - \bar{y} = 0$. ∎

Example 8.1.1. Find Green's function of the equation $y'' = -f$ with the boundary conditions $y(0) = 0$, $y'(L) = 0$, and solve the equation.

Solution. The general solution of the homogeneous equation is $y(x) = A + Bx$. The solution which satisfies $y(0) = 0$ is $y_1(x) = x$, up to a constant multiple. The solution which satisfies $y'(L) = 0$ is $y_2(x) = 1$, up to a constant multiple. Their wronskian is $W(y) = y_1 y_2' - y_1' y_2 = -1$. Green's function is

$$G(x, \xi) = \begin{cases} \xi & 0 \leq \xi \leq x \\ x & x \leq \xi \leq L \end{cases}$$

The solution of the inhomogeneous equation is $y(x) = \int_0^x \xi f(\xi) \, d\xi + x \int_x^L f(\xi) \, d\xi$. A direct verification shows that $y' = \int_x^L f(\xi) \, d\xi$, $y'' = -f$, $y(0) = 0$, $y'(L) = 0$. ●

To obtain a Fourier representation of Green's function in the generic case, let $\{\lambda_n\}_{n \geq 1}$ be the eigenvalues and $\{\varphi_n\}_{n \geq 1}$ a set of normalized eigenfunctions for the equation $(py')' + qy + \lambda y = 0$. Then Green's function is written as the series

$$G(x, \xi) = \sum_{n \geq 1} \frac{\varphi_n(x) \, \varphi_n(\xi)}{\lambda_n}$$

which is absolutely and uniformly convergent for $a \leq x \leq b$.

Example 8.1.2. Find a Fourier representation of Green's function of Example 8.1.1.

Solution. In this example we have $\lambda_n = (n - \frac{1}{2})^2 \pi^2 / L^2$, $\varphi_n(x) = \sqrt{2/L} \sin [(n - \frac{1}{2}) \pi x / L]$, $n \geq 1$, and Green's function is written $G(x, \xi) = (2L/\pi^2) \sum_{n \geq 2} \sin [(n - \frac{1}{2}) \pi x / L] \sin [(n - \frac{1}{2}) \pi \xi / L] / (n - \frac{1}{2})^2$. ●

The Exceptional Case: Modified Green's Function

If $\lambda = 0$ is an eigenvalue, then we lose existence and uniqueness. For example, if the equation is $y'' = -f$ and the boundary conditions are $y'(a) = 0 = y'(b)$, then we can solve the equation if and only if $\int_a^b f(x) \, dx = 0$. The homogeneous equation $y'' = 0$ has infinitely many solutions which satisfy the boundary condition, that is, $y(x) = $ constant, which destroys uniqueness. In these cases we may determine a unique solution by requiring in addition that the solution be orthogonal to the eigenfunction of the associated homogeneous problem. In the case just mentioned, this amounts to the requirement that $\int_a^b y(x) \, dx = 0$, leading to $y(x) \equiv 0$.

To formulate Green's function for this exceptional case, we consider the inhomogeneous equation

$$[p(x)y']' + q(x)y = -f(x) + \varphi(x) \int_a^b f(\xi)\varphi(\xi) \, d\xi$$

where $\varphi(x)$ is an eigenfunction which satisfies the homogeneous boundary conditions (8.1.2) and (8.1.3) and satisfies the normalization $\int_a^b \varphi(x)^2\, dx = 1$. We have written the equation so that the right side is orthogonal to the eigenfunction $\varphi(x)$. We look for a solution which satisfies the orthogonality condition $\int_a^b y(x)\varphi(x)\, dx = 0$. The solution is sought in the form $y(x) = \int_a^b G(x, \xi)f(\xi)\, d\xi$, and $G(x, \xi)$ is the *modified Green function*.

Example 8.1.3. Find the modified Green function for the equation $y'' = -f$ with the boundary conditions $y(0) = 0$, $y'(1) = y(1)$.

Solution. The general solution of the homogeneous equation is $y(x) = A + Bx$ which satisfies the boundary conditions with $A = 0$ and B arbitrary. A normalized eigenfunction is obtained by choosing $B = \sqrt{3}$. To solve the equation $y'' = -f + 3x \int_0^1 \xi f(\xi)\, d\xi$, we write

$$y'(x) = -\int_0^x f(\xi)\, d\xi + \frac{3x^2}{2} \int_0^1 \xi f(\xi)\, d\xi + C$$

$$y(x) = -\int_0^x (x - \xi)f(\xi)\, d\xi + \frac{1}{2}x^3 \int_0^1 \xi f(\xi)\, d\xi + Cx + D$$

The first boundary condition gives $D = 0$ while the second leaves C undetermined. To find C, we use the orthogonality condition $\int_0^1 y(x)\varphi(x)\, dx = 0$ to obtain $C = \int_0^1 (1 - 9\xi/5 + \frac{1}{2}\xi^2)f(\xi)\, d\xi$, and we have the modified Green function

$$G(x, \xi) = \begin{cases} \xi - x + \dfrac{1}{2}x^3\xi + \left(1 - \dfrac{9\xi}{5} + \dfrac{\xi^3}{2}\right)x & \xi < x \\[3mm] \dfrac{x^3\xi}{2} + \left(1 - \dfrac{9\xi}{5} + \dfrac{\xi^3}{2}\right)x & \xi \geq x \end{cases}$$

●

The modified Green function can be represented as a series of eigenfunctions in the form

$$G(x, \xi) = \sum_{n \geq 1} \frac{\varphi_n(x)\varphi_n(\xi)}{\lambda_n}$$

where the sum is over all *nonzero* eigenvalues $\{\lambda_n\}$ of the equation $(py')' + qy + \lambda y = 0$ with the given boundary conditions. For example, with the boundary conditions $y'(0) = 0$, $y'(L) = 0$ for the equation $y'' = -f$, we have $\lambda_n = (n\pi/L)^2$, $\varphi_n(x) = \sqrt{2/L} \cos(n\pi x/L)$, and Green's function is written as $G(x, \xi) = (2L/\pi^2)\sum_{n \geq 1} [\cos(n\pi x/L) \cos(n\pi\xi/L)]/n^2$.

The Fredholm Alternative

The above discussion leads us to the following dichotomy regarding the inhomogeneous equation (8.1.1) with the boundary conditions (8.1.2) and (8.1.3), known as the *Fredholm alternative* and stated as follows:

Either $\lambda = 0$ is not an eigenvalue of the homogeneous equation. Then the inhomogeneous equation (8.1.1) has a unique solution which satisfies the boundary conditions (8.1.2) and (8.1.3).

Or $\lambda = 0$ is an eigenvalue of the homogeneous equation with eigenfunction $\varphi(x)$. Then the inhomogeneous equation (8.1.1) has a solution if and only if the right-side $f(x)$ satisfies the orthogonality condition $\int_a^b f(x)\varphi(x)\ dx = 0$.

The solution is unique provided that we require that $\int_a^b y(x)\varphi(x)\ dx = 0$.

The reader familiar with the theory of linear equations will recognize a familiar principle at work. Let a system of linear equations be written in the form $Ay = -f$, where A is a square symmetric matrix. *Either* det A is nonzero, and we have a unique solution y for every vector f; *or* det A is zero, and we can solve the equation only if the right side is orthogonal to the set of solutions of $\{\varphi: A\varphi = 0\}$. In the case of a square matrix, det A is zero if and only if $\lambda = 0$ is an eigenvalue of A. In the case of differential equations, we cannot use the determinant to distinguish between the two cases. Nevertheless the existence or nonexistence of eigenfunctions for $\lambda = 0$ still makes sense and is a valid criterion to distinguish between the two cases.

EXERCISES 8.1

1. Show that Green's function for $y'' = -f$, $y(0) = 0 = y(L)$ is uniquely determined by properties 1 to 4.
2. Find Green's function for the equation $y'' = -f$ with the boundary conditions $y(0) = 0$, $y'(L) + hy(L) = 0$, where h is a positive constant.
3. Find Green's function for the equation $y'' = -f$ with the boundary conditions $y'(0) = hy(0)$, $y'(L) + hy(L) = 0$, where h is a positive constant.
4. Find Green's function for the equation $y'' - ky = -f$ with the boundary conditions $y(0) = 0$, $y(L) = 0$, where k is a positive constant.
5. Find Green's function for the equation $y'' - ky = -f$ with the boundary conditions $y(0) = 0$, $y(L) = 0$, where k is a negative constant

$$k \neq -\left(\frac{n\pi}{L}\right)^2 \qquad n = 1, 2, \ldots$$

6. Find the modified Green function for the equation $y'' = -f$ with the boundary conditions $y'(0) = 0$, $y'(L) = 0$.
7. Find the modified Green function for the equation $y'' = -f$ with the (periodic) boundary conditions $y(0) = y(L)$, $y'(0) = y'(L)$.

8.2 GREEN'S FUNCTION FOR THE THREE-DIMENSIONAL POISSON EQUATION

In this section and the following sections, we extend the concept of Green's function to certain second-order partial differential equations. Since the geometry and types of equations differ for each case, it is most efficient to consider separately the elliptic, parabolic, and hyperbolic problems.

The boundary-value problem for Poisson's equation is

$$\nabla^2 u = -h \qquad P \in D$$

$$u = f \qquad P \in \partial D$$

Here P is a point of two- or three-dimensional space, and D is a region whose boundary is denoted ∂D. For example, if D is the three-dimensional ball $\{P: |P| < R\}$, then ∂D is the sphere $\{P: |P| = R\}$.

By the superposition principle, the solution may be obtained in the form $u = u_f + u_h$, where u_f is the solution of the homogeneous equation $\nabla^2 u = 0$ with $u = f$ on ∂D and u_h is the solution of the Poisson equation $\nabla^2 u = -h$ with $u = 0$ on ∂D. The latter problem is the analog of the problem for ordinary differential equations which was solved in Sec. 8.1 by using Green's function. We now develop this idea for Poisson's equation.

Newtonian Potential Kernel

We begin with the case of the entire three-dimensional space $D = \mathbb{R}^3 = \{P = (x, y, z): x \in \mathbb{R}, y \in \mathbb{R}, z \in \mathbb{R}\}$. We proceed heuristically by Fourier transforms and then verify that we have obtained a rigorous solution. The Fourier representation of the solution and right member are written

$$u(P) = u(x, y, z) = \iiint e^{i\langle P, K\rangle} U(K)\, dK$$

$$= \iiint e^{i(kx+ly+mz)} U(k, l, m)\, dk\, dl\, dm$$

$$h(P) = h(x, y, z) = \iiint e^{i\langle P, K\rangle} H(K)\, dK$$

$$= \iiint e^{i(kx+ly+mz)} H(k, l, m)\, dk\, dl\, dm$$

where the integrals are over all $\mathbb{R}^3$. Differentiating formally, we have $\nabla^2 u = \iiint -|K|^2 e^{i\langle P, K\rangle} U(K)\, dK$. The equation $\nabla^2 u = -h$ requires that $|K|^2 U(K) = H(K)$. If H is zero in a small neighborhood of $K = 0$, this can be solved by $U(K) = H(K)/|K|^2$. We substitute in the integral for u and interchange orders of integration:

$$u(P) = \iiint \frac{e^{i\langle P, K\rangle} H(K)}{|K|^2}\, dK$$

$$= \iiint \frac{e^{i\langle P, K\rangle}}{|K|^2}\, dK \iiint \left(\frac{1}{2\pi}\right)^3 e^{-i\langle Q, K\rangle} h(Q)\, dQ$$

$$= \iiint h(Q)\, dQ \left(\frac{1}{2\pi}\right)^3 \iiint \frac{e^{i\langle K, P-Q\rangle}}{|K|^2}\, dK$$

The inner integral may be evaluated in spherical polar coordinates as

$$\iiint \frac{e^{i\langle K, P-Q\rangle}}{|K|^2}\, dK = \int_0^{2\pi} \int_0^\pi \int_0^\infty e^{i|P-Q|\rho \cos\theta} \sin\theta\, d\rho\, d\theta\, d\varphi$$

where $\rho = |K|$, θ is the (polar) angle between K and $P - Q$, and φ is the corresponding azimuthal angle. The θ integral is elementary, and the φ integral gives a factor of 2π, leading to

$$\iiint \frac{e^{i\langle K,\, P-Q\rangle}}{|K|^2}\, dK = 2\pi \int_0^\infty \frac{2\,\sin \rho|P-Q|}{\rho|P-Q|}\, d\rho$$

$$= 2\pi \frac{2}{|P-Q|}\frac{\pi}{2}$$

Recalling the factor $[1/(2\pi)]^3$ from the Fourier inversion, we are led to the explicit representation

$$u(P) = \frac{1}{4\pi}\iiint \frac{h(Q)}{|P-Q|}\, dQ \qquad (8.2.1)$$

It remains to verify that this is a (rigorous) solution of $\nabla^2 u = -h$. The function $P \to G(P, Q) := 1/(4\pi|P-Q|)$ is the *newtonian potential kernel*. It has the following basic properties:

1. $\nabla_P G = -\nabla_Q G = -\dfrac{1}{4\pi}\dfrac{(P-Q)}{|P-Q|^3} \qquad P \neq Q$

2. $\nabla_P^2 G = \nabla_Q^2 G = 0 \qquad P \neq Q$

3. $G(P, Q) \to 0 \qquad P \to \infty \text{ or } Q \to \infty$

We leave the verification of these as an exercise.

Theorem 8.2.1. Let $u = u(P)$ be the solution of $\nabla^2 u = -h$ in all $\mathbb{R}^3$ where $u(P) \to 0$, $|P||\nabla_P u| \to 0$ when $P \to \infty$. Then u is represented by (8.2.1) in the sense that $u(P) = [1/(4\pi)] \lim_{\epsilon \downarrow 0,\, R \uparrow \infty} \iiint_{\{Q:\ \epsilon < |Q-P| < R\}} [h(Q)/|P-Q|]\, dQ$.

Proof. We use Green's identity in the form

$$\iiint_{D_{\epsilon,R}} (u\nabla^2 G_P - G_P\nabla^2 u)\, dQ = \iint_{\partial D_{\epsilon,R}} \left(u\frac{\partial G_P}{\partial n} - G_P\frac{\partial u}{\partial n}\right) dS_Q$$

where $D_{\epsilon,R} = \{Q \in \mathbb{R}^3 : \epsilon < |Q-P| < R\}$ and $G_P(Q) = G(P, Q)$. The right member consists of two surface integrals, over the spheres of radius ϵ and radius R centered at P; the outward normal derivative is defined by $\partial G_P/\partial n = \langle \nabla_Q G_P, n\rangle$. To analyze the integral on the left, we note that $\nabla^2 G_P = 0$ in $D_{\epsilon,R}$ and $\nabla^2 u = -h$, by hypothesis. The integrals on the right are as follows:

$$\iint_{|Q-P|=\epsilon} u\frac{\partial G_P}{\partial n}\, dS_Q$$

$$= \frac{1}{4\pi}\int_0^{2\pi}\int_0^\pi u(x + \epsilon \sin\theta \cos\varphi,\, y + \epsilon \sin\theta \sin\varphi,\, z + \epsilon \cos\theta) \sin\theta\, d\theta\, d\varphi \qquad (8.2.2)$$

$$\iint\limits_{|Q-P|=R} u \frac{\partial G_P}{\partial n} \, dS_Q$$

$$= \frac{1}{4\pi} \int_0^{2\pi} \int_0^\pi u(x + R \sin \theta \cos \varphi, y + R \sin \theta \sin \varphi, z + R \cos \theta) \sin \theta \, d\theta \, d\varphi \quad (8.2.3)$$

$$\iint\limits_{|Q-P|=\epsilon} G_P \frac{\partial u}{\partial n} \, dS_Q$$

$$= \frac{1}{4\pi\epsilon} \int_0^{2\pi} \int_0^\pi \frac{\partial u}{\partial n} (x + \epsilon \sin \theta \cos \varphi, y + \epsilon \sin \theta \sin \varphi, z + \epsilon \cos \theta)\epsilon^2 \sin \theta \, d\theta \, d\varphi$$
$$(8.2.4)$$

$$\iint\limits_{|Q-P|=R} G_P \frac{\partial u}{\partial n} \, dS_Q$$

$$= \frac{1}{4\pi R} \int_0^{2\pi} \int_0^\pi \frac{\partial u}{\partial n} (x + R \cos \theta, y + R \sin \theta \cos \varphi, z + R \sin \theta \sin \varphi)R^2 \sin \theta \, d\theta \, d\varphi$$
$$(8.2.5)$$

The function u is assumed to be twice-differentiable, in particular, continuous. Therefore the integral (8.2.2) tends to $u(P)$ when $\epsilon \downarrow 0$. Integral (8.2.3) tends to zero when $R \uparrow \infty$ by the assumption on u. Integral (8.2.4) tends to zero by the continuity of $\partial u/\partial n$. Finally integral (8.2.5) tends to zero by the hypothesis on $|\nabla u|$. Putting these facts together, we see that the right side tends to $u(P)$ when $\epsilon \downarrow 0$, $R \uparrow \infty$. Therefore the left side tends to a limit, and we have proved that $u(P) = [1/(4\pi)] \lim_{\epsilon \downarrow 0, \, R \uparrow \infty} \iiint_{\epsilon < |P-Q| < R} h(Q)/|P - Q| \, dQ$, as required. ∎

Single- and Double-Layer Potentials

The newtonian potential integral (8.2.1) may be considered in its own right, without reference to Poisson's equation. In the theory of electrostatics, the function $P \to 1/(4\pi|P - Q|)$ is interpreted as the potential energy necessary to bring a particle of unit charge from infinity to point Q. The gradient $\nabla_P G$ is the force felt at point P due to the unit charge at point Q. If, instead of a unit charge, we have charge distributed according to a continuous density, then the resultant potential energy is given by the superposition integral (8.2.1). The force is obtained as the negative gradient of this integral. But we can also consider more general superpositions of charge, e.g., on surfaces, lines, or points.

A surface distribution of charges with surface density $h(Q)$ is written as the integral

$$\iint\limits_S \frac{h(Q)}{4\pi|P - Q|} \, dS_Q$$

This is called a *single-layer potential*.

A linear distribution of charges with linear density $h(Q)$ may be represented by the line integral

$$\int_C \frac{h(Q)}{4\pi|P - Q|} \, dl_Q$$

where the integration is over the parameterized curve C, for example, a straight-line segment.

It may be shown that each of the above integrals is a solution of Laplace's equation $\nabla^2 u = 0$ on the set of P for which $h(P) = 0$.†

The simplest charge distribution is that of a finite aggregate of point charges of the form

$$\sum_{i=1}^{N} \frac{C_i}{4\pi|P - Q_i|}$$

for a finite set of points Q_i with corresponding charges C_i, which may be positive or negative. In particular, we may take two nearby points $Q_0 = (x_0, y_0, z_0)$ and $Q_\epsilon = Q_0 + \epsilon n$, where n is a unit vector and the charges are of strength $1/\epsilon$ and $-1/\epsilon$. The total charge is zero, but in the limit we obtain a nonzero potential, since

$$\lim_{\epsilon \downarrow 0} \frac{1}{4\pi\epsilon} \left(\frac{1}{|P - Q_0|} - \frac{1}{|P - Q_\epsilon|} \right) = \left(\frac{\partial}{\partial n} \right) \left(\frac{1}{4\pi|P - Q|} \right)$$

$$= -(P - Q_0) \frac{n}{4\pi|P - Q_0|^3}$$

This is the *dipole potential*. It satisfies Laplace's equation for $P \neq Q_0$. A superposition of dipole potentials over a surface is called a *double-layer potential* and is written in the form

$$u(P) = \iint_S (P - Q) \frac{n}{4\pi|P - Q|^3} h(Q) \, dS_Q$$

where n is the unit normal to surface S. We will see below that double-layer potentials provide an explicit representation for many solutions of Laplace's equation.

We now return to the discussion of Green's function.

Green's Function of a Bounded Region

In the case of a smoothly bounded region D in three dimensions, Green's function is defined as the function $G(P, Q)$ with the following properties:

$$\nabla^2_Q G(P, Q) = 0 \quad \text{for } P, Q \in D, P \neq Q \tag{8.2.6}$$

$$G(P, Q) = 0 \quad \text{for } Q \in \partial D, P \in D \tag{8.2.7}$$

$$G(P, Q) - \frac{1}{4\pi|P - Q|} \text{ is a smooth function in all } D \tag{8.2.8}$$

Let's show that such a function is satisfactory for the solution of Poisson's equation.

---

† I. G. Petrovsky, *Partial Differential Equations*, Wiley-Interscience, New York, 1964, pp. 219–223.

Theorem 8.2.2. Suppose that $G(P, Q)$ satisfies (8.2.6) through (8.2.8) in the smoothly bounded three-dimensional region D. Then the solution of Poisson's equation $\nabla^2 u = -h$ in D with $u = 0$ on ∂D has the explicit representation

$$u(P) = \iiint_D G(P, Q)h(Q) \, dQ \tag{8.2.9}$$

Proof. We again use Green's identity, with $D_\epsilon = \{Q \in D: |Q - P| > \epsilon\}$, $G_P(Q) = G(P, Q)$. We have

$$\iiint_{D_\epsilon} (u\nabla^2 G_P - G_P\nabla^2 u) \, dQ = \iint_{\partial D_\epsilon} \left(u \frac{\partial G_P}{\partial n} - G_P \frac{\partial u}{\partial n} \right) dS_Q$$

In D_ϵ we have $\nabla^2 G_P = 0$ and $\nabla^2 u = -h$; thus the left member is $\iiint_{D_\epsilon} G_P h \, dQ$. The right member consists of two integrals, one on the outer boundary ∂D and one on the inner boundary $\{Q: |Q - P| = \epsilon\}$. On the outer boundary both G_P and u are zero, hence this integral is zero. On the inner boundary we can replace G_P by $1/(4\pi|P - Q|)$, since the difference is a smooth function, which will contribute zero to the integral in the limit $\epsilon \downarrow 0$. But the proof of Theorem 8.2.1 shows that for this choice of G, the integral $\iint_{|Q-P|=\epsilon} u \, (\partial G/\partial n) \, dS_Q$ tends to $u(P)$ and the integral $\iint_{|Q-P|=\epsilon} G_P (\partial u/\partial n) \, dS_Q$ tends to zero, for any smooth function u. Therefore we may let $\epsilon \downarrow 0$ to obtain the required representation. ∎

One may note that the integral (8.2.9) is absolutely convergent, for any continuous function $h(Q)$. This may be seen by writing the relevant part of the integral in a polar coordinate system about P, where r_0 is chosen small enough:

$$\iiint_{|Q-P|<r_0} \frac{h(Q)}{|P - Q|} \, dQ = \int_0^{2\pi} \int_0^\pi \int_0^{r_0} h(x + r \sin \theta \cos \varphi,$$

$$y + r \sin \theta \sin \varphi, z + r \cos \theta)r \sin \theta \, dr \, d\theta \, d\varphi$$

This means that $\lim_{\epsilon \downarrow 0} \iiint_{D_\epsilon} G(P, Q)h(Q) \, dQ = \iiint_D G(P, Q)h(Q) \, dQ$, an absolutely convergent integral.

We now show, as in the case of ordinary differential equations, that Green's function is symmetric.

Theorem 8.2.3. For any two points P_1, P_2 we have $G(P_1, P_2) = G(P_2, P_1)$.

Proof. We apply Green's identity with $u(Q) = G_{P_1}(Q) = G(P_1, Q)$, $v(Q) = G_{P_2}(Q) = G(P_2, Q)$ in the region $D_\epsilon = \{Q \in D: |Q - P_1| > \epsilon, |Q - P_2| > \epsilon\}$ where $\epsilon < \frac{1}{2}|P_1 - P_2|$. Since both functions satisfy Laplace's equation in D_ϵ, we have $0 = \iiint_{D_\epsilon} (u\nabla^2 v - v\nabla^2 u) \, dQ = \iint_{\partial D_\epsilon} (u \, \partial v/\partial n - v \, \partial u/\partial n) \, dS_Q$. The surface integral on the outer boundary ∂D is zero. It remains to analyze the following four integrals on the inner boundaries:

I: $$\iint_{|Q-P_1|=\epsilon} G_{P_1} \frac{\partial G_{P_2}}{\partial n} \, dS_Q$$

II:
$$\iint_{|Q-P_1|=\epsilon} G_{P_2} \frac{\partial G_{P_1}}{\partial n} dS_Q$$

III:
$$\iint_{|Q-P_2|=\epsilon} G_{P_1} \frac{\partial G_{P_2}}{\partial n} dS_Q$$

IV:
$$\iint_{|Q-P_2|=\epsilon} G_{P_2} \frac{\partial G_{P_1}}{\partial n} dS_Q$$

The first and fourth integrals tend to zero when $\epsilon \downarrow 0$, since each integrand is $O(1/\epsilon)$ and the surface area is $O(\epsilon^2)$ when $\epsilon \downarrow 0$. For the second and third integrals, we repeat the analysis used in the proof of Theorem 8.2.2 to conclude that II $\rightarrow$ $G_{P_2}(P_1)$ and III $\rightarrow G_{P_1}(P_2)$ when $\epsilon \downarrow 0$. The proof is complete. ∎

We now turn to some applications of these ideas.

Example 8.2.1. Find Green's function of the ball $D = \{Q \in \mathbb{R}^3 : |Q| < R\}$.

Solution. This can be obtained from the case $D = R^3$ by the *method of images*. We look for Green's function in the form

$$G(P, Q) = \frac{1}{4\pi|P - Q|} - \frac{C}{|P - Q'|}$$

where the image point Q' is suitably chosen with $Q' \notin D$ and C is a constant. The above combination satisfies (8.2.6) and (8.2.8). It remains to satisfy the boundary condition (8.2.7). To do this requires that $|P - Q| = |P - Q'|/C$ for all $P \in \partial D$. Specifically we choose the image point as $Q' = Q(R^2/|Q|^2)$; this is the point along ray OQ whose distance from O satisfies $|Q| |Q'| = R^2$. To choose the constant C, we write for $P \in \partial D$

$$|P - Q|^2 = |P|^2 - 2\langle P, Q \rangle + |Q|^2$$
$$= R^2 - 2\langle P, Q \rangle + |Q|^2$$
$$|P - Q'|^2 = |P|^2 - 2\langle P, Q' \rangle + |Q'|^2$$
$$= R^2 - 2\frac{R^2}{|Q|^2}\langle P, Q \rangle + \frac{R^4}{|Q|^2}$$
$$= \frac{R^2}{|Q|^2}(|Q|^2 - 2\langle P, Q \rangle + R^2)$$

Therefore $|P - Q'|/|P - Q| = R/|Q|$ if $|P| = R$. This leads to the choice $C = R/|Q|$ and Green's function in the form

$$G(P, Q) = \frac{1}{4\pi}\left(\frac{1}{|P - Q|} - \frac{R}{|Q|}\frac{1}{|P - Q'|}\right)$$ ●

We now return to the theory.

Solution of the Dirichlet Problem

Green's function of a region D also can be used to solve the Dirichlet problem $\nabla^2 u = 0$ in D with $u = f$ on ∂D. For this purpose we again write Green's identity

$$\iiint_{D_\epsilon} (G\nabla^2 u - u\nabla^2 G) \, dQ = \iint_{\partial D_\epsilon} \left(G \frac{\partial u}{\partial n} - u \frac{\partial G}{\partial n} \right) dS_Q$$

The left side is identically zero. The right member contributes $-\iint f \, (\partial G/\partial n)$ on the outer boundary ∂D; on the inner boundary we have $\iint u \, (\partial G/\partial n) \to -u(P)$. We have proved the following result.

Theorem 8.2.4. The solution of the Dirichlet problem $\nabla^2 u = 0$ in D with $u = f$ on ∂D has the representation

$$u(P) = \iint_{\partial D} \frac{\partial G}{\partial n} f(Q) \, dS_Q$$

where $G(P, Q)$ is Green's function of the region D. ∎

We can combine this with the particular solution of Poisson's equation to find the solution of Poisson's equation with general boundary conditions.

Example 8.2.2. Find the explicit representation of the solution of Poisson's equation $\nabla^2 u = -h$ in D with the boundary condition $u = f$ on ∂D.

Solution. By the superposition principle, the solution may be obtained in the form

$$u(P) = \iiint_D G(P, Q)h(Q) \, dQ + \iint_{\partial D} \frac{\partial G}{\partial n} f(Q) \, dS_Q \qquad ●$$

We can use Green's function for the ball to find a suitable form of Poisson's integral formula in three dimensions (cf. Sec. 3.2 for the two-dimensional case). We have $G(P, Q) = [1/(4\pi)][1/|P - Q| - (R/|Q|)(1/|P - Q'|)]$. We take a system of polar coordinates for which

$$P = (r \cos \varphi \sin \theta, \, r \sin \varphi \sin \theta, \, r \cos \theta)$$

$$Q = (\tau \cos \psi \sin \alpha, \, \tau \sin \psi \sin \alpha, \, \tau \cos \alpha)$$

$$Q' = \frac{R}{\tau} (R \cos \psi \sin \alpha, \, R \sin \psi \sin \alpha, \, R \cos \alpha)$$

A straightforward differentiation yields

$$\left. \frac{\partial G}{\partial r} \right|_{r=R} = \frac{R^2 - \rho^2}{4\pi R(R^2 + \rho^2 - 2R\rho \cos \gamma)^{3/2}}$$

where $\cos \gamma = \cos \alpha \cos \theta + \sin \alpha \sin \theta \cos (\psi - \varphi)$. The solution of Laplace's equation $\nabla^2 u = 0$ in the ball with the boundary condition $u = f$ is written

$$u(P) = \frac{1}{4\pi R} \iint\limits_{|Q|=R} \frac{R^2 - \rho^2}{(R^2 + \rho^2 - 2R\rho \cos \gamma)^{3/2}} f(Q) \, dS_Q$$

EXERCISES 8.2

1. Show that the newtonian potential kernel satisfies properties 1 through 3.
2. Let $g = g(r)$ be a radial solution of Laplace's equation $\nabla^2 g = 0$ defined for $r > 0$ with the properties that $g(r) \to 0$ when $r \to \infty$ and that

   $$\lim_{r \downarrow 0} \int_0^{2\pi} \int_0^\pi u(r \sin \theta \cos \varphi, r \sin \theta \sin \varphi, r \cos \theta) (\partial g/\partial r) r^2 \sin \theta \, d\theta \, d\varphi$$
   $$= -u(0, 0, 0)$$

   for every continuous function $u = u(x, y, z)$. Prove that $g(r) = 1/(4\pi r)$, the newtonian potential kernel.
3. Let $G(P, Q) = 1/(4\pi|P - Q|) - R/(4\pi|Q||P - Q'|)$ be Green's function of the ball $|P| < R$. Prove that $\partial G/\partial r = (R^2 - r^2)/[4\pi R(R^2 + r^2 - 2Rr \cos \gamma)^{3/2}]$ is the Poisson kernel of the ball.

Exercises 4 to 8 provide some of the basic properties of solutions of Laplace's equation. They must be done in the order given.

4. Modify the proof of Theorem 8.2.1 to prove the following result: Any solution of Laplace's equation $\nabla^2 u = 0$ in the three-dimensional ball $|P| < R$ can be written as

   $$u(P) = \iint\limits_{\partial D_{0,R}} \left(u \frac{\partial G_P}{\partial n} - G_P \frac{\partial u}{\partial n} \right) dS_Q$$

   where $G_P(Q)$ is the newtonian potential kernel and the integration is over the surface of the sphere.
5. Use Exercise 4 to prove the mean-value theorem for harmonic functions: For every solution of $\nabla^2 u = 0$ in a three-dimensional ball, we have

   $$u(P) = \frac{1}{4\pi R^2} \iint\limits_{|Q-P|=R} u \, dS_Q$$

   [*Hint:* First show that on the surface we have $G_P = 1/(4\pi R)$, $\partial G_P/\partial n = -[1/(4\pi R^2)]$, and $\iint_{|Q-P|=R} (\partial u/\partial n) \, dS_Q = 0$.]
6. Extend Exercise 5 to show that every solution of $\nabla^2 u = 0$ has the *solid mean-value property* in the form

   $$u(P) = \frac{3}{4\pi R^3} \iiint\limits_{|Q-P|<R} u \, dQ$$

   where the integral is over the solid ball of radius R.
7. Use Exercise 6 to prove the *local maximum principle*: If u is a solution of $\nabla^2 u = 0$ in the ball $\{Q: |Q - P| < \delta\}$ such that $u(P) \geq u(Q)$ for all Q and $|Q - P| < \delta$, then $u(Q) = u(P)$ for all Q such that $|Q - P| < \delta$.
8. Extend Exercise 7 to prove the *global maximum principle*: If u is a solution of $\nabla^2 u = 0$ in a connected and smoothly bounded region, for which $\max_{P \in D} u(P)$ is

attained at an interior point P_0, then u is constant throughout P. [*Hint:* If there is an interior global maximum at $P_0 \in D$, then connect an arbitrary point $P \in D$ to P_0 by means of a polygonal path C. Along this path draw a finite sequence of spheres $\{B(x_i, \delta_i)\}_{i=0}^{N}$ such that $x_0 = P_0$, $x_N = P$, and $x_i \in C$, $x_i \in B(x_{i-1}, \delta_{i-1})$ for $i = 1, \ldots, N$. By Exercise 7, u is constant throughout $B(x_0, \delta_0)$, especially at x_1. Now apply Exercise 7 to $B(x_1, \delta_1)$ and continue the process inductively until you reach $P = x_N$, to conclude that $u(x_N) = u(x_0)$.]

9. Formulate and prove a local minimum principle and global minimum principle for solutions of Laplace's equation $\nabla^2 u = 0$.

Exercises 10 and 11 are designed to establish directly the properties of the newtonian potential operator $h \rightarrow u$, where $u(P) = [1/(4\pi)] \iiint h(Q)/|P - Q| \, dQ$ is the newtonian potential of the function $h(Q)$.

10. Suppose that h is a continuous function with $h(Q) = 0$ for $|Q| > r$. Prove that u is a solution of Laplace's equation $\nabla^2 u = 0$ for $|P| > r$.

11. Suppose that h is a differentiable function with $h(Q) = 0$ for $|Q| > r$.
    (*a*) Prove that the first derivatives of u can be computed by the formula

$$4\pi \frac{\partial u}{\partial x_i} P = \iiint \frac{h(Q)(P - Q)_i}{|Q - P|^3} \, dQ = - \iiint \frac{\partial h}{\partial Q_i} Q \frac{1}{|Q - P|} \, dQ$$

(*b*) Prove that the second derivatives of u can be computed by the formula

$$4\pi \frac{\partial^2 u}{\partial x_i, \partial x_j} = \frac{1}{4\pi} \iiint \frac{\partial h}{\partial Q_i} \frac{(Q - P)_j}{|P - Q|^3} \, dQ$$

(*c*) Transform the last integral to spherical polar coordinates to prove that

$$4\pi \, \nabla^2 u(x, y, z)$$

$$= \int_0^{2\pi} \int_0^{\pi} \int_0^{\infty} \left(\frac{\partial}{\partial r} \right) h(x + r \cos \varphi \sin \theta, y + r \sin \varphi \sin \theta, z + r \cos \theta) \sin \theta \, dr \, d\theta \, d\varphi$$

$$= -4\pi \, h(x, y, z).$$

8.3 TWO-DIMENSIONAL PROBLEMS

One may try to replicate the three-dimensional theory in two dimensions. In the case of a bounded region, the theory is entirely parallel, although the formulas are slightly different. For the basic case of the entire plane, we encounter the fundamental difficulty that Green's function cannot be uniquely obtained that satisfies all the previous conditions.

The Logarithmic Potential

To find a suitable Green function for two dimensions, we abstract the fundamental properties of the newtonian potential kernel in three dimensions. These are that $\nabla_Q^2 G_P = 0$ for $Q \neq P$ and $\iint_{|Q-P|=\epsilon} (\partial G_P / \partial n) u \, dS_Q \rightarrow -u(P)$, for every smooth function u. To replicate this in two dimensions, we begin with the radial solutions of Laplace's equation

$$0 = \nabla^2 G = G_{rr} + \frac{1}{r} G_r = \frac{1}{r} (rG_r)_r$$

leading to $rG_r = B$, $G = A + B \log r$ for suitable constants A, B. We can safely set $A = 0$, since we are looking for a particular solution. To determine B, we compute $\partial G/\partial r = B/r$, whose "surface integral" on the circle $|Q - P| = \epsilon$ is

$$\int_{|Q-P|=\epsilon} \frac{\partial G}{\partial r} u \, dS_Q = \int_{-\pi}^{\pi} \frac{B}{\epsilon} u(x + \epsilon \cos \theta, y + \epsilon \sin \theta)\epsilon \, d\theta \to 2\pi B u(x, y)$$

$$\epsilon \to 0$$

Therefore we choose $B = -1/(2\pi)$, and we have the *logarithmic potential kernel*

$$G(P, Q) = \frac{1}{2\pi} \log \frac{1}{|P - Q|}$$

This leads to the following explicit representation of the solution of Poisson's equation:

$$u(P) = \frac{1}{2\pi} \int\int \log \left(\frac{1}{|P - Q|} \right) h(Q) \, dQ$$

One can show[†] that this formula provides a solution of Poisson's equation. However, it is difficult to obtain a simple uniqueness criterion for the Poisson equation in the entire plane. For this reason we turn to some problems with unique solutions.

Green's Function of a Bounded Plane Region

Green's function of a bounded two-dimensional region is defined by the following requirements:

$$\nabla_Q^2 G = 0 \qquad Q \in D, Q \neq P \tag{8.3.1}$$

$$G(P, Q) = 0 \qquad Q \in D, P \in \partial D \tag{8.3.2}$$

$$G(P, Q) - \frac{1}{2\pi} \log \frac{1}{|P - Q|} \text{ is a smooth function} \tag{8.3.3}$$

We have the following theorem, exactly as in the three-dimensional case.

Theorem 8.3.1. Let u be a solution of the Poisson equation $\nabla^2 u = -h$ in the region D satisfying the boundary condition that $u = 0$ on ∂D. Then we have the explicit representation

$$u(P) = \int\int_D G(P, Q)h(Q) \, dQ$$

---

† Petrovsky, loc. cit.

Proof. We follow the proof of the three-dimensional case, beginning with Green's formula applied to $D_\epsilon = \{Q \in D: |Q - P| > \epsilon\}$:

$$\iint_{D_\epsilon} (u\nabla^2 G - G\nabla^2 u)\, dQ = \int_{\partial D_\epsilon} \left(u\frac{\partial G}{\partial n} - G\frac{\partial u}{\partial n} \right) dS_Q$$

The integral on the left is over the interior of the plane region D_ϵ, and the integral on the right is over the boundary curves which define ∂D_ϵ. In D_ϵ, G satisfies Laplace's equation and $\nabla^2 u = -h$, so that the right side becomes $\iint_{D_\epsilon} Gh\, dQ$. The right side is analyzed as before: The outer boundary ∂D contributes zero, while the inner boundary $\{Q: |Q - P| = \epsilon\}$ contributes $[1/(2\pi)] \int_0^{2\pi} u(x + \epsilon \cos\theta, y + \epsilon \sin\theta)\, d\theta$. In the limit $\epsilon \downarrow 0$, this gives $u(x, y) = u(P)$, which was to be proved. ∎

Example 8.3.1. Find Green's function of the circular disk $D = \{(x, y): x^2 + y^2 < a^2\}$, and use this to solve Poisson's equation $\nabla^2 u = -h$ in D with $u = 0$ on ∂D.

Solution. This can be solved by the method of images, as in the three-dimensional case. We look for Green's function in the form

$$G(P, Q) = \frac{1}{2\pi} \left(\log \frac{1}{|P - Q|} - C_1 \log \frac{C_2}{|P - Q'|} \right)$$

where Q' is the image point $Q' = Q(a^2/|Q|^2)$ and the constants C_1, C_2 are to be determined. As we showed in Sec. 8.2, the image point Q' satisfies $|P - Q|/|P - Q'| = a/|Q|$ when $|P| = a$. Therefore we choose $C_1 = 1$, $C_2 = a/|Q|$. The Poisson equation is solved by the explicit representation

$$u(P) = \frac{1}{2\pi} \int_{|Q|=a} \left(\log \frac{1}{|P - Q|} - \log \frac{a}{|P - Q'||Q|} \right) h(Q)\, dS_Q \qquad ●$$

Example 8.3.2. Find Green's function for the half-plane $D = \{(x, y): y > 0\}$, and solve the Poisson equation $\nabla^2 u = -h$ with $u = 0$ on ∂D.

Solution. In this case we may again use the method of images. If $Q = (\xi, \eta)$, the image point is $Q' = (\xi, -\eta)$; we have $|P - Q| = |P - Q'|$ if $P \in \partial D$. Green's function is

$$G(P, Q) = \frac{1}{2\pi} \log \frac{1}{|P - Q|} - \frac{1}{2\pi} \log \frac{1}{|P - Q'|}$$

$$= -\frac{1}{4\pi} \log \frac{(x - \xi)^2 + (y - \eta)^2}{(x - \xi)^2 + (y + \eta)^2}$$

and the solution of Poisson's equation is given by $u(P) = \iint_D G(P, Q)h(Q)\, dQ$. ●

Solution of the Dirichlet Problem

The two-dimensional Green function also can be used to solve the Dirichlet problem for Laplace's equation $\nabla^2 u = 0$ in a smoothly bounded plane region

D with given boundary data f. As in the three-dimensional case, we begin with Green's identity

$$\iint_{D_\epsilon} (u\nabla^2 G - G\nabla^2 u)\, dQ = \int_{\partial D_\epsilon} \left(u\,\frac{\partial G}{\partial n} - G\,\frac{\partial u}{\partial n} \right) dS_Q$$

where $D_\epsilon = \{Q \in D : |Q - P| > \epsilon\}$. The left side is zero. The right side gives $\iint \partial G/\partial n$ on the outer boundary ∂D, while on the inner boundary we have $\int u\, \partial G/\partial n \to -u(P)$. Thus we have proved the following result.

Theorem 8.3.2. The solution of the Dirichlet problem $\nabla^2 u = 0$ in the smoothly bounded plane region with $u = f$ on ∂D is given by the double-layer potential

$$u(P) = \int_{\partial D} \frac{\partial G}{\partial n}\, f\, dS_Q \qquad\blacksquare$$

Example 8.3.3. Solve the Dirichlet problem in the ball $|P| < a$.

Solution. In this case we have Green's function

$$G(P,\, Q) = \frac{1}{2\pi} \log\left(\frac{1}{|P - Q|} \right) - \frac{1}{2\pi} \log \frac{a}{|Q||P - Q'|}$$

with

$$\frac{\partial G}{\partial r} = \frac{a^2 - r^2}{2\pi[a^2 + r^2 - 2ar\cos(\theta - \varphi)]}$$

and we retrieve the Poisson integral formula

$$u(P) = \frac{1}{2\pi} \int_{-\pi}^{\pi} \frac{a^2 - r^2}{a^2 + r^2 - 2ar\cos(\theta - \varphi)}\, f(\theta)\, d\theta$$

This formula was first derived in Sec. 3.1 from the Fourier series of separated solutions of Laplace's equation. Now we see that it can be done directly, without any appeal to Fourier series or separated solutions. $\qquad\bullet$

Green's Functions and Separation of Variables

If we compare the Fourier representation of the solution with the explicit representation by Green's function, we can obtain a representation of the Green function in terms of the eigenfunctions of the associated homogeneous problem.

Suppose that we want to solve Poisson's equation $\nabla^2 u = -h$ with the condition that $u = 0$ on the boundary of the smoothly bounded region D. Let $\{\varphi_n\}$ be a complete system of eigenfunctions satisfying $\nabla^2 \varphi_n + \lambda_n \varphi_n = 0$ in D with $\varphi_n = 0$ on the boundary and normalized so that $\langle \varphi_n,\, \varphi_n \rangle = \int_D \varphi_n(P)^2\, dP = 1$. Here n is a "multi-index," depending on the dimension of the space. The Fourier representation of the solution is

$$u(P) = \sum_n \frac{1}{\lambda_n} \langle \varphi_n,\, h \rangle\, \varphi_n(P)$$

However, Green's function satisfies the boundary conditions and can be expanded in a series of eigenfunctions as $G(P, Q) = \sum_n c_n(Q)\varphi_n(P)$, where the Fourier coefficients are obtained as $c_n(Q) = \int_D G(P, Q)\varphi_n(P)\, dP$. Substituting in the explicit representation of u by Green's function and proceeding formally, we have

$$u(P) = \int_D G(P, Q)h(Q)\, dQ = \sum_n \varphi_n(P)\langle c_n, h\rangle$$

Comparing the two formulas for u leads to the identification $\langle c_n, h\rangle = \langle \varphi_n, h\rangle/\lambda_n$ for every h, or $c_n(Q) = \varphi_n(Q)/\lambda_n$, and we are led to the formula

$$G(P, Q) = \sum_n \frac{\varphi_n(P)\varphi_n(Q)}{\lambda_n}$$

This statement is known as *Mercer's theorem*. Although easily remembered, it may not be an efficient method for computation of Green's function. In particular, we do not expect that the series converges for $P = Q$. Rather than explore this in general, we consider the rigorous validity in each case.

Example 8.3.4. Find the Fourier representation of Green's function of ∇^2 in the rectangle $0 < x < a$, $0 < y < b$ with the boundary condition that $u = 0$ on all four sides.

Solution. The normalized eigenfunctions of this problem are

$$\varphi_{mn}(x, y) = \sqrt{\frac{4}{ab}} \sin \frac{m\pi x}{a} \sin \frac{n\pi y}{b}$$

with $\lambda_{mn} = (m\pi/a)^2 + (n\pi/b)^2$. The above formula for Green's function is

$$G(P, Q) = G(x, y; \xi, \eta)$$

$$= \frac{4}{ab} \sum_{mn} \frac{\sin(m\pi x/a) \sin(n\pi y/b) \sin(m\pi\xi/a) \sin(n\pi\eta/b)}{(m\pi/a)^2 + (n\pi/b)^2}$$

Although this series is not absolutely convergent, it can be shown to be conditionally convergent for $(x, y) \neq (\xi, \eta)$. ●

Example 8.3.5. Find the Fourier representation of Green's function of ∇^2 in the circle $x^2 + y^2 < a^2$ with zero boundary conditions.

Solution. In polar coordinates the normalized eigenfunctions in complex form are

$$\varphi_{mn}(\rho, \varphi) = C_{mn} J_m \left(\frac{z_{mn}\rho}{a}\right) e^{im\varphi}$$

where the eigenvalues are $\lambda_{mn} = (z_{mn}/a)^2$ and the normalizing constants are $C_{mn}^2 = 2/J_{m+1}(z_{mn})^2$. Green's function is

$$G(P, Q) = G(r, \varphi; \rho, \theta)$$

$$= \sum_{mn} C_{mn}^2 J_m\left(\frac{z_{mn}r}{a}\right) J_m\left(\frac{z_{mn}\rho}{a}\right) e^{im(\varphi-\theta)}$$

It may be shown, as in Example 8.3.4, that this series is conditionally convergent.

●

The above representations by double Fourier series are not absolutely convergent. To obtain more effective series representations, we use a Fourier series in one variable only, where the coefficient functions are obtained as Green's function of a closely related ordinary differential equation. We illustrate this for the Poisson equation $u_{xx} + u_{yy} = -f$ in the rectangle $0 < x < a$, $0 < y < b$. We write each term as a single Fourier sine series:

$$f(x, y) = \sum_{m=1}^{\infty} \sin\left(\frac{m\pi x}{a}\right) F_m(y) \qquad F_m(y) = \frac{2}{a} \int_0^a f(\xi, y) \sin\frac{m\pi\xi}{a} \, d\xi$$

$$u(x, y) = \sum_{m=1}^{\infty} \sin\left(\frac{m\pi x}{a}\right) U_m(y) \qquad U_m(y) = \frac{2}{a} \int_0^a u(\xi, y) \sin\frac{m\pi\xi}{a} \, d\xi$$

where U_m satisfies the equation $U_m'' - (m\pi/a)^2 U_m = -F_m$ with the boundary condition $U_m(0) = 0 = U_m(b)$. This ordinary differential equation is solved by the one-dimensional Green function

$$U_m(y) = \int_0^b G_m(y, \eta) F_m(\eta) \, d\eta$$

where

$$G_m(y, \eta) = \begin{cases} \dfrac{\sinh(m\pi y/b) \sinh[m\pi(b-\eta)/b]}{W_m(y)} & 0 \le y \le \eta \\[3mm] \dfrac{\sinh(m\pi(b-y)[b \sinh(m\pi\eta/b)]}{W_m(y)} & \eta \le y \le b \end{cases}$$

and the wronskian is

$$W_m(y) = \frac{m\pi}{b}\left[\sinh\frac{m\pi y}{b}\cosh\frac{m\pi(b-y)}{b} + \cosh\frac{m\pi y}{b}\sinh\frac{m\pi(b-y)}{b}\right]$$

Green's function for the problem is written as

$$G(P, Q) = G(x, y; \xi, \eta) = \frac{2}{L}\sum_{m=1}^{\infty} \sin\frac{m\pi x}{a} \sin\left(\frac{m\pi\xi}{a}\right) G_m(y, \eta)$$

If $y \neq \eta$, this series is absolutely convergent together with all its derivatives. This can be seen easily from the form of the coefficient $G_m(y, \eta)$: When

$m \to \infty$, the hyperbolic sine function satisfies $\sinh m\theta \simeq \frac{1}{2}e^{m\theta}$. If we make this substitution everywhere, we have for $y < \eta$

$$G_m(y, \eta) \simeq \frac{e^{m\pi y/b}e^{m\pi(b-\eta)/b}}{2(m\pi/b)e^{m\pi}}$$

$$= \frac{e^{(m\pi/b)(y-\eta)}}{2(m\pi/b)}$$

This tends to zero exponentially when $m \to \infty$. A similar computation applies to the case $y > \eta$.

EXERCISES 8.3

1. Use the method of images to find Green's function and to solve Poisson's equation $\nabla^2 u = -h$ in the quarter-plane $D = \{(x, y): x > 0, y > 0\}$ with the boundary condition that $u = 0$ on both axes. Use three image points.

2. Use the method of images to find Green's function and to solve Poisson's equation $\nabla^2 u = h$ in the quarter-plane $D = \{(x, y): x > 0, y > 0\}$ with the boundary conditions $u(x, 0) = 0$, $(\partial u/\partial x)(0, y) = 0$.

3. Show that the logarithmic potential can be obtained directly from the newtonian potential kernel by the following "renormalization procedure": Let $u_M(x, y) = \int_{-M}^{M} dz/(4\pi\sqrt{x^2 + y^2 + z^2})$ be the newtonian potential of a uniform line charge on the segment $[-M, M]$ of the z axis.
   (a) Show that this integral can be evaluated directly as $[1/(2\pi)] \sinh^{-1}(M/r)$ where $r = \sqrt{x^2 + y^2}$.
   (b) Use the behavior of $\sinh^{-1} x$ when $x \to \infty$ to find a constant c_M such that $\lim_{M\uparrow\infty} [u_M(x, y) - c_M] = [1/(2\pi)] \log(1/\sqrt{x^2 + y^2})$.
   (c) Show that the "potential difference" can be expressed directly as

$$\lim_{M\uparrow\infty} [u_M(x_1, y_1) - u_M(x_2, y_2)] = \frac{1}{2\pi}\left(\log\frac{1}{\sqrt{x_1^2 + y_1^2}} - \log\frac{1}{\sqrt{x_2^2 + y_2^2}}\right)$$

4. Find a Fourier representation for Green's function of ∇^2 in the rectangle $0 < x < a$, $0 < y < b$ with the boundary conditions that $u = 0$ on the bottom and two vertical sides while $u_y = 0$ on the top side $y = b$, $0 < x < a$.

5. Find a Fourier representation of Green's function of ∇^2 of the disk $x^2 + y^2 < a^2$ in the form $G(r, \theta; \rho, \varphi) = \Sigma G_m(r, \rho)e^{im(\theta-\varphi)}$, where G_m is a suitable one-dimensional Green function for the ordinary differential operator $u_{rr} + (1/r)u_r - (m^2/r^2)u$.

6. Show that Green's function of ∇^2 for the infinite strip $-\infty < x < \infty$, $0 < y < a$ can be written in the form

$$G(x, y; \xi, \eta) = \frac{2}{a}\sum_{m=1}^{\infty} \sin\left(\frac{m\pi y}{a}\right) \sin\left(\frac{m\pi\eta}{a}\right) g_n(x, \xi)$$

where g_n is a suitably defined Green's function for the ordinary differential equation $g_n'' - (n\pi/a)^2 g_n = 0$, $-\infty < x < \infty$.

7. Use the method of images to show that Green's function of ∇^2 for the strip $-\infty < x < \infty$, $0 < y < a$ can be written in the form

$$G(x, y; \xi, \eta)$$

$$= \frac{1}{2\pi} \sum_n \left[\log \frac{1}{(x - \xi)^2 + (y - \eta - 2na)^2} - \log \frac{1}{(x - \xi)^2 + (y + \eta - 2na)^2} \right]$$

8. Let D be the region described in polar coordinates by $\{(r, \theta): a < r < b, \alpha < \theta < \beta\}$ with $a > 0$, $0 < \alpha < \beta < 2\pi$. Show that Green's function can be written in the form

$$G(r, \theta; \rho, \varphi) = \sum \sin \left[\frac{n\pi(\theta - \alpha)}{(\beta - \alpha)} \right] \sin \left[\frac{n\pi(\varphi - \alpha)}{(\beta - \alpha)} \right] f_n(r; \rho)$$

where f_n is a suitable Green function for the ordinary differential equation $f'' + (1/r)f' - [n\pi/(\beta - \alpha)^2]f = 0$ on the interval $a < r < b$.

9. Let D be the region described in polar coordinates by $\{(r, \theta): a < r < b, \alpha < \theta < \beta\}$ with $a > 0$, $0 < \alpha < \beta < 2\pi$. Show that Green's function of ∇^2 can be written in the form $G(r, \theta; \rho, \varphi) = \sum_{n \geq 1} \sin[n\pi \log(r/a)/\log(b/a)] \sin[n\pi \log(\rho/a)/\log(b/a)] g_n(\theta, \varphi)$, where g_n is a suitable Green function for the ordinary differential equation $g_n'' - [n\pi/\log(b/a)]^2 g_n = 0$ on the interval $\alpha < \varphi < \beta$.

8.4 GREEN'S FUNCTION FOR THE HEAT EQUATION

Inhomogeneous Heat Equation

We can obtain a particular solution of the inhomogeneous heat equation

$$u_t - K u_{xx} = h \qquad 0 < t < T, \ -\infty < x < \infty \tag{8.4.1}$$

by suitably transforming the Fourier representation. To do this, we write

$$u(x; t) = \int_{-\infty}^{\infty} U(\xi; t)e^{i\xi x}\, d\xi \qquad h(x; t) = \int_{-\infty}^{\infty} H(\xi; t)e^{i\xi x}\, d\xi$$

$$U(\xi; t) = \frac{1}{2\pi} \int_{-\infty}^{\infty} u(x; t)e^{-i\xi x}\, dx \qquad H(\xi; t) = \frac{1}{2\pi} \int_{-\infty}^{\infty} h(x; t)e^{-i\xi x}\, dx$$

and transform equation (8.4.1) to the ordinary differential equation $U_t + K\xi^2 U = H$. A particular solution with $U(\xi; 0) = 0$ is found by means of the integrating factor $e^{K\xi^2 t}$, and we obtain

$$U(\xi; t) = \int_0^t e^{-K\xi^2(t-s)} H(\xi; s)\, ds$$

Substituting the above Fourier integral of $H(\xi; t)$, we obtain

$$u(x; t) = \int_{-\infty}^{\infty} e^{i\xi x} \, d\xi \int_0^t e^{-K\xi^2(t-s)} \left[\frac{1}{2\pi} \int h(y; s)e^{-iy\xi} \, dy\right] ds$$

$$= \int_0^t ds \int_{-\infty}^{\infty} h(y; s) \, dy \int e^{i\xi(x-y)} e^{-K\xi^2(t-s)} \, d\xi$$

The final integral is recognized from Sec. 5.2 as the heat kernel

$$G(x, y; \tau) = (4\pi K\tau)^{-1/2} e^{-(x-y)^2/(4K\tau)}$$

and we have the explicit representation

$$u(x; t) = \int_0^t \int_{-\infty}^{\infty} G(x, y; t - s)h(y; s) \, dy \, ds \tag{8.4.2}$$

Theorem 8.4.1. Let $h(y; s)$ be a bounded continuous function in the strip $0 \leq t \leq T$, $-\infty < x < \infty$. Then integral (8.4.2) defines a solution of the inhomogeneous heat equation (8.4.1) with $u(x; 0) = 0$.

Proof. We recall from Sec. 5.2 that $G(x, y; t)$ satisfies the heat equation $G_t - KG_{xx} = 0$ and that for any arbitrary bounded continuous function h, $\lim_{\tau \downarrow 0} \int_{-\infty}^{\infty} G(x, y; \tau)h(y) \, dy = h(x)$. We use this to compute the derivative as follows:

$$u(x; t + \Delta t) - u(x; t) = \int_0^{t+\Delta t} \int_{-\infty}^{\infty} G(x, y; t + \Delta t - s)h(y; s) \, dy \, ds$$

$$- \int_0^t \int_{-\infty}^{\infty} G(x, y; t - s)h(y; s) \, dy \, ds$$

$$= \int_0^t \int_{-\infty}^{\infty} [G(x, y; t + \Delta t - s) - G(x, y; t - s)]h(y; s) \, dy \, ds$$

$$+ \int_t^{t+\Delta t} \int_{-\infty}^{\infty} G(x, y; t + \Delta t - s)h(y; s) \, dy \, ds$$

When we divide by Δt and take the limit, the first integral tends to $\int_0^t \int_{-\infty}^{\infty} G_t(x, y; t - s)h(y; s) \, dy \, ds$ and the second integral tends to $h(x, t)$. Similarly, when we compute u_{xx}, the derivatives can be put directly onto G to obtain the corresponding integral with G replaced by G_{xx}. But G satisfies the heat equation $G_t = KG_{xx}$ from which we conclude that $u_t - Ku_{xx} = h$, as required. ∎

To solve the inhomogeneous heat equation with general initial data, we apply the superposition principle. The function

$$u(x; t) = \int_{-\infty}^{\infty} G(x, y; t)f(y) \, dy + \int_0^t \int_{-\infty}^{\infty} G(x, y; t - s)h(y; s) \, dy \, ds$$

satisfies the equation $u_t - Ku_{xx} = h$ with the initial condition $u(x; 0) = f(x)$.

To solve inhomogeneous heat equations with boundary conditions, we can use the method of images to suitably modify $G(x, y; t)$.

Example 8.4.1. Find an explicit representation of the solution of the equation $u_t - Ku_{xx} = h$ in the half-space $0 < x < \infty$, $0 < t < T$ satisfying the boundary condition $u(0; t) = 0$ and the initial condition $u(x; 0) = 0$.

Solution. The corresponding Green function for this case is obtained by using the image point $Q' = -y$, which leads to

$$G(P, Q; t) = (4\pi Kt)^{-1/2} \left(e^{-(x-y)^2/(4Kt)} - e^{-(x+y)^2/(4Kt)} \right)$$

and the solution

$$u(x; t) = \int_0^t \int_{-\infty}^{\infty} G(x, y; t - s)h(y; s) \, dy \, ds \qquad \bullet$$

Example 8.4.2. Find the solution of the inhomogeneous heat equation $u_t - K\nabla^2 u = h$ in the region $R^3 \times [0, T]$ and satisfying the initial condition $u(P; 0) = 0$.

Solution. We simply modify the previous one-dimensional construction to the case of three dimensions. Green's function for this case is

$$G(P, Q; t) = \int\int\int e^{-K|\xi|^2 t} e^{i\langle \xi, P-Q \rangle} \, d\xi$$

$$= (4\pi Kt)^{-1/2} e^{-|P-Q|^2/(4Kt)}$$

and the solution of Poisson's equation is

$$u(P; t) = \int_0^t \int\int\int G(P, Q; t - s)h(Q; s) \, dQ \, ds \qquad \bullet$$

To solve inhomogeneous heat equations in higher dimensions, we can suitably modify $G(x, y; t)$. For example, the three-dimensional equation

$$u_t - K\nabla^2 u = h \qquad P \in R^3, \, 0 \le t \le T$$

has the particular solution

$$u(P; t) \frac{1}{[4\pi K(t - s)]^{3/2}} \int_0^t \int\int\int e^{-|P-Q|^2/[4K(t-s)]} h(Q; s) \, dQ \, ds$$

The One-Dimensional Heat Kernel and the Method of Images

A closely related problem is the initial-value problem for the homogeneous heat equation

$$u_t = Ku_{xx} \qquad (t > 0, \, -\infty < x < \infty) \qquad u(x; 0) = f(x)$$

which was solved in Sec. 5.2. There we obtained the Gauss-Weierstrass explicit representation

$$u(x; t) = \frac{1}{\sqrt{4\pi Kt}} \int_{-\infty}^{\infty} e^{-(x-y)^2/(4Kt)} f(y) \, dy$$

This can be used to find explicit representations of the solution of initial-boundary-value problems on a finite interval with homogeneous boundary conditions, leading to some interesting identities involving infinite series. To do this, we make a suitable application of the method of images. We illustrate with an example.

Example 8.4.3. Find an explicit representation of the solution of the heat equation $u_t = Ku_{xx}$ on the interval $0 < x < L$ with the boundary conditions $u(0; t) = 0$, $u(L; t) = 0$ and the initial conditions $u(x; 0) = f(x)$, a piecewise smooth function.

Solution. We extend $f(x)$ as an odd $2L$-periodic function by setting $f_{odd}(x) = -f(-x)$ for $-L < x < 0$ and $f_{odd}(x + 2L) = f_{odd}(x)$ for all x. We apply the Gauss-Weierstrass formula to $f_{odd}(x)$ and transform as follows:

$$\sqrt{4\pi Kt}\ u(x; t) = \int_{-\infty}^{\infty} e^{-(x-y)^2/(4Kt)}\ f_{odd}(y)\ dy$$

$$= \sum_{-\infty < m < \infty} \left\{ \int_{2mL}^{(2m+1)L} + \int_{(2m+1)L}^{(2m+2)L} \right\} e^{-(x-y)^2/(4Kt)}\ f_{odd}(y)\ dy$$

But for each $m = 0, +1, -1, \ldots$ we can write

$$\int_{2mL}^{(2m+1)L} e^{-(x-y)^2/(4Kt)}\ f_{odd}(y)\ dy = \int_{odd}^{L} e^{-(x-y-2mL)^2/(4Kt)} f(y)\ dy$$

$$\int_{(2m+1)L}^{(2m+2)L} e^{-(x-y)^2/(4Kt)}\ f_{odd}(y)\ dy = -\int_{odd}^{L} e^{-[x+y-(2m+2)L]^2/(4Kt)} f(y)\ dy$$

Thus

$$u(x; t) = \sum_{-\infty < m < \infty} \int_0^L \left(e^{-(x-y-2mL)^2/(4Kt)} - e^{-[x+y-(2m+2)L]^2/(4Kt)} \right) \frac{f(y)\ dy}{\sqrt{4\pi Kt}}$$

which is the required representation. ●

It is instructive to compare the result of Example 8.4.3 with the Fourier representation obtained in Sec. 2.2. According to that method, the solution is written in the form

$$u(x; t) = \sum_{n \geq 1} A_n \left(\sin \frac{n\pi x}{L} \right) e^{-(n\pi/L)^2 Kt}$$

where the Fourier coefficient is

$$A_n = \frac{2}{L} \int_0^L f(y) \sin \left(\frac{n\pi y}{L} \right) dy$$

Interchanging the orders of integration and summation, we have

$$u(x; t) = \frac{2}{L} \int_0^L \left[\sum_{n \geq 1} \sin \frac{n\pi x}{L} \left(\sin \frac{n\pi y}{L} \right) e^{-(n\pi/L)^2 Kt} \right] f(y)\ dy$$

We have represented the function $u(x; t)$ in two different ways. Since this holds for all piecewise smooth functions $f(x)$, $0 < x < L$, we infer that the integrands are equal. Thus

$$\frac{2}{L} \sum_{n \geq 1} \left(\sin \frac{n\pi x}{L} \right) \left(\sin \frac{n\pi y}{L} \right) e^{-(n\pi/L)^2 Kt}$$

$$= \frac{1}{\sqrt{4\pi Kt}} \sum_{-\infty < m < \infty} \left(e^{-(x-y-2mL)^2/(4Kt)} - e^{-[x-y-(2m+2)L]^2/(4Kt)} \right)$$

EXERCISES 8.4

1. Find the solution of the heat equation $u_t - Ku_{xx} = h$ for $0 < x < \infty$ satisfying the boundary conditions $u_x(0; t) = 0$, $u(x; 0) = 0$.
2. Find the solution of the heat equation $u_t - Ku_{xx} = h$ for $0 < x < L$ satisfying the boundary conditions $u(0; t) = 0$, $u(L; t) = 0$, $u(x; 0) = 0$.
3. Find a particular solution of the two-dimensional heat equation $u_t - K(u_{xx} + u_{yy}) = h$ in the entire plane $-\infty < x < \infty$, $-\infty < y < \infty$, $0 \leq t \leq T$.
4. Find an explicit representation of the solution of the heat equation $u_t = Ku_{xx}$ on the interval $0 < x < L$ with the boundary conditions $u_x(0; t) = 0$, $u_x(L; t) = 0$ and the initial condition $u(x; 0) = f(x)$, a piecewise smooth function.
5. Find an explicit representation of the solution of the heat equation $u_t = Ku_{xx}$ on the interval $0 < x < L$ with the (periodic) boundary conditions $u(0; t) = u(L; t)$, $u_x(0; t) = u_x(L; t)$ and the initial condition $u(x; 0) = f(x)$, a piecewise smooth function.

8.5   GREEN'S FUNCTION FOR THE WAVE EQUATION

We now seek a Green function representation for the solution of the three-dimensional wave equation problem

$$u_{tt} - c^2 \nabla^2 u = h(P; t) \qquad P \in \mathbb{R}^3, \ t > 0$$

$$u(P; 0) = 0 \qquad u_t(P; 0) = 0 \qquad P \in \mathbb{R}^3$$

This will generalize the newtonian potential kernel of Sec. 8.2, which is the case in which $u = u(P)$, $h = h(P)$, independent of time.

Derivation of the Retarded Potential

To find the explicit representation by Green's function, we begin with the Fourier-transformed equation

$$U_{tt} + c^2 |\mu|^2 U = H(\mu; t)$$

$$U(\mu; 0) = 0 \qquad U_t(\mu; 0) = 0$$

where

$$u(P; t) = \iiint e^{i\langle \mu, P \rangle} U(\mu; t) \, d\mu$$

$$U(\mu; t) = \left(\frac{1}{2\pi}\right)^3 \iiint e^{-i\langle \mu, P \rangle} u(P; t) \, dP$$

$$h(P; t) = \iiint e^{i\langle \mu, P \rangle} H(P; t) \, d\mu$$

$$H(\mu; t) = \left(\frac{1}{2\pi}\right)^3 \iiint e^{-i\langle \mu, P \rangle} h(P; t) \, dP$$

The Fourier-transformed equation is a second-order ordinary differential equation which can be solved by the method of variation of parameters:

$$U(\mu; t) = \int_0^t \left[\frac{\sin |\mu| cs}{|\mu| c} \right] H(\mu; t - s) \, ds$$

It remains to invert the Fourier transform.

For this purpose we recall from Sec. 5.3 the surface integral representation of the sine function as a mean value:

$$\frac{\sin |\mu| R}{|\mu| R} = \frac{1}{4\pi R^2} \iint_{|\xi| = R} e^{i\langle \mu, \xi \rangle} \, d\xi$$

Inserting this into the Fourier representation of $u(P; t)$ and interchanging the order of integration, we have

$$u(P; t) = \iiint e^{i\langle \mu, P \rangle} \, d\mu \int_0^t H(\mu; t - s) \frac{1}{4\pi c^2 s} \, ds \iint_{|\xi| = cs} e^{i\langle \mu, \xi \rangle} \, d\xi$$

$$= \frac{1}{4\pi c^2} \int_0^t \frac{1}{s} \, ds \iint_{|\xi| = cs} d\xi \iiint e^{i\langle \mu, x + \xi \rangle} H(\mu; t - s) \, d\mu$$

$$= \frac{1}{4\pi c^2} \int_0^t \frac{1}{s} \, ds \iint_{|\xi| = cs} h(P + \xi; t - s) \, d\xi \tag{8.5.1}$$

This is the *retarded potential representation* of the solution $u(P; t)$. It can be written as a solid integral purely in terms of the spatial variables by writing $Q = P + \xi$ and

$$u(P; t) = \frac{1}{4\pi c^2} \int_0^t ds \iint_{|P - Q| = cs} \frac{h(Q, t - (1/c)|P - Q|) \, dQ}{|P - Q|}$$

$$= \frac{1}{4\pi c^2} \iiint_{|P - Q| < ct} \frac{h(Q, t - (1/c)|P - Q|)}{|P - Q|} \, dQ \tag{8.5.2}$$

This formula reduces to the newtonian potential in the special case of time-independent problems. To see this, let $h(P; t) = c^2 h(P)$ and let $c \to \infty$ in the above representation. We obtain

$$\lim_{c \uparrow \infty} h(P; t) = \frac{1}{4\pi} \iiint \frac{h(Q)}{|P - Q|} \, dQ$$

The representation formula (8.5.1) can be expressed in terms of the mean-value operator which was used to solve the Cauchy problem for the homogeneous wave equation in Sec. 5.3, $M_R f(P) = [1/(4\pi R^2)] \iint_{|\xi|=R} f(x + \xi) \, d\xi$. This can be written as an integral over the unit sphere in the form

$$M_R f(P) = \frac{1}{4\pi} \iint_{|\omega|=1} f(P + R\omega) \, d\omega$$

This form has the advantage that the dependence on radius R appears inside the integral and not in the region of integration. (The first form is preferable when we apply the divergence theorem.) To use this above, we write $h^{(t)}(P) = h(P; t)$, and the inner integral in (8.5.1) is written as $M_{cs}(sh^{(t-s)})(P)$ so that the solution takes the form

$$u(P; t) = \int_0^t M_{cs}(sh^{(t-s)})(P) \, ds = \int_0^t M_{c(t-s)}(sh^{(s)})(P) \, ds$$

From Sec. 5.3, the integrand is the solution of the Cauchy problem $u_{tt} - c^2 \nabla^2 u = 0$ with $u(P; 0) = 0$, $u_t(P; 0) = h^{(s)}(P)$. This connection makes it possible to first study the homogeneous problem and then apply the results obtained to the inhomogeneous problem.

First we turn to the proof of the representation formula (5.3.11) for the homogeneous equation, which was left open in Chap. 5.

Theorem 8.5.1. Suppose that $f(P)$, $P \in R^3$, is a real-valued continuous function with two continuous partial derivatives. Then the formula $u(P; t) = t M_{ct} f(P)$ defines a twice-differentiable function which satisfies the conditions $u_{tt} = c^2 \nabla^2 u$, $u(P; 0) = 0$, $u_t(P; 0) = f(P)$.

Proof. We have

$$u(P; t) = \frac{1}{4\pi c^2 t} \iint_{|\xi|=ct} f(P + \xi) \, d\xi = \frac{t}{4\pi} \iint_{|\omega|=1} f(P + ct\omega) \, d\omega$$

Clearly $u(P; 0) = 0$. The first derivative is given by

$$u_t(P; t) = \frac{1}{4\pi} \iint_{|\omega|=1} f(P + ct\omega) \, d\Omega + \frac{t}{4\pi} \iint_{|\omega|=1} \nabla f(P + ct\omega) \cdot c\omega \, d\omega$$

In particular, $u_t(P; 0) = f(P)$. To proceed further, we transform the second integral by the divergence theorem:

$$\frac{t}{4\pi} \iint\limits_{|\omega|=1} \nabla f(P + ct\omega) \cdot c\omega \, d\omega = \frac{1}{4\pi ct} \iint\limits_{|\xi|=ct} \left(\frac{\partial f}{\partial n}\right) (P + \xi) \, d\xi$$

$$= \frac{1}{4\pi ct} \iiint\limits_{|\xi|<ct} (\nabla^2 f)(P + \xi) \, d\xi$$

so that we may write

$$u_t(P; t) = \frac{1}{4\pi} \iint\limits_{|\omega|=1} f(P + ct\omega) \, d\omega + \frac{1}{4\pi ct} \int_0^{ct} ds \iint\limits_{|\xi|=cs} (\nabla^2 f)(P + \xi) \, d\xi$$

Differentiating again with respect to t and applying the divergence theorem, we have

$$u_{tt}(P; t) = \frac{c}{4\pi} \iint\limits_{|\omega|=1} \nabla f(P + ct\omega) \cdot \omega - \frac{1}{4\pi ct^2} \int_0^{ct} ds \iint\limits_{|\xi|=cs} (\nabla^2 f)(P + \xi) \, d\xi$$

$$+ \frac{1}{4\pi t} \iiint\limits_{|\xi|<ct} (\nabla^2 f)(P + \xi) \, d\xi$$

$$= \frac{1}{4\pi ct^2} \iiint\limits_{|\xi|<ct} (\nabla^2 f)(P + \xi) \, d\xi - \frac{1}{4\pi ct^2} \iiint\limits_{|\xi|<ct} (\nabla^2 f)(P + \xi) \, d\xi$$

$$+ \frac{1}{4\pi t} \iiint\limits_{|\xi|<ct} (\nabla^2 f)(P + \xi) \, d\xi$$

$$= \frac{c^2 t}{4\pi} \iint\limits_{|\omega|=1} \nabla^2 f(P + ct\omega) \, d\omega$$

But if we compute the spatial derivatives of u by differentiating the integral defining $u(P; t)$, we obtain the formulas

$$u_{x_i}(P; t) = \frac{t}{4\pi} \iint\limits_{|\omega|=1} f_{x_i}(P + ct\omega) \, d\omega \qquad 1 \leq i \leq 3$$

$$u_{x_i x_j}(P; t) = \frac{t}{4\pi} \iint\limits_{|\omega|=1} f_{x_i x_j}(P + ct\omega) \, d\omega \qquad 1 \leq i, j \leq 3$$

In particular,

$$\nabla^2 u(P; t) = u_{x_1 x_1} + u_{x_2 x_2} + u_{x_3 x_3} = \frac{t}{4\pi} \iint\limits_{|\omega|=1} (\nabla^2 f)(P + ct\omega) \, d\omega = \left(\frac{1}{c^2}\right) u_{tt}$$

which was to be proved. ∎

We can now use this theorem to solve the general Cauchy problem for the homogeneous wave equation.

Theorem 8.5.2. Suppose that $f(P)$, $P \in \mathbb{R}^3$, is a real-valued function with three continuous partial derivatives. Then the formula $v(P; t) = (\partial/\partial t)[tM_{ct}f(P)]$ defines a twice-differentiable function which satisfies $v_{tt}(P; t) = c^2\nabla^2 v$ with $v(P; 0) = f(P)$, $v_t(P; 0) = 0$.

Proof. Both f and ∇f are twice-differentiable, so that $v = u_t = (\partial/\partial t)(tM_{ct}g)$, given by the first line in the proof of Theorem 8.5.1, is twice-differentiable. Since u satisfies the wave equation (by Theorem 8.5.1 also), we can write

$$\left(\frac{\partial^2}{\partial t^2} - c^2\nabla^2\right)u_t = \frac{\partial}{\partial t}(u_{tt} - c^2\nabla^2 u) = 0$$

so that v also satisfies the wave equation. From the proof of Theorem 8.5.1 we have $v(P; 0) = u_t(P; 0) = f(P)$, $v_t(P; 0) = u_{tt}(P; 0) = 0$.

Combining the results of the above two theorems by the superposition principle, we have obtained the rigorous proof of the *Poisson formula*

$$u(P; t) = \frac{\partial}{\partial t}(tM_{ct}f_1)(P) + tM_{ct}f_2(P)$$

for the solution of the wave equation $u_{tt} = c^2\nabla^2 u$ with the initial conditions $u(P; 0) = f_1(P)$, $u_t(P; 0) = f_2(P)$. ∎

We can now return to the inhomogeneous equation and prove that formula (8.5.1) gives the rigorous solution of the Cauchy problem for the inhomogeneous wave equation. To do this, we write

$$u(P; t) = \int_0^t v^{(s)}(P; t - s) \, ds$$

where $v^{(s)}$ is the solution of the homogeneous wave equation $v_{tt} = c^2\nabla^2 v$, $v^{(s)}(P; 0) = 0$, $v_t^{(s)}(P; 0) = h^{(s)}$. By Theorem 8.5.1 the solution is $v^{(s)}(P; t) = tM_{ct}h^{(s)}$. Clearly $u(P; 0) = 0$. Now we can differentiate under the integral to obtain

$$u_t(P; t) = \left(\frac{\partial}{\partial t}\right)\int_0^t v^{(s)}(P; t - s) \, ds = v^{(s)}(P; 0) + \int_0^t v_t^{(s)}(P; t - s) \, ds$$

$$u_{tt}(P; t) = \left(\frac{\partial^2}{\partial t^2}\right)\int_0^t v^{(s)}(P; t - s) \, ds = v_t^{(s)}(P; 0) + \int_0^t v_{tt}^{(s)}(P; t - s) \, ds$$

Immediately $u_t(P; 0) = v^{(s)}(P; 0) = 0$, by construction. From the preceding calculations we know that $v_{tt} = c^2\nabla^2 v$, and hence the final integral in u_{tt} equals $c^2\nabla^2 u$ while the first term of u_{tt} is $h(P; t)$ by construction. We have proved that u satisfies the inhomogeneous wave equation $u_{tt} - c^2\nabla^2 u = h$.

This is summarized as follows.

Theorem 8.5.3. Suppose that $h(P; t)$ is a continuous function for which $P \rightarrow H(P; t)$ has two continuous partial derivatives. Then the solution of the wave equation $u_{tt} - c^2\nabla^2 u = h(P; t)$ with $u(P; 0) = 0$, $u_t(P; 0) = 0$ is given by the retarded potential

$$u(P; t) = \int_0^t M_{c(t-s)}(sh^{(s)})(P) \, ds$$

$$= \frac{1}{4\pi c^2} \iiint\limits_{|P-Q|<ct} \frac{h(Q, t - (1/c)|P - Q|)}{|P - Q|} \, dQ$$

Green's Function for the Helmholtz Equation

In case the right side has a harmonic time dependence, we can give a direct treatment of the wave equation in terms of the *Helmholtz equation*. Specifically, we assume that the right side is written in the complex form $h(P; t) = h(P)e^{i\omega t}$. We look for a solution in the form $u(P; t) = u(P)e^{i\omega t}$. This leads us to the inhomogeneous Helmholtz equation

$$\nabla^2 u + k^2 u = -h \qquad k = \frac{\omega}{c}$$

We may look for solutions in the entire three-dimensional space or in a region with a smooth boundary.

To formulate Green's function for entire three-dimensional space, we begin by looking for radial solutions with the characteristic singularity at $r = 0$. In polar coordinates we have $0 = \nabla^2 u + k^2 u = u_{rr} + (2/r)u_r + k^2 u = (1/r)[(ru)_{rr} + k^2(ru)]$ with the general solution $u = (1/r)(A \cos kr + B \sin kr)$. To achieve the proper singularity at $r = 0$, we choose $A = 1/(4\pi)$; the value of B is undetermined, since that part of the solution is smooth at $r = 0$ and is not determined by the singularity of Green's function or by requiring that $u \to 0$ at infinity since $u \to 0$ for any choice of B.

To determine a unique Green function for the Helmholtz equation, we write the solution in complex form:

$$u = \frac{1}{r} (Ae^{ikr} + Be^{-ikr})$$

which redefines the constants A, B. To have the proper singularity at $r = 0$, we must have $A + B = 1/(4\pi)$. To determine both constants, we look at the corresponding complex separated solutions of the wave equation:

$$u(P; t) = \frac{1}{r} (Ae^{i(kr+\omega t)} + Be^{i(-kr+\omega t)})$$

This describes an *outgoing spherical wave* if $A = 0$. It describes an *incoming spherical wave* if $B = 0$. We use these notions to determine a unique Green function.

The function $u = e^{-ikr}/r$ satisfies the first-order differential equation $(ru)_r = -ike^{-ikr} = -ikru$, which entails $ru_r + u = -ikru$ or $u_r + iku = -e^{-ikr}/r^2$. This equation motivates the concept of the *outgoing radiation condition* which is the statement that

$$u_r + iku = O\left(\frac{1}{r^2}\right) \qquad r \to \infty$$

It is satisfied, in particular, for the function $u = e^{ikr}/r$. Similarly, the *incoming radiation condition* is the statement that $u_r - iku = O(1/r^2), r \to \infty$. It is satisfied, in particular, for the function $u = e^{-ikr}/r$. Either may be used to determine a unique Green function for the Helmholtz equation. We formulate the following theorem.

Theorem 8.5.4. Let u be a solution of the Helmholtz equation $\nabla^2 u + k^2 u = -h$ in the entire three-dimensional space and satisfying the outgoing radiation condition and the condition that $u \to 0$ when $r \to \infty$. Then we have the representation

$$u(P) = \iiint \frac{e^{-ik|P-Q|}}{4\pi|P-Q|} h(Q) \, dQ$$

Proof. We begin with Green's identity, written in the form

$$\iiint_{D_{\epsilon,R}} [u(\nabla^2 + k^2)G - G(\nabla^2 + k^2)u] = \iint_{\partial D_{\epsilon,R}} \left(u \frac{\partial G}{\partial n} - G \frac{\partial u}{\partial n}\right) dS_Q$$

where we have added and subtracted the term $k^2 uG$ on the left side. This integral is analyzed exactly as for the case of Poisson's equation in Sec. 8.2. For the right side, the integral on $|Q - P| = \epsilon$ is analyzed as before. For the integral on $|Q - P| = R$, we note that both u and G satisfy the outgoing radiation condition, and hence we can replace $\partial u/\partial n$ and $\partial G/\partial n$ by $-iku + O(1/r^2)$ and (respectively) $-ikG + O(1/r^2)$. Two of the terms cancel, and the resulting integrals tend to zero when $R \to \infty$, giving the required result. ∎

Application to the Telegraph Equation

In Sec. 5.4 we found the Fourier representation of the solution of the telegraph equation $u_{tt} + 2\beta u_t + \alpha u = c^2 u_{xx}$. The form of the solution depends on whether $\alpha < \beta^2$, $\alpha = \beta^2$, or $\alpha > \beta^2$. In the second case, the explicit representation is available from d'Alembert's formula, and there is nothing further to do. We now treat the other two cases, using a *method of descent* beginning with the two-dimensional wave equation.

As a preliminary simplification, we recall that the definition $v(x; t) = u(x; t)e^{\beta t}$ transforms the telegraph equation to a special form with $\beta = 0$, that is, the equation $v_{tt} = c^2 v_{xx} \pm k^2 v$, where either $k^2 = \beta^2 - \alpha$ or $k^2 = \alpha - \beta^2$, whichever is positive, corresponding to case 1 or case 3. First we examine the case where $\alpha < \beta^2$, which is the initial-value problem for

$$v_{tt} - c^2 v_{xx} = k^2 v$$

with

$$v(x; 0) = f_1(x) \qquad v_t(x; 0) = f_2(x)$$

To relate this system to the two-dimensional wave equation, we introduce a new independent variable y and consider the function

$$w(x, y; t) = v(x; t)e^{(ky/c)}$$

For this new function, we have $w_{tt} = v_{tt}e^{(ky/c)}$, $w_{xx} = x_{xx}e^{(ky/c)}$, $w_{yy} = (k/c)^2ve^{(ky/c)}$, and the equation

$$w_{tt} - c^2(w_{xx} + w_{yy}) = 0$$

with

$$w(x, y; 0) = f_1(x)e^{(ky/c)} \qquad w_t(x, y; 0) = f_2(x)e^{(ky/c)}$$

This two-dimensional wave equation is solved by formula (5.3.11):

$$w(x, y; t) = \frac{d}{dt}(tM_{ct}F_1) + tM_{ct}F_2$$

where $F_1(x, y) = f_1(x)e^{(ky/c)}$, $F_2(x, y) = f_2(x)e^{(ky/c)}$.

The mean-value operator is expressed (from 5.3.12) as

$$M_{ct}F_i(x, y) = \frac{1}{2\pi ct} \iint_{|\xi|<ct} \frac{F_i(x + \xi_1, y + \xi_2)\, d\xi_1\, d\xi_2}{\sqrt{(ct)^2 - \xi_1^2 - \xi_2^2}} \qquad i = 1, 2$$

$$= \frac{1}{2\pi ct} \iint_{|\xi|<ct} \frac{f_i(x + \xi_1)e^{(k/c)(y+\xi_2)}}{\sqrt{(ct)^2 - \xi_1^2 - \xi_2^2}}\, d\xi_1\, d\xi_2$$

To evaluate the ξ_2 integral, we make the substitution $\xi_2 = \sqrt{(ct)^2 - \xi_1^2}\cos\theta$, $0 < \theta < \pi$, to find

$$\int_{|\xi_2|<\sqrt{(ct)^2-\xi_1^2}} \frac{e^{(k/c)(y+\xi_2)}}{\sqrt{(ct)^2 - \xi_1^2 - \xi_2^2}}\, d\xi_2 = e^{(ky/c)}\int_0^\pi e^{(k/c)\sqrt{(ct)^2-\xi_1^2}\cos\theta}\, d\theta$$

The final integral was seen in Sec. 3.2 as the integral representation of the Bessel function with imaginary argument, that is, $I_0\left((k/c)\sqrt{(ct)^2 - \xi_1^2}\right)$. Therefore we have

$$M_{ct}F_i(x, y) = e^{(ky/c)} \frac{1}{2ct} \int_{-ct}^{ct} f_i(x + \xi_1)I_0\left(\frac{k}{c}\sqrt{(ct)^2 - \xi_1^2}\right) d\xi_1$$

Canceling the factor $e^{(ky/c)}$, we have found the solution of the telegraph equation with $f_1 = 0$. To find the solution in general, we need to differentiate this integral with respect to t:

$$\frac{d}{dt}(tM_{ct}f_i) = \frac{1}{2c}[f_i(x + ct) + f_i(x - ct)]$$

$$+ \frac{1}{2c}\int_{-ct}^{ct} f_i(x + \xi_1)\frac{d}{dt}\left\{I_0\left[\frac{k}{c}\sqrt{(ct)^2 - \xi_1^2}\right]\right\} d\xi_1$$

But the derivative of the Bessel function I_0 is the Bessel function I_1: $I_0' = I_1$.

Therefore we conclude that

$$\frac{d}{dt}(tM_{ct}f_i) = \frac{1}{2c}[f_i(x + ct) + f_i(x - ct)]$$

$$+ \frac{k}{2}\int_{-ct}^{ct} f_i(x + \xi)I_1\left[\frac{k}{c}\sqrt{(ct)^2 - \xi^2}\right]d\xi$$

We can summarize these calculations in the following form.

Theorem 8.5.5. The explicit representation of the solution of the telegraph equation $v_{tt} - c^2 v_{xx} = k^2 v$ with $v(x; 0) = f_1(x)$, $v_t(x; 0) = f_2(x)$ is given by

$$v(x; t) = \frac{1}{2c}\int_{-ct}^{ct} f_2(x + \xi)I_0\left[\frac{k}{c}\sqrt{(ct)^2 - \xi^2}\right]d\xi$$

$$+ \frac{1}{2c}[f_1(x + ct) + f_1(x - ct)] + \frac{k}{2}\int_{-ct}^{ct} f_1(x + \xi)I_1\left[\frac{k}{c}\sqrt{(ct)^2 - \xi^2}\right]d\xi$$

This concludes the derivation of the solution. ∎

In the exercises we solve the telegraph equation in case 3, for $v_{tt} - c^2 v_{xx} = -k^2 v$, leading to the replacement of the modified Bessel function by the usual Bessel function J_0.

EXERCISES 8.5

1. Find the explicit representation of the solution of the two-dimensional wave equation $u_{tt} - c^2(u_{xx} + u_{yy}) = h(x, y; t)$ with $u(x, y; 0) = 0$, $u_t(x, y; 0) = 0$. [*Hint:* Apply the method of Theorem 8.5.3 with the two-dimensional mean-value formula (5.3.12).]
2. Find the explicit representation of the solution of the one-dimensional wave equation $u_{tt} - c^2 u_{xx} = h(x; t)$ with $u(x; 0) = 0$, $u_t(x; 0) = 0$. [*Hint:* Apply the method used in Theorem 8.5.3 with the one-dimensional mean-value operator $M_{ct}f = [1/(2ct)]\int_{x-ct}^{x+ct} f(z)\,dz$.
3. Consider the one-dimensional Helmholtz equation $u'' + k^2 u = -h(x)$.
   (*a*) Find a Fourier representation of the solution.
   (*b*) Show that this can be rewritten as the explicit representation

   $$u(x) = -\int_{-\infty}^{\infty} \frac{e^{ik|x-y|}}{2ik} h(y)\,dy$$

   (*c*) Suppose that h is continuous and satisfies $h(x) = O(1/|x|^2)$ when $|x| \to \infty$. Prove that $u'(x)$, $u''(x)$ [defined in (*b*)] exist and that u satisfies the one-dimensional Helmholtz equation.
4. Find the explicit representation of the solution of the telegraph equation $v_{tt} - c^2 v_{xx} = -k^2 v$ with $u(x; 0) = f_1(x)$, $v_t(x; 0) = f_2(x)$. [*Hint:* The function $w(x, y; t) := e^{i(ky/c)}v(x; t)$ satisfies the two-dimensional wave equation $w_{tt} = c^2(w_{xx} + w_{yy})$.]
5. Find the explicit representation of the solution of the two-dimensional telegraph equation $v_{tt} - c^2(v_{xx} + v_{yy}) = k^2 v$ with $v(x, y; 0) = f_1(x, y)$, $v_t(x, y; 0) = f_2(x, y)$. [*Hint:* The function $w(x, y, z; t) := e^{(kz/c)}v(x, y; t)$ satisfies the three-dimensional wave equation $w_{tt} = c^2(w_{xx} + w_{yy} + w_{zz})$.]

CHAPTER
9

APPROXIMATE SOLUTIONS

Introduction: Perturbation Methods and Variational Methods

In this chapter we explore various methods for obtaining approximate solutions of certain problems involving partial differential equations. Although the treatment is mostly at a formal, heuristic level, all the methods can be rigorously justified by more advanced methods of analysis.

The methods fall into two main categories: perturbation methods and variational methods. In the first category we consider a problem which contains a small parameter ϵ. In the limiting case $\epsilon = 0$, the problem is supposed to be exactly solvable in terms of previously known methods. The solution of the problem for $\epsilon > 0$ is sought in the form of a power series $u = \sum_{n \geq 0} u_n \epsilon^n$, where the coefficients u_n are to be determined as the solution of some auxiliary problems. The series may not converge for any value of $\epsilon > 0$. Nevertheless a fixed number of terms provides a good approximation of the solution if ϵ is very small.

In the second category of methods, we characterize the exact solution of a problem as the solution of a suitable minimization problem. Then we look for methods to obtain an approximate minimum, which leads to an approximation of the exact solution of the problem. In specific cases this is described by the approximation methods of Ritz, Kantorovich, and Galerkin. In each method we reduce a given problem to a system of ordinary differential equations by looking for an approximate solution in terms of a finite combination of functions of one variable.

In Sec. 9.1 we develop the perturbation method for the vibrating string with variable density $\rho(s)$, supposed to be close to a constant: $\rho(s) \simeq \rho_0 + \epsilon\rho_1(s)$. In Sec. 9.2 we extend the perturbation method to treat the corresponding problem for the vibrating drumhead with variable density $\rho(x, y) \simeq \rho_0 + \epsilon\rho_1(x, y)$ and other problems of this type. In the remaining sections we develop the variational method and the related approximation methods of Ritz, Kantorovich, and Galerkin.

9.1 PERTURBATIVE SOLUTION OF THE VIBRATING STRING WITH VARIABLE DENSITY

We now turn to the vibrating string with nonconstant density. This can be solved by a method of *perturbation theory,* where we look for solutions which are close to the solutions of a known solvable problem.

Problem Formulation

Equation (2.4.1) for transverse vibrations with no external forces is written in the form $\rho(s) \, \partial^2 y/\partial t^2 = T_0 \, \partial^2 y/\partial s^2$, where $\rho(s)$ is the (variable) density and T_0 is the (constant) tension. We look for separated solutions in the form $y(s; t) = f(s)e^{i\omega t}$ which are zero at the extremities; this leads to the following Sturm-Liouville boundary-value problem for $f(s)$:

$$T_0 f''(s) + \omega^2 \rho(s) f(s) = 0 \qquad 0 < s < L \tag{9.1.1}$$
$$f(0) = 0 = f(L)$$

For a general density $\rho(s)$ this problem is not explicitly solvable. However, if $\rho(s)$ is close to a constant ρ_0, we expect the solution to be close to the solution of the equation $T_0 f_0'' + \omega_0^2 \, \rho_0 f_0 = 0$. To explore this further, we hypothesize that the density, frequency, and profile have power series expansions in terms of a suitable small parameter ϵ. In detail, we write

$$\rho(s) = \rho_0 + \sum_{k \geq 1} \epsilon^k \rho_k(s)$$

$$\omega^2 = \omega_0^2 + \sum_{k \geq 1} \epsilon^k (\omega^2)_k$$

$$f(s) = f_0(s) + \sum_{k \geq 1} \epsilon^k f_k(s)$$

The function $\rho_1(s)$, assumed to be continuous and piecewise smooth, measures (to the first order of approximation) the extent of nonuniform density of the string. The coefficient $(\omega^2)_1$ measures the first-order change in square frequency, while $f_1(s)$ measures the first-order change in the string profile. The higher terms have similar interpretations. The computations below can be rigorously justified by known methods of analysis. It is our purpose here to exhibit the formulas.

To obtain these, we insert these expansions into the basic equation (9.1.1) and equate like powers of ϵ. This leads to the following equations:

$$T_0[f_0''(s) + \sum_{k\geq 1} \epsilon^k f_k''(s)] \tag{9.1.2}$$
$$+ [\omega_0^2 + \sum_{k\geq 1} \epsilon^k(\omega^2)_k] [\rho_0 + \sum_{k\geq 1} \epsilon^k \rho_k(s)] [f_0(s) + \sum_{k\geq 1} \epsilon^k f_k(s)] = 0$$

$$T_0 f_0''(s) + \omega_0^2 \rho_0 f_0(s) = 0$$

$$T_0 f_1''(s) + \omega_0^2 \rho_0 f_1(s) + [\omega_0^2 \rho_1(s) + (\omega^2)_1 \rho_0] f_0(s) = 0$$

and for $k \geq 2$

$$T_0 f_k''(s) + \sum_{k_1+k_2+k_3=k} (\omega^2)_{k_1} \rho_{k_2}(s) f_{k_3}(s) = 0$$

These are to be solved with the boundary conditions

$$f_k(0) = 0 = f_k(L) \qquad k = 0, 1, 2, \ldots$$

To simplify the computations below, we make the normalization

$$\int_0^L f_0(s)^2 \, ds = 1$$

This uniquely determines f_0 up to a $\pm$ sign.

First-Order Frequency Correction

For $k = 0$ we have the known solutions

$$f_{0,\, n}(s) = \sqrt{\frac{2}{L}} \sin \frac{n\pi s}{L} \qquad \omega_{0,n}^2 = \frac{T_0}{\rho_0} \left(\frac{n\pi}{L}\right)^2 \qquad n \geq 1$$

We now fix the integer parameter $n = 1, 2, \ldots$ which gives the higher frequencies; it has nothing to do with the index $k = 1, 2, \ldots$ which gives the higher corrections due to nonconstant density. To obtain the first frequency correction $(\omega^2)_1$, we integrate the $k = 1$ equation against the profile $f_0(s)$ to obtain

$$\int_0^L f_0(s) [T_0 f_1''(s) + \omega_0^2 \rho_0 f_1(s) + [\omega_0^2 \rho_1(s) + (\omega^2)_1 \rho_0] f_0(s)] \, ds = 0$$

The first integral is transformed by repeated integration by parts and use of the boundary condition as follows:

$$\int_0^L f_0 f''_1 \, ds = f_0 f'_1 \big|_0^L - \int_0^L f'_1 \, f'_0 \, ds$$

$$= (f_0 f'_1 - f'_0 f_1) \big|_0^L + \int_0^L f''_0 f_1 \, ds$$

$$= \int_0^L f''_0 f_1 \, ds$$

The first two integrals are written as

$$\int_0^L f_0(s) \, [T_0 f''_1(s) + \omega_0^2 \, \rho_0 f_1(s)] \, ds = \int_0^L f_1(s) \, [T_0 f''_0(s) + \omega_0^2 \, \rho_0 f_0(s)] \, ds$$

$$= 0$$

Therefore we obtain the following equation for the first-order change in frequency:

$$(\omega^2)_1 = -\frac{\omega_0^2}{\rho_0} \int_0^L \rho_1(s) f_0(s)^2 \, ds \qquad (9.1.3)$$

The integral is a weighted average of the perturbation $\rho_1(s)$ with respect to the squared profile $f_0(s)^2$. The change of frequency has the opposite sign from the perturbation.

First-Order Profile Correction

To obtain the first-order correction to the profile, we write the equation for $k = 1$:

$$T_0 f''_1(s) + \omega_0^2 \, \rho_0 f_1(s) = -[\omega_0^2 \, \rho_1(s) + (\omega^2)_1 \, \rho_0] f_0(s) \qquad (9.1.4)$$

To solve this inhomogeneous equation with the boundary conditions and normalization, we look for the solution as a Fourier sine series

$$f_1(s) = \sum_{m \geq 1} A_k \sin \frac{m \pi s}{L}$$

The perturbation also has an expansion which is written as

$$\rho_1(s) \sin \frac{n \pi s}{L} = \sum_{m \geq 1} C_{mn} \sin \frac{m \pi s}{L}$$

Equivalently, $C_{mn} = (2/L) \int_0^L \rho_1(s) \sin (m \pi s / L) \sin (n \pi s / L) \, ds$. In this notation we have, for example, $(\omega^2)_1 = -\omega_0^2 C_{nn} / \rho_0$.

Equating coefficients in (9.1.4), we obtain

$$\left[-T_0 \left(\frac{m \pi}{L} \right)^2 + \omega_0^2 \, \rho_0 \right] A_m = -\sqrt{\frac{2}{L}} \, [\omega_0^2 \, C_{mn} - (\omega^2)_1 \, \rho_0 \delta_{mn}]$$

(The Kronecker δ symbol $\delta_{mn} := 1$ for $m = n$ and 0 otherwise.) The coefficient A_n is not determined from (9.1.4); without loss of generality we may take $A_n = 0$.

Thus we have the first-order correction to the profile in the form

$$f_1(s) = -\omega_0^2 \sqrt{\frac{2}{L}} \sum \frac{C_{mn} \sin{(m\pi s/L)}}{\omega_0^2 \rho_0 - T_0 (m\pi/L)^2}$$

Here the summation is extended over $m \neq n$.

Example 9.1.1. Find the perturbed solution to the vibrating string in the case $\rho(s) = \rho_0 + \epsilon \rho_1 s$, where ρ_1 is a constant.

Solution. In this case we have the integrals

$$C_{mn} = \begin{cases} \dfrac{2}{L} \displaystyle\int_0^L \rho_1 s \sin\left(\dfrac{m\pi s}{L}\right) \sin\left(\dfrac{n\pi s}{L}\right) ds = \dfrac{L}{\pi} \rho_1 & m = n \\[4mm] \dfrac{L}{\pi^2} \left[\dfrac{1 - (-1)^{m-n}}{(m-n)^2} - \dfrac{1 - (-1)^{m+n}}{(m+n)^2} \right] \rho_1 & m \neq n \end{cases}$$

The first-order correction in square frequency is

$$(\omega^2)_1 = -\frac{\omega_0^2}{\rho_0} \frac{2}{L} \int_0^L \rho_1 s \sin^2 \frac{n\pi s}{L} ds = -\frac{\rho_1 L}{\pi \rho_0} \omega_0^2$$

The first-order correction to the profile is given by substituting the present value of C_{mn} in the above formula for $f_1(s)$.   ●

Second-Order Frequency Correction

Returning to the theory, we now determine the second-order corrections to the frequency and profile. To do this, we introduce the Fourier coefficients of the second-order density correction:

$$D_{mn} = \frac{2}{L} \int_0^L \rho_2(s) \sin\left(\frac{n\pi s}{L}\right) \sin\left(\frac{m\pi s}{L}\right) ds$$

The equation for the second-order profile correction is

$$T_0 f_2''(s) + \omega_0^2 \rho_0 f_2(s) + (\omega^2)_2 \rho_0 f_0(s) + \omega_0^2 \rho_2(s) f_0(s)$$
$$+ (\omega^2)_1 \rho_1(s) f_0(s) + (\omega^2)_1 \rho_0(s) f_1(s) + \omega_0^2 \rho_1(s) f_1(s) = 0$$

Multiplying by $f_0(s)$ and integrating, we find, as before, that the first two terms contribute zero, and we are left with

$$(\omega^2)_2 \rho_0 + \omega_0^2 \int_0^L \rho_2(s) f_0(s)^2 ds + (\omega^2)_1 \int_0^L \rho_1(s) f_0(s)^2 ds$$
$$+ (\omega^2)_1 \rho_0 \int_0^L f_1(s) f_0(s) ds + \omega_0^2 \int_0^L \rho_1(s) f_1(s) f_0(s) ds = 0$$

We need to compute the four integrals which appear above.

The first is given by the expansion of the second-order density correction from above:

$$\int_0^L \rho_2(s) f_0(s)^2 \, ds = D_{nn}$$

The second integral can be written in terms of the first-order frequency correction as

$$\int_0^L \rho_1(s) f_0(s)^2 \, ds = -\frac{\rho_0(\omega^2)_1}{\omega_0^2} = C_{nn}$$

The third integral is zero by virtue of the choice of f_1.

The fourth integral can be expressed through the expansion coefficients by writing

$$\rho_1(s) f_0(s) = \sqrt{\frac{2}{L}} \sum_{m \geq 1} C_{mn} \sin \frac{m \pi s}{L}$$

Comparing this with the formula for $f_1(s)$ and using Parseval's theorem, we have

$$\int_0^L \rho_1(s) f_0(s) f_1(s) \, ds = -\omega_0^2 \sum_{m \geq 1} \frac{C_{mn}^2}{\omega_0^2 \rho_0 - T_0 (m\pi/L)^2}$$

Combining these calculations, we arrive at the second-order frequency correction in the form

$$(\omega^2)_2 = -\frac{\omega_0^2}{\rho_0} D_{nn} + \omega_0^2 \frac{C_{nn}^2}{\rho_0^2} + \frac{\omega_0^4}{\rho_0^2} \sum_m \frac{C_{mn}^2}{\omega_0^2 \rho_0 - T_0(m\pi/L)^2}$$

EXERCISES 9.1

1. For the string with density function $\rho(s) = \rho_0 + \epsilon \rho_1 s$, obtain the second-order correction in frequency ω_2^2.
2. For the string with density function $\rho(s) = \rho_0 + \epsilon \rho_1 \sin(\pi s/L)$, find the first- and second-order corrections in square frequency.
3. For the string with density function $\rho(s) = \rho_0 + \epsilon \rho_1 \sin(\pi s/L)$, show that $C_{mn} = 0$ if $m - n$ is an odd integer.
4. Consider the mathematical problem of the perturbation of the eigenvalues of the Sturm-Liouville problem $f'' + [\mu - V(x)]f = 0$ with the boundary condition $f(0) = 0 = f(L)$. Assume that the *potential function* $V(x)$ and the solutions $(f(x), \mu)$ have expansions of the form $V(x) = V_0(x) + \sum_{k \geq 1} \epsilon^k V_k(x)$, $f(x) = f_0(x) + \sum_{k \geq 1} \epsilon^k f_k(x)$, $\mu = \mu_0 + \sum_{k \geq 1} \epsilon^k \mu_k$, where $f_0''(x) + [\mu_0 - V_0(x)]f_0(x) = 0$ and $f_0(0) = f_0(L)$. The problem is to show that $\mu_1 = \int_0^L f_0(x)^2 V_1(x) \, dx$.

9.2 PERTURBATIVE SOLUTION OF THE VIBRATING DRUMHEAD WITH VARIABLE DENSITY

Problem Formulation

The vibrating drumhead with nonconstant density can be treated by the method of perturbation theory, used for the vibrating string in Sec. 9.1. The equation for small transverse vibrations with no external forces is written

$$D(x, y)u_{tt}(x, y) = T_0(u_{xx}(x, y) + u_{yy}(x, y)) = T_0\nabla^2 u(x, y)$$

where the tension T_0 is a constant and the density $D(x, y)$ is assumed to have an expansion in a small parameter ϵ in the form

$$D(x, y) = D_0 + \sum_{k\geq 1} \epsilon^k D_k(x, y)$$

We look for separated solutions of the form $u(x, y; t) = f(x, y) e^{i\omega t}$, where $f(x, y)$ is a solution of the boundary-value problem

$$T_0\nabla^2 f(x, y) + \omega^2 D(x, y)f(x, y) = 0 \qquad x^2 + y^2 < R^2$$
$$f(x, y) = 0 \qquad x^2 + y^2 = R^2 \tag{9.2.1}$$

For a general density $D(x, y)$ this problem is not explicitly solvable. To obtain useful information, we look for solutions of the form

$$f(x, y) = f_0(x, y) + \sum_{k\geq 1} \epsilon^k f_k(x, y)$$

$$\omega^2 = \omega_0^2 + \sum_{k\geq 1} \epsilon^k (\omega^2)_k$$

To obtain the correction terms $f_k(x, y)$ and $(\omega^2)_k$, we insert these expansions into the basic equation (3.3.11) together with the assumed form of $D(x, y)$, which leads to the following equations:

$$T_0\left(\nabla^2 f_0 + \sum_{k\geq 1} \epsilon^k \nabla^2 f_k\right) + \left(\omega_0^2 + \sum_{k\geq 1} \epsilon^k(\omega^2)_k\right)\left(D_0 + \sum_{k\geq 1} \epsilon^k D_k\right)\left(f_0 + \sum_{k\geq 1} \epsilon^k f_k\right) = 0$$

$$T_0\nabla^2 f_0 + \omega_0^2 D_0 f_0 = 0$$

$$T_0\nabla^2 f_1 + \omega_0^2 D_0 f_1 + [\omega_0^2 D_1 + (\omega^2)_1 D_0]f_0 = 0 \qquad x^2 + y^2 < R^2$$

and for $k \geq 2$

$$T_0\nabla^2 f_k + \sum_{k_1+k_2+k_3=k} (\omega^2)_{k_1} D_{k_2} f_{k_3} = 0 \qquad x^2 + y^2 < R^2$$

These are to be solved with the boundary conditions

$$f_k(x, y) = 0 \qquad x^2 + y^2 = R^2$$

and the normalization

$$\iint_{x^2+y^2<R^2} f_0(x, y)^2 \, dx \, dy = 1$$

For $k = 0$ we have the known solutions (3.3.11)

$$f_0(x, y) = J_m \left(\frac{\rho x_n^{(m)}}{R} \right) (A \cos m\varphi + B \sin m\varphi) \qquad m = 0, 1, \ldots; n = 1, 2, \ldots$$

$$\omega_0^2 = \frac{T_0}{D_0} \left(\frac{x_n^{(m)}}{R} \right)^2$$

First-Order Frequency Correction

We now fix a pair (n, m) and compute the first-order correction to the square frequency. The analysis differs for the cases $m = 0$ and $m \neq 0$, which we consider separately. Specifically, in the first case $(\omega^2)_1$ depends on only ω_0^2 while in the second case it depends also on the choice of f_0, which is ambiguous.

$m = 0$ To do this case, we integrate the equation for f_1 against the profile f_0 to obtain

$$\iint\limits_{x^2+y^2<R^2} f_0\{T_0\nabla^2 f_1 + \omega_0^2 D_0 f_1 + [\omega_0^2 D_1 + (\omega^2)_1 D_0]f_0\} \, dx \, dy = 0$$

Applying the divergence theorem and using the equation for f_0 and the boundary conditions for f_0 and f_1, we see that the first two terms contribute zero. The final integral may be evaluated by using the normalization for $f_0(x, y)$. Thus we are left with

$$(\omega^2)_1 D_0 + \omega_0^2 \iint\limits_{x^2+y^2<R^2} D_1 f_0^2 \, dx \, dy = 0$$

which is solved to yield

$$\boxed{(\omega^2)_1 = -\frac{\omega_0^2}{D_0} \iint\limits_{x^2+y^2<R^2} D_1(x, y) f_0(x, y)^2 \, dx \, dy}$$

This has the same structure as the perturbed solution of the vibrating string; the first-order change in square frequency is negatively proportional to a weighted average of the first-order density with respect to the unperturbed solution.

$m \neq 0$ In this case, the frequency corresponds to two linearly independent solutions which we write in terms of the following basis, where the coefficients A, B are to be determined:

$$f_0 = J_m \left(\frac{\rho x_n^{(m)}}{R} \right) (A \cos m\varphi + B \sin m\varphi)$$

$$\hat{f}_0 = J_m \left(\frac{\rho x_n^{(m)}}{R} \right) (B \cos m\varphi - A \sin m\varphi)$$

These two functions are orthogonal to each other and can be normalized as in the case $m = 0$. We write the equation for f_1 and integrate against both f_0 and $\hat{f}_0$ to obtain

$$\omega_0^2 \iint D_1 f_0^2 \, dx \, dy + (\omega^2)_1 D_0 \iint f_0^2 \, dx \, dy = 0$$

$$\omega_0^2 \iint D_1 f_0 \hat{f}_0 \, dx \, dy = 0$$

where we have used the equations for f_0, $\hat{f}_0$, and their orthogonality. This second equation gives a quadratic equation for the coefficients A and B which must be satisfied, leading to two choices of the pair (A, B). For the corresponding choice of f_0 we substitute in the first equation to determine the value of $(\omega^2)_1$. Again we are led to the formula $(\omega^2)_1 = -(\omega_0^2/D_0)\iint D_1 f_0^2 \, dx \, dy$ but with a different interpretation: We must choose f_0 as determined by the quadratic equation. This is an example of *degenerate perturbation theory*.

> **Example 9.2.1.** Find the possible solutions for f_0 if $m = 1$ for the perturbation $D_1 = \rho^2 \sin 2\varphi$.
>
> **Solution.** In this case the quadratic equation for (A, B) is determined from the second condition above as
>
> $$\int_{-\pi}^{\pi} \sin 2\varphi (A \cos \varphi + B \sin \varphi)(B \cos \varphi - A \sin \varphi) \, d\varphi = 0$$
>
> By using orthogonality relations from Fourier series, this reduces to $A^2 - B^2 = 0$. Therefore the possible solutions are of the form
>
> $$f_0 = A J_1 \left(\frac{\rho x_n^{(1)}}{R} \right) (\cos \varphi + \sin \varphi)$$
>
> $$\hat{f}_0 = A J_1 \left(\frac{\rho x_n^{(1)}}{R} \right) (\cos \varphi - \sin \varphi) \qquad \bullet$$

First-Order Profile Correction

Returning to the theory, we now obtain the first-order correction in profile in case $m = 0$. We introduce the Fourier coefficients of the perturbation in the form

$$\rho_1(x, y)\psi_{mn}(x, y) = \sum_{kl} C_{kl,mn}\psi_{kl}(x, y)$$

where we have used the complex notation $\psi_{mn}(x, y) = (\text{const.}) \, J_m(\rho x_n^{(m)}/R)e^{im\varphi}$ for the normalized eigenfunctions of the unperturbed problem. The solution is sought in the form

$$f_1(x, y) = \sum_{kl} A_{kl}\psi_{kl}(x, y)$$

where we may take $A_{mn} = 0$, due to the normalization of f_1. Substituting in the equation and using the equation for ψ_{kl} give

$$\sum_{kl} A_{kl}\left[-T_0\left(\frac{x_l^{(k)}}{R}\right)^2 + \omega_0^2\right]\psi_{kl}(x, y) = -\omega_0^2\sum_{kl} C_{kl,mn}\psi_{kl}(x, y)$$

Thus we choose $A_{kl} = \omega_0^2 C_{kl,mn}/[\omega_0^2 - T_0(x_l^{(k)}/R)^2]$ and the solution

$$f_1(x, y) = \omega_0^2 \sum_{kl \neq mn} \frac{C_{kl,mn}\psi_{kl}(x, y)}{\omega_0^2 - T_0(x_l^{(k)}/R)^2}$$

The solution for the case $m \neq 0$ is left for the exercises.

Perturbation of the Boundary

We now consider the perturbation of frequency due to a small change in the geometry of the drumhead. Specifically, we suppose that the periphery of the drumhead is given in polar coordinates by $\rho = R + \epsilon g(\theta)$, where ϵ is a small parameter and $g(\theta)$ is a given piecewise smooth function. (Higher-order corrections may also be considered, if necessary.) The solution is sought in the form

$$f(\rho, \theta) = f_0(\rho, \theta) + \sum_{k \geq 1} \epsilon^k f_k(\rho, \theta)$$

$$\omega^2 = \omega_0^2 + \sum_{k \geq 1} \epsilon^k(\omega^2)_k$$

where $f(\rho, \theta)$ satisfies equation (3.4.11) in polar coordinates and the boundary condition is written in the form

$$f(R + \epsilon g(\theta), \theta) = 0 \qquad -\pi \leq \theta \leq \pi$$

Expanding this in powers of ϵ and setting $\epsilon = 0$, we obtain for the first two terms

$$f_0(R, \theta) = 0 \qquad -\pi \leq \theta \leq \pi$$

$$f_1(R, \theta) + g(\theta)\left(\frac{\partial f_0}{\partial r}\right)(R, \theta) = 0 \qquad -\pi \leq \theta \leq \pi$$

The differential equations for $f_0, f_1, \ldots$ are as before with $\rho = R_0$. As before we analyze separately the cases of simple and multiple frequencies.

$m = 0$ In this case we integrate the f_1 equation against f_0 to obtain

$$-T_0 R \int_{-\pi}^{\pi} f_1 \left(\frac{\partial f_0}{\partial r}\right)(R, \theta)\, d\theta + (\omega^2)_1 D_0 \int\int f_0^2\, dx\, dy = 0$$

Using the boundary conditions and normalizations, we get

$$(\omega^2)_1 = -R\frac{T_0}{D_0} \int_{-\pi}^{\pi} g(\theta) \left(\frac{\partial f_0}{\partial r}\right)^2 (R, \theta)\, d\theta$$

The first-order frequency change is negatively proportional to a weighted average of the variation of the periphery.

$m \neq 0$ In this case we integrate the f_1 equation against each of f_0 and $\hat{f}_0$ as for the variation-of-density problem. Using the boundary conditions and normalizations, we get the two equations

$$(\omega^2)_1 = -R \frac{T_0}{D_0} \int_{-\pi}^{\pi} g(\theta) \left(\frac{\partial f_0}{\partial r}\right)^2 (R, \theta) \, d\theta$$

$$0 = \int_{-\pi}^{\pi} g(\theta) \frac{\partial f_0}{\partial r} (R, \theta) \left(\frac{\partial \hat{f}_0}{\partial r}\right) (R, \theta) \, d\theta$$

The latter is a quadratic equation for the constants A, B leading to two choices for the basic solution f_0. For each of these we may substitute in the first equation to determine the corresponding value of the first-order square-frequency correction $(\omega^2)_1$.

EXERCISES 9.2

1. For the perturbed solution of the vibrating drumhead with $m = 0$, find the second-order square-frequency correction $(\omega^2)_2$.
2. For the perturbed solution of the vibrating drumhead with $m \neq 0$, find the first-order correction to the profile f_1.
3. Consider a circular membrane which is simultaneously perturbed by a small change in periphery and a small change in density. Show that the first-order change in square frequency is obtained as the *sum* of the two terms obtained in the separate cases above.
4. For the perturbed solution of the vibrating drumhead with $m \neq 0$, suppose that $\int_{-\pi}^{\pi} D_1(\rho, \varphi) \cos m\varphi \sin m\varphi \, d\varphi = 0$. Show that we may choose $A = 0$ or $B = 0$ in the choice of f_0.
5. For the perturbed solution of the vibrating drumhead with $m \neq 0$, suppose that $\int_{-\pi}^{\pi} D_1(\rho, \varphi) \cos^2 m\varphi \, d\varphi = \int_{-\pi}^{\pi} D_1(\rho, \varphi) \sin^2 m\varphi \, d\varphi$. Show that we may choose $A = B$ or $A = -B$ in the choice of f_0.

9.3   VARIATIONAL METHODS

Many solutions of partial differential equations which come from physics can be characterized as solutions of *variational problems,* where one is required to minimize a certain *functional* with boundary conditions. This immediately suggests many constructive approaches for finding approximate solutions.

Variational Formulation of Poisson's Equation

Consider the solution of Poisson's equation $\nabla^2 u = -\rho$ in the interior of a bounded three-dimensional region D with the boundary condition $u = 0$. We claim that this can be obtained as the solution of the following problem:

Minimize $\displaystyle\iiint_D [\tfrac{1}{2}(u_x^2 + u_y^2 + u_z^2) - \rho u]\, dx\, dy\, dz$

subject to $\quad u = 0$ on ∂D, the boundary of D.

Indeed, suppose that there exists a smooth function $u_0(x, y, z)$ which achieves the stated minimum, in competition with all smooth functions satisfying the boundary condition. We will prove that u_0 also satisfies Poisson's equation. Since Poisson's equation is known to have a unique solution, we will have found the required variational characterization of the solution.

To prove this, let $v(x, y, z)$ be any smooth function which satisfies the boundary condition $v = 0$. Then the function $u_0 + \epsilon v$ also satisfies the boundary condition for any real constant ϵ; by hypothesis

$$\iiint_D [\tfrac{1}{2}|\nabla(u_0 + \epsilon v)|^2 - \rho(u_0 + \epsilon v)]\, dV \geq \iiint_D (\tfrac{1}{2}|\nabla u_0|^2 - \rho u_0)\, dV$$

(Here we have used vector notation for brevity.) Expanding the left side and simplifying, we have the inequality

$$\iiint_D (\epsilon \nabla u_0 \cdot \nabla v + \tfrac{1}{2}\epsilon^2 |\nabla v|^2 - \epsilon \rho v)\, dV \geq 0$$

This quadratic function of ϵ has a minimum at $\epsilon = 0$; therefore the derivative at $\epsilon = 0$ is zero, which is written as

$$\iiint_D (\nabla u_0 \cdot \nabla v - \rho v)\, dV = 0$$

We can use the divergence theorem and the boundary condition $v = 0$ to transform the first term as follows. We integrate the identity $\text{div}(v \cdot \nabla u_0) = v\nabla^2 u_0 + (\nabla v \cdot \nabla u_0)$ to obtain

$$0 = \iint_{\partial D} v(\nabla u_0 \cdot N)\, dS = \iiint_D \text{div}\, (v \cdot \nabla u_0)\, dV$$

$$= \iiint_D (v\nabla^2 u_0 + \nabla v \cdot \nabla u_0)\, dV$$

We have shown that for every smooth function v which satisfies $v = 0$ on the boundary,

$$0 = \iiint_D v(\nabla^2 u_0 + \rho)\, dV$$

From this we conclude that $\nabla^2 u_0 + \rho = 0$ everywhere in D, that is, u_0 satisfies

Poisson's equation. This final step depends on the following general principle, which we will use repeatedly.

Fundamental Lemma of the Calculus of Variations

Theorem 9.3.1. Let $w(x, y, z)$ be a continuous function in a region D such that

$$\iiint_D w(x, y, z)\varphi(x, y, z)\, dx\, dy\, dz = 0$$

for every once-differentiable function φ for which $\varphi = 0$ on ∂D. Then $w(P) \equiv 0$ in D.

Proof. Given $P_0 \in D$, we must prove that $w(P_0) = 0$. Suppose that $w(P_0) \neq 0$; then either $w(P_0) > 0$ or $w(P_0) < 0$. In the first case there exists $\delta > 0$ such that $\delta < \text{dist}\,(P_0, \partial D)$ and $w(P) \geq \frac{1}{2}w(P_0) > 0$ for $|P - P_0| < \delta$, $P \in D$. Let $\varphi(P) = (\delta^2 - |P - P_0|^2)^2$ for $|P - P_0| < \delta$ and $\varphi(P) = 0$ elsewhere. Then φ is once-differentiable with $\varphi = 0$ on ∂D. Applying the hypothesis, we have

$$0 = \iiint_D \varphi(x, y, z)w(x, y, z)\, dx\, dy\, dz$$

$$= \iiint_{|P-P_0|<\delta} \varphi(x, y, z)w(x, y, z)\, dx\, dy\, dz$$

$$\geq \iiint_{|P-P_0|<\frac{1}{2}\delta} \varphi(x, y, z)w(x, y, z)\, dx\, dy\, dz$$

$$\geq \frac{1}{2}|w(P_0)| \iiint_{|P-P_0|<\frac{1}{2}\delta} \varphi(x, y, z)\, dx\, dy\, dz$$

But $\varphi(x, y, z) < 0$ when $|P - P_0| < \frac{1}{2}\delta$. Therefore the final integral is positive, and we have a contradiction, so we conclude that $w(P_0) = 0$. In case $w(P_0) < 0$, we apply the above argument to $-w(P_0)$. ∎

More-General Variational Problems

We have treated the above example of Poisson's equation at length to illustrate the connection between a minimization problem and the solution of a partial differential equation. More generally we may consider the problem to minimize a functional of the form

$$I(u) := \iiint_D F(x, y, z, u, u_x, u_y, u_z)\, dx\, dy\, dz$$

with the condition that on the boundary $u = f$, a given function. The corresponding partial differential equation is the *equation of Euler and Lagrange*

$$\partial_x F_5(x, y, z, u, u_x, u_y, u_z) + \partial_y F_6(x, y, z, u, u_x, u_y, u_z)$$
$$+ \partial_z F_7(x, y, z, u, u_x, u_y, u_z) = F_4(x, y, z, u, u_x, u_y, u_z)$$

Here F_4 (respectively F_5, F_6, F_7) refers to the partial derivative of F with respect to the fourth (respectively fifth, sixth, seventh) variable; ∂_x refers to the derivative of the composite function with respect to x (respectively ∂_y, ∂_z). In the case of Poisson's equation we have $F = \frac{1}{2}(u_x^2 + u_y^2 + u_z^2) - \rho u$, and thus $F_4 = -\rho$, $F_5 = u_x$, $F_6 = u_y$, $F_7 = u_z$.

Example 9.3.1. Consider the functional

$$I(u) = \int\int\int_D \left[\tfrac{1}{2}(u_x^2 + u_y^2 + u_z^2) + auu_x + buu_y + cuu_z + \tfrac{1}{2}du^2\right] dx\, dy\, dz$$

where a, b, c, d are C^1 functions in D. Find the Euler-Lagrange equation.

Solution. We have $\partial F/\partial u_x = u_x + au$, $\partial F/\partial u_y = u_y + bu$, $\partial F/\partial u_z = u_z + cu$, and $\partial F/\partial u = du + au_x + bu_y + cu_z$. The Euler-Lagrange equation is

$$\frac{\partial}{\partial x}(u_x + au) + \frac{\partial}{\partial x}(u_y + bu) + \frac{\partial}{\partial z}(u_z + cu) = du + au_x + bu_y + cu_z$$

which can be simplified to $\nabla^2 u + (a_x + b_y + c_z - d)u = 0$. ●

We have illustrated the Euler-Lagrange equation and its connection with variational problems in the case of three independent variables. The same connections apply in fewer variables, in particular for problems in one and two variables, especially for ordinary differential equations.

Rigorous Equivalence of Poisson's Equation with a Variational Problem

The above treatment, although somewhat heuristic, has outlined the salient features of variational techniques. In this subsection, we give the formal statement and proof of the variational formulation of a partial differential equation. The following theorem demonstrates the equivalence of the two approaches for the Dirichlet problem for Poisson's equation. In case $\rho = 0$ this is known as *Dirichlet's principle*.

Theorem 9.3.2. Let D be a bounded three-dimensional region and $\rho \in C(D)$, $f \in C(\partial D)$.

1. Suppose that there exists $u \in C^2(D)$ such that $u = f$ on ∂D and $\nabla^2 u = -\rho$ in D. Then for any $v \in C^1(D)$ with $v = f$ on ∂D we have

$$\int\int\int_D (\tfrac{1}{2}|\nabla v|^2 - \rho v)\, dx\, dy\, dz \geq \int\int\int_D (\tfrac{1}{2}|\nabla u|^2 - \rho u)\, dx\, dy\, dz$$

2. Conversely, suppose that there exists $u \in C^2(D)$ with $u = f$ on ∂D such that for all $v \in C^1(D)$ with $v = f$ on ∂D we have

$$\iiint_D (\tfrac{1}{2}|\nabla v|^2 - \rho v)\, dx\, dy\, dz \geq \iiint_D (\tfrac{1}{2}|\nabla u|^2 - \rho u)\, dx\, dy\, dz$$

Then $\nabla^2 u = -\rho$ in D.

Proof. We begin with the identity

$$\operatorname{div}(\zeta \nabla u) = \nabla \zeta \cdot \nabla u + \zeta \nabla^2 u \tag{9.3.1}$$

valid for any $u \in C^2$, $\zeta \in C^1$. Writing $v = u + \epsilon\zeta$, we have by hypothesis

$$\iiint_D [\tfrac{1}{2}|\nabla u + \epsilon\nabla\zeta|^2 - \rho(u + \epsilon\zeta)]\, dx\, dy\, dz \geq \iiint_D (\tfrac{1}{2}|\nabla u|^2 - \rho u)\, dx\, dy\, dz$$

for any real number ϵ and $\zeta \in C^1(D)$, $\zeta = 0$ on ∂D. Subtracting the right side from both sides of this inequality, dividing by ϵ, and letting $\epsilon \downarrow 0$ and $\epsilon \uparrow 0$, we conclude that

$$\iiint_D (\nabla\zeta\nabla u - \zeta\rho)\, dx\, dy\, dz = 0 \tag{9.3.2}$$

In view of identity (9.3.1), we can write $\iiint_D \zeta(\nabla^2 u + \rho)\, dx\, dy\, dz = \iiint_D [\operatorname{div}(\zeta\nabla u) + \zeta\rho - \nabla\zeta\nabla u]\, dx\, dy\, dz = 0$, where we have used the divergence theorem, the boundary condition on ζ, and (9.3.2). Since this is true for all $\zeta \in C^1(D)$, we conclude from the fundamental lemma of the calculus of variations that $\nabla^2 u + \rho = 0$, as required.

Conversely, suppose that $u \in C^2(D)$ solves $\nabla^2 u = -\rho$ in D. Then for any ζ with $\zeta = 0$ on ∂D we can apply identity (9.3.1) to conclude that $\iiint_D (\nabla u \nabla \zeta - \rho u)\, dx\, dy\, dz = 0$. Writing $v = u + \zeta$, we have

$$\iiint_D (\tfrac{1}{2}|\nabla v|^2 - \rho v)\, dx\, dy\, dz = \iiint_D [\tfrac{1}{2}|\nabla u|^2 + \nabla u\nabla\zeta + \tfrac{1}{2}|\nabla\zeta|^2\rho(u + \zeta)]\, dx\, dy\, dz$$

$$= \iiint_D (\tfrac{1}{2}|\nabla u|^2 + \tfrac{1}{2}|\nabla\zeta|^2 - \rho u)\, dx\, dy\, dz$$

$$\geq \iiint_D (\tfrac{1}{2}|\nabla u|^2 - \rho u)\, dx\, dy\, dz$$

which was to be proved. ■

If we try to replicate Theorem 9.3.2 for other variational problems, we encounter some difficulties. On the one hand, it is true quite generally that a smooth minimum of a smooth functional $I(u)$ is indeed a solution of the appropriate Euler-Lagrange equation. On the other hand, it is difficult to show that a solution of a Euler-Lagrange equation is necessarily a minimum of the associated functional $I(u)$. For example, in the space of two variables (x, t) the functional $I(u) = \iint_D (c^2 u_x^2 - u_t^2)\, dx\, dt$ has for the Euler-Lagrange equation the wave equation $u_{tt} = c^2 u_{xx}$. A minimum (or maximum) of $I(u)$ is necessarily a solution of the wave equation, but the converse is not true. All that we can say is that the solution u is a *stationary point* of the functional $I(u)$, meaning that for any $v \in C^1(D)$ and satisfying the boundary conditions, we have $(d/dt)I(u + tv)|_{t=0} = 0$. This is often used as a basic postulate for classical mechanics, under the name *Hamilton's principle*. This acts as a substitute for Newton's laws of motion which are a basis for the elementary approach to classical mechanics. In the variational formulation, the emphasis is on writing down the appropriate form of the functional $I(u)$ whose stationarity is sought. In various guises this is called the *lagrangian* or *action functional*, and the variational problem is referred to (incorrectly) as the *principle of least action*. For detailed examples, see Exercises 10 to 12 at the end of this section.

Variational Formulation of Eigenvalue Problems

As a second illustration of variational methods, consider the problem of determining the *vibrating frequencies* of a three-dimensional region D. These are the numbers λ_n that are obtained in the solution of the boundary-value problem $\nabla^2 u + \lambda u = 0$ with the condition that $u = 0$ on the boundary. To establish the connection, we consider the functional

$$I(u) = \frac{\iiint_D |\nabla u|^2\, dV}{\iiint_D u^2\, dV}$$

Suppose that $I(u)$ has a minimum when $u = u_0$, in competition with all smooth functions which are zero on the boundary. We will show that $\nabla^2 u_0 + \lambda u_0 = 0$, where $\lambda = I(u_0)$.

To prove this, we write the hypothesis in the form

$$\iiint_D |\nabla u_0 + \epsilon \nabla v|^2\, dV \geq \lambda \iiint_D |u_0 + \epsilon v|^2\, dV$$

where v is an arbitrary smooth function and ϵ is an arbitrary real number. Expanding and simplifying, we have

$$2\epsilon \iiint_D (\nabla u_0 \cdot \nabla v - \lambda u_0 v)\, dv + \epsilon^2 \iiint_D (|\nabla v|^2 - \lambda v^2)\, dV \geq 0$$

This quadratic function has a minimum at $\epsilon = 0$, so the first integral is zero. Transforming the first term of this by using the divergence theorem and the boundary condition $v = 0$, we obtain, exactly as before, the condition that

$$0 = \iiint_D v(\nabla^2 u_0 + \lambda u_0) \, dV$$

If this happens for all smooth functions which satisfy the boundary condition, we conclude from the fundamental lemma of the calculus of variations that the term in parentheses is identically zero, that is, $\nabla^2 u_0 + \lambda u_0 = 0$, which was to be proved.

We remark that the minimum just obtained is the *smallest* frequency of the three-dimensional region. Indeed, if μ is another such, associated with the solution w of the equation $\nabla^2 w + \mu w = 0$, then by the divergence theorem and the boundary conditions we must have $\mu = \iiint_D |\nabla w^2| \, dV / \iiint_D w^2 \, dV$. But this quotient is greater than or equal to the absolute minimum λ discussed above. Therefore we have $\mu \geq \lambda$ for any other frequency μ.

EXERCISES 9.3

1. Let u_0 be the solution of Poisson's equation $\nabla^2 u_0 = -\rho$ with the condition that $u_0 = 0$ on the boundary of a two-dimensional region D. Let u be any other smooth function which satisfies the boundary condition. Show that $I(u) \geq I(u_0)$, where

$$I(u) = \iint_D (\tfrac{1}{2}\nabla u^2 - \rho u) \, dx \, dy$$

2. Let

$$I(u) = \iiint_D [\tfrac{1}{2}(Au_x^2 + Bu_y^2 + Cu_z^2 + Du^2) - \rho u] \, dx \, dy \, dz$$

Suppose u_0 is a smooth function that minimizes $I(u)$ in competition with all smooth functions which are zero on the boundary of the three-dimensional region D. Prove that u_0 satisfies the equation $(Au_x)_x + (Bu_y)_y + (Cu_z)_z + Du = -\rho$ in D.

3. Let

$$I(u) = \iint_G [\tfrac{1}{2}(Au_x^2 + 2Bu_x u_y + Cu_y^2 + Du^2) - \rho u] \, dx \, dy$$

where G is a smooth two-dimensional region. Suppose that u_0 is a smooth function which minimizes $I(u)$ in competition with all once-differentiable functions that are zero on the boundary of G. Prove that u_0 satisfies the equation $(Au_x + Bu_y)_x + (Bu_x + Cu_y)_y + Du = -\rho$ in G.

4. Let

$$I(u) = \iint_D \sqrt{1 + u_x^2 + u_y^2} \, dx \, dy$$

where G is a smooth two-dimensional region. Suppose that u_0 is a smooth function which minimizes $I(u)$ in competition with all once-differentiable functions u for which $u = f$ on ∂G. Prove that u_0 satisfies the equation

$$(1 + u_y^2)u_{xx} + (1 + u_x^2)u_{yy} - 2u_x u_y u_{xy} = 0$$

5. Let

$$I(u) = \iint_G [\tfrac{1}{2}(u_{xx} + u_{yy})^2 + \rho(x, y)u] \, dx \, dy$$

where G is a smooth two-dimensional region. Suppose that u_0 is a function which minimizes $I(u)$ in competition with all twice-differentiable functions for which $u = 0$, $\partial u/\partial n = 0$ on ∂G. Prove that u_0 satisfies the equation $\nabla^2\nabla^2 u = -\rho$ in G.

6. Let
$$I(u) = \iint_G [\tfrac{1}{2}(u_x^2 + u_y^2) - \rho(x, y)u] \, dx \, dy$$

where G is a smooth two-dimensional region. Suppose that u_0 is a smooth function which minimizes $I(u)$ in competition with all once-differentiable functions u (assuming no boundary conditions).

(a) Prove that u_0 satisfies the equation $\nabla^2 u = -\rho$ in G together with the boundary condition $\partial u/\partial n = 0$ on ∂G.

(b) Conclude that $\iint_G \rho(x, y) \, dx \, dy = 0$.

7. Let
$$I(u) = \iint_G [\tfrac{1}{2}(u_x^2 + u_y^2) - \rho(x, y)u] \, dx \, dy - \int_{\partial G} f(s)u(s) \, ds$$

Suppose that u_0 is a smooth function which minimizes $I(u)$ in competition with all once-differentiable functions u (assuming no boundary conditions). Prove that u_0 satisfies the equation $\nabla^2 u = -\rho$ in G with the boundary condition $\partial u/\partial n = f(s)$ on ∂G.

8. Let
$$I(u) = \iint_G [\tfrac{1}{2}(u_x^2 + u_y^2) - \rho(x, y)u] \, dx \, dy + \int_{\partial G} (fu - gu^2) \, ds$$

where $\rho(x, y)$ is a given function on ∂G and $f(s)$, $g(s)$ are given on ∂G. Suppose that u_0 is a smooth function which minimizes $I(u)$ in competition with all once-differentiable functions on G. Find the partial differential equation and boundary condition satisfied by u_0.

9. Repeat Exercise 8 for the functional

$$I(u) = \iint_G \varphi(x, y, u, u_x, u_y) \, dx \, dy + \int_{\partial G} \psi\left(s, u, \frac{\partial u}{\partial s}\right) \, ds$$

The following exercises illustrate examples in which the solution of the Euler-Lagrange equation is not a maximum or minimum of the functional whose stationariness is sought. This clarifies the "principle of least action" in mechanics. Exercise 10 gives a positive result for the simple harmonic oscillator.

10. Let
$$I_T(u) = \int_0^T [u'(t)^2 - u(t)^2] \, dt \qquad u(0) = A, \, u(T) = B$$

(a) Show that the Euler-Lagrange equation is $u'' + u = 0$.

(b) If T is not of the form $\pi, 2\pi, \ldots$, show that there is a unique solution $v(t)$, $0 \le t \le T$, of the Euler-Lagrange equation satisfying the boundary conditions.

(c) Show that $I_T(u)$ can be written as

$$I_T(u) = \int_0^T \left[u'(t) - \frac{v'(t)}{v(t)}u(t) \right]^2 dt - \frac{v}{v'} u^2 \Big|_0^T$$

if $v(t)$, $0 \le t \le T$, is any solution of $v'' + v = 0$ with $v(t) \ne 0$.

(d) Conclude that if $T < \pi$ and $A > 0$, $B > 0$, then we have $I_T(u) \ge I_T(v)$ for any function u which satisfies the boundary conditions $u(0) = A$, $u(T) = B$ where $v(t)$, $0 \le t \le T$, is the solution obtained in (b). Equality is obtained if and only if $u = v$. Hence we have a unique minimum in this case.

11. (a) With the same choice of $I_T(u)$, compute its value for the function $u(t) = Bt/T + a_1 \sin(\pi t/T) + \cdots + a_n \sin(n\pi t/T)$, where $a_1, \ldots, a_n$ are unspecified constants.

(b) Using the formula obtained in (a), show that if $T > \pi$, then sup $I_T(u) = +\infty$ and inf $I_T(u) = -\infty$ over the indicated class of functions. Hence we have no minimum or maximum in this case.

12. Let

$$I_T(u) = \int_0^T \int_0^\pi (u_t^2 - u_x^2) \, dx \, dt$$

with the boundary conditions $u(0; t) = 0$, $u(\pi; t) = 0$, $u(x; 0) = 0$, $u(x; T) = B \sin x$.

(a) Show that the Euler-Lagrange equation is the wave equation $u_{tt} = u_{xx}$, and solve it if T is not of the form $\pi, 2\pi, 3\pi, \ldots$.

(b) Compute the value of $I_T(u)$ for a function of the form $u(x; t) = (Bt/T) \sin x + a \sin Nx \sin(k\pi t/T)$, where N, k are integers and a is any constant.

(c) By suitable choice of k, N, a, show that we have inf $I_T(u) = -\infty$, sup $I_T(u) = +\infty$ over the indicated class of functions in (b) for *any* $T > 0$, *no matter how small*. Hence we have no maximum or minimum in this case.

9.4 APPROXIMATE METHODS OF RITZ, KANTOROVICH, AND GALERKIN

Having established the theoretical connection between a partial differential equation and a minimization problem in Sec. 9.3, we turn to some approximate methods of solution. All such methods involve the use of *trial solutions*, or functions that satisfy the boundary conditions and are determined by additional conditions which are intended to make them close to the desired minimum. The additional conditions may be obtained by means such as

1. Solution of a minimum problem in finitely many variables
2. Orthogonality conditions in finitely many variables
3. Solution of a related ordinary differential equation
4. Combinations of 1, 2, and 3

The Ritz Method: Rectangular Regions

The general idea behind this method is to look for a trial solution in the form

$$u(x, y, z) = U(x, y, z; c_1, c_2, \ldots, c_n)$$

which satisfies the boundary condition identically where $c_1, \ldots, c_n$ are parameters which may be adjusted. If we substitute this into the minimization problem, we obtain

$$\Phi(c_1, \ldots, c_n) = \iiint_D F(x, y, z; u, u_x, u_y, u_z) \, dx \, dy \, dz$$

We minimize this function of n variables by looking for a critical point, i.e., an n-tuple $(c_1, \ldots, c_n)$ which satisfies the equations

$$\frac{\partial \Phi}{\partial c_i}(c_1, \ldots, c_n) = 0 \qquad 1 \le i \le n$$

We then substitute these values into the trial solution.

Example 9.4.1. Find the approximate solution of Poisson's equation $u_{xx} + u_{yy} = -\rho$ in the square $|x| < a$, $|y| < a$ with the boundary condition $u = 0$ where ρ is a constant. Use trial solutions of the form $u = c(x^2 - a^2)(y^2 - a^2)$.

Solution. We have $u_x = 2cx(y^2 - a^2)$, $u_y = 2cy(x^2 - a^2)$, $\frac{1}{2}(u_x^2 + u_y^2) = 2c^2[x^4y^2 + x^2y^4 - 4a^2x^2y^2 + a^4(x^2 + y^2)]$. This leads to $\iint [\frac{1}{2}(u_x^2 + u_y^2)\, dx\, dy = \frac{128}{45} a^8c^2$. Likewise $\iint \rho u\, dx\, dy = \frac{16}{9} \rho ca^8$. Thus we are required to minimize the function $\Phi(c) = \frac{128}{45} a^8c^2 - \frac{16}{9} \rho ca^6$. The required minimum is attained at $c = \frac{5}{16} \rho/a^2$, which gives the required trial solution with $\Phi(c_0) = -\frac{5}{18} \rho^2 a^4 \approx -0.277\rho^2 a^4$. ●

Example 9.4.2. Find the approximate solution of Poisson's equation $u_{xx} + u_{yy} = -\rho$ in the square $|x| < a$, $|y| < a$ with the boundary condition $u = 0$, where ρ is a constant. Use trial solutions of the form $u = c \cos[\pi x/(2a)] \cos[\pi y/(2a)]$ and compare with Example 9.4.1.

Solution. We have $u_x = -[c\pi/(2a) + \sin[\pi x/(2a)] \cos[\pi y/(2a)]$, $u_y = -[c\pi/(2a)] \cos[\pi x/(2a)] \sin[\pi y/(2a)]$. This leads to $\iint \frac{1}{2}(u_x^2 + u_y^2)\, dx\, dy = \pi^2 c^2/4$. Likewise $\iint \rho u\, dx\, dy = \rho c(4a/\pi)^2$. Thus we are required to minimize the functional $\Phi(c) = c^2\pi^2/4 - \rho c(4a/\pi)^2$; this leads to a minimum at $c = 32\rho a^2/\pi^4$. The associated value is $\Phi(c) = -\rho^2 a^4(256/\pi^6)$, about $-0.266\rho^2 a^4$ slightly larger than the minimum obtained in Example 9.4.1. ●

We note here that the Ritz method applies in any number of variables. In particular, it may be applied to variational problems for *ordinary* differential equations (see the Exercises).

The Kantorovich Method: Rectangular Regions

In this method we begin with a trial solution that contains one or more arbitrary functions of one variable and identically satisfies the boundary condition. These functions are chosen as solutions of the ordinary differential Euler-Lagrange equations which are obtained when the composite functional is minimized.

Example 9.4.3. Find the approximate solution of Poisson's equation $u_{xx} + u_{yy} = -\rho$ in the square $|x| < a$, $|y| < a$ with the boundary condition $u = 0$, where ρ is a constant. Use trial solutions of the form $u = (a^2 - y^2)c(x)$, where $c(x)$ is an arbitrary function.

Solution. We have $u_x = (a^2 - y^2)c'(x)$, $u_y = -2yc(x)$, and thus

$$\iint \left[\frac{1}{2}(u_x^2 + u_y^2) - \rho u\right] dx\, dy = \int_{-a}^{a} \left[\frac{16a^5}{15} c'(x)^2 + \frac{8a^3}{3} c(x)^2 - \frac{4\rho a^3}{3} c(x)\right] dx$$

The Euler-Lagrange equation for this functional is $c''(x) = [5/(2a^2)]c(x) - 5\rho/8$; the solution with the boundary conditions $c(-a) = 0 = c(a)$ is $c(x) = \frac{1}{4}\rho a^2(1 - \cosh kx/\cosh ka)$, where $k = \sqrt{5/(2a^2)}$. ●

Example 9.4.4. Find the approximate solution of Poisson's equation $u_{xx} + u_{yy} = -\rho$ in the square $|x| < a$, $|y| < a$ with the boundary condition $u = 0$, where ρ is a constant. Use trial solutions of the form $u = \sum_{j=0}^{N} c_j(x) \cos [(j + \frac{1}{2})(\pi y/a)]$.

Solution. For this trial solution we have

$$u_x = \sum_{j=0}^{N} c_j'(x) \cos \left[\left(j + \frac{1}{2}\right) \left(\frac{\pi y}{a}\right) \right]$$

$$u_y = \sum_{j=0}^{N} \left(j + \frac{1}{2}\right) \left(\frac{\pi}{a}\right) c_j(x) \sin \left[\left(j + \frac{1}{2}\right) \left(\frac{\pi y}{a}\right) \right]$$

$$\int\int \left[\frac{1}{2}(u_x^2 + u_y^2) - \rho u \right] dx\, dy$$

$$= \sum_{j=0}^{N} \int_{-a}^{a} \left[ac_j'(x)^2 + \left(j + \frac{1}{2}\right)^2 \left(\frac{\pi}{a}\right)^2 c_j(x)^2 - 2\rho c_j(x) \frac{(1)^j}{(j + \frac{1}{2})(\pi/a)} \right] dx$$

The Euler-Lagrange equations are

$$c_j''(x) = \left(j + \frac{1}{2}\right)^2 \left(\frac{\pi}{a}\right)^2 c_j(x) - \frac{\rho(1)^j}{(j + \frac{1}{2})(\pi/a)}$$

with the solutions

$$c_j(x) = \left[\frac{\rho(-1)^j}{(j + \frac{1}{2})(\pi/a)} \right] \left\{ 1 - \frac{\cosh (j + \frac{1}{2})(\pi x/a)}{\cosh [(j+\frac{1}{2})\pi]} \right\}$$

Note that the resulting function $u(x, y)$ is the Nth partial sum of the Fourier representation of the solution. $\bullet$

Unlike the Ritz method, the Kantorovich method can be applied only in two or more variables. It is specifically suited to *partial* differential equations and has no counterpart for ordinary differential equations.

The Galerkin Method: Rectangular Regions

In this method we look for a trial solution of the form

$$u(x, y, z) = \sum_{1 \leq i \leq N} c_i \Phi_i(x, y, z)$$

where $\{\Phi_i\}_{1 \leq i \leq N}$ are given functions which satisfy the boundary condition and the c_i are adjustable constants. Writing the differential equation in the form $Lu = 0$, we determine the constants $\{c_i\}_{1 \leq i \leq N}$ by the requirement

$$\int\int\int_D Lu(x, y, z)\Phi_i(x, y, z)\, dV = 0 \qquad 1 \leq i \leq N$$

Indeed, if u were the true solution, we would have $Lu = 0$ identically; the method of Galerkin requires that Lu be orthogonal to the linear span of

$\{\Phi_i\}_{1\leq i\leq N}$. Note that this method does *not* require that we formulate the differential equation in terms of a variational problem.

Example 9.4.5. Find the approximate solution of the Poisson equation $\nabla^2 u = -\rho$ in the rectangle $|x| < a$, $|y| < b$ with zero boundary conditions. Use Galerkin's method with trial solutions of the form

$$u(x, y) = (a^2 - x^2)(b^2 - y^2)(A_{00} + A_{10}x^2 + A_{01}y^2 + \cdots + A_{ij}x^{2i}y^{2j})$$

Solution. We have

$$Lu = [-2(a^2 - x^2) - 2(b^2 - y^2)](A_{00} + \cdots + A_{ij}x^{2i}y^{2j})$$

$$+ (a^2 - x^2)(b^2 - y^2)[2A_{10} + 2A_{01} + \cdots + 2i(2i - 1)(2j)(2j-1)x^{2i-1}y^{2j-1}] + \rho$$

$$- 2x(b^2 - y^2)(2A_{10}x + \cdots) - 2y(a^2 - x^2)(2A_{01}y + \cdots)$$

Applying the method for $(i, j) = (0, 0)$ gives the orthogonality condition

$$0 = \int_{-a}^{a}\int_{-b}^{b} \{[-2(a^2 - x^2) - 2(b^2 - y^2)]A_{00} + \rho\}(a^2 - x^2)(b^2 - y^2) \, dx \, dy$$

Integrating and solving for A_{00} lead to the value $A_{00} = 5/[4(a^2 + b^2)]$ and the trial solution

$$u(x, y) = \frac{5}{4} \frac{(a^2 - x^2)(b^2 - y^2)}{a^2 + b^2}$$

The determination of A_{ij} for $(i, j) \neq (0, 0)$ is left to the exercises. ●

Example 9.4.6. Find the approximate solution of the Poisson equation $\nabla^2 u = -\rho$ in the rectangle $|x| < a$, $|y| < b$ with zero boundary conditions. Use Galerkin's method with trial solutions of the form

$$u(x, y) = \sum_{m,n \text{ odd}} A_{mn} \cos\left[\frac{m\pi x}{(2a)}\right] \cos\left[\frac{n\pi y}{(2b)}\right]$$

Solution. We have

$$Lu = \rho - \sum_{m,n \text{ odd}} A_{mn}\left[\left(\frac{m\pi}{2a}\right)^2 + \left(\frac{n\pi}{2b}\right)^2\right] \cos\frac{m\pi x}{2a} \cos\frac{n\pi y}{2b}$$

When we integrate this against the orthogonal functions $\cos(m\pi x/a)\cos(n\pi y/b)$, we obtain the equations

$$\frac{\rho(16ab)}{\pi^2 mn(-1)^{(m+n)/2-1}} - A_{mn}\, ab\left[\left(\frac{m\pi}{2a}\right)^2 + \left(\frac{n\pi}{2b}\right)^2\right] = 0$$

This leads to the trial solution

$$u(x, y) = \frac{128a^2b^2}{\pi^4} \sum_{m,n \text{ odd}} (-1)^{(m+n)/2-1} \frac{\cos\left[m\pi x/(2a)\right] \cos\left[n\pi y/(2b)\right]}{mn(b^2m^2 + a^2n^2)}$$

We now apply Galerkin's method to find an approximation to the fundamental frequency of a circular drumhead. This problem was solved exactly in Sec. 3.4 by using Bessel functions.   ●

Example 9.4.7. Find the approximate solution of $\nabla^2 u + \lambda u = 0$ in the circular disk $0 < \rho < a$ in the form $u = A \cos [\pi\rho/(2a)]$.

Solution. In polar coordinates, we have $\nabla^2 u = u_{\rho\rho} + (1/\rho)u_\rho + (1/\rho^2)u_{\varphi\varphi}$. In this case u is independent of φ, and we have

$$u_\rho = -A \left(\frac{\pi}{2a}\right) \sin \frac{\pi\rho}{2a} \qquad u_{\rho\rho} = -\left(\frac{\pi}{2a}\right)^2 A \cos \frac{\pi\rho}{2a}$$

$$0 = \iint\limits_{x^2+y^2\le a^2} (\lambda u + \nabla^2 u) \cos \frac{\pi\rho}{2a} \, dx \, dy$$

$$= 2\pi \int_0^a \left[\lambda A \cos^2 \frac{\pi\rho}{2a} - A \left(\frac{\pi}{2a}\right)^2 \cos^2 \frac{\pi\rho}{2a} - A \frac{\pi}{2a} \sin \frac{\pi\rho}{2a} \cos \frac{\pi\rho}{2a} \right] \rho \, d\rho$$

Doing the required integrals and canceling the common factor of A, we have $\frac{1}{4}\pi^2$ $(\frac{1}{2} + 2\pi^{-2}) - \lambda a^2 (\frac{1}{2} - 2\pi^{-2}) = 0$, or $\lambda = 5.832/a^2$. This approximate solution is to be compared with the exact solution $u = J_0 (\rho z/a)$ which leads to $\lambda = 5.779/a^2$.   ●

Nonrectangular Regions

We now turn to the problems in nonrectangular regions. Consider the region D in the xy plane which is bounded by the vertical lines $x = a$ and $x = b$; the horizontal boundaries are defined by the curves $y = \pm \varphi(x)$, where φ is a smooth function (the case $\varphi =$ constant corresponds to the rectangle already studied). We look for a trial solution of Poisson's equation $u_{xx} + u_{yy} = -\rho$ in the form

$$u(x, y) = [y^2 - \varphi(x)^2]f(x)$$

Here ρ may be a function of x, and $f(x)$ is to be determined from a suitable ordinary differential equation (method of Kantorovich). To find this equation, we compute the variational integral as follows:

$$u_x = -2\varphi\varphi' f + f'(y^2 - \varphi^2) \qquad u_y = 2yf$$

$$\frac{1}{2}(u_x^2 + u_y^2) - \rho u$$

$$= 2\varphi^2\varphi'^2 f^2 - 2\varphi\varphi' ff'(y^2 - \varphi^2) + \frac{1}{2}f'^2(y^2 - \varphi^2)^2 + 2y^2 f^2 - \rho(y^2 - \varphi^2)f$$

Performing the y integrations directly, we have for the variational integral

$$I(u) = \int_a^b \int_{-\varphi(x)}^{\varphi(x)} [\frac{1}{2}(u_x^2 + u_y^2) - \rho u] \, dy \, dx$$

$$= \int_a^b (4\varphi^3\varphi'^2 f^2 - \frac{8}{3}\varphi^4\varphi' ff' + \frac{8}{15}\varphi^5 f'^2 + \frac{4}{3}f^2\varphi^3 + \frac{4}{3}\rho f\varphi^3) \, dx$$

The Euler-Lagrange ordinary differential equation for the minimization of this functional is of the form $(d/dx)(\partial\Phi/\partial f') = \partial\Phi/\partial f$; written in detail, this is

$$\frac{\partial\Phi}{\partial f'} = \frac{8}{3}\varphi^4\varphi'f + \frac{16}{15}f'\varphi^5$$

$$\frac{\partial\Phi}{\partial f} = 8\varphi^3\varphi'^2f + \frac{8}{3}\varphi^4\varphi'f' + \frac{8}{3}f\varphi^3 + \frac{4}{3}\rho\varphi^3$$

Performing the d/dx and canceling a common factor of φ^3 give the differential equation for the unknown function f:

$$\varphi^2f'' + (5\varphi\varphi')f' + \tfrac{5}{2}[\varphi\varphi'' - 1 + (\varphi')^2]f = \frac{5\rho}{4}$$

We solve this with the condition that $f(x) = 0$ for $x = a$ and $x = b$.

Example 9.4.8. Find the approximate solution of Poisson's equation $u_{xx} + u_{yy} = -\rho$ in the trapezoid defined by the inequalities $a < x < b$, $-kx < y < kx$, where ρ is a constant and $a > 0$.

Solution. In this case we have $\varphi(x) = kx$ with $\varphi' = k$, $\varphi'' = 0$; the differential equation for $f(x)$ is $k^2x^2f'' + 5k^2xf' + \tfrac{5}{2}(k^2 - 1)f = 5\rho/4$. This is a differential equation of the Euler type, whose general solution is of the form $f(x) = C_0 + C_1x^{\nu_1} + C_2x^{\nu_2}$, where C_0, C_1, C_2 are constants determined from ρ and the boundary conditions, while ν_1, ν_2 are roots of the indicial equation $k^2\nu(\nu - 1) + 5k^2\nu + \tfrac{5}{2}(k^2 - 1) = 0$. The roots are $\nu = -2 \pm \tfrac{1}{2}\sqrt{6 + 10k^{-2}}$, and the constant C_0 is determined from the right side by $C_0 = 5\rho/[4(k^2 - 1)]$. Constants C_1 and C_2 are determined from the boundary conditions by solving the system of simultaneous equations $0 = C_0 + C_1a^{\nu_1} + C_2a^{\nu_2}$, $0 = C_0 + C_1b^{\nu_1} + C_2b^{\nu_2}$. ●

We consider the following limiting case.

Example 9.4.9. Find the approximate solution of Poisson's equation $u_{xx} + u_{yy} = -\rho$ in the isosceles triangle bounded by the lines $y = \pm kx$, $x = b$, where ρ is a constant.

Solution. In this case we may dispense with the boundary condition at $x = a$ and determine $f(x)$ by the equation $k^2x^2f'' + 5kxf' + \tfrac{5}{2}(k^2 - 1)f = 5\rho/4$ with the boundary condition $f(b) = 0$. We use the larger exponent $\nu = -2 + \tfrac{1}{2}\sqrt{6 + 10k^2}$. The solution which satisfies the boundary condition is written in the form $f(x) = \rho[1 - (x/b)^\nu]/[2(k^2 - 1)]$. Since $\nu > -2$, we note that the product $f(x)(y^2 - k^2x^2)$ is bounded in the triangle and tends to zero as $(x, y) \to (0, 0)$. In the case of an equilateral triangle, we have $k = 1/\sqrt{3}$; thus $\nu = 1$, and the approximate solution is a polynomial: $u(x, y) = -\tfrac{3}{4}[1 - (x/b)(y^2 - \tfrac{1}{3}x^2)]$. You can check that in this case we have found the exact solution. ●

EXERCISES 9.4

1. Use the Ritz method to find an approximate minimum of the functional $I(u) = \int_0^1 [u'(t)^2 - u(t)^2 - 2tu(t)]\, dt$ with the boundary conditions $u(0) = u(1) = 0$. Use the following trial functions.
   (a) $U(t) = ct(1 - t)$
   (b) $U(t) = t(1 - t)(c_1 + c_2 t)$
   (c) $U(t) = t(1 - t)(c_1 + c_2 t + c_3 t^2)$
   (d) $U(t) = \sum_{n=1}^{N} a_n \sin n\pi t$

2. Use the Ritz method to find an approximate minimum of the functional $I(u) = \int_0^2 [u'(t)^2 + u(t)^2 - 2tu(t)]\, dt$ with the boundary conditions $u(0) = u(2) = 0$. Use the following trial functions.
   (a) $U(t) = ct(2 - t)$
   (b) $U(t) = t(2 - t)(c_1 + c_2 t)$
   (c) $U(t) = t(2 - t)(c_1 + c_2 t + c_3 t^2)$
   (d) $U(t) = \sum_{n=1}^{N} a_n \sin n\pi t$

3. Use the Ritz method to find an approximate minimum of the functional $I(u) = \iint [\frac{1}{2}(u_x^2 + u_y^2) - xyu]\, dx\, dy$, where the integration is over the square $-1 \le x \le 1$, $-1 \le y \le 1$ and the boundary conditions are that $u(x, y) = 0$ if $x = -1$, $x = 1$ or $y = -1$, $y = 1$. Use the following trial functions.
   (a) $U(x, y) = c(1 - x^2)(1 - y^2)$
   (b) $U(x, y) = c(1 - x^2)(1 - y^2)(1 + ax + by)$
   (c) $U(x, y) = c(1 - x^2)(1 - y^2)(1 + ax^2 + by^2)$

4. Use the Ritz method to find an approximate minimum of the functional $I(u) = \int_{-\pi}^{\pi} \int_0^1 [u_\rho^2 + (1/\rho^2)u_\varphi^2]\rho\, d\rho\, d\varphi$ with the boundary conditions that $u(\rho; \varphi) = 0$ when $\rho = 1$. Use the following trial functions.
   (a) $U(\rho) = c(1 - \rho^2)$
   (b) $U(\rho) = c(1 - \rho^2)(1 + a\rho)$
   (c) $U(\rho) = c(1 - \rho^2)(1 + a\rho + b\rho^2)$

5. Consider the Sturm-Liouville eigenvalue problem $\varphi'' + \lambda\varphi = 0$ for $-1 \le x \le 1$ with the boundary conditions $\varphi(-1) = 0$, $\varphi(1) = 0$.
   (a) Find (exactly) the smallest eigenvalue.
   (b) Use the Ritz method with the trial function $U(x) = 1 - x^2$ to find a first approximation to the smallest eigenvalue, and compare with the exact result obtained in (a).
   (c) Refine the approximation of (b) by using the trial function $U(x) = (1 - x^2)(1 + ax^2)$, and show that the error obtained is less than 1 percent.

6. Use the Kantorovich method to find an approximate minimum of the functional $I(u) = \iint [\frac{1}{2}(u_x^2 + u_y^2) - xyu]\, dx\, dy$, where the integration is over the unit square $-1 \le x \le 1$, $-1 \le y \le 1$ and the boundary conditions are that $u(x, y) = 0$ when $x = -1$, $x = 1$ or $y = -1$, $y = 1$. Use the following trial functions.
   (a) $U(x, y) = (1 - y^2)[c_1(x) + y^2 c_2(x)]$
   (b) $U(x, y) = \sum_{n=0}^{N} \cos[(n + \frac{1}{2})\pi y]\, c_n(x)$

7. Use the Kantorovich method to find an approximate minimum of the functional $I(u) = \int_{-\pi}^{\pi} \int_0^1 \{\frac{1}{2}[u_\rho^2 + (1/\rho^2)u_\varphi^2] - u \cos \varphi\}\rho \, d\rho \, d\varphi$ with the boundary condition that $u(\rho; \varphi) = 0$ when $\rho = 1$. Use the following trial functions.

   (a) $U(\rho; \varphi) = (1 - \rho^2)c_1(\varphi)$

   (b) $U(\rho; \varphi) = J_0(\rho x_1)c_1(\varphi) + J_0(\rho x_2)c_2(\varphi)$ ($x_i = i$th zero of J_0)

8. Use Galerkin's method to find an approximate solution of the ordinary differential equation $u'' + xu' + u = 2x$ on the interval $-1 \le x \le 1$ with the boundary conditions that $u(-1) = 0$, $u(1) = 0$, choosing a trial function of the form $U(x) = (1 - x^2)(c_1 + c_2x + c_3x^2)$.

9. Use Galerkin's method to find an approximate solution of the ordinary differential equation $(xu')' + u = x$ on the interval $0 \le x \le 1$ with the boundary conditions $u(0) = 0$, $u(1) = 1$. Use a trial function of the form $U(x) = x + x(1 - x)(c_1 + c_2x)$.

Using Mathematica[1]

Alfred Gray

The traditional approach to partial differential equations and boundary value problems uses analytical methods that produce graphical and numerical illustrations of the solutions. For example, the methods of Fourier analysis and separation of variables lead to series and integral representations of solutions of the heat equation. In many cases these lead to accurate numerical approximations and/or informative graphs of the solution. However, the effort required to produce such results by hand is frequently prohibitive.

Thanks to new computer technology it is now possible to obtain high-speed representations of solutions. Of the several programs available we have chosen **Mathematica**, which has the capabilities of symbolic manipulation and graphical illustration as well as traditional numerical computations. This program will factor polynomials, differentiate the standard functions of calculus and produce accurate graphs of the resultant symbolic calculations. **Mathematica** is available for many different computers, including 386-PC's, MacSE's, MacII's, Sun Workstations, **NeXT**'s and Iris workstations.

Mathematica excels in its ability to do graphics. For example, complicated Bessel functions can be graphed with ease using **Mathematica**. Because it uses the printing language Postscript the graphics of **Mathematica** can be easily printed. Graphics displayed on a monitor are ephemeral, but print outs can be studied at one's leisure.

In this appendix we illustrate the use of **Mathematica** in the following areas. (Section numbers refer to the main part of the book.)

- Computing Fourier coefficients .... (Section 1.1)

- Graphing Fourier series .... (Sections 1.1, 1.3 and 1.4)

- Computing and graphing Legendre polynomials ....(Section 4.2)

- Graphing 3-dimensional plots of heat flow .... (Section 2.1)

- Computing and graphing Bessel Functions.............. (Section 3.2)

---

[1]Mathematica is a registered trademark of Wolfram Research, Inc., P. O. Box 6059, Champaign IL 61821.

404

- Graphing and animating 3-dimensional plots of a
  vibrating drum ....(Sections 2.5 and 3.3)

- Bifurcation of the wave equation .... (Section 2.4)

Implementations of Mathematica

Of the several interfaces probably the most useful are those with notebooks. Notebooks can be used on MacSE's, MacII's and **NeXT**'s. A notebook is a kind of extended file that can display both text and graphics. You can also type in **Mathematica** commands and evaluate them. Such commands can do symbolic or numerical calculations or display graphics, either 2- or 3-dimensional. The usefulness of a notebook lies in its ability to combine theoretical discussions with the immediate execution of commands, especially graphics commands. At this writing the notebooks on Mac's have more features, such as color and extensive fonts, than do **NeXT**'s. However, because the operating system on a **NeXT** is based on Unix, it is easier on a **NeXT** to transfer calculations and graphics to other applications. Furthermore, the **NeXT** printer (400 dpi) has sharper graphics than the Apple LaserWriter (300 dpi).

Unix based workstations such as those of Sun and Iris do not have notebooks in the current version of **Mathematica**. However, windows can be used effectively. A good system is to use one window for **Mathematica** and another for text. The text can be transferred from the text window to the **Mathematica** window using a mouse. Each graphic is displayed in its own window. Both Sun and Iris workstations, especially if they have a great deal of memory, are much faster than a Mac and somewhat faster than a **NeXT**.

The most spectacular **Mathematica** screen graphics can be obtained on an Iris workstation. Using the **Live** command a standard **Mathematica** 3-dimensional graphic can be displayed using the Iris's graphics hardware. This image can then be rotated in real time.

The most primitive interface for **Mathematica** is the one on a 386-PC because windows are unavailable. But that should change soon. At any rate all of the standard **Mathematica** commands and graphics are available, but a great deal of memory is needed, especially for graphics.

The essential references for **Mathematica** are the well-written books

Mathematica by Steven Wolfram

Programming in Mathematica by Roman Maeder.

Both books are published by Addison-Wesley.

Getting Started

Here are simple instructions for getting started with **Mathematica** on a **NeXT** workstation; the instructions for a MacII or a MacSE are quite similar. First you need to logon. To bring up **Mathematica** you should click on the **Mathematica** icon (it is a small polyhedron). This brings up a window called "Untitled"; it is actually a notebook. To enter a formula just start typing. For example, the polynomial $x^{15} - 1$ can be factored by typing

```
Factor[x^15-1]
```

Mathematica on a **NeXT** workstation (as on a MacII) uses an interface called a "notebook". Notebooks are divided into cells. Each formula is entered into a "cell", recognizable by a bracket on the right side of the window. After typing a formula into a cell, you can evaluate the formula by hitting the key "Enter". (Notice that "Enter" is different from "Return"; the latter is used to go to the next line of a cell.

Cells can be used to display either text or graphics and can be printed individually. To print a whole notebook, click the mouse on the "print" icon. To print an individual cell, first you need to use the mouse to move the cursor to a cell bracket and click on it. You should see a long dark line encompassing the cell. Then move the cursor to the "Print Selection" icon and click the mouse button. This will bring up a dialog box that allows you to send the graphics to the printer.

More complete information can be found in the documentation **Mathematica: Summary of the Front End**. It and other useful files can be found in the directory /NextLibrary/Documentation/Mathematica. This documentation is present on each **NeXT** and can be easily printed.

Graphing Fourier Series Using Mathematica

Let us find the first n terms in the Fourier series of the function defined by

$$f(x) = x \qquad \text{for } -\pi < x < \pi,$$

and plot it. The Fourier series of $f(x)$ is

$$2\sum_{k=1}^{\infty} \frac{(-1)^{k+1}\sin(kx)}{k}.$$

Enter the following commands in a cell:

```
F[x_,n_]:=2 Sum[(((-1)^(i+1))/i) Sin[i x],{i,1,n}];

F[n_]:=Plot[F[x,n],{x,-Pi,Pi}];
```

Then do "Enter". (You won't see anything immediately.) The first line defines a function F of two variables, and the second line defines a plot-valued function. It does not matter that the same symbol F is used for both functions. You can get a quick check whether F is defined correctly[2] by typing ?F. Here F[x,n] is the sum of the first n terms of the Fourier series for $f(x)$. For example, you can see the first three terms of the series by using the mouse to move the cursor to a new cell and typing F[x,3].

Don't forget "Enter". The result, which will appear in the cell immediately below the cell in which the command F[x,3] was entered, will be

```
                Sin[2 x]     Sin[3 x]
Out[2]= 2 (Sin[x] - -------- + --------)
                   2           3
```

This is **Mathematica**'s way of printing out

$$2\left(\sin(x) - \frac{\sin(2x)}{2} + \frac{\sin(3x)}{3}\right).$$

To graph the function $F[3,x]$ simply repeat the above instructions, but

---

[2]Two common mistakes to avoid are typing **sin** instead of **Sin** and ( ) instead of [ ].

type F[3] instead of F[x,3]. The result will be

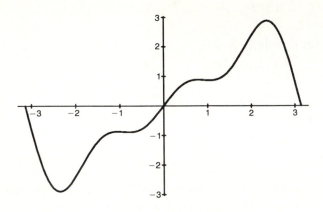

Computing Fourier Coefficients

Here are some commands to compute Fourier coefficients in **Mathematica**:

```
A[n_,L_,f_]:=(1/L) Integrate[f[x] Cos[n Pi x /L],{x,-L,L}]
A0[L_,f_]:=(1/(2L)) Integrate[f[x],{x,-L,L}]
B[n_,L_,f_]:=(1/L) Integrate[f[x] Sin[n Pi x /L],{x,-L,L}]
A[n_,f_]:=(1/Pi) Integrate[f[x] Cos[n x],{x,-Pi,Pi}]
A0[f_]:=(1/(2Pi)) Integrate[f[x] Cos[n x],{x,-Pi,Pi}]
B[n_,f_]:=(1/Pi) Integrate[f[x] Sin[n x],{x,-Pi,Pi}]
```

Before using these formulas we must define a function f via the command
f[x_]:=
On the right hand side of this equation you should type the formula that
defines $f(x)$ using proper **Mathematica** syntax. Then if L and n are specific
numbers, the command

```
A[n,L,f]
```

will yield the quantity which is denoted by Formula (1.1.6) as A_n. (**Mathematica** is able to distinguish between A[n,L,f] and A[n,f] because the first
is a function of 3 variables, and the second is a function of 2 variables.)

We use these formulas to compute the Fourier coefficients of the function[3]
f which has period $2L$ and is given by

---

[3]Problem 4 of Section 1.1

$$f(x) = e^x \quad \text{for } -L < x < L.$$

First we define the function f via the command

```
f[x_]:=E^x
```

Then we type in the commands above. Next typing

```
A[n,L,f] /. Sin[Pi n]->0 /. Sin[-Pi n]->0
/. Cos[Pi n]->(-1)^n /. Cos[-Pi n]->0
```

results in

```
             n  L
         (-1)   E
Out[2]= --------------
                2  2
              Pi   n
        L (1 + ------)
                 2
                 L
```

The "Out[2]" means output line number 2; it may be some other number on your machine. Here we have used the command /. which means "such that". Note also that -> consists of the two characters - and >. Without special programming **Mathematica** has to be told that $\sin(n\pi) = 0$; this is done via the command Sin[n Pi]->0. Similar remarks apply to the other substitutions.

We can simplify the equation "Out[2]" further with the command

```
Simplify[%2]
```

We get

```
             n  L
         (-1)   E  L
Out[9]= ------------
          2      2  2
         L  + Pi   n
```

Exercise. Use the same method in **Mathematica** to compute A_0 and the B_n's for the function $f(x) = e^x$ $(-L < x < L)$.

Gibbs' Phenomenon

Let f be a function defined by

$$f(x) = \begin{cases} -1 & \text{for } -\pi \leq x < 0, \\ 1 & \text{for } 0 \leq x < \pi. \end{cases}$$

It is easy to graph $f(x)$ using **Mathematica**. First we define a step function by means of the "**If**" function:

```
u[a_,x_]:=If[a<x,1,0]
```

(In **Mathematica** the syntax of "**If**" can be found by typing "**?If**". The function **If** takes three arguments. The first argument is a condition. The second argument is the value of **If** if the condition is satisfied, and the third argument is the value of **If** if the condition is not satisfied.) Discontinuous functions such as step functions can be graphed in **Mathematica** just as easily as continuous functions. First we define another function whose value is a graph:

```
U[a_]:=Plot[u[a,x],{x,a-1,a+1}]
```

Then, for example, the graph of u[2,x] can be obtained by typing U[2]:

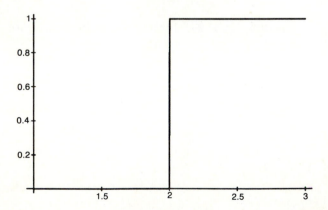

The function f can be written in terms of step functions as

```
f[x_]:=1 - 2 u[-Pi,x] + 2 u[0,x]
```

Now we extend f periodically to be a function defined on the whole real line. Then the Fourier series of f has sin terms only. The coefficients are given by

$$B_n = \frac{2}{\pi} \int_0^\pi \sin nx \, dx = \frac{2}{n\pi}(1 - (-1)^n).$$

Consequently, B_n is zero if n is even and

$$B_{2n-1} = \frac{4}{(2n-1)\pi}.$$

A partial sum for the Fourier series of f is given by

$$f_{2n-1}(x) = \sum_{k=1}^n B_{2k-1} \sin((2k-1)x) = \frac{4}{\pi} \sum_{k=1}^n \frac{\sin((2k-1)x)}{2k-1}.$$

In Mathematica we define a function of 2 variables to represent this partial sum:

```
f[n_,x_]:=(4/Pi) Sum[(1/(2 k-1)) Sin[(2 k-1) x],{k,1,n}]
```

Note that typing f[n,x] produces no new result, only

```
            Sin[(-1 + 2 k) x]
        4 Sum[------------------, {k, 1, n}]
              -1 + 2 k
Out[5]= --------------------------------------
                     Pi
```

whereas typing f[3,x] will result in

```
            Sin[3 x]   Sin[5 x]
        4 (Sin[x] + -------- + --------)
               3          5
Out[6]= --------------------------------
                   Pi
```

In other words, the expression f[n,x] is too general for **Mathematica** to expand, but if we substitute a specific number for n then **Mathematica** will expand it.

To graph the partial sum, we first define a function that assigns to each integer n the plot of f[n,x]:

```
fgraph[n_]:=Plot[f[n,x],{x,-2Pi,2Pi}]
```

You can type `fgraph[3]`, `fgraph[10]` or `fgraph[50]` to see how closely the partial sums approximate f. For example,

`fgraph[3]`

yields

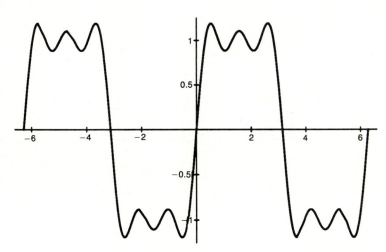

and

`fgraph[10]`

yields

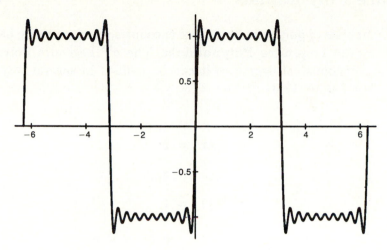

You can even type

```
Plot[{f[3,x],f[10,x]},{x,-2Pi,2Pi}]
```

to plot graphs simultaneously!

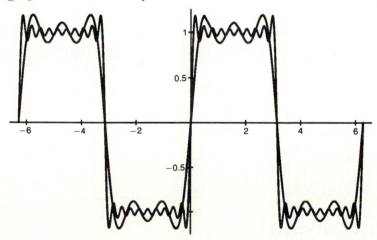

Legendre Polynomials

An important class of polynomials that we encounter when we study Laplace's Equation are the **Legendre Polynomials**. The nth Legendre Polynomial $P_n(x)$ is a polynomial of degree n; it can be defined in several ways. For example, if we require that

$$P_0(x) = 1,$$

$$P_1(x) = x,$$

$$P_n(1) = 1,$$

$$\int_{-1}^{1} P_n(x)P_m(x)\,dx = 0 \text{ for } m \neq n,$$

then the $P_n(x)$'s are determined uniquely. However, there are more explicit ways to find the Legendre Polynomials. There is a recursion relation to get P_n. Temporarily, we shall denote the Legendre Polynomial $P_n(x)$ in **Mathematica** as LP[n,x]. In **Mathematica** we type in the commands

```
LP[0,x_]:=1
LP[1,x_]:=x
LP[n_,x_]:=(1/n)((2n-1)x LP[n-1,x]-(n-1)LP[n-2,x])
```

We get a function LP of the two variables n and x. For example if we type LP[5,x] and simplify the resulting expression using **Simplify** we get

```
                2         4
        x (15 - 70 x  + 63 x )
Out[5]= ----------------------
                  8
```

To graph LP[5,x] from -1 to 1 we type

```
Plot[LP[5,x],{x,-1,1}];
```

(Notice that we type a semicolon at the end of a Plot command to keep

Mathematica from printing out "Graphics" below the plot.) The result is

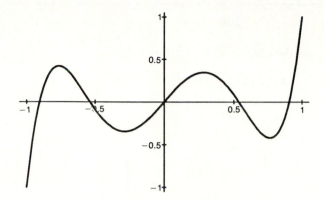

Next, we use **Mathematica** to compute the integral

$$\int_{-1}^{1} LP[n,x]^2 dx.$$

The most straightforward command, namely

`Integrate[LP[n,x]^2,{x,-1,1}]`

doesn't work because "`LP[n,x]`" is too general for **Mathematica**. However,

`Integrate[LP[5,x]^2,{x,-1,1}]`

yields

```
             2
Out[11]=  --
            11
```

Integrating `LP[n,x]^2` for a few other values of n leads us to the heuristic conclusion that

$$\int_{-1}^{1} LP[n,x]^2 dx = \frac{2}{2n+1}$$

for general n. This is the correct answer. [See eq. (4.2.16), page 227.]

Actually **Mathematica** has a built-in function for the nth Legendre Polynomial, namely `LegendreP[n,x]`. Computations using `LegendreP[n,x]` are much faster than those using `LP[n,x]`.

Exercise. Use **Mathematica** to verify that `LegendreP[10,x]` and `LP[10,x]` are the same function of x:

$$
\frac{-63 + 3465\ x^2 - 30030\ x^4 + 90090\ x^6 - 109395\ x^8 + 46189\ x^{10}}{256}
$$

This function may be plotted using the command

```
Plot[LegendreP[10,x],{x,-1,1}]
```

The graph is

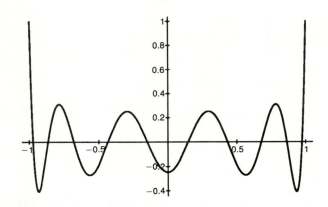

The **Legendre Equation** is

$$(1 - x^2)y'' - 2xy' + n(n+1)y = 0.$$

Mathematica can be used to verify that `Legendre[20,x]` is indeed a solution of the Legendre Equation for $n = 20$. For this we will need to use the differentiation operator **D**. To find the exact syntax type **?D**. Next define a function `y[x]` by

```
y[x_]:=LegendreP[20,x]
```

The first and second derivatives of `y[x]` are `D[y[x],x]` and `D[y[x],{x,2}]`. Then we compute

```
(1-x^2)D[y[x],{x,2}]-2 x D[y[x],x] + 20 21 y[x]
```

The result is complicated, but it reduces to 0 when we apply the command `Simplify`.

The Heat Equation

We can use **Mathematica** to do a 3-dimensional plot of the function $u(z, t)$ that satisfies the heat equation

$$u_t = K u_{zz} \qquad z > 0, \quad -\infty < t < \infty, \quad |u(t, z)| \leq M$$

with the boundary condition

$$u(0, t) = \cos \frac{2\pi t}{T}.$$

According to Example 2.1.3 the solution is

$$u(z; t) = e^{-c_1 z} \cos \left(\frac{2\pi t}{T} - z \sqrt{\frac{\pi}{KT}} \right)$$

where $c_1 = \sqrt{\dfrac{\pi}{KT}}$. This function can be defined in **Mathematica** using the command

```
u[z_,t_,K_,T_]:=E^(-z Sqrt[Pi/(K T)])*
        Cos[2Pi t/T-z Sqrt[Pi/(K T)]]
```

In order to be able to graph a specific example we assume that $T = K = 2$ and $0 \leq t \leq 2, 0 \leq z \leq 5$.

Then typing

```
Plot3D[u[z,t,2,2],{t,0,5},{z,0,2},PlotPoints->40,PlotRange->{-1,1}]
```

yields the actual plot:

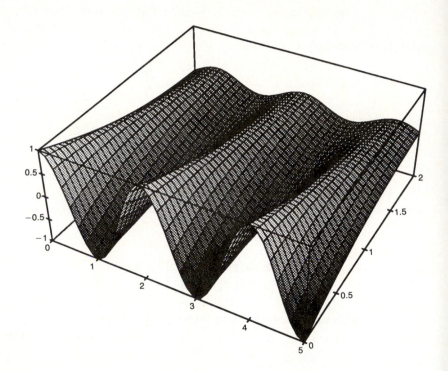

At the front of this graph we see the change of seasons at the surface of the earth, and at the back of the graph we see the change of seasons at a depth of 2 feet.

Graphing Bessel Functions

The **Mathematica** command for the Bessel function $J_n(x)$ is `BesselJ[n,x]`. To obtain the power series expansion for $J_0(x)$ at the origin we can use, for example,

```
Series[BesselJ[0,x],{x,0,7}]
```

The result is

$$\text{Out[2]} = 1 - \frac{x^2}{4} + \frac{x^4}{64} - \frac{x^6}{2304} + \frac{x^8}{147456} + O[x]^9$$

Bessel functions can be plotted in the same way that we plotted trigonometric functions and Legendre polynomials. In particular, `??Plot` lists all of the options associated with `Plot` together with their default values. Changing the default value of 25 of the option `PlotPoints` is useful for getting accurate graphs; for a complicated function such as `BesselJ[5,x]` the number 25 should be increased to 40. Thus, for example, $J_5(x)$ is nicely plotted with the **Mathematica** command

```
Plot[BesselJ[5,x],{x,-50,50},PlotPoints->40] .
```

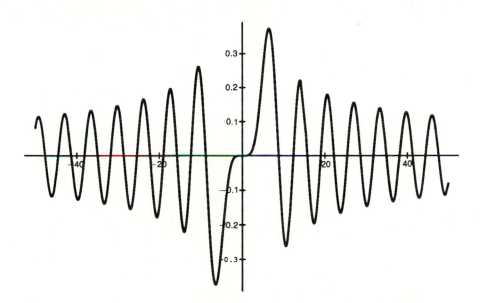

The graphing facilities of **Mathematica** can be used very effectively to find the zeros[4] of Bessel functions. Let us find the first zero of J_1, that is,

---

[4]**Mathematica** has a built in function for finding zeros called `FindRoot`. But at times it gives strange results. The graphing procedure described here has the advantage that at each stage one has a graph to see what is going on.

the first number x_{11} such that $x_{11} > 0$ and $J_1(x_{11}) = 0$. To do this we draw successive graphs of J_1 that narrow in on x_{11}. We shall find x_{11} correct to 4 decimal places.

First we define a plot-valued function `jj` by

```
jj[a_,b_]:=Plot[BesselJ[1,x],{x,a,b}]
```

Then `jj[0,5]` yields

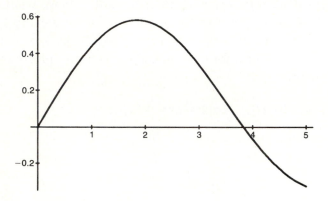

Now we have a good idea that x_{11} lies somewhere between 3.5 and 4. So next we type `jj[3.5,4]` and we get

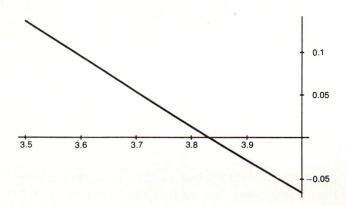

Consequently x_{11} lies between 3.8 and 3.85. So for the next approximation we type `jj[3.8,3.85]`. Continuing is this way we find that x_{11} is slightly

more than 3.8317.

It is possible to see just how well the function

$$\left(\frac{2}{\pi x}\right)^{\frac{1}{2}} \cos\left(x - \frac{\pi}{4}\right)$$

approximates $J_0(x)$ for large x. The **Plot** command can be used to graph two functions simultaneously. When we type

```
Plot[{BesselJ[0,x],(2/Pi x))^(1/2) Cos[x-Pi/4]},{x,.1,50}]
```

we get

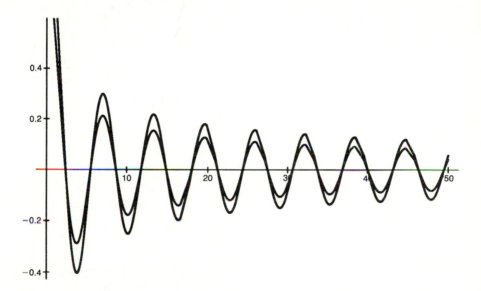

There is a Gibbs phenomenon for Bessel functions similar to that of trigonometric functions. To see it let us graph some of the partial sums of the Fourier-Bessel expansion of the function defined by

$$f(x) = \begin{cases} 1 & \text{for } 0 \le x \le \frac{1}{2}, \\ 0 & \text{for } \frac{1}{2} < x \le 1. \end{cases}$$

The Fourier-Bessel expansion of $f(x)$ is computed in Example 3.2.7:

$$f(x) = \sum_{n=1}^{\infty} \frac{J_1(\frac{1}{2}x_n)}{x_n J_1(x_n)^2} J_0(xx_n),$$

where x_n is the nth positive zero of the Bessel function $J_0(x)$. Working with Bessel functions is more complicated than working with trigonometric functions because the x_n's and the zeros of the other Bessel functions do not occur at regular intervals like $2\pi, 4\pi, 6\pi, \ldots$ The x_n's either have to be computed explicitly or looked up in a table.

We define a list X in **Mathematica** whose elements are the first 20 positive zeros of $J_0(x)$:

```
X={2.405,5.520,8.654,11.792,14.931,
    18.071,21.212,24.352,27.493,30.635,
    33.776,36.972,40.058,43.200,46.483,
    49.482,52.624,55.766,58.907,62.049}
```

Then X[[k]] is the kth element of X, that is, x_k. Next we define a function of two variables f[x,n] that represents the nth partial sum for the Fourier-Bessel expansion of $f(x)$:

```
f[x_,n_]:=Sum[
    BesselJ[1,Release[X[[k]]]/2] BesselJ[0,x Release[X[[k]]]]
    /(Release[X[[k]]] BesselJ[1,Release[X[[k]]]]^2),{k,n}]
```

Typing

```
Plot[f[x,10],{x,-1,2},PlotPoints->40]
```

and

```
Plot[f[x,10],{x,-1,2},PlotPoints->40]
```

yields the following graphs:

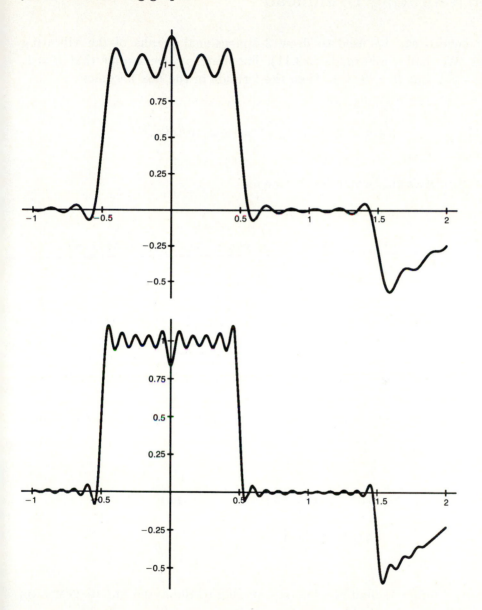

The Gibbs phenomenon is evident from these graphs, but with a new twist: the graph of the first 10 terms goes up at 0, but the graph of the first 20 terms goes down!

The Vibrating Drumhead

Mathematica can be used to draw 3-dimensional graphs of the vibrating drum. We shall use Formula (3.3.11). For simplicity we assume that $R = 1$, $A = \widetilde{A} = 1$ and $B = \widetilde{B} = 0$. Then the formula in the box becomes

$$u(\rho, \varphi; t) = J_m(\rho x_n^{(m)}) \cos(m\varphi) \cos(ctx_n^{(m)}).$$

In Mathematica this would be written as

```
u[m_,x_,rho_,phi_,t_]:=BesselJ[m,rho x]Cos[m phi]Cos[t x]
```

To graph u we need the command `CylindricalPlot3D`. This command is not a part of the standard **Mathematica** package; however, it is described on page 41 of **Programming in Mathematica**. The command `CylindricalPlot3D` is a part of the package `NewParametricPlot3D`. This package is described at the end of **Programming in Mathematica**; it can be obtained on diskettes from Wolfram Research.

Assuming the existence of `CylindricalPlot3D`, we define a plot-valued function uu via

```
uu[m_,x_,t_]:=CylindricalPlot3D[u[m,x,rho,phi,t],
            {rho,0,1},{phi,0,2Pi,Pi/15},
            Boxed->False]
```

Then the 3-dimensional plots corresponding to the nodal line diagrams on page 198 are generated by the commands

```
uu[0,2.40482,0]        uu[1,3.83171,0]        uu[2,5.13562,0]
uu[0,5.52007,0]        uu[1,7.01559,0]        uu[2,8.41724,0]
```

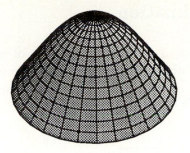

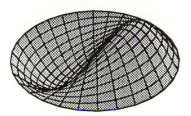

The Square Drumhead and Animation

It is even simpler to use **Mathematica** to draw 3-dimensional graphs of a rectangular drum. Furthermore, **Mathematica** does the plotting much more quickly, because `Plot3D` can be used instead of `CylindricalPlot3D`. The relevant function to plot is given in Formula (2.5.6); we write it as

$$v_{mn}(x, y; t) = \sin mx \sin ny \cos(\sqrt{m^2 + n^2}\, t),$$

where for simplicity we have taken $L_1 = L_2 = \pi$, $A = 1$ and $B = 0$. In Mathematica this function would be written as

```
v[m_,n_,x_,y_,t_]:=Sin[m x]Sin[n y]Cos[Sqrt[m^2+n^2]t]
```

No packages are necessary to plot v_{mn}; all we need to get 3-dimensional graphs of the diagrams of Figure 2.5.2 is the plot-valued function

```
vv[m_,n_,t_]:=Plot3D[Sin[m x]Sin[n y]Cos[Sqrt[m^2+n^2]t],
              {x,0,Pi},{y,0,Pi},PlotPoints->40]
```

Similarly, the plot-valued function

```
vV[m_,n_,t_]:=Plot3D[(Sin[m x]Sin[n y]-Sin[n x]Sin[m y])*
              Cos[Sqrt[m^2+n^2]t],{x,0,Pi},{y,0,Pi},PlotPoints->4(
```

yields the 3-dimensional versions of the diagrams on Figure 2.5.3. Here are some examples. They are obtained by typing the commands

```
vv[1,2,0]        vv[2,2,0]
vv[1,4,0]        vv[2,5,0]
vV[1,2,0]        vV[1,3,0]
```

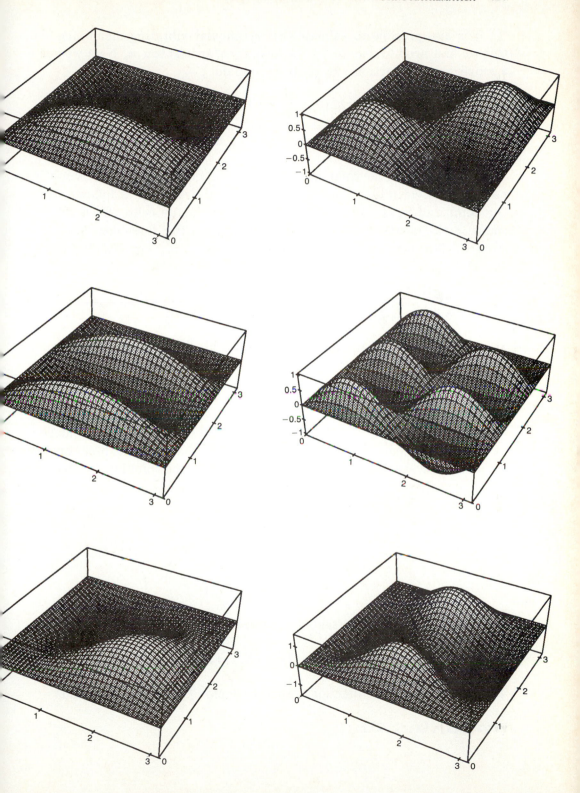

It is also possible to animate the rectangular vibrating membrane. For this we first need to read in the file **Graphics/Animation.m**. Next we define a movie-valued function **vvv** via the command

```
vvv[m_,n_]:=MoviePlot3D[Sin[m x]Sin[n y]Cos[Sqrt[m^2+n^2]t],
           {x,0,Pi},{y,0,Pi},
           {t,0,2Pi/Sqrt[m^2+n^2],.05},PlotRange->{-1,1}]
```

Then, for example, the command **vvv[5,2]** will create a movie of a vibrating membrane divided into 10 sections. The actual process is as follows. What the command **vvv[5,2]** does is create a great many cells. Each cell is a frame in the movie of the vibrating membrane. After these cells have been created, click in the cell bracket that surrounds all of them. The frames can then be put into motion by typing "command-y".

The Wave Equation and Bifurcation

The solution $y(s;t)$ to the wave equation with initial data f according to (2.4.8) can be written as

```
y[s_,t_,f_]:=(1/2)(f[s+t]+f[s-t])
```

Here f can be any piecewise smooth function; then $y(s;t)$ can be visualized as a surface in space. To understand the wave equation let us first consider a triangular wave. We represent it analytically by the function f defined by

$$f(x) = \begin{cases} 0 & \text{if } x < -1, \\ x+1 & \text{if } -1 \le x < 0, \\ 1-x & \text{if } 0 \le x < 1, \\ 0 & \text{if } x \ge 1 \end{cases}$$

This function can be represented in **Mathematica** by

```
f[x_]:=If[x<-1,0,If[x<0,x+1,If[x<1,1-x,0]]]
```

To see that this is the function we want we plot f with the command

```
Plot[f[x],{x,-2,2}]
```

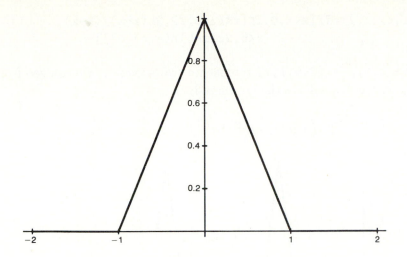

The physical interpretation of $y(s;t)$ is the the sum of two traveling waves. This can be seen very easily using Plot3D.

```
Plot3D[y[s,t,f],{s,-3,3},{t,0,3}]
```

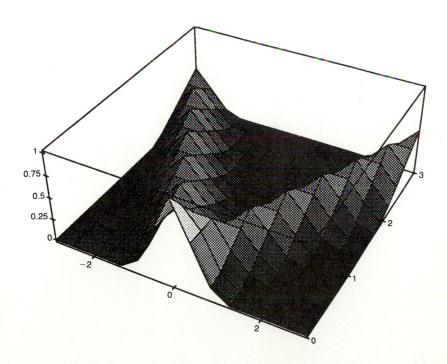

Next let us wee what happens when there are two waves. First we define a function which we can use to build waves.

```
v[a_,b_,c_,x_]:=If[x<a,0,If[x<(a+b)/2,2c(x-a)/(b-a),
                     If[x<b,2c(x-b)/(a-b),0]]]
```

Then `v[1,2,1,x]+v[3,4,1,x]` represents two waves, as we can see from its plot, which is obtained via the command:

```
Plot[v[1,2,1,x]+v[3,4,1,x],{x,0,6}]
```

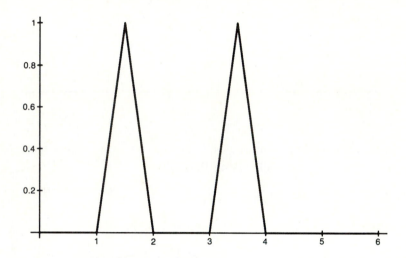

We define *g* by

```
g[x_]:=v[1,2,1,x]+v[4,5,1,x]
```

and use it as initial data for the wave equation. Not only does each of the two waves bifurcate, but also we get interference in the solution to the wave equation:

`Plot3D[y[s,t,g],{s,-1,7},{t,0,4},PlotPoints->40,PlotRange->{0,1}]`

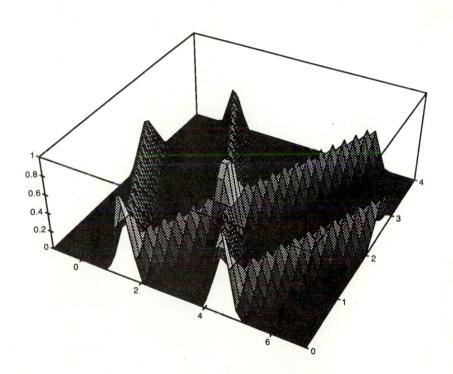

BIBLIOGRAPHY

Abramowitz, M., and I. Stegun: *Handbook of Mathematical Functions,* Dover, New York, 1965.

Bateman, H.: *Partial Differential Equations of Mathematical Physics,* Cambridge University Press, New York, 1959.

Berg, P., and J. MacGregor: *Elementary Partial Differential Equations,* Holden-Day, San Francisco, 1966.

Birkhoff, G., and G. C. Rota: *Ordinary Differential Equations,* Ginn, Lexington, Mass., 1962.

Churchill, R. V., and J. W. Brown: *Fourier Series and Boundary Value Problems,* 4th ed., McGraw-Hill, New York, 1987.

Courant, R., and D. Hilbert: *Methods of Mathematical Physics,* vol. 1, Wiley-Interscience, New York, 1953.

DuChateau, P., and D. Zachman: *Applied Partial Differential Equations,* Harper & Row, New York, 1989.

Dym, H., and H. P. McKean, Jr.: *Fourier Series and Integrals,* Academic Press, New York, 1972.

Epstein, B.: *Partial Differential Equations,* McGraw-Hill, New York, 1962.

Garabedian, P.: *Partial Differential Equations,* Wiley, New York, 1964.

Gelfand, I., and S. Fomin: *Calculus of Variations,* Prentice-Hall, Englewood Cliffs, N.J., 1963.

Guenther, R., and J. W. Lee: *Partial Differential Equations of Mathematical Physics and Integral Equations,* Prentice-Hall, Englewood Cliffs, N.J., 1988.

Haberman, R.: *Elementary Applied Partial Differential Equations,* 2d ed., Prentice-Hall, Englewood Cliffs, N.J., 1983.

Hildebrand, F.: *Advanced Calculus for Applications,* Prentice-Hall, Englewood Cliffs, N.J., 1976.

Kreith, F.: *Principles of Heat Transfer,* Harper & Row, New York, 1973.

Lebedev, N. N., I. P. Skalskaya, and Y. S. Ufyland: *Solved Problems in Applied Mathematics,* Dover, New York, 1965.

Marion, J.: *Classical Dynamics of Particles and Systems,* Academic Press, New York, 1970.

Petrovsky, I.: *Partial Differential Equations,* Wiley-Interscience, New York, 1954.

Sommerfield, A.: *Partial Differential Equations of Physics,* Academic Press, New York, 1949.

Stakgold, I.: *Boundary Value Problems of Mathematical Physics,* Macmillan, New York, 1967, two volumes.

Tikhonov A., and A. A. Samarski: *Equations of Mathematical Physics,* Pergamon Press, New York, 1963.

Weinberger, H., *A. First Course in Partial Differential Equations,* Ginn, Lexington, Mass., 1965.

Whittaker, E., and G. N. Watson: *Modern Analysis,* Cambridge University Press, New York, 1950.

Zachmanoglou, E., and D. Thoe: *Introduction to Partial Differential Equations with Applications,* Dover, New York, 1986.

Zygmund, A.: *Trigonometric Series,* Cambridge University Press, New York, 1959.

ANSWERS TO SELECTED EXERCISES

CHAPTER 0

Section 0.1

1. Hyperbolic
2. Hyperbolic for $x < 0$, parabolic for $x = 0$, elliptic for $x > 0$.
3. Hyperbolic 4. Elliptic
5. Hyperbolic for $y(1 - x^2) < 0$, elliptic for $y(1 - x^2) > 0$, parabolic for $y(1 - x^2) = 0$
6. $u_{xx} = -k^2 \sin kx \sinh ky$, $u_{yy} = +k^2 \sin kx \sinh ky$
7. $u_{xx} = -k^2(\sin kx)e^{-k^2 t}$, $u_t = -k^2(\sin kx)e^{-k^2 t}$
8. $u_x = f'(x)$, $u_{xy} = (\partial/\partial y)f'(x) = 0$ 9. Use the chain rule.
10. $u_t = -e^{-x} \sin (t - x)$, $u_{xx} = 2e^{-x} \sin (t - x)$

Section 0.2

1. (a) $f_1''(x) - \lambda f(x) = 0$ (b) $f_1''(x) + 2f_1'(x) - \lambda f_1(x) = 0$
     $f_2''(y) - 2\lambda f_2(y) = 0$ $f_2''(y) + \lambda f_2(y) = 0$
   (c) $x^2 f_1''(x) - \lambda f_1(x) = 0$ (d) $f_1''(x) + f_1'(x) - \lambda f_1(x) = 0$
     $2y f_2'(y) - \lambda f_2(y) = 0$ $f_2'(y) + (\lambda - 1)f_2(y) = 0$

3. (b) is a solution of Laplace's equation.
4. $\lambda < 0$: $u(x, y) = (A_1 e^{x(1-1+\sqrt{1-\lambda})} + A_2 e^{x(-1-\sqrt{1-\lambda})})(A_3 \cos y \sqrt{-\lambda} + A_4 \sin y \sqrt{-\lambda})$
   $\lambda = 0$: $u(x, y) = (A_1 + A_2 e^{-2x})(A_3 + A_4 y)$
   $0 > \lambda > 1$: $u(x, y) = (A_1 e^{x(-1-\sqrt{1-\lambda})} + A_2 e^{x(-1-\sqrt{1-\lambda})})(A_3 e^{y\sqrt{\lambda}} + A_4 e^{-y\sqrt{\lambda}})$
   $\lambda = 1$: $u(x, y) = (A_1 e^{-x} + A_2 xe^{-x})(A_3 e^y + A_4 e^{-y})$
   $\lambda > 1$: $u(x, y) = (A_1 e^{-x} \cos x \sqrt{\lambda - 1} + A_2 e^{-x} \sin x \sqrt{\lambda - 1})(A_3 e^{y\sqrt{\lambda}} + A_4 e^{-y\sqrt{\lambda}})$

6. $\lambda < 0$: $u(x, y) = [A_1 \cos (\sqrt{-\lambda} \ln |x|) + A_2 \sin (\sqrt{-\lambda} \ln |x|)](A_3 e^{y\sqrt{-\lambda}} + A_4 e^{-y\sqrt{-\lambda}})$
   $\lambda = 0$: $u(x, y) = (A_1 + A_2 \log |x|)(A_3 + A_4 y)$
   $\lambda > 0$: $u(x, y) = (A_1 |x|^{\sqrt{\lambda}} + A_2 |x|^{-\sqrt{\lambda}})(A_3 \cos y \sqrt{\lambda} + A_4 \sin y \sqrt{\lambda})$

8. $\lambda < 0$: $u(x, y) = (A_1 e^{x\sqrt{-\lambda}} + A_2 e^{-x\sqrt{-\lambda}})(1/|y|^{1+\lambda})$

   $\lambda = 0$: $u(x, y) = (A_1 + A_2 x)(1/y)$

   $\lambda > 0$: $u(x, y) = (A_1 \cos x \sqrt{\lambda} + A_2 \sin x \sqrt{\lambda})(1/|y|^{1+\lambda})$

10. $u_n(x, y) = A_n \cos (n\pi x/L) \sinh (n\pi y/L), \quad n = 1, 2, 3, \ldots$

11. $u_n(x, y) = A_n \sin (n\pi x/L) e^{-n\pi y/L}, \quad n = 1, 2, \ldots$

12. $u_n(x; t) = A_n \sin (n\pi x/L) \exp [-(n\pi/L)^2 t], \quad n = 1, 2, \ldots$

13. $u_n(x; t) = A_n \sin \dfrac{(n - \frac{1}{2})\pi x}{L} \exp \left[-\left(\dfrac{(n - \frac{1}{2})\pi}{L} \right)^2 t \right], \quad n = 1, 2, \ldots$

14. $u_n(x; t) = A_n \cos \dfrac{n\pi x}{L} \exp \left[-\left(\dfrac{n\pi}{L} \right)^2 t \right], \quad n = 0, 1, 2, \ldots$

Section 0.3

1. (a) $\langle \varphi_1, \varphi_2 \rangle = \frac{1}{2}$ (b) $\langle \varphi_1, \varphi_3 \rangle = \frac{1}{3}$ (c) $\|\varphi_1 - \varphi_2\|^2 = \frac{1}{3}$ (d) $|2\varphi_1 + 3\varphi_2\|^2 = 13$

2. $\langle \varphi_1, \varphi_3 \rangle = 0, \quad \langle \varphi_1, \varphi_4 \rangle = 0, \quad \langle \varphi_2, \varphi_3 \rangle = 0, \quad \langle \varphi_3, \varphi_4 \rangle = 0$ All others are non zero. Therefore (φ_1, φ_3) are orthogonal, (φ_1, φ_4) are orthogonal, (φ_2, φ_3) are orthogonal, and (φ_3, φ_4) are orthogonal.

3. $\psi_1(x) = 1, \psi_2(x) = x, \psi_3(x) = x^2 - \frac{1}{3}$

5. $2/\pi; d_{min}^2 = \frac{1}{2} - 4/\pi^2 = 0.0947, d_{min} = 0.3078$

6. $\frac{1}{2} + \frac{1}{2} \cos 2x$

7. (b) $x/(2|x|); d_{min}^2 = \frac{1}{6}, d_{min} = 0.4082$

10. (d) $\cos \theta = \sqrt{3}/2, \theta = \pi/6$

11. $1/\sqrt{2}, x\sqrt{\frac{3}{2}} (x^2 - \frac{1}{3}), \sqrt{\frac{45}{8}}$

Section 0.4

1. (a) $(\varphi')' + (\lambda/x^2)\varphi = 0, \rho(x) = (1/x^2)$ (b) $[(\sin x)\varphi']' + \lambda(\sin x)\varphi = 0, \rho(x) = \sin x$

   (c) In standard form; $\rho(x) = 1$ (d) $(e^{-x^2/2}\varphi')' + \lambda e^{-x^2/2}\varphi = 0, \rho(x) = e^{-x^2/2}$

   (e) $(e^{-x}\varphi')' + \lambda(e^{-x}\varphi) = 0, \rho(x) = e^{-x}$ (f) $(x^2\varphi')' + \lambda(x^2\varphi) = 0, \rho(x) = x^2$

2. $\varphi_n(x) = A_n \sin \dfrac{n\pi \log x}{\log 2}, \quad n = 1, 2, \ldots$

   $\lambda_n = \left(\dfrac{n\pi}{\log 2} \right)^2, \quad \log = \text{natural logarithm}$

3. (a), (b), (c), (f) satisfy the orthogonality conditions (0.4.2).

4. (a) $\lambda_n = (n\pi/L)^2, \quad \varphi_n(x) = B_n \cos (n\pi x/L), \quad n = 0, 1, 2, \ldots$

   (b) $\lambda_n = [(2n - 1)\pi/(2L)]^2, \quad \varphi_n(x) = A_n \sin [(2n - 1)\pi x/(2L)], \quad n = 1, 2, \ldots$

   (c) $\lambda_n = (2n\pi/L)^2, \quad \varphi_n(x) = A_n \sin (2n\pi x/L) + B_n \cos (2n\pi x/L), \quad n = 0, 1, 2, \ldots$

   (d) $\lambda_n = [(2n - 1)\pi/L]^2, \quad \varphi_n(x) = A_n \sin [(2n - 1)\pi x/L] + B_n \cos [(2n - 1)\pi x/L], \quad n = 1, 2, \ldots$

   (e) $\sqrt{\lambda_n} = -\tan L \sqrt{\lambda_n}, \quad \varphi_n(x) = A_n \sin x\sqrt{\lambda_n}, \quad$ determine λ_n graphically.

   (f) $\sqrt{\lambda_n} = \cot L \sqrt{\lambda_n}, \quad \varphi_n(x) = A_n(\sin x \sqrt{\lambda_n} + \sqrt{\lambda_n} \cos x \sqrt{\lambda_n}), \quad$ determine λ_n graphically.

5. (a), (b), (c), (d), (e), (f) satisfy the positivity criterion (0.4.3).

6. (a), (c), (e), and (f)

7. The positivity criterion is not satisfied.

Section 0.5

1. (*a*) diverges;   (*b*) converges;   (*c*) diverges;   (*d*) converges.

2. (*a*) $\sum_{n=1}^{\infty} (-1)^{n+1} \dfrac{x^{2n-1}}{(2n-1)!}$ (*b*) $2 \sum_{n=1}^{\infty} \dfrac{x^{2n-1}}{2n-1}$

   (*c*) $1 + \sum_{n=1}^{\infty} \dfrac{x^{2n}}{(2n)!}$ (*d*) $\sum_{n=1}^{\infty} \dfrac{x^{2n-1}}{(n-1)!}$ $(0! = 1)$

3. (*a*) $f'(x) = \dfrac{1}{x^2} e^{-1/x}$ and $f''(x) = \left(\dfrac{1}{x^4} - \dfrac{2}{x^3}\right) e^{-1/x}$ for $x > 0$

   (*d*) The Taylor series converges for all x, and its sum is zero. This equals $f(x) - f(0)$ only when $x \le 0$.

5. (*a*), (*c*), (*d*) can be differentiated term by term according to Proposition 0.1.3.

6. $\sum_{n=1}^{\infty} n^3 x^n = \dfrac{x(1 + 4x + x^2)}{(1-x)^4},$ $\sum_{n=1}^{\infty} n^4 x^n = \dfrac{x(1 + 11x + 11x^2 + x^3)}{(1-x)^5},$ $-1 < x < 1$

9. (*a*) diverges;   (*b*) converges;   (*c*) converges;   (*d*) converges.

10. (*a*) Can choose any $T > 0$, $M = 2$ (*b*) Can choose any $T > 0$, $M = 2^{10}(10!)$
    (*c*) Can choose $T = 1$, $M = 2$ (*d*) Can choose any $T > 0$, $M = 2$

12. (*a*) Use L'Hospital's rule.

    (*b*) $\int_0^1 n^2 x e^{-nx} dx = \int_0^n y e^{-y} dy = -(1+y)e^{-y}\Big|_0^n \to 1$ when $n \to \infty$

    (*c*) Choose $u_n(x) = n^2 x e^{-nx} - (n-1)^2 e^{-(n-1)x}$ for $n = 1, 2, \ldots$.

15. Not necessarily; for example, let $f_1(t) = \sin t$, $f_2(t) = \cos t$, $g(t) = 1$.

Section 0.6

1. $y(t) = c_1 \cos 2t + c_2 2t$ 2. $y(t) = c_1 e^{-2t} + c_2 t e^{-2t}$
3. $y(t) = c_1 e^{3t} + c_2 e^{-5t}$ 4. $y(t) = c_1 + c_2 e^{-3t}$
5. $y(t) = c_1 e^{-t/3} + c_2 e^{2t}$ 6. $y(t) = \sin 2t$
7. $y(t) = e^{-2t} + 2t e^{-2t}$ 8. $y(t) = \frac{13}{8} e^{3t} + \frac{3}{8} e^{-5t}$
9. $y(t) = \frac{4}{3} - \frac{4}{3} e^{3-3t}$ 10. $y(t) = \frac{3}{7} e^{-t/3} + \frac{4}{7} e^{2t}$
11. $y(t) = e^{-t^2}(t + c)$ 12. $y(t) = 1 + c/t$
13. $y(t) = e^{2t}/5 + C e^{-3t}$ 14. $y(t) = t^2/6 + C/t^4$

15. $y(t) = 1 + C \csc t$ 16. $y(t) = e^{-2t}\left(-\dfrac{1}{3}\cos \sqrt{2}t + \dfrac{2\sqrt{2}}{3}\sin \sqrt{2}t\right) + \dfrac{1}{3}$

17. $y(t) = \frac{1}{4} e^{2t} + \frac{1}{4} e^{-2t} - \frac{1}{2}$ 18. $y(t) = \frac{7}{8} - \frac{7}{8} e^{-4t} + \frac{1}{2} t$
19. $y(t) = 3 + 4t$ 20. $y(t) = -t + \ln t + 1$
21. $y_\infty = 3,\quad \tau = \frac{1}{2}$ 22. $y_\infty = 5,\quad \tau = 1$
23. $y_\infty = \frac{1}{2},\quad \tau = \frac{1}{2}$ 24. $y_\infty = 1, \tau = 1$

25. $y_\infty = 4,\quad \tau = 1$ 26. $y(t) = t + \sum_{n=1}^{\infty} \dfrac{(-4)^n \prod_{m=1}^{n} (3m-1)}{(3n+1)!} t^{3n+1}$

27. $y(t) = 1 + \sum_{n=1}^{\infty} \dfrac{(-4)^n \prod_{m=1}^{n} (3m-2)}{(3n)!} t^{3n}$

31. (a), (c), (d), (e) have regular singular points.

32. $r(r - 1) + 1 = 0$ $y(t) = t^{\frac{1 + \sqrt{5}}{2}}$

33. $r^2 - 1 = 0$ $y(t) = t$

34. $r^2 = 0$ $y(t) = 1 + \sum_{n=1}^{\infty} \frac{(-1)^n 3^n}{2^{2n}(n!)^2} t^{2n}$

35. $r(r - 1) + 1 = 0,\ y(t) = t^r \left\{ 1 + \sum_{n=1}^{\infty} \frac{(-3)^n t^{2n}}{\prod\limits_{j=1}^{n} [r(4j - 2) + 4j^2 - 2j + 2]} \right\},\ r = \frac{1 + \sqrt{5}}{2}$

CHAPTER 1

Section 1.1

1. $\dfrac{L^2}{3} + \sum\limits_{n=1}^{\infty} \dfrac{4L^2}{n^2 \pi^2} (-1)^n \cos \dfrac{n\pi x}{L}$

2. $2L^3 \sum\limits_{n=1}^{\infty} (-1)^n \left[\dfrac{6}{(n\pi)^3} - \dfrac{1}{n\pi} \right] \sin \dfrac{n\pi x}{L}$

3. $\dfrac{L^3}{4} + 2L^3 \sum\limits_{n=1}^{\infty} \left\{ \dfrac{3(-1)^n}{(n\pi)^2} + \dfrac{6[1 - (-1)^n]}{(n\pi)^4} \right\} \cos \dfrac{n\pi x}{L}$

4. $\dfrac{\sinh L}{L} \left[1 + 2 \sum\limits_{n=1}^{\infty} (-1)^n \dfrac{\cos (n\pi x/L) - (n\pi/L) \sin (n\pi x/L)}{1 + (n\pi/L)^2} \right]$

5. $\tfrac{1}{2} - \tfrac{1}{2} \cos 4x$

6. $\tfrac{1}{4} \cos 3x + \tfrac{3}{4} \cos x$

7. $\dfrac{1}{2} + \sum\limits_{n=1}^{\infty} \dfrac{1 - (-1)^n}{n\pi} \sin \dfrac{n\pi x}{L}$

8. $\dfrac{L}{4} + \sum\limits_{n=1}^{\infty} \left\{ \dfrac{L(-1)^{n+1}}{n\pi} \sin \dfrac{n\pi x}{L} - \dfrac{L}{(n\pi)^2} [1 - (-1)^n] \cos \dfrac{n\pi x}{L} \right\}$

9. $\dfrac{1}{\pi} + \dfrac{1}{2} \sin x - \dfrac{1}{\pi} \sum\limits_{n=2}^{\infty} \dfrac{\cos nx}{n^2 - 1} [1 - (-1)^n]$

10. $2 \dfrac{\sinh L}{L} \sum\limits_{n=1}^{\infty} (-1)^{n+1} \dfrac{n\pi/L \sin(n\pi x/L)}{1 + (n\pi/L)^2}$

14. $f_1(x) = \dfrac{f(x) - f(-x)}{2},\quad f_2(x) = \dfrac{f(x) + f(-x)}{2}$

15. (b), (c), (g), (h) are even;   (a), (e), (f) are odd;   (d) is neither.

16. (a) $\dfrac{2L}{\pi} \sum\limits_{n=1}^{\infty} \dfrac{(-1)^{n+1}}{n} \sin \dfrac{n\pi x}{L}$

(b) $2L^2 \sum\limits_{n=1}^{\infty} \left[\dfrac{(-1)^{n+1}}{n\pi} - \dfrac{2[1 - (-1)^n]}{(n\pi)^3} \right] \sin \dfrac{n\pi x}{L}$

(c) $\dfrac{2\pi}{L^2} \sum\limits_{n=1}^{\infty} n \left\{ \dfrac{1 - e^L(-1)^n}{1 + [L/(n\pi)]^2} \right\} \sin \dfrac{n\pi x}{L}$

(d) Same as Exercise 2.

17. (a) $\dfrac{L}{2} - \dfrac{2L}{\pi^2} \sum\limits_{n=1}^{\infty} \dfrac{1 - (-1)^n}{n^2} \cos \dfrac{n\pi x}{L}$

(b) Same as Exercise 1.

(c) $\dfrac{e^L - 1}{L} + \dfrac{2}{L} \sum\limits_{n=1}^{\infty} \dfrac{(-1)^n e^L - 1}{1 + (n\pi/L)^2} \cos \dfrac{n\pi x}{L}$

(d) Same as Exercise 3.

21. $1 = 1$; $\quad 1 = \dfrac{4}{\pi}\left(\sin x + \dfrac{1}{3} \sin 3x + \dfrac{1}{5} \sin 5x + \cdots \right)$;

$1 = \dfrac{4}{\pi}\left(\cos x - \dfrac{1}{3} \cos 3x + \dfrac{1}{5} \cos 5x - \cdots \right)$;

$1 = \dfrac{8}{\pi}\left(\dfrac{\sin 2x}{2} + \dfrac{\sin 6x}{6} + \dfrac{\sin 10x}{10} + \cdots \right), \quad 0 < x < \dfrac{\pi}{2}$

22. Yes, period is 2.   **23.** Yes, period is 2π.

24. Yes, period is π.   **25.** Not periodic.

26. Yes, period is 1.   **27.** Yes, period is π.

28. Yes, period is 2π.   **29.** Not periodic.

30. $-\dfrac{4L^2}{\pi^3} \sum\limits_{n=1}^{\infty} \dfrac{1 - (-1)^n}{n^3} \sin \dfrac{n\pi x}{L} = x^2 - Lx, \quad 0 \le x \le L$

31. $\dfrac{48L^4}{\pi^5} \sum\limits_{n=1}^{\infty} \dfrac{1 - (-1)^n}{n^5} \sin \dfrac{n\pi x}{L} = x^4 - 2Lx^3 + L^3x, \quad 0 \le x \le L$

Section 1.2

1. (b), (d) are piecewise smooth.   **9.** (a) 1   (b) $\frac{1}{2}$ (c) 1   (d) 0

10. $(1/\pi)(N + \frac{1}{2})$ **11.** $u = n\pi/(N + \frac{1}{2})$ for $n = \pm 1, \pm 2, \ldots$

12. For $N = 1$, $D_N'(u) = 0$ at $u = 0, \pm \pi$

     For $N = 2$, $D_N'(u) = 0$ at $u = 0, \pm \cos^{-1}(-\frac{1}{4}), \pm \pi$

16. (b) $\pi^2/12$ (c) $\pi^2/6$ (d) $\pi^2/8$

17. 0   **18.** $\cosh \pi = \frac{1}{2}(e^\pi + e^{-\pi})$ **22.** $n > 4$

23. $A_n = -\dfrac{1}{n\pi} \sum\limits_{-\pi < x_i \le \pi} (\sin nx_i)[f(x_i + 0) - f(x_i - 0)] + O\left(\dfrac{1}{n^2}\right), \quad n \to \infty$

$B_n = \dfrac{1}{n\pi} \sum\limits_{-\pi < x_i \le \pi} (\cos nx_i)[f(x_i + 0) - f(x_i - 0)] + O\left(\dfrac{1}{n^2}\right), \quad n \to \infty$

Section 1.3

2. $k = 2$, 1.42; $k = 3$, 1.67; $k = 4$, 1.49

11. $x^2 = \dfrac{4L^2}{\pi^2} \sum\limits_{n=1}^{\infty} \dfrac{(-1)^{n+1}}{n^2}\left(1 - \cos \dfrac{n\pi x}{L} \right) = \dfrac{L^2}{3} - \dfrac{4L^2}{\pi^2} \sum\limits_{n=1}^{\infty} \dfrac{(-1)^{n+1}}{n^2} \cos \dfrac{n\pi x}{L}, \quad 0 < x < L$

12. $\dfrac{x^3}{3} - \dfrac{L^2 x}{3} = \dfrac{4L^3}{\pi^3} \sum\limits_{n=1}^{\infty} \dfrac{(-1)^n}{n^3} \sin \dfrac{n\pi x}{L}, \quad 0 < x < L$

13. The series for x^2 and $x^3 - L^2x$ are uniformly convergent. The series for x is not uniformly convergent, since the sum of the series is discontinuous at $x = \pm L$.

Section 1.4

1. $\sigma_N^2 = \dfrac{2}{\pi^2} \sum\limits_{n=N+1}^{\infty} \dfrac{[(-1)^n - 1]^2}{n^2}$

2. $\sigma_N^2 = 8 \sum\limits_{n=N+1}^{\infty} \dfrac{1}{n^4}$

3. $\sigma_N^2 = 0$ for $N \geq 10$

4. $\pi^2/8 = 1 + \frac{1}{9} + \frac{1}{25} + \cdots$

5. $\pi^4/90 = 1 + \frac{1}{16} + \frac{1}{81} + \cdots$

8. (a) $\dfrac{4}{\pi} \sum\limits_{n=1}^{\infty} \dfrac{1 - (-1)^n}{n^3} \sin nx$ (b) $\dfrac{\pi^2}{6} - 2 \sum\limits_{n=1}^{\infty} \dfrac{1 + (-1)^n}{n^2} \cos nx$

(c) σ_N^2 (sine series) $= \dfrac{8}{\pi^2} \sum\limits_{n=N+1}^{\infty} \dfrac{[1 - (-1)^n]^2}{n^6} = O(N^{-5})$,

σ_N^2 (cosine series) $= 2 \sum\limits_{n=N+1}^{\infty} \dfrac{[1 + (-1)^n]^2}{n^4} = O(N^{-3})$

10. (a) $\dfrac{1}{a\pi} + \dfrac{2n}{\pi} \sum\limits_{n=1}^{\infty} \dfrac{(-1)^n}{a^2 - n^2} \cos nx$ (b) $\sigma_N^2 = O(N^{-3})$

16. $\dfrac{\sum\limits_{n=N+1}^{\infty} e^{-n}}{\displaystyle\int_N^{\infty} e^{-x}\, dx} = \dfrac{1}{e - 1}$

17. $\dfrac{P^2}{A} = 12\sqrt{3}$ **18.** $\dfrac{P^2}{A} = 16$ **19.** $\dfrac{P^2}{A} = 4n \tan \dfrac{\pi}{n}$

Section 1.5

3. $e^x = \sum\limits_{-\infty}^{\infty} (-1)^n \dfrac{L + in\pi}{L^2 + n^2\pi^2} (\sinh L) \exp\left(\dfrac{in\pi}{L} x\right)$, $-L < x < L$

4. $\dfrac{1}{1 - re^{ix}} = \sum\limits_{n=0}^{\infty} r^n e^{inx}$

CHAPTER 2

Section 2.1

1. $U(z) = T_1 + (z/L)(T_2 - T_1)$

2. $\Phi = -(k/L)(T_2 - T_1)$

3. $U(z) = \Phi_0(z - L) + T_0$

4. $U(z) = \{T_1(k + hz) + T_0[k + h(L - z)]\}/(2k + hL)$

5. $U(z) = T_3 + (T_2 - T_3)\dfrac{\sinh z \sqrt{\beta/K}}{\sinh L \sqrt{\beta/K}} + (T_1 - T_3)\dfrac{\sinh (L - z) \sqrt{\beta/K}}{\sinh L \sqrt{\beta/K}}$

6. $U(z) = T_1 + [r/(2K)](L^2 - z^2)$; $\Phi|_{z=L} = krL/K$

7. $U(z) = \dfrac{-rz^2}{2K} + \left(\dfrac{T_2 - T_1}{L} + \dfrac{rL}{2K}\right) z + T_1$

8. $U(z) = \dfrac{r_0 Lz}{6K}$, $0 < z < \dfrac{L}{3}$; $U(z) = \dfrac{r_0 z(L - z)}{2K} - \dfrac{r_0 L^2}{18K}$, $\dfrac{L}{3} < z < \dfrac{2L}{3}$;

$U(z) = \dfrac{r_0 L(L - z)}{6K}$, $\dfrac{2L}{3} < z < L$

9. 0.001792 cal/(s · cm²)

10. $u(z; t) = A_0 + A_1 \exp\left(-\sqrt{\dfrac{\pi}{KT_1}}\, z\right) \cos\left(\dfrac{2\pi t}{T_1} - \sqrt{\dfrac{\pi}{KT_1}}\, z\right)$

$$+ A_2 \exp\left(-\sqrt{\dfrac{\pi}{KT_2}}\, z\right) \cos\left(\dfrac{2\pi t}{T_2} - \sqrt{\dfrac{\pi}{KT_2}}\, z\right)$$

12. $u(z; t) = e^{\pm(1+i)\sqrt{\beta/(2Kz)}}\, e^{i\beta t}$

14. $u(z; t) = A_0\left(1 - \dfrac{z}{L}\right) + A_1 \,\mathrm{Re}\, \dfrac{e^{-c(z-L)}e^{i[\beta t - c(z-L)]} - e^{c(z-L)}e^{i[\beta t + c(z-L)]}}{e^{cL}e^{icL} - e^{-cL}e^{-icL}}, \quad \beta = \dfrac{2\pi}{T},$

$c = \sqrt{\dfrac{\beta}{2K}}$

15. $\sqrt{\pi KT} = 23.3$ cm

16. $u(z; t) = \dfrac{2}{\pi} \displaystyle\sum_{n=1}^{\infty} \exp\left(-z\sqrt{\dfrac{n\pi}{KT}}\right)\left[\dfrac{1 - (-1)^n}{n}\right] \sin\left(\dfrac{2n\pi t}{T} - z\sqrt{\dfrac{n\pi}{KT}}\right)$

17. $u(z; t) = -\dfrac{A_1}{2c}\, e^{-cz}[\cos(\beta t - cz) + \sin(\beta t - cz)], \quad c = \sqrt{\dfrac{\beta}{2K}}$

18. $u(z; t) = -A_1 e^{-cz}\, \dfrac{[(c + h)\cos(\beta t - cz) + c\sin(\beta t - cz)]}{c^2 + (c + h)^2} \quad c = \sqrt{\dfrac{\beta}{2K}}$

19. $u(z; t) = A_0 + A_1 \cos\left(\dfrac{2\pi t}{T} - z\sqrt{\dfrac{\pi}{KT}}\right)$

20. $U(z) = \dfrac{rL}{2Kh} + \dfrac{rz}{2K}(L - z) + \dfrac{T_1[1 + h(L - z)]}{2 + Lh} + \dfrac{T_2(1 + hz)}{2 + Lh}$

21. The constants must satisfy $K(\Phi_2 - \Phi_1) + rL = 0$.

Section 2.2

1. $u(z; t) = \dfrac{2L}{\pi} \displaystyle\sum_{n=1}^{\infty} \dfrac{(-1)^{n+1}}{n} \sin\dfrac{n\pi z}{L} \exp\left[-\left(\dfrac{n\pi}{L}\right)^2 Kt\right]$

2. $u(z; t) = \dfrac{2T}{\pi} \displaystyle\sum_{n=1}^{\infty} \dfrac{1}{n}\left(1 - \cos\dfrac{n\pi}{2}\right) \sin\dfrac{n\pi z}{L} \exp\left[-\left(\dfrac{n\pi}{L}\right)^2 Kt\right]$

3. $u(z; t) = 3 \sin\dfrac{\pi z}{2L} \exp\left[-\left(\dfrac{\pi}{2L}\right)^2 Kt\right] + 5 \sin\dfrac{3\pi z}{2L} \exp\left[-\left(\dfrac{3\pi}{2L}\right)^2 Kt\right]$

4. $u_n(z; t) = \cos\dfrac{n\pi z}{L} \exp\left[-\left(\dfrac{n\pi}{L}\right)^2 Kt\right], \, n = 0, 1, 2, \ldots$

5. $u(z; t) = \dfrac{L}{2} - \dfrac{4L}{\pi^2} \displaystyle\sum_{n=1}^{\infty} \dfrac{\cos[(2n - 1)\pi z/L]}{(2n - 1)^2} \exp\left[-\dfrac{(2n - 1)^2\pi^2 Kt}{L^2}\right]$

6. $u(z; t) = 3 + 4 \cos\dfrac{\pi z}{L} \exp\left[-\left(\dfrac{\pi}{L}\right)^2 Kt\right] + 7 \cos\dfrac{3\pi z}{L} \exp\left[-\left(\dfrac{3\pi}{L}\right)^2 Kt\right]$

9. $u(z; t) = \displaystyle\sum_{n=1}^{\infty} B_n \sin(z\sqrt{\lambda_n})e^{-\lambda_n Kt}, \quad \sqrt{\lambda_n} = -h\tan(L\sqrt{\lambda_n}),$

$B_n = \dfrac{4(1 - \cos L\sqrt{\lambda_n})}{2L\sqrt{\lambda_n} - \sin(2L\sqrt{\lambda_n})}$

10. $u_n(z; t) = [h \sin (z\sqrt{\lambda_n}) + \sqrt{\lambda_n} \cos (z\sqrt{\lambda_n})]e^{-\lambda_n Kt}$,
$\tan (L\sqrt{\lambda_n}) = 2h\sqrt{\lambda_n}/(\lambda_n - h^2)$

11. $u(z; t) = \sum\limits_{n=1}^{\infty} A_n u_n(z; t), \quad A_n = \dfrac{\displaystyle\int_0^L u_n(z; 0) \, dz}{\displaystyle\int_0^L u_n(z; 0)^2 \, dz}$

12. $\tau = L^2/(\pi^2 K)$ **13.** $\tau = L^2/(\pi^2 K)$ **14.** $\tau = 4L^2/(\pi^2 K)$

15. $\tau = \dfrac{1}{\lambda_1 K}$, $\lambda_1 = $ smallest root of the equation $\sqrt{\lambda} = -h \tan (L\sqrt{\lambda})$

16. $\tau = 1080$ s

18. $u_n(z; t) = \left(A_n \cos \dfrac{2n\pi z}{L} + B_n \sin \dfrac{2n\pi z}{L} \right) \exp \left[-\left(\dfrac{2n\pi}{L}\right)^2 Kt \right]; \quad n = 0, 1, 2, \ldots$

19. $u(z; t) = 50 + \dfrac{100}{\pi} \sum\limits_{n=1}^{\infty} \left[\dfrac{1 - (-1)^n}{n} \right] \sin \dfrac{2n\pi z}{L} \exp \left[-\left(\dfrac{2n\pi}{L}\right)^2 Kt \right]$

20. $\tau_{ring} = \dfrac{L^2}{4\pi^2 K} = \dfrac{1}{4} \tau_{slab}$

Section 2.3

1. $u(z; t) = T_1 + \Phi_2 z + \sum\limits_{n=1}^{\infty} A_n \sin \dfrac{(n - \frac{1}{2})\pi z}{L} \exp \left\{ -\left[\dfrac{(n - \frac{1}{2})\pi}{L} \right]^2 Kt \right\}$

$A_n = \dfrac{2(T_3 - T_1)}{(n - \frac{1}{2})\pi} - \dfrac{2L\Phi_2(-1)^{n+1}}{(n - \frac{1}{2})^2\pi^2}, \quad \tau = \dfrac{4L^2}{\pi^2 K}$

2. $u(z; t) = T_2 + \sum\limits_{n=1}^{\infty} A_n \cos (z\sqrt{\lambda_n})e^{-\lambda_n Kt}, \quad \sqrt{\lambda_n} = h \cot (L\sqrt{\lambda_n})$

$A_n = \dfrac{2(T_3 - T_2) \sin (L\sqrt{\lambda_n})}{L\sqrt{\lambda_n} + \sin (L\sqrt{\lambda_n}) \cos (L\sqrt{\lambda_n})}$

3. $u(z; t) = \dfrac{rz}{2K} (L - z) + T_1 + \dfrac{(T_2 - T_1)z}{L} + \sum\limits_{n=1}^{\infty} A_n \sin \dfrac{n\pi z}{L} \exp \left[-\left(\dfrac{n\pi}{L}\right)^2 Kt \right]$

$A_n = \dfrac{2(T_3 - T_1)}{\pi} \dfrac{1 - (-1)^n}{n} - \dfrac{2}{L} (T_2 - T_1) \dfrac{(-1)^{n+1}}{n} - \dfrac{2L^2 r}{K\pi^3} \dfrac{1 - (-1)^n}{n^3}$

4. $u(z; t) = 273 + \dfrac{768L}{\pi^2} \sum\limits_{n=1}^{\infty} \dfrac{1 - (-1)^n}{n^2} \cos \dfrac{n\pi z}{L} \exp \left[-\left(\dfrac{n\pi}{L}\right)^2 Kt \right]$

5. $u(z; t) = T_3 + \Phi(z - \frac{1}{2} L) + \dfrac{2L\Phi}{\pi^2} \sum\limits_{n=1}^{\infty} \dfrac{1 - (-1)^n}{n^2} \cos \dfrac{n\pi z}{L} \exp \left[-\left(\dfrac{n\pi}{L}\right)^2 Kt \right]$

6. $U(z) = (a/6)(L^2 z - z^3)$

$u(z; t) = U(z) + \dfrac{2aL^3}{\pi^3} \sum\limits_{n=1}^{\infty} \dfrac{(-1)^n}{n^3} \left(\sin \dfrac{n\pi z}{L} \right) e^{-(n\pi/L)^2 Kt}$

$\tau = \dfrac{L^2}{\pi^2 K}$

7. $u(z; t) = \dfrac{Azt}{L} + \dfrac{A}{6L} (L^2 z - z^3) + \dfrac{2AL^2}{\pi^3} \sum\limits_{n=1}^{\infty} \dfrac{(-1)^n}{n^3} \left(\sin \dfrac{n\pi z}{L} \right) e^{-(n\pi/L)^2 Kt}$

Section 2.4

1. $y(s; L/(2c)) = 0$ for $0 < s < L$

2. $B_{2n+1} = 0$ for $n = 0, 1, 2, \ldots$

3. $B_{3n+1} = 0,\ B_{3n+2} = 0$ for $n = 0, 1, 2, \ldots$

4. $E = \dfrac{L}{4} \displaystyle\sum_{n=1}^{\infty} \left[\rho \omega_n^2\, \tilde{B}_n^2 + T_0 \left(\dfrac{n\pi}{L} \right)^2 \tilde{A}_n^2 \right]$

6. $E = \dfrac{4T_0 L}{\pi^2} \left(1 + \dfrac{1}{9} + \dfrac{1}{25} + \cdots \right) = \dfrac{T_0\, L\pi}{2} = \displaystyle\sum_{n=1}^{\infty} E_n, \quad \displaystyle\sum_{n>1} E_n = 0.189\,E$

13. (a) $y(s; t) = \dfrac{2L}{\pi^2 c} \displaystyle\sum_{n=1}^{\infty} \dfrac{1 - (-1)^n}{n^2} \sin \dfrac{n\pi s}{L} \sin \dfrac{n\pi c t}{L}$

18. $u(x; t) = \dfrac{e^{-at}}{2} [g_1(x + ct) + g_1(x - ct)] + \dfrac{ae^{-at}}{2c} \displaystyle\int_{x-ct}^{x+ct} g_1(s)\, ds$

19. $u(x; t) = \dfrac{e^{-at}}{2c} \displaystyle\int_{x-ct}^{x+ct} g_2(s)\, ds$

20. $y(s; t) = \dfrac{2A}{\pi} \displaystyle\sum_{n=1}^{\infty} \dfrac{[(n\pi c/L)^2 - \omega^2] \cos \omega t + 2a\omega \sin \omega t}{[(n\pi c/L)^2 - \omega^2]^2 + (2a\omega)^2} \left(\sin \dfrac{n\pi s}{L} \right) \left[\dfrac{1 - (-1)^n}{n} \right]$

(Figure shows y vs s axes with labels $y(s; L/4c)$, $y(s; 3L/4c)$, and point $(L, 0)$.)

Section 2.5

1. $u(x, y; t) = \dfrac{4}{\pi^2} \displaystyle\sum_{m,n=1}^{\infty} \dfrac{\sin [(m - \frac{1}{2})(\pi x/L_1)]}{m - \frac{1}{2}} \dfrac{\sin [(n - \frac{1}{2})(\pi y/L_2)]}{n - \frac{1}{2}}\, e^{-\lambda_{mn} K t}$

$\lambda_{mn} = (m - \frac{1}{2})^2 (\pi/L_1)^2 + (n - \frac{1}{2})^2 (\pi/L_2)^2 \qquad \tau = [4/(\pi^2 K)][L_1^2 L_2^2/(L_1^2 + L_2^2)]$

2. $u(x, y; t) = \dfrac{2}{\pi} \displaystyle\sum_{m=1}^{\infty} \dfrac{1 - (-1)^m}{m} \sin \dfrac{m\pi x}{L_1} \exp \left[-\left(\dfrac{m\pi}{L_1} \right)^2 K t \right], \quad \tau = \dfrac{L_1^2}{\pi^2 K}$

3. $u_0(x, y) = Ay + B,\quad u_n(x, y) = \cos (n\pi x/L_1)[A \cosh n\pi y/L_1) + B \sinh (n\pi y/L_1)]$,

$n = 1, 2, \ldots$

4. $u(x, y) = \dfrac{yL_2}{2L_2} - \dfrac{2L_1}{\pi} \displaystyle\sum_{n=1}^{\infty} \dfrac{[1 - (-1)^n] \cos (n\pi x/L_1) \sinh (n\pi y/L_1)}{n^2 \sinh (n\pi L_2/L_1)}$

5. $u(x, y) = y/L_2$

6. $u(x, y) = yT_2/L_2 + (L_2 - y)T_2/L_2$

7. $u_{mn}(x, y, z) =$
$\sin [(m - \frac{1}{2})(\pi x/L)] \sin [(n - \frac{1}{2})(\pi y/L)](A \cosh \{\pi z/[L\sqrt{(m - \frac{1}{2})^2 + (n - \frac{1}{2})^2}]\}$
$+ B \sinh \{\pi z/[L\sqrt{(m - \frac{1}{2})^2 + (n - \frac{1}{2})^2}]\}), \qquad m, n = 1, 2, \ldots$

8. $u_{00}(x, y, z) = Az + B,\quad u_{mn}(x, y, z)$
$= \left(\cos \dfrac{m\pi x}{L} \right)\left(\cos \dfrac{n\pi y}{L} \right)\left[A \cosh \left(\dfrac{\pi z}{L} \sqrt{m^2 + n^2} \right) + B \sinh \left(\dfrac{\pi z}{L} \sqrt{m^2 + n^2} \right) \right]$,
$m, n = 0, 1, 2, \ldots$ with $m^2 + n^2 \neq 0$

9. $u(x, y)$
$= \dfrac{4}{\pi^2} \displaystyle\sum_{m,n=1}^{\infty} \dfrac{\sin [(m - \frac{1}{2})(\pi x/L)] \sin [(n - \frac{1}{2})(\pi y/L)] \sinh [(\pi z/L) \sqrt{(m - \frac{1}{2})^2 + (n - \frac{1}{2})^2}]}{(m - \frac{1}{2})(n - \frac{1}{2}) \sinh [\pi \sqrt{(m - \frac{1}{2})^2 + (n - \frac{1}{2})^2}]}$

10. $u(x, y, z) = 1$

11. $u(x, y; t) = \dfrac{T_2 y}{L_2} + \dfrac{(L_2 - y)T_1}{L_1} + \sum\limits_{n=1}^{\infty} A_n \sin \dfrac{n\pi y}{L_2} \exp\left[-\left(\dfrac{n\pi}{L} \right)^2 Kt \right]$

$A_n = \dfrac{2(T_3 - T_1)[1 - (-1)^n]}{n\pi} + \dfrac{2(T_1 - T_2)(-1)^n}{L_2\, n\pi}$

12. $u(x, y; t) = \dfrac{T_1 y}{L} - \dfrac{4T_1}{\pi^2 L^2} \sum\limits_{m,n=1}^{\infty} \dfrac{1 - (-1)^m}{m} \dfrac{(-1)^{n+1}}{n} \left(\sin \dfrac{m\pi x}{L} \right)\left(\sin \dfrac{n\pi y}{L} \right)(e^{-\lambda_{mn}Kt},)$

$\lambda_{mn} = \dfrac{(m^2 + n^2)\pi^2}{L^2}, \quad \tau = \dfrac{L^2}{2K\pi^2}$

13. $u(x, y; t) = 3 \sin (\pi x/L) \sin (2\pi y/L) \cos (\pi ct\ \sqrt{5}/L)$
  $+ 4 \sin (3\pi x/L) \sin (5\pi y/L) \cos (\pi ct\ \sqrt{34}/L)$

14. $u_{mn}(x, y; t) = \cos (m\pi x/L) \cos (n\pi y/L) \sin [(\pi ct/L)\ \sqrt{m^2 + n^2}]$,
  $m, n = 0, 1, 2, \ldots$

15. $\dfrac{\pi c}{L}, \dfrac{\pi c}{L}, \dfrac{\pi c}{L} \sqrt{2}, \dfrac{\pi c}{L} \sqrt{4}, \dfrac{\pi c}{L} \sqrt{4}, \dfrac{\pi c}{L} \sqrt{5}, \dfrac{\pi c}{L} \sqrt{5}, \dfrac{\pi c}{L} \sqrt{8}, \dfrac{\pi c}{L} \sqrt{9}, \dfrac{\pi c}{L} \sqrt{9}$

17. $\dfrac{\pi c}{L} \sqrt{5}, \dfrac{\pi c}{L} \sqrt{10}, \dfrac{\pi c}{L} \sqrt{13}, \dfrac{\pi c}{L} \sqrt{17}, \dfrac{\pi c}{L} \sqrt{20}, \dfrac{\pi c}{L} \sqrt{25}, \dfrac{\pi c}{L} \sqrt{26}, \dfrac{\pi c}{L} \sqrt{29}, \dfrac{\pi c}{L} \sqrt{34},$
  $\dfrac{\pi c}{L} \sqrt{41}$

CHAPTER 3

Section 3.1

1. $12\rho^2 \cos 2\varphi$ **2.** 0 **3.** $n^2\rho^{n-2}$ **4.** $(n^2 - m^2)\rho^{n-2} \cos m\varphi$
5. $e^\rho \cos \varphi + (1/\rho)e^\rho \cos \varphi - (1/\rho^2)e^\rho \cos \varphi$
6. 1, 2, 3 if n is even;   4 if $n \geq m$ and $n - m$ is even
9. $f(\rho) = A \ln \rho + B,\ \rho \neq 0$ **10.** $f(\rho) = -\tfrac{1}{4}\rho^2 + A \ln \rho + B,\ \rho \neq 0$

11. $f(\rho) = \dfrac{2}{\ln 2} \ln \rho + 3$ **12.** $f(\rho) = -\dfrac{1}{4}\rho^2 + \dfrac{7 \ln \rho}{4 \ln 2} + \dfrac{1}{4}$

13. $u(\rho, \varphi) = 1 + \left(\dfrac{\rho}{R} \right)^2 \cos 2\varphi + 3\left(\dfrac{\rho}{R} \right)^3 \sin 3\varphi$

14. $u(\rho, \varphi) = \dfrac{\ln \rho}{\ln 2} - \dfrac{\rho^2}{15} \cos 2\varphi + \dfrac{16 \cos 2\varphi}{15\rho^2}$

15. $u(\rho, \varphi) = \dfrac{1}{2} + \dfrac{1}{\pi} \sum\limits_{n=1}^{\infty} \dfrac{\rho^n - \rho^{-n}}{2^n - 2^{-n}} \dfrac{1 - (-1)^n}{n} \sin n\varphi$

16. $u(\rho, \varphi) = 3 + 4\left(\dfrac{R}{\rho} \right)^2 \cos 2\varphi + 5\left(\dfrac{R}{\rho} \right)^3 \sin 3\varphi$

18. $u_n(\rho, \varphi) = \rho^{2n} \sin 2n\varphi,\ n = 1, 2, 3, \ldots$

19. $u(\rho, \varphi) = \dfrac{2}{\pi} \sum\limits_{n=1}^{\infty} \dfrac{1 - (-1)^n}{n} \rho^{2n} \sin 2n\varphi$

20. $u_n(\rho, \varphi) = \rho^n \cos n\varphi,\ n = 0, 1, 2, 3, \ldots$

21. $u(\rho, \varphi) = \dfrac{\pi^2}{6} + 4 \sum\limits_{n=1}^{\infty} \dfrac{(-1)^{n+1}}{n^2} \rho^n \cos n\varphi$

Section 3.2

8. $J_0(x) = 1 - \dfrac{x^2}{4} + \dfrac{x^4}{64} - \dfrac{x^6}{2304} + \cdots \qquad J_1(x) = -\dfrac{x}{2} + \dfrac{x^3}{16} - \dfrac{x^5}{384} + \dfrac{x^7}{18{,}432} - \cdots$

28. $A_n = \dfrac{128/x_n^5 - 16/x_n^3}{J_1(x_n)}$

29. $P_6(x) = \dfrac{19 - 27x^2 + 9x^4 - x^6}{4608}$

30. $(1 - x^2)^3 = 4608 P_6 - 768 P_4 \qquad A_n = \dfrac{4608/x_n^7 - 768/x_n^5}{J_1(x_n)}$

31. $P_8 = \dfrac{211 - 304x^2 + 108x^4 - 16x^6 + x^8}{294{,}912}$

$P_{10} = \dfrac{3651 - 5275x^2 + 1900x^4 - 300x^6 + 25x^8 - x^{10}}{29{,}491{,}200}$

32. $A_n = 2/[x_n J_2(x_n)]$ **33.** $F_3 = \rho(1 - \rho^2)/16$ **34.** $F_5 = \rho(1 - \rho^2)(2 - \rho^2)/384$

Section 3.3

1. $U(\rho) = [g/(4c^2)](\rho^2 - R^2)$

4. $u(\rho, \varphi; t) = \displaystyle\sum_{n=1}^{\infty} A_n J_0\!\left(\dfrac{\rho x_n}{R}\right) \cos \dfrac{ctx_n}{R}, \quad J_0(x_n) = 0,$

$A_n = \dfrac{2x_n^2}{J_1(x_n)^2} \displaystyle\int_0^R F_1(\rho) J_0\!\left(\dfrac{\rho x_n}{R}\right)\rho\, d\rho$

5. $u(\rho, \varphi, t) = \displaystyle\sum_{n=1}^{\infty} A_n J_0\!\left(\dfrac{\rho x_n}{R}\right) \sin \dfrac{ctx_n}{R}, \quad A_n = \dfrac{2Rx_n}{cJ_1(x_n)^2} \displaystyle\int_0^R F_2(\rho) J_0\!\left(\dfrac{\rho x_n}{R}\right)\rho\, d\rho$

6. $u(\rho, \varphi; t) = \dfrac{2R}{c} \displaystyle\sum_{n=1}^{\infty} \dfrac{J_0(\rho x_n/R)}{x_n^2 J_1(x_n)} \sin \dfrac{ctx_n}{R}, \quad J_0(x_n) = 0$

7. $u(\rho, \varphi; t) = \dfrac{8R^3}{c} \displaystyle\sum_{n=1}^{\infty} \dfrac{J_0(\rho x_n/R)}{x_n^4 J_1(x_n)} \sin \dfrac{ctx_n}{R}, \quad J_0(x_n) = 0$

8. $u(\rho, \varphi; t) = [R/(cx_1^{(3)})]J_3(\rho x_1^{(3)}/R) \sin (ctx_1^{(3)}/R) \cos 3\varphi$

Section 3.4

1. $u(\rho, \varphi; t) = 8R^2 \displaystyle\sum_{n=1}^{\infty} \dfrac{J_0(\rho x_n/R)}{x_n^3 J_1(x_n)} \exp\left(-\dfrac{x_n^2 Kt}{R^2}\right)$, where $J_0(x_n) = 0$

2. $u(\rho, \varphi; t) = 1 - 2 \displaystyle\sum_{n=1}^{\infty} \dfrac{J_0(\rho x_n/R)}{x_n J_1(x_n)} \exp\left(-\dfrac{x_n^2 Kt}{R^2}\right)$, where $J_0(x_n) = 0$; $\tau = 0.1736 \dfrac{R^2}{K}$

3. $u(\rho, \varphi; t) = 1 + \dfrac{\rho}{2R} \cos \varphi - 2 \displaystyle\sum_{n=1}^{\infty} \dfrac{J_0(\rho x_n^{(0)}/R)}{x_n^{(0)} J_1(x_n^{(0)})} \exp\left\{-\dfrac{(x_n^{(0)})^2 Kt}{R^2}\right\}$

$- \cos \varphi \displaystyle\sum_{n=1}^{\infty} \dfrac{J_1(\rho x_n^{(1)}/R)}{x_n^{(1)} J_2(x_n^{(1)})} \times \exp\left[-(x_n^{(1)})^2 \dfrac{Kt}{R^2}\right]$, where $J_0(x_n^{(0)}) = 0$, $J_1(x_n^{(1)}) = 0$

4. $u(\rho, \varphi; t) = T_1 + \dfrac{\sigma(R^2 - \rho^2)}{4K} + \displaystyle\sum_{n=1}^{\infty} A_n J_0\!\left(\dfrac{\rho x_n}{R}\right) \exp\left(-\dfrac{x_n^2 Kt}{R^2}\right)$, where $J_0(x_n) = 0,$

$A_n = \dfrac{8[T_2 - \sigma R^2/(4K)]}{x_n^3 J_1(x_n)} + \dfrac{2(T_2 - T_1)}{x_n J_1(x_n)}$

5. $u_{mn} = J_m(\rho x_n^{(m)}/R) \sin m\varphi \exp(-(x_n^{(m)})^2 Kt/R^2]$, where $m = 1, 2, \ldots$
   $J_m(x_n^{(m)}) = 0$

6. $u(\rho, \varphi; t) = \sum\limits_{m,n=1}^{\infty} A_{mn} J_m\left(\dfrac{\rho x_n^{(m)}}{R}\right) \sin m\varphi \exp\left[-(x_n^{(m)})^2 \dfrac{Kt}{R^2}\right]$

   $A_{mn} = \dfrac{4(x_n^{(m)})^2(1 - (-1)^m]}{m\pi J_{m+1}(x_n^{(m)})^2} \int_0^R J_m\left(\dfrac{\rho x_n^{(m)}}{R}\right) f(\rho)\rho\, d\rho$

7. $U(\rho) = 100 \dfrac{\ln(\rho/3)}{\ln 5}$ $\qquad \tau = \dfrac{390}{K}$

8. $u(\rho; t) = 100 - \sum\limits_{n=1}^{\infty} A_n J_0\left(\dfrac{\rho x_n}{2}\right) \exp\left[-(x_n)^2 \dfrac{Kt}{4}\right]$, $\quad A_n = \dfrac{100}{x_n J_1(x_n)} + \dfrac{50 J_1(x_n/2)}{x_n J_1(x_n)^2}$,
   $J_0(x_n) = 0$

9. $u_{mn}(\rho, \varphi; t) = J_m\left(\dfrac{\rho x_n^{(m)}}{R}\right)(A \cos m\varphi + B \sin m\varphi) \exp\left[-(x_n^{(m)})^2 \dfrac{Kt}{R^2}\right]$,
   $J_m'(x_n^{(m)}) = 0$, $m = 0, 1, 2, \ldots$; $n = 1, 2, \ldots$

10. $u(\rho, \varphi, z) = J_m(\rho x_n^{(m)}/R)(A \cos m\varphi + B \sin m\varphi)[C \sinh(zx_n^{(m)}/R) + D \cosh(zx_n^{(m)}/R)]$ $\quad$ where $m = 0, 1, 2, \ldots$, $\quad n = 1, 2, \ldots$, $\quad J_m(x_n^{(m)}) = 0$

11. $u(\rho, \varphi, z) = 2\sum\limits_{n=1}^{\infty} \dfrac{J_0(\rho x_n/R) \sinh(zx_n/R)}{x_n J_1(x_n) \sinh(Lx_n/R)}$, where $J_0(x_n) = 0$, $\quad n = 1, 2, \ldots$

12. $u(\rho, \varphi, z) = 2T_2 \sum\limits_{n=1}^{\infty} \dfrac{J_0(\rho x_n/R) \sinh(zx_n/R)}{x_n J_1(x_n) \sinh(Lx_n/R)} + 2T_1 \sum\limits_{n=1}^{\infty} \dfrac{J_0(\rho x_n/R) \sinh[(L - z)x_n/R]}{x_n J_1(x_n) \sinh(Lx_n/R)}$,
   where $J_0(x_n) = 0$

13. $u(\rho, \varphi, z) = I_m(n\pi\rho/L)(A \cos m\varphi + B \sin m\varphi) \sin(n\pi z/L)$, where $m = 0, 1, 2, \ldots$, $\quad n = 1, 2, \ldots$

14. $u(\rho, \varphi, z) = \dfrac{2}{\pi} \sum\limits_{n=1}^{\infty} \dfrac{I_0(n\pi\rho/L)}{I_0(n\pi R/L)} \dfrac{1 - (-1)^n}{n} \sin \dfrac{n\pi z}{L}$

16. $u(\rho, \varphi; t) = 2R \sin \varphi \sum\limits_{n=1}^{\infty} \dfrac{J_1(\rho x_n^{(1)}/R)}{x_n J_2(x_n)} \exp\left[-(x_n)^2 \dfrac{Kt}{R^2}\right]$, $\quad J_1(x_n) = 0$

17. $u(\rho, \varphi; t) = 2 \sum\limits_{n=1}^{\infty} \dfrac{J_2(\rho x_n/R)}{x_n J_3(x_n)} \cos 2\varphi \exp\left[\left(-x_n^2 \dfrac{Kt}{R^2}\right)\right]$, where $J_2(x_n) = 0$

18. $u(\rho, \varphi; t) = \sum\limits_{n=1}^{\infty} A_n J_0\left(\dfrac{\rho x_n}{R}\right) \exp\left(-x_n^2 \dfrac{Kt}{R^2}\right)$,

   $A_n = \dfrac{1}{\{1 + [x_n k/(hR)^2\} J_1(x_n)}\left(\dfrac{8T_2}{x_n^3} - \dfrac{2T_1}{x_n}\right)$, $2x_n^{(m)} J_m'(x_n^{(m)}) + hRJ_m(x_n^{(m)}) = 0$

CHAPTER 4

Section 4.1

1. $12r$	2. $3 \sin^3 \theta + 9 \sin \theta \cos^2 \theta$	3. $2/r$
4. $(\cot \theta)/r^2$	5. 0	6. $(9 + 6/r)e^{3r}$
7. $n(n + 1)r^{n-2}$	11. $f(r) = (a^2 - r^2)/6$	12. $f(r) = (a^4 - r^4)/20$

13. $f(r) = (a^6 - r^6)/42$

14. $u(r; t) = 3 \dfrac{a}{r} \operatorname{Re}\left\{\exp[c_1(r - a)(1 + i)]e^{2it} \dfrac{1 - \exp[-2c_1 r(1 + i)]}{1 - \exp(-2c_1 a(1 + i)]}\right\}$,

   where $c_1 = \sqrt{\dfrac{1}{K}}$

17. $u(r; t) = \dfrac{\sigma(a^2 - r^2)}{6K} + T_1 + \dfrac{1}{r} \sum\limits_{n=1}^{\infty} A_n \sin \dfrac{n\pi r}{a} \exp\left[-\left(\dfrac{n\pi}{a}\right)^2 Kt\right],$

where $A_n = \dfrac{2(T_2 - T_1)}{n\pi}(-1)^{n+1} - \dfrac{2\sigma a^3(-1)^{n+1}}{\pi^3 K n^3}, \quad \tau = \dfrac{a^2}{\pi^2 K}$

18. $u(r; t) = \dfrac{1}{r} \sum\limits_{n=1}^{\infty} A_n \sin \dfrac{n\pi r}{2a} \exp\left[-\left(\dfrac{n\pi}{2a}\right)^2 Kt\right] + T_1,$

where $A_n = T_2\left[\dfrac{-4a}{(n\pi)^2} \sin \dfrac{n\pi}{2} - \dfrac{2a}{n\pi} \cos \dfrac{n\pi}{2}\right] + T_1\left[\dfrac{4a(-1)^n}{n\pi}\right]$

20. $u_n(r; t) = \dfrac{1}{r} \sin (r\sqrt{\lambda_n})e^{-\lambda_n Kt}$, where $a\sqrt{\lambda_n} \cot (a\sqrt{\lambda_n}) = 1$

22. $u(r; t) = T_1 + \dfrac{1}{r} \sum\limits_{n=1}^{\infty} A_n \sin \dfrac{(n - \frac{1}{2})\pi r}{a} \exp\left\{-\left[\dfrac{(n - \frac{1}{2})\pi}{a}\right]^2 Kt\right\},$

where $A_n = \dfrac{2(T_2 - T_1)(-1)^n a}{\pi^2(n - \frac{1}{2})^2}$

23. $\tau = \dfrac{4a^2}{\pi^2 K} \cong 2 \text{ min}$ **26.** $u(r; t) = \dfrac{Aa}{nr\pi c} \sin \dfrac{n\pi r}{a} \sin \dfrac{n\pi ct}{a}$

Section 4.2

1. $1, 0, -\frac{1}{2}, 0$ **2.** $0, 1, 0, -\frac{3}{2}$

3. $P_5(z) = \dfrac{63z^5 - 70z^3 + 15z}{8}, \quad P_6(z) = \dfrac{231z^6 - 315z^4 + 105z^2 - 5}{16}$

10. $P_1(0) = 0, \quad P_2\left(\pm\dfrac{1}{\sqrt{3}}\right) = 0, \quad P_3(0) = 0, \quad P_3(\pm\sqrt{\frac{3}{5}}) = 0, \quad P_4\left(\pm\sqrt{\dfrac{30 \pm \sqrt{30}}{70}}\right) = 0$

12. $\dfrac{1}{2} + \sum\limits_{l=1}^{\infty} \dfrac{2l + 1}{2l(l + 1)} P_l'(0)P_l(z) = \frac{1}{2}[f(z - 0) + f(z + 0)], \quad -1 < z < 1$

13. $\sum\limits_{l=1}^{\infty} \dfrac{2l + 1}{l(l + 1)} P_l'(0)P_l(z) = \frac{1}{2}[f(z - 0) + f(z + 0)], \quad -1 < z < 1$

14. $\frac{1}{2} + \frac{45}{8} P_2(z) - \frac{135}{28} P_4(z) - \frac{5733}{1024}P_6(z)$

15. $P_{41}(z) = \sqrt{1 - z^2}(140z^3 - 90z), \quad P_{42}(z) = (1 - z^2)(420z^2 - 90),$
$P_{43}(z) = (1 - z^2)^{3/2}(840z), \quad P_{44}(z) = 840(1 - z^2)^2$

19. $z^2 = \frac{1}{3}P_0(z) + \frac{2}{3}P_2(z), \quad z^3 = \frac{3}{5}P_1(z) + \frac{2}{5}P_3(z), \quad z^4 = \frac{7}{35}P_0(z) + \frac{20}{35}P_2(z) + \frac{8}{35}P_4(z)$

Section 4.3

1. $u(r, \theta) = \frac{11}{3} P_0(\cos \theta) + 4(r/a)P_1(\cos \theta) + \frac{4}{3}(r/a)^2 P_2(\cos \theta)$

2. $u(r, \theta) = \frac{2}{3}P_0(\cos \theta) + \frac{4}{3}(r/a)^2 P_2^2(\cos \theta)$

3. $u(r, \theta) = \sum\limits_{n=0}^{\infty} \dfrac{1}{2^{n+1}} \left(\dfrac{r}{a}\right)^n P_n(\cos \theta)$

4. $u(r, \theta) = \dfrac{1}{2} + \sum\limits_{l=1}^{\infty} \dfrac{(2l + 1)P_l'(0)}{2l(l + 1)} \left(\dfrac{r}{a}\right)^l P_l(\cos \theta)$

5. $u(r, \theta) = \sum\limits_{l=1}^{\infty} \dfrac{(2l + 1)P_l'(0)}{l(l + 1)} \left(\dfrac{r}{a}\right)^l P_l(\cos \theta), \quad u\left(r, \dfrac{\pi}{2}\right) = 0$

7. $u(r, \theta) = \frac{26}{5}(r/a)P_1(\cos \theta) + \frac{4}{5}(r/a)^3 P_3(\cos \theta)$

8. See Exercise 5.

9. $u(r, \theta) = \frac{37}{35}(a/r)P_0(\cos \theta) + 2(a/r)^2 P_1(\cos \theta) + \frac{10}{35}(a/r)^3 P_2(\cos \theta) + \frac{8}{35}(a/r)^5 P_4(\cos \theta)$

10. $u(r, \theta) = \sum_{l=1}^{\infty} \frac{(2l + 1)P_l'(0)}{l(l + 1)} \left(\frac{a}{r}\right)^{l+1} P_l(\cos \theta)$

11. $u(r, \theta) = -\frac{7}{5}(a^3/r^2)P_1(\cos \theta) - \frac{3}{10}(a^5/r^4)P_3(\cos \theta)$

12. $u(r, \theta) = \dfrac{3a^3}{a - 2} \dfrac{\cos \theta}{r^2} \quad (a \neq 2)$

13. $u(r, \theta) = (r/a) \sin \theta \cos \varphi + (r/a)^2 \sin^2 \theta \sin 2\varphi$

14. $u(r, \theta) = (a/r)^2 \sin \theta \cos \varphi + (a/r)^3 \sin^2 \theta \sin 2\varphi$

CHAPTER 5

Section 5.1

1. $F(\mu) = \dfrac{\sin 2\mu}{\pi\mu}$

2. $F(\mu) = \dfrac{4}{i\pi\mu}(1 - \cos \mu)$

3. $F(\mu) = \dfrac{1}{2\pi}\left(\dfrac{1}{2 - i\mu} + \dfrac{1}{3 + i\mu}\right)$

4. $F(\mu) = \dfrac{2\mu}{i\pi(1 + \mu^2)^2}$

5. $F(\mu) = \dfrac{1}{2\pi}\left[\dfrac{1}{1 + (1 + \mu)^2} + \dfrac{1}{1 + (1 - \mu)^2}\right]$

6. $F(\mu) = \dfrac{1}{4\pi}\left[\dfrac{2}{1 + \mu^2} + \dfrac{1}{1 + (\mu - 2)^2} + \dfrac{1}{1 + (\mu + 2)^2}\right]$

7. $F(\mu) = \dfrac{-i\mu}{4} e^{-|\mu|}$

8. $F(\mu) = \dfrac{1}{2\pi} \exp\left[-\dfrac{1}{2}\left(\mu - \dfrac{3i}{2}\right)^2\right]$

9. $F(\mu) = \dfrac{1}{4\pi}\left\{\exp\left[-\dfrac{1}{2}(1 + \mu)^2\right] + \exp\left[-\dfrac{1}{2}(1 - \mu)^2\right]\right\}$

10. $F(\mu) = -\dfrac{i\mu e^{-\mu^2/2}}{2\pi}$

13. $F_c(\mu) = \dfrac{2}{\pi}\dfrac{1 - \mu^2}{(1 + \mu^2)^2} \quad F_s(\mu) = \dfrac{4\mu}{\pi(1 + \mu^2)^2}$

17. $F(\mu) = \pi e^{-3i\mu}e^{-|\mu|}$

18. $F(\mu) = \dfrac{1}{2\pi}e^{-2i\mu}e^{-\mu^2/2}$

19. $F(\mu) = 3e^{-2i\mu}/[\pi(9 + \mu^2)]$

20. $F(\mu) = (2 - i\mu)/[\pi(4 + \mu^2)]$

Section 5.2

6. $u(x; t) = \dfrac{1}{\sqrt{4\pi Kt}} \int_0^{L_1} \left\{\exp\left[-\dfrac{(x - \xi)^2}{4Kt}\right] - \exp\left[-\dfrac{(x + \xi)^2}{4Kt}\right]\right\} d\xi$

$|u(x; t)| \leq \dfrac{xL_1^2}{4K\sqrt{\pi K}} t^{-3/2}$

7. $u(x; t) = \dfrac{1}{\sqrt{4\pi Kt}} \int_0^{L_1} \left\{\exp\left[-\dfrac{(x - \xi)^2}{4Kt}\right] + \exp\left[-\dfrac{(x + \xi)^2}{4Kt}\right]\right\} d\xi$

$|u(x; t)| \leq \dfrac{2L_1}{\sqrt{4\pi Kt}}$

11. $u(x; t) = T_1\left[1 - \Phi\left(\dfrac{x}{\sqrt{2Kt}}\right)\right]$; $C = 10$ then $t = (0.61)x^2$
$\qquad\qquad\qquad\qquad\qquad\qquad\quad C = 30$ then $t = (3.6)x^2$
$\qquad\qquad\qquad\qquad\qquad\qquad\quad C = 50$ then $x = 0$

12. $\tau^* = 1.81x^2/K$

13. $u(x; t) = T_1\left[\Phi\left(\dfrac{x + L}{\sqrt{2Kt}}\right) - \Phi\left(\dfrac{x}{\sqrt{2Kt}}\right)\right] + T_2\left[\Phi\left(\dfrac{x}{\sqrt{2Kt}}\right) - \Phi\left(\dfrac{x - L}{\sqrt{2Kt}}\right)\right]$
$\qquad \lim_{t\to\infty} u(x; t) = 0$

Section 5.3

1. $y(x; t) = 3 \sin 2x \cos 2ct$ $\qquad\qquad$ **2.** $y(x; t) = 4 \cos 5x \cos 5ct$

4. $y(x; t) = \dfrac{1}{2c}\displaystyle\int_{ct-x}^{ct+x} g(\xi)\,d\xi$ for $0 < x < ct$; $\quad y(x; t) = \dfrac{1}{2c}\displaystyle\int_{x-ct}^{x+ct} g(\xi)\,d\xi$ for $x > ct$

5. $y(x; t) = \dfrac{1}{2c}\displaystyle\int_{ct-x}^{x+ct} g(\xi)\,d\xi + s\left(t - \dfrac{x}{c}\right)$ for $0 < x < ct$

$\qquad y(x; t) = \dfrac{1}{2c}\displaystyle\int_{x-ct}^{x+ct} g(\xi)\,d\xi$ for $x > ct$

7. $u(r; t) = \dfrac{1}{2r}[(r + ct)f_1(r + ct) + (r - ct)f_1(r - ct)] + \dfrac{1}{2cr}\displaystyle\int_{r-ct}^{r+ct} \xi f_2(\xi)\,d\xi$

8. $u(r; t) = \begin{cases} Tt & 0 < ct < a - r \\ [T/(2cr)](a^2 - (r - ct)^2) & 0 < a - r < ct < a + r \\ 0 & ct > a + r \end{cases}$

9. $u(x, y) = \dfrac{2}{\pi}\left(\tan^{-1}\dfrac{4 - x}{y} + \tan^{-1}\dfrac{4 + x}{y}\right)$

10. $u(x, y) = \dfrac{1}{\pi}\displaystyle\int_0^\infty \left[\dfrac{y}{y^2 + (x - \xi)^2} - \dfrac{y}{y^2 + (x + \xi)^2}\right] f(\xi)\,d\xi$

11. $u(x, y) = \dfrac{1}{\pi}\displaystyle\int_0^\infty \left[\dfrac{y}{y^2 + (x - \xi)^2} + \dfrac{y}{y^2 + (x + \xi)^2}\right] f(\xi)\,d\xi$

12. $u(x, y) = \dfrac{1}{\pi}\displaystyle\int_0^\infty \left[\dfrac{y}{y^2 + (x - \xi)^2} + \dfrac{x}{x^2 + (y - \xi)^2}\right.$
$\qquad\qquad\qquad\qquad\qquad\qquad\left. - \dfrac{y}{y^2 + (x + \xi)^2} - \dfrac{x}{x^2 + (y + \xi)^2}\right] f(\xi)\,d\xi$

14. $u(x, y) = \displaystyle\sum_{n=1}^\infty B_n \sin\left(\dfrac{n\pi x}{L}\right)e^{-(n\pi y/L)}$ $\qquad B_n = \dfrac{2}{L}\displaystyle\int_0^L f(x)\sin\dfrac{n\pi x}{L}\,dx$

15. $u(x, y) = \displaystyle\int_{-\infty}^\infty B(\lambda)\sinh(\lambda x)e^{i\lambda y}\,d\lambda$ $\qquad B(\lambda)\sinh \lambda L = \dfrac{1}{2\pi}\displaystyle\int_{-\infty}^\infty g(y)e^{-i\lambda y}\,dy$

CHAPTER 6

Section 6.1

5. $n = 1$, 0.922; $n = 10$, 3.599×10^6; $n = 100$, 9.325×10^{157}; $n = 1000$, 4.027×10^{2567}

Section 6.2

1. $f(t) = (e^t/t)[\sin 1 + O(1/t)], \quad t \to \infty$ **2.** $f(t) = (e^{-t}/t)[\frac{1}{2} + O(1/t)], \quad t \to \infty$

3. $f(t) = (e^{+t}/t)[1 + O(1/t)], \quad t \to \infty$ **4.** $f(t) = (1/t)[1 + O(1/t)], \quad t \to \infty$

5. $f(t) = (1/t)[1 + O(1/t)], \quad t \to \infty$

6. $u(x; t) = \begin{cases} 100 + O(-x^2/(4Kt)) & x > 0, t \to 0 \\ O(e^{-x^2/(4Kt)}) & x < 0, t \to 0 \\ 50 + O(1/\sqrt{t}) & t \to \infty \end{cases}$

7. $u(at; t) = 100 + O(e^{+a^2t/(4K)}) \quad t \to \infty, a > 0$

Section 6.3

1. $f(t) = 2\sqrt{\pi/t}[1 + O(1/\sqrt{t})], \quad t \to \infty$ **2.** $f(t) = 5\sqrt{\pi/t}[1 + O(1/\sqrt{t})], \quad t \to \infty$

3. $f(t) = e^{3t/4}\sqrt{8\pi/(15t)}[1 + O(1/\sqrt{t})], \quad t \to \infty$

4. $I_0(t) = e^t\sqrt{1/(2\pi t)}[1 + O(1/\sqrt{t})], \quad t \to \infty$

7. $u(x; t) = O(1/t), \quad u_x(x; t) = O(1/t), \quad t \to \infty$

8. $u(x; t) = \dfrac{1}{\sqrt{\pi Kt}}\left[\displaystyle\int_0^\infty f(x)\,dx + O\left(\dfrac{1}{\sqrt{t}}\right)\right], \quad u_x(x; t) = O\left(\dfrac{1}{t}\right), \quad t \to \infty$

9. $u(x; t) = 50\sqrt{\dfrac{\pi}{Kt}}\left[1 + O\left(\dfrac{1}{\sqrt{t}}\right)\right], \quad u_x(x; t) = O\left(\dfrac{1}{t}\right), \quad t \to \infty$

10. $u(x; t) = \dfrac{1}{\sqrt{\pi Kt}}\left[\dfrac{1}{2} + O\left(\dfrac{1}{t}\right)\right], \quad u_x(x; t) = O\left(\dfrac{1}{t}\right), \quad t \to \infty$

Section 6.4

4. $f(t) = e^{-t}\left[\dfrac{1}{t} - \dfrac{1}{t^2} + \dfrac{2}{t^3} - \dfrac{6}{t^4} + \cdots + (-1)^n\dfrac{n!}{t^{n+1}} + O\left(\dfrac{1}{t^{n+2}}\right)\right], \quad t \to \infty$

6. $u(x; t) = \sqrt{\dfrac{\pi}{Kt}}\left[F(0) - \dfrac{2x^2 + 1}{4Kt}\right] + O\left(\dfrac{1}{t^{5/2}}\right), \quad t \to \infty$

Section 6.5

1. $f(t) = \sqrt{\pi/t}\,e^{i\pi/4}[1 + O(1/\sqrt{t})], \quad t \to \infty$

2. $f(t) = \sqrt{2\pi/t}\,e^{it}e^{-(i\pi/4)}[1 + O(1/\sqrt{t})], \quad t \to \infty$

3. $f(t) = \sqrt{\pi/(2t)}\,e^{it}e^{-(i\pi/4)}[1 + O(1/\sqrt{t})], \quad t \to \infty$

4. $f(t) = 2\sqrt{2\pi/t}\,\cos(t - \pi/4) + O(1/t), \quad t \to \infty \quad m \text{ even}$

     $f(t) = 2i\sqrt{2\pi/t}\,\sin(t - \pi/4) + O(1/t), \quad t \to \infty \quad m \text{ odd}$

CHAPTER 7

Section 7.1

1. (a) 0.04996   (b) −0.1000   (c) −0.4986   (d) −0.9975

2. (a) −0.04996   (b) 0.1000   (c) 0.4082   (d) 0.9975

3. (a) 0   (b) 0   (c) -0.01126 (d) 0
5. (a) 0   (b) -0.0618 (c) -0.0008 (d) 2.000
8. $u_1 = 0.0990$ $u_2 = 0.1951$ $u_3 = 0.2865$ $u_4 = 0.3717$ $u_5 = 0.4496$
    $u_6 = 0.5193$ $u_7 = 0.5806$ $u_8 = 0.6333$ $u_9 = 0.6778$ $u_{10} = 0.7146$
9. $\left|u(\tfrac{1}{2}) - u_5\right| \le 0.05$
10. $u_1 = 0.1942$ $u_2 = 0.3719$ $u_3 = 0.5246$ $u_4 = 0.6470$ $u_5 = 0.7365$
     $u_6 = 0.7937$ $u_7 = 0.8212$ $u_8 = 0.8236$ $u_7 = 0.8059$ $u_{10} = 0.7738$

CHAPTER 8

Section 8.1

2. $G(x, z) = \begin{cases} \dfrac{1 + h(L - z)}{1 + hL}\, x & 0 \le x \le z \\[2mm] \dfrac{1 + h(L - x)}{1 + hL}\, z & z \le x \le L \end{cases}$

3. $G(x, z) = \begin{cases} \dfrac{1 + h(L - z)}{h(2 + hL)}\,(1 + hx) & 0 \le x \le z \\[2mm] \dfrac{1 + hz}{h(2 + hL)}\,[1 + h(L - x)] & z \le x \le L \end{cases}$

4. $G(x, z) = \begin{cases} \dfrac{\sinh[(L - z)\sqrt{k}]\,\sinh(x\sqrt{k})}{\sinh(L\sqrt{k})} & 0 \le x \le z \\[2mm] \dfrac{\sinh(z\sqrt{k})\,\sinh[(L - x)\sqrt{k}]}{\sinh(L\sqrt{k})} & z \le x \le L \end{cases}$

6. $G(x, z) = \begin{cases} \dfrac{\sin[(L - z)\sqrt{-k}]\,\sin(x\sqrt{-k})}{\sin(L\sqrt{-k})} & 0 \le x \le z \\[2mm] \dfrac{\sin(z\sqrt{-k})\,\sin[(L - z)\sqrt{-k}]}{\sin L\sqrt{-k}} & z \le x \le L \end{cases}$

7. $G(x, z) = \begin{cases} -\dfrac{(z - L)^2}{2L} & 0 \le x \le z \\[2mm] x - z - \dfrac{(z - L)^2}{2L} & z \le x \le L \end{cases}$

8. $G(x, z) = \begin{cases} \dfrac{(L - z)(z - 2x)}{2L} & 0 \le x \le z \\[2mm] (x - z) + \dfrac{(L - z)(z - 2x)}{2L} & z \le x \le L \end{cases}$

Section 8.3

1. $G(P, Q) = \dfrac{1}{4\pi} \log \dfrac{(x - \xi)^2 + (y - \eta)^2}{(x - \xi)^2 + (y + \eta)^2} - \dfrac{1}{4\pi} \log \dfrac{(x + \xi)^2 + (y - \eta)^2}{(x + \xi)^2 + (y + \eta)^2},$
    $P = (x, y), \ Q = (\xi, \eta)$

2. $G(P, Q) = \dfrac{1}{4\pi} \log \dfrac{(x - \xi)^2 + (y + \eta)^2}{(x - \xi)^2 + (y + \eta)^2} + \dfrac{1}{4\pi} \log \dfrac{(x + \xi)^2 + (y - \eta)^2}{(x + \xi)^2 + (y + \eta)^2},$
    $P = (x, y), \ Q = (\xi, \eta)$

4. $G(P, Q) = \sum\limits_{m,n=1}^{\infty} \dfrac{\sin(m\pi x/a)\,\sin[(n-\frac{1}{2})\pi y/b]\,\sin(m\pi\xi/a)\,\sin[(n-\frac{1}{2})\pi\eta/b]}{(m\pi/a)^2 + [(n-\frac{1}{2})\pi/b]^2}\,\dfrac{4}{ab}$

$P = (x, y),\ Q = (\xi, \eta)$

Section 8.4

1. $u(x;\,t) = \displaystyle\int_0^t\int_0^\infty \dfrac{e^{-(x-\xi)^2/[4K(t-s)]} + e^{-(x+\xi)^2/[4K(t-s)]}}{\sqrt{4\pi K(t-s)}}\,h(\xi;\,s)\,d\xi\,ds$

2. $u(x;\,t) = \displaystyle\sum_{m=\infty}^{\infty}\int_0^t\int_0^L \dfrac{e^{-(x-y-2mL)^2/[4K(t-s)]} - e^{-[x+y-(2m+2)L]^2/[4K(t-s)]}}{\sqrt{4\pi K(t-s)}}\,h(y,\,s)\,dy\,ds$

3. $u(x,\,y;\,t) = \displaystyle\int_0^t\int_{-\infty}^{\infty}\int_{-\infty}^{\infty} \dfrac{e^{-[(x-\xi)^2+(y-\eta)^2]/[4K(t-s)]}}{\sqrt{4\pi K(t-s)}}\,h(\xi,\,\eta,\,s)\,d\xi\,d\eta\,ds$

4. $u(x;\,t) = \displaystyle\sum_{n=-\infty}^{\infty}\int_0^t\int_0^L \dfrac{e^{-(x-y-2mL)^2/(4K(t-s))} - e^{-[x+y-(2m+2)L]^2/(4K(t-s))}}{\sqrt{4\pi K(t-s)}}\,h(y,\,s)\,dy\,ds$

5. $u(x;\,t) = \displaystyle\sum_{m=-\infty}^{\infty}\int_0^t\int_0^L \dfrac{e^{-(x-y-2mL)^2/[4K(t-s)]}}{\sqrt{4\pi K(t-s)}}\,h(y;\,s)\,dy\,ds$

CHAPTER 9

Section 9.1

1. $(\omega^2)_2 = \dfrac{\omega_0^2}{\rho_0^2}\left(\dfrac{L\rho_1}{\pi}\right)^2 + \dfrac{\omega_0^4}{\rho_0^2}\displaystyle\sum_{m \neq n}\dfrac{C_{mn}^2}{\omega_0^2\rho_0^2 - T_0(m\pi/L)}$

2. $(\omega^2)_1 = -\dfrac{\omega_0^2}{\rho_0}\dfrac{\rho_1 L}{\pi}\dfrac{4n^2}{4n^2 - 1}$

Section 9.4

1. (a) $c = \frac{5}{18}$, $I(U_{\min}) = -\frac{5}{216} \approx -0.023$
   (b) $c_1 = 0.533$, $c_2 = 0.048$, $I(U_{\min}) \approx -0.0032$
   (d) $a_1 = 0.0057$, $I(a_1 \sin \pi t) \approx 0.0057$
        $a_2 = -0.00414$, $I(a_1 \sin \pi t + a_2 \sin 2\pi t) \approx 0.0050$
3. (a) $c = \frac{1}{144}$, $I(u_{\min}) = -0.00035$

INDEX

Abel's theorem, 41, 120
Abramowitz, M., 207, 299
Acoustics, 136ff.
Amplitude variation, 105
Angular index, 174
Antman, S., 132
Apples, cooling of, 215ff.
Associated Legendre function, 229ff.
Asymptotic amplitude, 187
Asymptotic behavior, 33, 121, 124, 127, 183
Asymptotic estimate, 86ff.
Asymptotic expansion, 308ff.
Asymptotic phase, 187
Asymptotic solution, 294ff.
Azimuthal angle, 25, 209

Bessel functions, 174ff., 197ff., 201ff., 314
  differentiation formula for, 180ff.
  integral representation of, 178
  modified, 177, 376
  recurrence formula for, 181
  spherical, 231ff.
  zeros of, 183ff., 197ff., 201ff.
Bessel's equation, 23ff., 174ff.
  power series solution of, 174ff.
  second solution of, 181ff.
Bessel's inequality, 15ff.
Big-O notation, 43ff.
Birkhoff, G., 31
Blackman, E. D., 137
Boas, R. P., 318